Systems Architecture

Hardware and Software in Business Information Systems

Second Edition

Stephen D. Burd
University of New Mexico

COURSE
TECHNOLOGY

ONE MAIN STREET , CAMBRIDGE, MA 02142

an International Thomson Publishing company I(T)P®

Cambridge • Albany • Bonn • Boston • Cincinnati • London • Madrid • Melbourne • Mexico City
New York • Paris • San Francisco • Singapore • Tokyo • Toronto • Washington

Systems Architecture, Second Edition is published by Course Technology.

Managing Editor	Kristen Duerr
Senior Product Manager	Jennifer Normandin
Associate Product Manager	Lisa Ayers
Production Editor	Debbie Masi
Book Design, Composition and Art	GEX, Inc.
Cover Illustrator	Cuttriss & Hambleton

© 1998 by Course Technology — I T P®

For more information contact:

Course Technology
One Main Street
Cambridge, MA 02142

ITP Europe
Berkshire House 168-173
High Holborn
London WCIV 7AA
England

Nelson ITP, Australia
102 Dodds Street
South Melbourne, 3205
Victoria, Australia

ITP Nelson Canada
1120 Birchmount Road
Scarborough, Ontario
Canada M1K 5G4

International Thomson Editores
Seneca, 53
Colonia Polanco
11560 Mexico D.F. Mexico

ITP GmbH
Königswinterer Strasse 418
53227 Bonn
Germany

ITP Asia
60 Albert Street, #15-01
Albert Complex
Singapore 189969

ITP Japan
Hirakawacho Kyowa Building, 3F
2-2-1 Hirakawacho
Chiyoda-ku, Tokyo 102
Japan

ISBN 0-7600-4960-2

Printed in the United States of America

1 2 3 4 5 6 7 8 9 10 BH 02 01 00 99 98

DEDICATION

To Andy Whinston and Rao Vemuganti, thank you for showing me the many facets of scholarship.

CONTENTS

CHAPTER 2
Computer Hardware

CHAPTER 3
Computer Software

CHAPTER 4
Data Representation

CHAPTER 5
Processor Technology and Architecture

CHAPTER 6
Data Storage Technology

CHAPTER 7
System Integration & Performance

CHAPTER 8
Data and Network Communication Technology

CHAPTER 9
Networks and Distributed Systems

CHAPTER 10
Input/Output Technology

CHAPTER 11
Applications Development

CHAPTER 12
Operating Systems

CHAPTER 13
Mass Storage Access and Management

CHAPTER 14
Operating System Input/Output

CHAPTER 15
System Administration

PREFACE

Intended Audience

The purpose of this text is to provide a broad technical description of computer hardware and system software. It is intended as a reference for information systems (IS) professionals and as an undergraduate text for students majoring or concentrating in information systems. The text covers a broad range of hardware and system software technology in an integrated manner. Topics that are most useful to IS students and professionals are stressed at an appropriate level of detail. This book provides a technical foundation for systems design, systems implementation, hardware and software procurement, and computing resource management. In some areas, the reader will gain a sufficient depth of technical knowledge to solve technical problems alone. In other areas, the reader will gain a sufficient depth of technical knowledge to allow effective communication with technical specialists.

There is a substantial difference between traditional curriculums in computer science and information systems. Students of computer science are exposed to a great deal of computer hardware and system software technology in their normal course work. The topics covered in this text are covered in depth in several different undergraduate computer science courses (e.g. computer hardware, operating systems, and data communications). There are many texts designed for such computer science courses that focus specifically on a subset of the topics in this book.

Exposure to computer hardware and system software in an IS curriculum is usually limited. A brief overview may be provided in an introductory IS course and some specific technical topics may be covered in other courses (e.g., system design and computer networks). However, the primary emphasis of most information systems curricula is application development, and courses in that area comprise the bulk of the program. There is little time to incorporate technical computer science courses and, thus, there is generally one IS course devoted to covering many topics in hardware and system software technology.

The topical coverage presented here closely matches the requirements for IS'97.4—Information Technology Hardware and Software in the most recent recommended IS curriculum.[1] This text is specifically designed for that course but may also be used as supplemental reading in courses on system design and computer resource management. With respect to system design, the text covers many technical topics that must be addressed in the selection and configuration

---

[1] *IS'97 Model Curriculum and Guidelines for Undergraduate Degree Programs in Information Systems*, published by the Association for Computing Machinery, Association for Information Systems, and Association of Information Technology Professionals, 1997.

of computer hardware and system software. With respect to computer resource management, this book provides the broad technical foundation necessary to effectively manage computer technology.

READER BACKGROUND KNOWLEDGE

IS'97.4 is placed early in the recommended IS curriculum. Prerequisite and corequisite courses cover personal productivity software, an overview of IS theory and practice, and introductory computer programming. Thus, this book makes few assumptions of reader prerequisite knowledge. The reader is not assumed to have an extensive background in mathematics, physics, or engineering, as is the case in many computer science texts. Where necessary, background information in these areas is presented in sufficient depth to understand subsequent material.

The reader is not assumed to know any particular programming language. However, classroom or practical experience with at least one programming language is helpful to fully comprehend the coverage of operating systems and application development software. Programming examples are provided in several different programming languages and in psuedocode. All of these examples should be understandable to readers not familiar with the particular programming language used.

The reader is not assumed to have detailed knowledge of any particular operating system. However, as with programming experience, exposure to at least one operating system is helpful. Lengthy examples from specific operating systems are purposely avoided. However, there are some short examples from MS-DOS, UNIX, Windows 95, Windows NT, and MVS. These examples are designed to be understandable by readers not familiar with the particular operating system used in the example.

The reader is not assumed to possess any knowledge of low-level machine actions or assembly language programming. Low-level machine actions are discussed in detail in the hardware section of this book. Assembly language programs are discussed in several chapters, but no specific assembly language is used and no detailed coverage of assembly language program construction is provided.

CHANGES IN THE SECOND EDITION

The goals of the second edition are to match the requirements of IS'97.4, modernize the topical coverage, and improve the pedagogical approach. These goals led to a number of changes from the first edition including:

► Revision of most of the first edition material to update its technological coverage.

- Expansion of the number of "Technology Focus" sections to approximately two per chapter. Each presents a modern in-depth example to illustrate concepts covered in the chapter.
- Addition of a "Business Focus" section (a mini-case) to most chapters.
- Addition of topical coverage for semiconductor technology, network hardware devices, network software, graphical interfaces, system administration, and security.
- Increased coverage depth for many topics including microprocessors, optical storage, caching, networks, audio and graphic I/O, and object-oriented programming and system development.
- Elimination of the chapters covering assembly language programming, advanced computer architecture, and non-procedural programming. The coverage of assembly language programming has been substantially reduced in scope and distributed throughout the text. The material on advanced computer architectures and non-procedural programming has been eliminated (however, this material will be available on the accompanying Web site).
- Revision and expansion of end-of-chapter material including additional vocabulary exercises, review questions, exercises, and research problems.

SUPPLEMENTS

Systems Architecture, Second Edition includes teaching supplements to support professors in the classroom. The ancillary materials consist of an Instructor's Manual and Course Test Manager, an electronic testing engine. The Instructor's Manual provides materials to help professors make their classes informative and interesting. The Manual includes teaching tips, discussion topics, and solutions to the end-of-chapter materials. Course Test Manager is a powerful testing and assessment package that enables instructors to create and print tests designed specifically to accompany *Systems Architecture*.

WORLD WIDE WEB SITE

A support site for this book is located at http://averia.unm.edu. The site will be more active than the first edition site and will include:

- An Instructor's Manual.
- Interactive end of chapter material.
- Web resource links for most text topics and research problems.
- Text updates and errata.

One of the problems with any text covering hardware and system software is that is goes out of date before it leaves the printer. The Web site will be used to address that problem by providing regular updates and additions to text material.

ORGANIZATION OF THE TEXT

This text is organized into three groups of chapters. The first group consists of three introductory chapters containing a general overview, a hardware overview, and a software overview. The second group consists of seven chapters covering hardware technology. The third group consists of five chapters covering details of software technology and system administration.

Although the chapters were designed to be covered in a linear sequence, a number of alternative orderings are possible. In particular, some instructors may want to intermix coverage of hardware and related software topics. The prerequisite material for each chapter is given in the chapter descriptions below. Alternate orderings may be constructed based upon these prerequisite relationships. For any alternative ordering, it is suggested that Chapters 2 through 4 be covered first. Several chapters (e.g., 1, 7, 10, and 13) and chapter sections may be skipped without loss of continuity.

There should be time to cover between 8 and 10 chapters in a three credit undergraduate course. This text purposely overshoots this number to provide flexibility in course coverage. The material in some chapters may be covered in other courses in a specific curriculum (e.g., Chapters 8 and 9 in a networking class, Chapter 13 in a database management class, and Chapter 15 in a system administration class). Thus, the instructor can choose specific chapters to best match the overall curriculum design and teaching preferences.

CHAPTER DESCRIPTIONS

Chapter 1: Systems Architecture: An Introduction. This chapter briefly describes the uses of technical knowledge about computer hardware and system software. These uses are described in terms of structured system development life cycle phases and the persons who participate in that life cycle. This chapter also describes various sources of knowledge about computer hardware and system software and provides a list of recommended periodicals.

Chapter 2: Computer Hardware. This chapter provides an overview of computer hardware technology and architecture. The primary components and functions of a computer system are described. Various methods of computer system performance measurement and evaluation are also discussed. This chapter also describes classes of computer systems and the characteristics that distinguish them.

Chapter 3: Computer Software. This chapter begins by describing the role and function of software. Classes of software including application, application development, and application support software are defined. Operating system components and functions are also discussed. A layered model of software is stressed throughout the chapter. Chapter 2 is a recommended prerequisite.

Chapter 4: Data Representation. This chapter describes data representation in terms of coding formats and digital signals. Binary, octal, and hexadecimal numbering systems are described. Primitive CPU data types are described along with common coding conventions for each type. A discussion of data structures is also included. Chapter 2 is a recommended prerequisite.

Chapter 5: Processor Technology and Architecture. This chapter describes the architecture of the central processing unit and of primary storage. The description of processor functions includes discussions of primitive data transformations, data movement, and sequence control. Various architectural features of a von Neumann processor are discussed including instruction/execution cycles, word size, clock rate, registers, and instruction formats. Semiconductor and microprocessor fabrication technology and operating limits are also described. Chapter 4 is a necessary prerequisite and Chapter 2 is a recommended prerequisite.

Chapter 6: Data Storage Technology. This chapter describes primary and secondary storage implementation with electrical and optical technologies. The general operating principles of electrical, magnetic and optical storage technologies are described. Specific storage are described including RAM, ROM, magnetic disk, magnetic tape, and optical disk. A brief discussion of primary storage addressing and allocation is also provided. Chapters 4 and 5 are necessary prerequisites. Chapter 2 is a recommended prerequisite.

Chapter 7: System Integration & Performance. This chapter describes communication among computer system components and various methods of performance enhancement. The chapter begins with a discussion of the system bus and bus protocols followed by a description of device controllers, mainframe channels, and interrupt processing. Performance enhancement methods are described including buffering, caching, and compression. Chapters 5 and 6 are required prerequisites and Chapters 2, 3, and 4 are recommended prerequisites.

Chapter 8: Data and Network Communication Technology. This chapter describes data communication technology. Basic technological issues discussed include signals, signal propagation media, and data coding methods. Various communication and protocol topics are discussed including serial and parallel transmission, synchronous and asynchronous transmission, channel sharing methods, and error detection and correction. The first section of Chapter 4 is a necessary prerequisite. Chapters 2 and 5 are recommended prerequisites.

Chapter 9: Networks and Distributed Systems. This chapter describes the architectural and hardware aspects of computer networks. The chapter begins with a general discussion of distributed computing. Next, specific modes of resource distribution are described including shared I/O devices, file sharing, distributed processing, and client/server architecture. This chapter then describes various low-level networking issues including physical and logical topology, media access control methods, routing and addressing, and network hardware devices. The chapter concludes with a discussion of IEEE and OSI networking standards. Chapters 4, 5, and 8 are necessary prerequisites. Chapters 2, 3, and 6 are recommended prerequisites.

Chapter 10: Input/Output Technology. This chapter describes I/O devices including keyboards, pointing devices, printers and plotters, video display terminals, graphic displays, optical scanners, and audio I/O devices. This chapter also covers more general I/O topics such as fonts, image representation, color representation, and image description languages. Chapter 4 is a necessary prerequisite. Chapters 2, 6, and 7 are recommended prerequisites.

Chapter 11: Applications Development. This chapter describes the development of application programs using higher level languages and other application development tools. This chapter begins with a discussion of various approaches to application development and the application development software required to support each approach. A detailed discussion of third generation, fourth generation, and object-oriented programming languages is then provided. This is followed by a detailed discussion of compilation, interpretation, and the use of support libraries. The final sections discuss various forms of advanced application development tools including CASE tools, code generators, and programmers' workbenches. Chapters 2, 4, and 5 are necessary prerequisites. Chapter 3 is a recommended prerequisite.

Chapter 12: Operating Systems. This chapter provides an overview of operating systems and detailed descriptions of resource allocation, processor control, process and thread management, and memory management. This chapter begins with a discussion of software layers with particular attention to operating system layers. The role of the operating system as a resource allocator is then discussed. This is followed by detailed discussions of the mechanisms by which processor and memory resources are allocated to processes and threads. Chapters 3, 5, and 6 are necessary prerequisites.

Chapter 13: Mass Storage and Access Management. This chapter provides a detailed description of mass storage device organization and file management. This chapter begins with a description of logical and physical secondary storage accesses. File content, structure, and manipulation are then discussed. Physical considerations for device I/O, storage allocation, and directory structure are described next. Other topics in the chapter include file protection and security, file sharing, and file system administration. Chapters 3 and 6 are necessary prerequisites. Chapters 7, 11, and 12 are recommended prerequisites.

Chapter 14: Operating System Input/Output. This chapter describes operating system facilities for user interface, interactive command and control, batch job control, and access to distributed resource access. This chapter begins with a discussion of I/O support services interfaces to interactive I/O devices. This chapter then covers interactive control via command language, form-based, and window-based interfaces, Interactive and batch process control using command and job control languages is discussed next. This chapter concludes with a discussion of network protocol stacks and mechanisms for locating and accessing distributed resources. Chapters 2, 9, and 10 are necessary prerequisites. Chapters 3, 4, 12, and 13 are recommended prerequisites.

Chapter 15: System Administration. This chapter discusses various issues of computer system administration. This chapter begins with systems administration tasks and the strategic role of hardware and software resources in an organization. The normal acquisition process is then discussed. Next, a detailed discussion of requirements determination and performance modeling is given. The next section describes the implementation of system security models. The final section discusses various physical aspects of computer operation. Chapters 2, 3, 9, and 13 are necessary prerequisites.

ACKNOWLEDGMENTS

The first edition of this text originated from a planned revision of another text entitled *System Architecture: Software and Hardware Concepts*, by Leigh and Ali. The project evolved to encompass a broader range of topics and a more specific information system emphasis. Many of the figures and some of the textual material are taken or adapted from that text. I am indebted to Leigh and Ali and the boyd & fraser publishing company (now merged with Course Technology) for providing a starting point for the first and second editions of this text.

I would like to thank everyone who contributed to the development of this edition. This includes the editors and staff of Course Technology and Masi Book Productions. Thanks also to the peer reviewers of the project proposal and manuscript drafts: Charles Bilbrey, James Madison University; Jeff Butterfield, University of Idaho; Richard Coppins, Virginia Commonwealth University; Rahul De, Bowie State University; Donna Dufner, University of Springfield; Alexis Koster, San Diego State University; Michael Moorman, St. Leo College; Philip Pfeiffer, East Tennessee State University; John Reynolds, Missouri Baptist College. Thanks also go to the students of the undergraduate MIS concentration at the Anderson Schools of Management, University of New Mexico. Several drafts of the first and second editions have been used in the Software and Hardware Concepts course at the Anderson Schools during the last seven years. Comments from students in these classes were invaluable for improving the book.

Chapter 1

Systems Architecture:
An Introduction

Chapter Goals

▶ Describe the activities of information systems professionals.

▶ Describe the technical knowledge of computer hardware and systems software needed to develop and manage information systems.

▶ Identify additional sources of information for continuing education in computer hardware and systems software.

COMPUTER TECHNOLOGY: YOUR NEED TO KNOW

To be widely accessible, technical devices must appear to be simple. Our daily lives contain many examples of hidden technical complexity. For example, most people know little about the inner workings of a car, a home stereo system, a microwave oven, and most other common technologically complex devices. If detailed knowledge about such devices were required to use them, then they would be out of reach for the average person.

The use of computers is no exception to this technological reality. The earliest computers required that users (or, operators) have substantial training and knowledge to accomplish even the simplest tasks. Over time, computers have become increasingly more complex and powerful machines, and, yet, the prerequisite knowledge has considerably decreased. At least partly in response to this so-called simplification at the user interface level, computers have proliferated beyond their original scientific applications and moved into business, the classroom, and the home. Why, then, do you need to know anything about a computer's innermost workings?

Acquiring and Configuring Technological Devices

The knowledge necessary to select and configure technically complex devices is far greater than that required to effectively use these machines. For example, consider the operation of a home stereo system, which is relatively simple. You need to know how to plug in the system, the location of the power switch, how to select input (e.g., tape or compact disc), how to adjust volume, and so forth. The average person can gain this knowledge quickly by reading the operating manual or making deductions from past experiences with similar machines.

But what happens when someone goes into a stereo store? The customer is immediately confronted with a wide range of choices, such as watts of power, size of speakers, and whether to get an integrated system versus a combination of separate components. The prerequisite knowledge to advise a customer includes a thorough understanding of the purchaser's desires and requirements (e.g., musical taste, sensitivity to distortion, size of listening rooms, etc.) as well as insight into available alternatives.

To determine compatibility and best fit, a purchaser must know about the technical details of the alternatives and their relationship to specific customer needs. A purchaser who has a high degree of sensitivity to distortion may decide to get "high-end," specialized equipment.

A typical purchaser is not familiar with technical terminology (e.g., what is signal-to-noise ratio? and how is it measured?). Further complicating the purchaser's decision is the large number of similar products, the availability of alternative technologies to meet the same need (e.g., analog cassette tapes, compact discs, and digital audiotape), and compatibility issues.

Acquiring and Configuring Computer Systems

The purchase of a computer system is an even more complex endeavor than that of a home stereo system. For example, a customer must decide which basic type of computer, storage capacity, processor type and speed, input/output devices, and so forth to purchase. Another area of decision involves software. What operating system is required? Is a database manager needed? What about a word processor or spreadsheet package? Does the purchaser need tools to develop software from scratch? What is the optimum combination of software and hardware components to meet a particular customer's needs?

A home computer purchaser may deal with these issues infrequently and with a relatively simple set of requirements. An information system (IS) professional, on the other hand, may confront these issues daily and for varying requirements. Customer needs might even change during the selection process.

IS DEVELOPMENT: TECHNOLOGICAL KNOWLEDGE

Within the context of IS development, what knowledge of computer hardware and software is required and when is it required? To answer this question, briefly examine the steps in the process of IS development outlined in the following sections. For each step, the technical knowledge needed to complete the step is described.

The steps necessary to develop an IS are commonly called a **systems development life cycle (SDLC)**. A number of different SDLCs have been used over the history of automated information processing. One commonly used approach to develop large and complex ISs is the **structured SDLC**. The structured SDLC specifies a set of steps to be followed in systems development as well as tools and procedures for each step. The phases of the structured SDLC are depicted in Figure 1-1.

Figure 1-1

Steps in the structured systems development life cycle (SDLC)

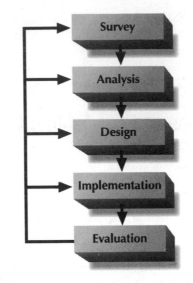

Systems Survey

The **systems survey** is a process that briefly examines user information needs, existing ways in which those needs are being addressed (if any), and the feasibility of developing or acquiring an IS to address those needs. Normally, the survey is conducted within a relatively short time frame (days or weeks) and is intended to determine the existence and nature of unmet (or poorly met) needs. If unmet needs are found, the next step in the survey is to briefly analyze the potential costs and benefits of alternative systems that might address those needs. This data is used by systems analysts and users to decide whether any of the alternatives appears promising enough to perform a full systems analysis.

An IS specialist requires technical knowledge of computer hardware and systems software to assess the degree to which user needs are currently being met and to estimate the resources required to address unmet needs. For example, an analyst surveying a point of sale system in a retail store might pose questions such as:

► How much time is required to process a sale?

► Is the existing system easy for a salesperson to use?

► Is sufficient information being gathered (e.g., for marketing purposes)?

► Can the existing hardware handle peak sales volumes (e.g., Christmas)?

► Will the existing hardware be sufficient three years from now?

► Can the existing system be expanded easily?

► What are the current hardware operating costs?

► Are there cheaper hardware alternatives?

All of these questions directly or indirectly require technical knowledge of hardware and systems software.

Processing time and ease of use depend on the hardware and software capabilities and on the user interface structure. The system's ability to handle peak demands requires detailed knowledge of the processing and storage capabilities of hardware, operating systems, and application software. This same knowledge is required to determine future sufficiency. Expansion possibilities depend on unused capacity in the hardware and limitations in both the hardware and software. Determining whether or not cheaper alternatives exist requires all of the aforementioned knowledge—not only with regard to the current system but also with respect to a wide range of currently available hardware and software options.

The systems survey requires a broad range of technical knowledge. However, the time limit inherent in the survey usually precludes detailed use of this knowledge. For example, it is rare to see exacting calculations of processor performance or storage capacity made during this phase. Similarly, cost estimates are rarely made "to the dollar."

Systems Analysis

Systems analysis is concerned primarily with the detailed examination of user needs and the extent to which they are being met. This phase of the SDLC may also be called **needs analysis**, or **requirements analysis**. Information requirements are examined in greater depth; processing requirements are precisely defined; and performance requirements for the current and future systems are determined.

Many of the questions posed during the systems survey are examined during systems analysis. Ease of use and time constraints lead to specifications for interface style and processing speed. Detailed estimates of current and future transaction volumes are used to derive both processing and storage requirements. Other needs for the data determine requirements for integrating this system with other ISs and lead to detailed specifications for hardware, software, and database compatibility.

The detailed specification of requirements serves two primary functions. First, it serves as a basis for systems design. Detailed requirements allow the analyst to precisely determine cost and desirability of various systems implementations (combinations of hardware and software). The requirements specification also serves as a contract with the user. It is a statement of the acceptance criteria for the system that will be delivered ultimately.

Systems Design

During the **systems design** phase, the exact configuration of all hardware and software components is specified (see Figure 1-2). Specific design phase tasks include:

► Select computer hardware (e.g., processing, storage, input/output, and network components).
► Select system software (e.g., an operating system and/or network communication software).
► Select application program development tools (e.g., programming language and database management system).

Figure 1-2

Detailed system design tasks in the structured systems development life cycle (SDLC)

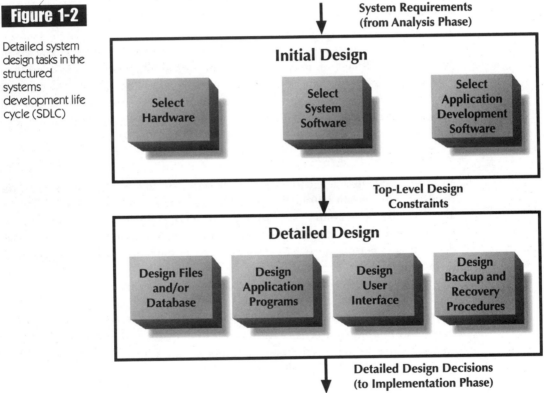

Because these design decisions have a broad impact on the IS, they are made early in the design phase.

Other design decisions are more limited and depend on earlier decisions. Examples of these include:

► Design files and/or databases (e.g., the grouping of data elements into records and files, indexing, and sorting).
► Design individual programs.
► Design user interfaces (e.g., input screen formats, report formats, and command dialogues).
► Design system backup and recovery mechanisms.

The output of these decisions is fully documented and forms a blueprint for detailed implementation tasks.

During the early stages of systems design, the analyst needs considerable technical knowledge of computer hardware and systems software. Choosing hardware components requires a detailed knowledge of product capabilities and limitations. The integration of multiple hardware components into a single system requires that all of the devices be compatible. The analyst's choice of operating system is constrained by the hardware system and by the overall system performance requirements. Compatibility of both the hardware and the systems software with other systems within the organization must be determined during this phase.

The choice of application program development tools depends on both the requirements of the IS and the limitations of the hardware and operating system. An IS analyst must select and integrate programming languages, window managers, database managers, and other tools. All of these tools (and the software components built with them) vary widely in their efficiency, power, and compatibility. Because such tools are often expensive, these choices will affect future systems development projects.

Systems Implementation

During the **systems implementation** phase, the IS analyst implements the plans formulated during systems design. Hardware and systems software are acquired, installed, configured, and tested. Application programs and files are developed, installed, and tested. The entire system is checked to ensure that all of its components work together and that the system correctly meets the user's needs as defined during the analysis phase. Generally, there is some overlap of tasks during the design and implementation phases. For example, the installation and testing of hardware and systems software is often performed concurrently with the detailed design of files and application programs.

A great deal of technical knowledge is required to perform many of the system implementation tasks. The installation and configuration of hardware and systems software is a highly specialized task that requires a great deal of understanding of both the components being installed and the purposes for which they will be used. Both hardware and software usually must be configured to a particular operating environment, which requires adjustments to the components and follow-up testing. This process can require many iterations and much time. More mundane related tasks, such as formatting storage devices, setting up system security, installing application programs, and establishing accounting and auditing controls also require time and technical expertise.

Evaluation and Maintenance

Over time, various shortcomings in the system may emerge. Errors or problems that escaped detection during installation and testing may appear. The system may become overloaded due to inadequate estimates of processing volume. Additional information needs may become apparent, thereby necessitating additional data collection, storage, and/or processing.

Normally, minor system changes (e.g., correcting program errors or minor processing changes) are handled as maintenance changes. Such changes may or may not require extensive technical knowledge, but some technical knowledge may be required to properly classify a proposed change as major or minor. Will new or additional processing requirements overtax the hardware or software capacity? Do the proposed changes require programming tools that are incompatible with the current system design or configuration? The answers to questions such as these determine whether the existing system will be modified or replaced by an entirely new system.

If the existing system is to be modified, the programs and/or files to be changed must be identified, modified, and tested. The specific technical knowledge requirements rely heavily on the specific hardware and software components affected by the change. In the event that an entirely new system is required, a new SDLC will be initiated with all of the technical knowledge requirements described previously.

MANAGEMENT OF COMPUTER RESOURCES

Thus far, the need for technological knowledge has been discussed in the context of developing a single IS. Consider, though, the complexities and knowledge required to manage a large organization's computer resources. In such an environment, existing ISs may number in the hundreds or thousands, and many new development projects or major upgrades of existing systems may be in progress at any one time.

Such an environment requires increased attention to two key technological issues: compatibility and future trends. Compatibility and integration have become increasingly important as the number of computer systems and applications has increased. Few ISs are standalone systems; rather, most are integrated with other systems through both software and hardware. For example, accounts payable and accounts receivable programs may share a common hardware platform and operating system. Data from both of these systems may be input to a financial reporting system on an entirely different computer, and data from many sources within the organization may be available over a communications network.

The manager of such a disparate collection of ISs must contend with significant technical complexity. One challenge is to ensure that each new system not only operates correctly by itself, but in conjunction with all of the other systems in the organization. The manager also must ensure that acquisitions of hardware and software are cost-effective. Current acquisitions also must provide a sound basis for the implementation of future systems.

Given the rapid pace of change in computer technology, the manager must have a broad understanding of both the current technology and its likely future trends. For example, will the computer purchased today be compatible with the hardware available three years from now? Will today's communications network have sufficient capacity to meet future needs? Should the organization invest only in proven technologies, or should it risk the purchase of cutting edge technologies in hopes of greater performance or competitive advantage?

Answering these questions requires great depth of technical knowledge, which is far more than any one person can provide. Typically, a manager confronted by such questions will rely on experts and other sources of information. Even so, the responsible manager must have a sufficient base of technical knowledge to understand and evaluate the information and advice provided.

ROLES AND JOB TITLES

A large number of people can be loosely classified as "computer professionals," and an even larger number of workers use computers. There is a bewildering array of job titles, specializations, and professional certifications to accompany this wide range of roles. The following sections attempt to classify these workers into groups, describe some of their common characteristics, and describe the knowledge of computer hardware and systems software required by each group.

Users

By definition, anyone who comes into contact (directly or indirectly) with a computer system can be considered a user. The term **user** is generally applied to persons who interact directly with **application software**. Application software is software designed to accomplish a specific purpose. Examples of application software include payroll programs, financial reporting programs, and programs to input and edit business transactions (e.g., invoices or customer orders). Some people who use such software include data entry clerks, customer service representatives, and managers.

In the 1980s, the number of users grew as general purpose **application tools** came into widespread use. Typical users of these tools included secretaries, accountants, and managers at a variety of organizational levels. Examples of such tools include spreadsheet programs (e.g., Excel and Quatro Pro), database managers (e.g., Access and Paradox), and word processors (e.g., WordPerfect and Microsoft Word). Although this class of software is not geared toward a particular need such as payroll or customer billing, it focuses on specific types of processing needs including financial calculation and document preparation.

Application Development Personnel

A large class of computer professionals are employed to create application software for specific processing needs. These professionals have many different job titles, including **programmer**, **programmer/analyst**, **systems analyst**, and **systems designer**, among others. Each role contributes to a different part of the SDLC. Systems analysts are primarily responsible for conducting surveys, determining feasibility, and defining and documenting user requirements. Systems designers are primarily responsible for procuring hardware and procuring or designing application software. **Application programmers** are responsible mainly for software implementation and testing.

Many professionals engage in activities that don't neatly fit into these job definitions. For example, a person with the job title "systems analyst" commonly has responsibility for survey, analysis, design, and management of a development project. Programmers often are responsible for analysis and design tasks as well.

Adding further confusion are the types of software applications that can be developed and the training required for each type. Applications can be loosely classified into two types: information processing and scientific/technical. People who develop information processing applications typically develop software geared toward the processing of business transactions or the provision of information to managers. These developers usually have college or technical degrees in management or business specializing in information processing[1].

---

[1] Other names for the field include management information systems, data processing, business computer systems, and so forth.

People who develop scientific/technical applications typically have degrees in **computer science** or some branch of engineering. The applications they develop are oriented toward scientific pursuits such as astronomy, meteorology, and physics. Technical applications are often involved with the control of hardware devices. Examples include applications for control of robots, flight navigation, and scientific instrumentation.

Systems Software Personnel

Another large class of computer professionals is responsible for the development of systems software. The typical job title for such a person is **systems programmer**. This includes software such as operating systems, compilers, database management systems, and general purpose application tools. These professionals typically have degrees in computer science or computer engineering and do not develop specific applications. Organizations with large amounts of computer equipment and software frequently employ systems programmers as hardware/software consultants and to perform hardware and software installation and configuration. Larger numbers of systems programmers are employed by organizations that develop and market systems software and/or application tools (e.g., Microsoft).

Hardware Personnel

People who interact with computer hardware include those who operate it, install it, and design or build it. A **computer operator** is responsible for executing application software and for other operational tasks such as backing up files to tape, starting and stopping the system, and creating user accounts. A computer operator usually has a technical degree or has been through vendor-specific training. The proliferation of small- and medium-sized computers has caused the role of user and computer operator to be combined in many cases. In the past, users relied almost exclusively on operators for tasks such as starting programs and copying files. Today, users of personal and small computers must usually perform such operations themselves.

People responsible for hardware installation and maintenance typically are employed by vendors of computer hardware. Lower-level personnel usually have a technical degree and/or vendor-specific training while higher-level personnel often have a degree in computer science or **computer engineering**. Computer systems designers typically have degrees in computer science or a related branch of engineering.

Systems Management Personnel

The proliferation of computer hardware, networks, and software applications over the last four decades has created a need for large numbers of managers and administrators. The job descriptions for such positions vary widely from one organization to another because of differences in organizational structure and in the amount and types of computing and information resources employed by the organization. Some of the more common job titles in this category are **computer operations manager**, **network administrator**, **database administrator**, and **chief information officer**. Many variations on these titles also exist.

A computer operations manager is responsible for the operation of a large information-processing facility. Typically, such facilities have one or more large computer systems and all related peripheral equipment housed in a central location. Large databases, thousands of application programs, dozens or hundreds of employees, and significant amounts of batch processing characterize such installations. Examples of organizations with this type of computing facility include large banks, credit-reporting bureaus, the Social Security Administration, and the Internal Revenue Service. The management of day-to-day operations in such an environment is extremely complex. Scheduling, staffing, security, system backups, maintenance, and upgrades are some of the key responsibilities of a computer operations manager.

The title of network administrator typically implies one of two possible roles. The first is responsibility for the network infrastructure of a large distributed organization. A multinational corporation with dozens or hundreds of worldwide locations typically will possess a complex network to tie together its databases, application software, and computer hardware. The design, operation, and maintenance of such a network is a complex task requiring substantial technical expertise in computer hardware, telecommunications, and systems software. The role of network administrator in this environment is a high-level position.

In smaller organizations, the title of network administrator usually implies the management of a local area network (LAN). Such networks connect anywhere from a half dozen to a few hundred computers (mostly microcomputers) and provide access to one or more shared databases. Especially in smaller organizations, the network administrator may be responsible for many tasks other than operating and maintaining the network. These tasks might include the installation and maintenance of end-user software, hardware installation and configuration, training of users, and assisting management in the selection and acquisition of software and hardware resources. The position of network administrator represents one of the fastest-growing areas of employment for IS professionals and is one of the most demanding positions in terms of breadth and depth of required skills and technical knowledge.

Since the early 1980s, many organizations have realized the economic and strategic importance of the data stored in their computer systems. This data may be exploited for many purposes, including coordinating organizational subunits and developing effective marketing strategies. The technology by which large collections of data (databases) are managed and accessed is highly complex. This complexity, combined with managerial recognition of the importance of data resources, has resulted in the creation of many positions with the title of "database administrator." The title implies more than just technical expertise; it also suggests a manager who is responsible for helping the organization realize the potential benefits of its investment in stored data.

As the name suggests, the role of chief information officer is a high-level management position that typically exists only in large organizations with substantial investments in computer and software technology. Many of the previously defined positions (e.g., database administrator, network administrator, and computer operations manager) report to a company's chief information officer. The responsibilities of this position span the breadth of an organization's computers, network, and software. As with a database administrator, the position of chief information officer has a responsibility for strategic planning and the effective use of information and computing technology in the broadest sense of those terms.

SOURCES OF INFORMATION ABOUT COMPUTER TECHNOLOGY

This text and other study materials will provide you with a technical knowledge foundation for a career in IS development or management. Unfortunately, that foundation will quickly erode due to the rapid pace of change in computer and information technology, which means that you will need to constantly update your knowledge to stay current.

Numerous means are available for this purpose. For example, newspapers and periodicals can provide a wealth of information on current trends and technologies. Training courses offered by hardware and software vendors can teach the specifics of current products. Additional course work and self-study can keep you abreast of technologies and trends less specifically geared toward particular products. By far, the most important of these activities consists of reading the periodical literature.

Periodical Literature

The volume of literature available on computer topics is vast, often making it difficult to determine the most important and/or appropriate sources for IS professionals. Much of the problem arises from differences in training among IS professionals, computer scientists, and computer engineers. IS professionals are extensively trained in application development tasks (e.g., systems analysis), managerial tasks (e.g., project management), and functional areas of business (e.g., accounting and marketing). The extensive coverage of these topics leaves less time available for detailed coverage of computer hardware and systems software.

Computer scientists and engineers are more extensively trained in hardware and systems software than are IS professionals, but they have less training in applications development, management, and functional business areas. Thus, many sources of information that are oriented toward one computer or technical specialty are difficult reading for the other. For example, a detailed discussion of user interaction and validation of requirements models may be beyond the training of a computer scientist or engineer. Similarly, detailed descriptions of the theory of optical telecommunications may be beyond an IS professional's background.

The periodicals listed below are recommended as a good source of technical information for the IS professional. They avoid excessive levels of technical depth while still providing sufficient detail.

► **ACM Computing Surveys.** An excellent source of information on the latest research trends in computer software and hardware. The articles are written as in-depth summaries of technologies or trends geared toward a readership with a moderate level of familiarity with computer hardware and software.

► **Byte Magazine.** An excellent source of information on computer hardware, software, and trends primarily oriented toward small computers. Coverage is provided of both specific products and of general trends and technologies. Comparative reviews of specific products are provided regularly. Tutorials and theme articles on basic technologies are also regular features.

► **ComputerWorld.** A weekly magazine primarily geared toward computer "news items." Coverage of product releases, trade shows, and occasional coverage of technologies and trends is provided.

► **Communications of the ACM.** A source of information about research topics in computer science. Many of the articles are highly technical and specialized to research but some are geared toward a less research-oriented audience.

► **Computer.** A publication of the Institute of Electrical and Electronics Engineers (IEEE) devoted to coverage of computer hardware and software. Many of the articles are research-oriented, but occasional coverage of technologies and trends is provided for a less technical audience.

The following periodicals are also good sources of information on computer hardware and systems software. With the exception of *Science* and *Scientific American*, they tend to be more specialized than the periodicals listed above:

AI Magazine	*PC Magazine*
Data Base	*PC World*
Datamation	*Personal Computing*
EDN	*Science*
Electronics Week	*Scientific American*
IEEE Spectrum	*UNIX Review*
InfoWorld	

World Wide Web

Since the early 1990s, the **World Wide Web** (or, simply, the **Web** or **WWW**) has become an increasingly important source of information for IS professionals. Several important categories of information sources are increasingly found on the Web, including periodical information, vendor-specific information, information from various standard-setting organizations, and information produced and distributed by computer-oriented professional societies. The Web allows relatively easy access to a much larger body of information than could previously be accessed through traditional indices and paper-based publications.

Web-Based Periodicals

Most publishing is still based on prepaid subscriptions and the distribution of printed materials. Web-based publishing has languished due to the lack of simple and standardized methods of electronic payment and distribution. Once these have been defined, electronic publishing will likely rapidly supplant paper publishing as the predominant means of information distribution. Thus, the periodicals described earlier may not be available in printed form within a few years.

Most periodical publishers have established a **Web site** and are rapidly developing their Web publishing capabilities (see Figure 1-3). In general, these sites include full or partial content from back issues of the printed periodical. This may be supplemented by partial content from the current issue and by additional information that doesn't appear in printed form. Search capabilities are typically an integral part of the Web site, which reduces the need for publishers to build and distribute indexes for specific periodicals.

Figure 1-3

A sample Web
periodical: the
main page for
the periodical
ComputerWorld

A significant advantage of using Web-based periodicals is the ability to utilize Web links to article references and other related material. For example, an article that refers to another article in a back issue will frequently contain a hypertext link to the back issue. This also may be true for material in an article's reference list or bibliography. Thus, the task of accessing background and reference material is reduced to the click of a mouse button. Similarly, reviews of specific software or hardware will typically contain links to product information and specifications maintained on vendor or manufacturer Web sites.

Vendor and Manufacturer Web Sites

Many vendors and manufacturers of computer hardware and software products have rushed to create a Web presence because many purchasers of computer hardware and software are already online. Vendor Web pages are typically oriented to sales, but they usually contain detailed information on products either directly or as links to manufacturer Web sites. Manufacturer Web sites typically contain detailed information on their products. Technical support and other customer support services are also provided on these Web sites.

It is important to remember that vendor and manufacturer Web sites exist primarily as marketing and customer-support tools, which results in good and bad points. The benefit is the wealth of technical information provided for specific products. This information is far more extensive and current than was ever available through paper-based product brochures and technical documents. The breadth and depth of this information potentially allows IS professionals to make faster, better, and more informed choices.

The negative side of manufacturer Web sites is that the information they provide is often biased to the vendors' products and the technologies on which they are based. As when viewing any information source, a reader must consider the motives and objectivity of the information provider. Hardware and software manufacturers aren't typically in the business of providing unbiased information; rather, they are in the business to sell what they produce. Thus, a reader should expect the content of a manufacturer or vendor Web site to be biased toward the products it produces or sells.

In the best case, biased content may manifest itself as "marketing hype" that a reader must wade through or filter out to get to the "real" information. In the worst case, the information content may be purposely biased by content, omission, or both. Many sites contain technology overviews and similar information that masquerade as independent reviews or research, whereas others contain links to a biased subset of supporting reviews or research.

It is the readers' responsibility to balance potentially biased vendor and manufacturer information with information from unbiased sources, which is present on the Web although it isn't always easy to find. The old saying that "you get what you pay for" certainly applies to the Web. High-quality, unbiased information is typically the product of intensive research. Some organizations do this as a public service (e.g., some computer professional societies and governmental organizations), but much of the "good" information is produced for a profit. Expect to pay for such information and don't think it will be readily available on the Web until a standardized method of electronic payment has been adopted.

Finding Information on the Web

The biggest challenge for Web users is locating relevant, useful information. Because the Web is not owned or managed by any central authority, there is no centralized index to its content. Due to its ever-changing content, any Web index is quickly obsolete unless it is continuously updated.

Many organizations offer free indexing and search services through the Web (see Figure 1-4). As with manufacturer and vendor Web sites, a user must pay careful attention to the potential for bias. Most index and search services are in the business of selling marketing services to information providers. This may show up in the searching and indexing services in many ways including:

► Advertisements displayed with the search results.

► Ordering of search results that favors the Web pages of organizations that have paid a fee to the search provider.

► Omission of information from organizations that haven't paid a fee to the search provider.

► Omission of information that is counter to the interests of the search provider's financial backers.

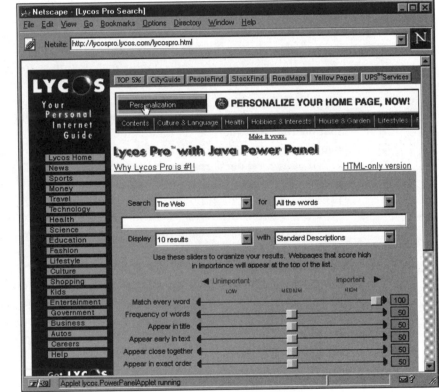

It is not always obvious to the reader which, if any, of these biases are present in the searching and indexing service being used.

A simple strategy for finding information provided by specific organizations is a trial and error approach to guessing the organization's **Internet site name**. A Web access that uses only the site name will typically display a default page for the organization. This default page, in turn, usually contains a table of contents and/or an indexing and searching service for the entire Web site. Most Web site names consist of three parts separated by periods:

1. A leading "www" indicating an Internet site name that offers World Wide Web services

2. A middle name derived from the name of the providing organization (e.g., a full name, abbreviation, or acronym)

3. A trailing identifier that indicates the type of organization

The four most common organization types are .com (commercial), .edu (educational), .gov (governmental), and .org (other, such as a professional society). You can make reasonable guesses at Web site names based on this structure and the set of organization types, as the following examples illustrate:

American National Standards Institute	www.ansi.org
Association for Computing Machinery	www.acm.org
Byte Magazine	www.byte.com
Los Alamos National Laboratories	www.lanl.gov
Microsoft Corporation	www.microsoft.com
University of New Mexico	www.unm.edu

Since Web search services are generally provided by commercial organizations, you can make similar guesses based on the service name (e.g., www.yahoo.com and www.lycos.com).

Sites outside of the United States generally use three- or four-part Web site names. The final part is generally a two-letter abbreviation that indicates the country in which the site is registered. Country abbreviations include .ca for Canada, .uk for the United Kingdom, and .de for Germany (Deutschland). The domain structure (com, gov, edu, etc.) outside of this country is not the same as for U.S. Web sites.

An exhaustive list of Web sources is purposely omitted from this text because it would likely be outdated before the text left the printer. You can access a set of Web source links through the Web site that supports this text.

Professional Societies

There are several professional societies that are excellent sources of information about computer technology, including:

▶ **Association for Information Technology Professionals (AITP).** The **AITP** was known as the Data Processing Management Association (DPMA) before 1996. The membership consists primarily of IS managers and application developers. AITP has local chapters throughout the country and publishes several periodicals, including *Information Executive*. Their Web site is www.aitp.org.

▶ **Association for Computing Machinery (ACM).** The **ACM** is a well-established organization with a primary emphasis on computer science. ACM has dozens of special-interest groups. ACM sponsors hundreds of research conferences each year. ACM publishes dozens of periodicals and many technical books. The membership represents a broad cross-section of the computer community, including hardware and software manufacturers, educators, researchers, and students. Their Web site is www.acm.org.

▶ **Institute for Electrical and Electronics Engineers (IEEE) Computer Society.** The **IEEE Computer Society** is a subgroup of the IEEE that specializes in computer and data communications technology. The membership is largely composed of engineers with an interest in computer hardware, but there are many members with software interests and many with academic and other backgrounds. The IEEE Computer Society sponsors many conferences (many jointly with the ACM). Its publications include several periodicals (e.g., *IEEE Computer*) and a large collection of technical books and standards. Their Web site is www.computer.org.

SUMMARY

The development of ISs requires technical knowledge of computer hardware and systems software. The depth and breadth of required knowledge differs among phases of the structured systems development life cycle (SDLC). Breadth of knowledge is required in the survey phase to evaluate feasibility and related issues. Depth of knowledge is most required during systems design, when detailed specifications for hardware and systems software are derived.

IS managers also require technical knowledge. Particular attention must be given to computer resource compatibility and future trends. Compatibility is important because computer hardware and systems software are typically shared among organizational units and subsystems. Future trends must be considered in computer acquisitions because of the long-term nature of hardware and software investments.

Technical knowledge must be constantly updated to incorporate changes in the hardware and software technology. IS professionals must engage in continuing education and study to keep pace with these changes. Training is available through vendors, educational organizations, and self-study, which relies heavily on periodical literature. IS professionals must exercise care in choosing periodical literature because of the differences among them in intended audience and required background and training.

Information about computer hardware and software is readily available on the World Wide Web (WWW). Sites of interest include those maintained by publishers, vendors, manufacturers, and professional organizations. The range of available information is substantial, but IS professionals must exercise care when relying on the Web as a source of technical information because much of the available information is biased and/or incomplete.

Key Terms

application programmer
application software
application tool
Association for Computing Machinery (ACM)
Association for Information Technology Professionals (AITP)
chief information officer
computer engineering
computer operations manager
computer operator
computer science

database administrator
Institute for Electrical and Electronics Engineers (IEEE) Computer Society
Internet site name
needs analysis
network administrator
programmer
programmer/analyst
requirements analysis
structured SDLC
systems analysis
systems analyst

systems design
systems designer
systems development life cycle (SDLC)
systems implementation
systems programmer
systems survey
Web
user
Web site
World Wide Web (WWW)

Vocabulary Exercises

1. Students of _____ generally focus on system software whereas students of ISs generally focus on _____.

2. Configuration of hardware and system software occurs during the _____ phase of the structured SDLC.

3. Professional organizations with which an IS student or professional should be familiar include _____, _____, and _____.

4. Selection of hardware and system software occurs in the _____ phase of the structured SDLC.

5. A(n) _____ is typically responsible for a large computer center and all of the related software.

6. Computer specialties most concerned with hardware and the hardware/software interface are _____ and _____.

7. An initial estimate of hardware and system software requirements is made during the _____ phase of the structure's SDLC.

8. The job titles of persons directly responsible for developing application software include _____, _____, and _____.

Review Questions

1. In what way(s) is the knowledge needed to operate complex devices different from the knowledge needed to acquire and configure them?

2. What knowledge of computer hardware and systems software is necessary to successfully complete the survey phase of the systems development life cycle (SDLC)?

3. What knowledge of computer hardware and systems software is necessary to successfully complete the design phase of the SDLC?

4. What additional technical issues must be addressed when managing a computer center or local area network (LAN) as compared to developing a single IS application?

Research Problem

The Bureau of Labor Statistics (BLS) compiles employment statistics for a variety of job categories and industries. BLS also produces predictions of employment trends by job category and industry. Most of the information produced by the BLS is available on the Web. Access the BLS Web site (stats.bls.gov) and investigate the current and expected employment prospects for IS professionals.

Chapter 2

Computer Hardware

Chapter Goals

► Describe the functions and capabilities of automated computing devices and the means by which to implement them.

► Describe the functions, capabilities, and operation of a computer processor.

► Describe the components of a computer system and the function of each.

► Describe the classes of computer systems and their distinguishing characteristics.

► Describe the ways in which computer system power is measured and compared.

AUTOMATED COMPUTATION

Automated computational devices share three capabilities:

► To perform computation.

► To store data.

► To accept input and return output (i.e., communicate).

Each of these functions may be implemented by many methods and/or devices. In automated devices, computations may be implemented by many methods and/or devices, including mechanics[1], electronics, and optics. For example, a computer may do calculations electronically (e.g., using transistors within a microprocessor), store data optically (e.g., using a laser and the reflective coating on an optical disk), and communicate using a combination of electronics and mechanics (e.g., the mechanical and electrical components of a printer).

Mechanical Implementation

Early mechanical computational devices were designed to perform repetitive mathematical calculations. The most famous of these machines, called the "difference engine," was built by Charles Babbage in 1821. This entirely mechanical device computed logarithms through the movements of various gears and other mechanical components. Many other mechanical computational machines (e.g., mechanical adding machines used by bookkeepers) were developed well into the Twentieth Century and were used as recently as the 1960s.

The common element in all early computational devices was a mechanical representation of a mathematical calculation. Consider the simple example of a mechanical clock driven by a spring and pendulum. Each swing of the pendulum allows a gear to move one step under pressure from the spring. As time passes, the pendulum swings repetitively, and the gears advance the hands of the clock.

A clock is a relatively simple adding machine in that each swing of the pendulum performs addition by advancing the position of the gears. The current position of the gears is the sum of their initial position at some time in the past and the accumulated (added) value of the time that has passed since the clock was started. A set of hands attached to one or more of the gears displays this position in relation to the numbers printed on the face of the clock.

A more modern example of simple mechanical calculation is an automobile odometer. The wheels of the car are connected to the odometer through a series of gears and mechanical links. As the wheels of the car move forward, the physical movement is transmitted by the gears and links and is accumulated by

---

[1] The term **mechanics** refers to the use of interconnected (or interacting) moving parts.

the odometer in terms of the current position of its display wheels. Each of the display wheels has numbers printed upon it, which provides a display of the current accumulated value (miles the car has driven).

More complex computational functions also can be represented mechanically. Multiplication of whole numbers, for example, can be implemented mechanically as repeated addition. Thus, a machine capable of addition can perform multiplication by executing the addition function multiple times (e.g., six times three can be calculated by adding six to six and then adding six to the result).

Although a variety of computational functions can be performed mechanically, mechanical computation has inherent shortcomings. The first of these is the difficulty in designing and building mechanical devices. Automated computation through the use of gears and wheels requires a complex set of components that must be designed, manufactured, and assembled to exacting specifications. As the complexity of the computational function increases, the complexity of the mechanical device that performs it also increases, thus exacerbating these manufacturing problems.

Mechanical devices are also subject to wear and breakdown and tend to require frequent maintenance. In a mechanical clock, for example, the internal gears are subject to friction-induced wear. As a result, the size of the gear teeth decreases, and the gears interact with less and less precision. Thus, the clock starts to run fast or slow and requires adjustment. This problem can be minimized by regular cleaning and lubrication, but that can only slow down the wear, not eliminate it. Eventually, the clock will either cease working or become so unreliable as to become unusable.

Another problem with mechanical devices is the speed with which they perform their function. The speed of any device that uses physical movement to implement a function is limited by the maximum possible speed of the moving parts. A car's speed, for example, is limited not only by the power of the engine but by the physical limits of transmission gears and engine components. Causing any of these components to move too quickly (e.g., revving the engine to 20,000 revolutions per minute) will cause rapid breakdown of the device.

Electrical Implementation

Much as the era of mechanical clocks gave way to the era of electrical clocks, the era of mechanical computation eventually gave way to electrical computers. The change to electricity was motivated by increasing knowledge of how to effectively use it and a desire to overcome the shortcomings of mechanical computation. The major impetus in the era of electrical computing came just before and during World War II. Many military challenges (e.g., navigation, breaking codes, etc.) required massive amounts of complex computation. Because the mechanical devices of the time were not up to the task, the quest for electrically based computation began.

In an electrical computer, the movement of electrons performs essentially the same functions as those performed by the gears and wheels of earlier mechanical computers. Storage of numerical values electrically is accomplished by the storage of magnetic charges rather than by the position of gears and wheels. Where necessary, electricity can be used to cause physical movement (e.g., the hands of an electrical clock, the pins of a dot matrix printer, etc.).

Electrical computers addressed most of the shortcomings of mechanical computation. They were inherently faster due to the relatively high speed at which electrons move. With improvements in the design and construction of electrical devices, they became more reliable and easier to build than their mechanical counterparts. Electrical computers made it possible to perform complex calculations at a speed previously thought impossible, which allowed the addressing of larger and more complex problems and simplified the solution process.

Optical Implementation

Light also may be used as a basis for implementing computation. In direct parallel with electrical implementation, light can be considered a particle (i.e., a photon) that moves at a high rate of speed. As with electricity, the energy of the moving optical particles can be harnessed to perform computational work. Light can be transmitted over appropriate conductors (e.g., fiberoptic cables). Data can be represented as pulses of light and stored either directly (e.g., an image stored as a hologram) or indirectly by materials that reflect (or don't reflect) a light source (e.g., the bits of information on an optical disk).

Optical implementations are at the leading edge of computer hardware technology. Optical data communication is now common in computer networks that cover relatively large distances. Currently, optical devices are in use for storing and retrieving large amounts of data. Some input/output devices (e.g., laser printers and optical scanners) are based on optical technologies and devices. Purely optical and hybrid electro-optical devices have been developed for communication among computer system components and for computer processing. These devices are largely experimental; however, it is expected that optical and electro-optical technologies will find much wider application in the computer hardware of the next decade and beyond.

COMPUTER PROCESSORS

A **processor** is a device that performs data manipulation and/or transformation functions. These functions are usually computational in nature but may include other functions (e.g., comparison and the movement of data among storage locations). A processor is usually capable of many different functions, including addition, subtraction, multiplication, division, and comparison. An

instruction is a signal (or command) to a processor to perform one of its functions. When a processor performs a function in response to an instruction, it is said to be **executing** that instruction.

Each of the functions that a processor can perform is simple, or primitive. An example of a simple processor function is the addition of two numbers. Complex functions are performed as a sequence of simple functions. For example, a typical processor cannot add a set of 10 numbers together in response to a single instruction. Rather, it must be instructed to add the first number to the second and to temporarily store the result. Then it must be instructed to add the stored result to the third number and to temporarily store that result. Individual instructions must be issued and executed until all 10 numbers have been added.

Most useful computational tasks (e.g., recalculating a spreadsheet) require that a long sequence of instructions be executed in a **program**. Usually, a program is a complex mixture of different processing operations. For example, payroll calculation requires a complex series of input, computation, storage, and output functions to be performed on many data items. Some programs (e.g., the addition example above) require the repetitive execution of similar instructions.

A processor can be classified as either general purpose or special purpose. A **general purpose processor** can execute many different instructions in many different sequences or combinations. The task such a processor performs can be altered by changing the program that directs its actions. Thus, a general purpose processor can be instructed to do many different tasks (e.g., payroll calculation, text processing, scientific calculation, etc.) simply by supplying it with an appropriate program.

A **special purpose processor** is designed to perform only one specific task. Although this processor may be capable of executing many types of instructions, it can only execute them in one sequence. This is analogous to a general purpose processor that has only one program. Many commonly used devices such as simple calculators, microwave ovens, video cassette recorders, and computer printers contain special purpose processors. Although such processors (or the devices that contain them) can be called computers, the term **computer** is more often used to refer to a device containing a general purpose processor that can execute any program supplied to it.

COMPUTER SYSTEM CAPABILITIES

Although a computer system is an automated computing device, all automated computing devices are not computer systems. However, the distinction between the two is not absolute. The primary characteristics that distinguish a computer system from simpler automated computation devices include:

▶ A general purpose processor capable of performing computation, data movement, comparison, and branching functions.

- Storage capacity sufficient to hold large numbers of program instructions and data.
- Flexible communication capability through the use of multiple communication media and devices.

Using these characteristics and criteria, a device such as an IBM personal computer (PC) system would be classified as a computer, whereas devices such as adding machines and calculators would not.

Processor Functions

Some processing tasks require little more than the computational functions of a processor. Examples of these tasks include single-step and multiple-step computations, such as the following calculation:

Profit = Sales - Expenses - Taxes

To compute profit, the processor needs the ability to perform subtraction and to store a single temporary result. Thus, expenses may be subtracted from sales and the result stored in a temporary location, and taxes are then subtracted from the temporarily stored value. The result is profit. Processing tasks such as this one are called **formulaic problems**. They can be expressed as a formula and solved by a processor using a sequential series of computational operations.

The vast majority of processing tasks that computers are expected to perform are not formulaic problems. Although many processing tasks contain subtasks that can be solved by formulaic approaches, additional processor capabilities are required, including the ability to compare data items and to alter the sequence of instruction execution (i.e., **branching**). Programming languages implement this capability with an if-then or if-then-else statement. A problem-solving procedure (or program) that uses comparison and branching functions is called an **algorithm**.

For example, consider the computation of taxes within a payroll program. In the United States, most income taxes are computed at progressively higher rates on higher levels of income. Figure 2-1 shows a typical computational method for income taxes. Note that different levels of income require the use of different formulas in order to calculate the correct amount of tax. Thus, a program that computes tax based on this table must be capable of using different formulas depending upon the income level supplied to the program. For a program to exercise this capability, the processor that executes the program must be capable of both comparison and branching functions.

Figure 2-1

Single: Schedule 1

If gross pay is: **The tax is:**

A payroll tax table

Over:	But not over:		Of the amount over:
$0	$2,465	15%	$0
2,465	5,975	$369.75 + 28%	2,465
5,975	12,465	1,352.55 + 31%	5,975
12,465	27,105	3,364.45 + 36%	12,465
27,105	—	8,634.85 + 39.6%	12,105

Figure 2-2 illustrates a program that uses comparison and branching functions to calculate taxes. In statements 20, 50, 80, and 110, an explicit comparison of GROSS_PAY to a constant is made to determine the next appropriate instruction to execute. Depending on the results of the comparison, the program may execute the next instruction in the sequence or alter the sequence by issuing a GOTO instruction.

Figure 2-2

A program to calculate federal income taxes. Comparison and branching functions are used to select the proper formula.

```
10      INPUT GROSS PAY
20      IF GROSS PAY > 2465 THEN GOTO 50
30      TAX = GROSS PAY * 0.15
40      GOTO 150
50      IF GROSS PAY > 5975 THEN GOTO 80
60      TAX = 369.75 + (GROSS_PAY - 2465) * 0.28
70      GOTO 150
80      IF GROSS_PAY > 12465 THEN GOTO 110
90      TAX = 1352.55 + (GROSS PAY - 5975) * 0.31
100     GOTO 150
110     IF GROSS PAY > 27105 THEN GOTO 140
120     TAX = 3364.45 + (GROSS_PAY - 12465) * 0.36
130     GOTO 150
140     TAX = 8634.85 + (GROSS_PAY - 27105) * 0.396
150     OUTPUT TAX
160     END
```

The ability to do comparison and alter program execution sequence is the basis for implementing intelligent behavior within a computer processor. The comparison instructions are sometimes called **logic instructions**, which implies their relationship to intelligent behavior. Much of human intelligence is based on the ability to recognize patterns, conditions, and other aspects of the environment and to (re)act accordingly.

Humans can compare complex objects and phenomena and handle uncertainty in the resulting conclusions. A general purpose processor within a computer is substantially more restricted in its comparative abilities. It can only perform simple comparisons (e.g., equality, less than, and greater than) with numeric data where the results are either completely true or completely false. Despite these limitations, comparison and branching are the basic capabilities by which the processor may solve complex problems. These capabilities also distinguish a computer processor from the processors of lesser automated computation devices such as adding machines and calculators.

Storage Functions

The storage capacity of a computer system must be large for a variety of reasons, including the need to store:

▶ Intermediate processing results.

▶ Currently executing programs.

▶ Programs that aren't currently being executed.

▶ Data used by currently executing program(s).

▶ Data that will be needed by programs in the future.

The need to store intermediate results arises from the need to perform problem-solving tasks using primitive operations (e.g., a single addition operation within a multistep formula). Many programs that solve real world problems require the storage of hundreds, thousands, or millions of intermediate results during program execution. To avoid processing delays, intermediate results should be stored in devices that operate at speeds comparable to the processor.

A typical program consists of many instructions, which may number in the thousands for simple programs to the millions for extremely complex programs. In modern computer systems, the processor may be executing many programs simultaneously. A computer processor reads and executes these instructions one at a time. As with intermediate results, instructions should be stored for rapid access to ensure fast program execution.

Data must also be stored for present and/or future use. A user may require storage of (and access to) thousands (or millions) of data items. A large organization may require storage of (and access to) billions of data items. This data may be used by currently executing programs, held for future processing needs, or archived.

Communication Functions

A computer's communication capability must be flexible and encompass a variety of media and devices. A typical small computer has several communication devices including a video display, keyboard, mouse, printer, and modem or network interface. A large computer typically has many more communication devices and may use devices of substantially more power and complexity than a smaller computer.

COMPUTER SYSTEM COMPONENTS

The components of a **computer system** are illustrated in Figure 2-3. Each component addresses one of the computer capabilities described in the previous section. The number, implementation, complexity, and power of these components may vary substantially from one computer system to another, but, the functions generally are similar.

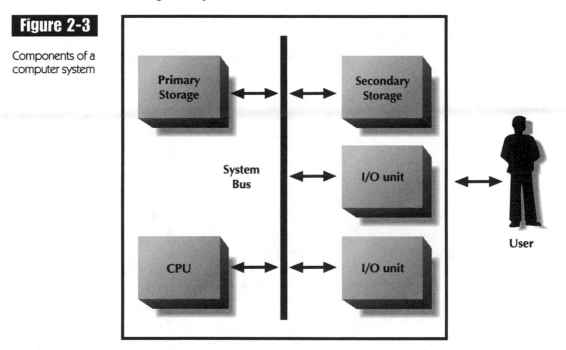

Figure 2-3

Components of a computer system

The "heart and brain" of the computer system is the **central processing unit (CPU)**. The CPU is a general purpose processor, as described earlier. Each of the other computer system components is attached to the CPU by the **system bus,** which serves as a general purpose communication channel.

Instructions and data from currently executing programs flow to and from **primary storage**. **Secondary storage** holds programs that are not currently being executed as well as groups of data items that are too large to fit in primary storage. Secondary storage may be composed of several different devices (e.g., multiple disk drives) although only one device is shown in Figure 2-3.

The CPU is also connected to a number of **input-output (I/O) units**, which allow the CPU (and the computer system as a whole) to communicate with the outside world (e.g., to a user through a video display and keyboard). Two such units are shown in the diagram, although the actual number of I/O units can vary widely from one computer system to another.

Central Processing Unit

Figure 2-4 illustrates the components of central processing unit (CPU). In most computers, these are:

▶ Arithmetic logic unit (ALU).

▶ Registers.

▶ Control unit.

Figure 2-4

Components of the central processing unit (CPU)

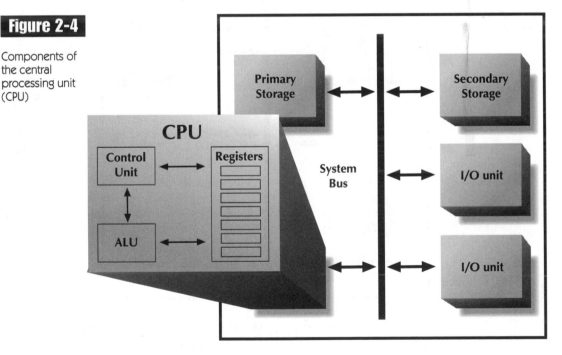

Computer Hardware

The **arithmetic logic unit (ALU)** is composed of electronic circuits that perform the processor's computational (e.g., addition and subtraction) and logical (e.g., comparison) functions. Different portions of the circuitry correspond to different functions. The execution of a math or logic instruction causes data (electronic signals) to flow through the appropriate portion(s) of the circuitry. The typical arithmetic instructions include addition, subtraction, multiplication, and division. More advanced computational functions such as exponentiation and logarithms also may be implemented. Logic instructions include various forms of comparison (e.g., equality, greater than, less than) and other instructions that will be discussed in Chapter 5.

A small set of temporary storage locations (or cells) are located within the CPU. The cells are called **registers** and can typically hold a single instruction or data item. Registers are used to store data or instructions that are needed immediately, quickly, and/or frequently. For example, two numbers that are about to be added together are each stored in a register. The ALU reads these numbers from the registers and stores the result of the addition in another register. Because they are located within the CPU, the contents of registers can be accessed quickly by the other CPU components.

The **control unit** is responsible primarily for the movement of data and instructions to and from CPU registers and for controlling the ALU. As program instructions and data are needed, they are moved from primary storage to registers by the control unit. The control unit examines incoming instructions to determine how they should be processed. Computation and logic instructions are routed to the ALU for processing. Data movement instructions to/from primary storage, secondary storage, and I/O devices are processed by the control unit itself.

A complex chain of events occurs when a computer executes a program. To start, the first instruction of the program is read from primary storage by the control unit. The control unit stores the instruction in a register and, if necessary, reads data inputs from primary storage and stores them in registers. If the instruction is a computational or comparison instruction, then the control unit signals the ALU what function to perform, where the input data is located, and where to store the output data. Other types of instructions (e.g., input and output to secondary storage or I/O devices) are executed by the control unit itself. When the first instruction has been executed, the next instruction is read and executed and so forth until the final instruction of the program has been executed.

The steps required to process each instruction can be divided into two groups: the **instruction cycle** and the **execution cycle**. The instruction cycle also may be referred to as the **fetch cycle**, or simply as a **fetch**. The operation of the instruction and execution cycles for computation and comparison instructions is depicted in Figure 2-5. The sequence of events performed during the instruction cycle is as follows:

1. The control unit retrieves an instruction from primary storage and increments a pointer showing the location of the next instruction.

2. The instruction is separated into its components: the instruction code (or number) and the data inputs to the instruction. Each of these components is stored in a register.

3. The control unit generates an internal signal to the ALU to execute the instruction.

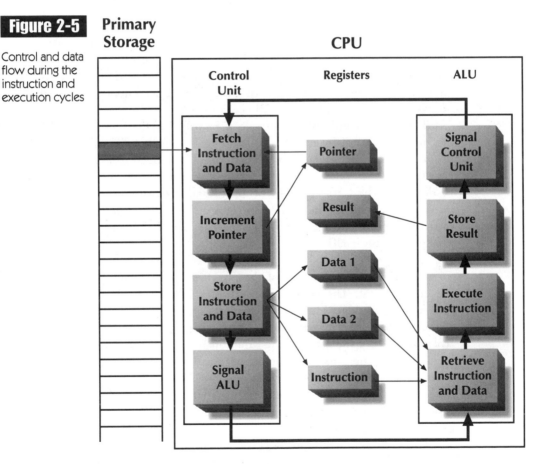

Figure 2-5

Control and data flow during the instruction and execution cycles

The sequence of events during the execution cycle is as follows:

1. The ALU accesses register contents to determine what function (instruction) to perform and to obtain the input data for the instruction.

2. Data is passed through circuitry to perform the function (instruction).

3. The processing result is placed in a register or returned to the control unit to be written to memory.

At the conclusion of the execution cycle, a new instruction cycle is started. Thus, a central processing unit executes a program by constantly alternating between the instruction and execution cycles. This switching is precisely regulated by an internal clock that allocates fixed intervals of time to each cycle.

System Bus

The system bus is a set of parallel communication lines that serves as the primary pathway for data transmission among computer system devices. This channel's capacity is a critical factor in the overall performance of the computer system. A powerful CPU requires a high-capacity system bus to keep it supplied with instructions and data from primary storage. Bus capacity is also critical to high-performance secondary storage and I/O devices.

von Neumann Computer Architecture

In a broad sense, the underlying architecture of computers has not changed since the early 1950s. This architecture was first described by a committee at the University of Pennsylvania. A mathematician, John von Neumann, contributed many of the essential architectural concepts including the stored program. The essential characteristics of this type of architecture are those discussed in the previous sections:

▶ Use of a single general purpose processor.

▶ Use of stored programs.

▶ Sequential processing of instructions.

▶ Alternating instruction and execution cycles.

Computers designed around these basic concepts came to be called **von Neumann machines**, although the term is no longer widely used.

Until recently, virtually all computers used for information processing were von Neumann machines. The increases in speed and power that occurred over the last several decades were primarily a result of improvements in the design and construction of computer circuits and components, not of changes in the basic architecture. Some changes to von Neumann architecture have found their way into current computers.

Primary Storage

Primary storage consists of storage cells that hold programs (and their associated data) that are currently being executed. Primary storage is usually referred to as **main memory** or, simply, as **memory**. Program execution requires continual movement of both instructions and data between main memory and the CPU. Because the CPU is a relatively fast device, it is desirable to implement main memory using devices capable of rapid access.

Primary storage can be thought of as a sequence of contiguous (or adjacent) memory cells as shown in Figure 2-6. Each cell is large enough to contain a single instruction or data item, and each has a unique address corresponding to its physical location in the sequence. The CPU uses these addresses to specify which instructions are to be loaded into the CPU (during the instruction cycle) and what data is to be read or written. Because the CPU may require instructions or data from many different parts of memory, it is necessary to implement primary storage with a device capable of **direct access** (or **random access**). These terms refer to the ability to specify that a specific location be read or written, ignoring any other locations that may occur before or after it.

Figure 2-6

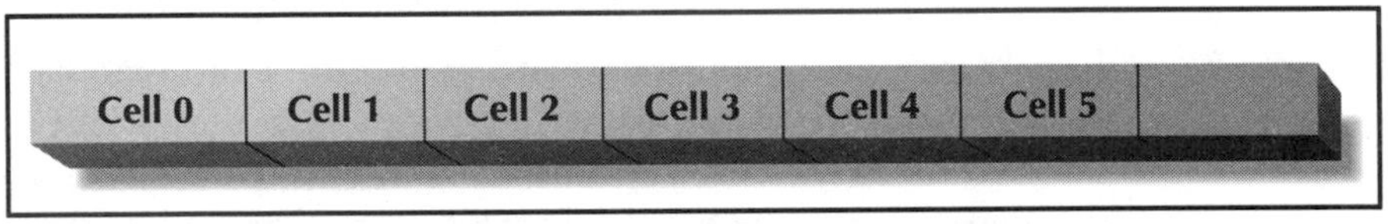

Depiction of the sequential organization of cells within primary storage

| Cell 0 | Cell 1 | Cell 2 | Cell 3 | Cell 4 | Cell 5 | |

In current computer hardware, main memory is implemented with silicon-based semiconductor devices commonly referred to as **random access memory (RAM)**. These devices provide the access speed required by the CPU and allow the CPU to read or write to a specific memory cell by referring to its location. Unfortunately, RAM is relatively costly, which limits the amount of main memory that can be included in a computer system. Another problem with RAM is that it does not provide permanent storage. When power to RAM is turned off, its contents are lost. This characteristic of RAM (or any other storage device) is called **volatility**. Any type of storage device that cannot retain data values indefinitely is said to be volatile; in contrast, storage devices that permanently retain data values are said to be nonvolatile. Due to both the volatility and limited capacity of primary storage, a computer system requires auxiliary means of storing data and programs over long periods.

Secondary Storage

The term secondary storage refers to storage devices that are nonvolatile and able to store large volumes of data. Secondary storage devices are used to store programs that are not currently being executed. They also store data that is not needed by currently executing programs as well as data that cannot fit into primary storage. For various architectural and efficiency-related reasons, the contents of secondary storage are first moved to primary storage before they are accessed by the CPU.

Within a typical IS, the number of programs and amount of data is large, which means that a typical computer system must contain a large amount of secondary storage as compared to its primary storage. For example, a typical mid-sized computer might contain 64 million primary storage locations and billions of secondary storage locations. Differences in utilization and implementation of the various types of storage in a computer system are summarized in Table 2-1.

TABLE 2-1	Storage Type	Implementation	Contents
A summary of the differences in implementation and usual content among various types of storage in a computer system	Registers	Very high-speed electrical devices within the CPU	Currently executing instruction, a few items of related data
	Primary Storage	High-speed electrical devices (RAM) outside of the CPU	Entire programs currently being executed, small amounts of data
	Secondary Storage	Low-speed electromagnetic or optical devices (e.g., magnetic and optical disk)	Programs not currently being executed, large amounts of data

To keep the total cost of secondary storage within acceptable limits it is necessary to use storage devices that provide relatively slow access speeds and/or limited access methods as compared to primary storage devices. The most common devices used to implement secondary storage are magnetic disks, optical disks, and magnetic tape. Disk storage provides relatively fast access (compared to magnetic tape) and allows each storage location to be accessed directly through its address (i.e., direct or random access). Magnetic tapes provide a slower, but cheaper, method of storage.

Input/Output Devices

To be useful to humans, a computer system must have the ability to communicate with them. It is also desirable to allow the computer system to communicate directly with other computers or processing devices. This role is filled by a general class of devices called I/O devices, which are implemented with a wide range of technologies, depending on the exact nature of the communication they support.

Examples of input devices for human use include keyboards, pointing devices (e.g., a mouse), and voice-recognition devices. The purpose of such devices is to accept input on human terms (e.g., voice or keystroke input) and convert that input into the computer's language (electrical signals). Output devices for human use include video displays, printers, plotters, and speech and sound output devices. All of these devices perform a conversion from electronic signals to a communication medium that humans can understand (e.g., pictures, words, or sound).

Other input/output devices include modems, network interface units, and multiplexers. These devices provide communication capabilities among computers or between a computer and a distant I/O device.

CLASSES OF COMPUTER SYSTEMS

Computer systems can be loosely classified into the following categories based on performance and capabilities:

► Microcomputers.
► Minicomputers.
► Mainframes.
► Supercomputers.

A **microcomputer** is a computer system designed to meet a single user's information processing needs. A microcomputer may also be called a **personal computer (PC)**, or **workstation**. The term personal computer often implies use in a home, and the term workstation often suggests use in a work setting. The hardware capabilities of a microcomputer are designed to meet the typical computing needs of a single user. Examples of processing performed with microcomputers include word processing, computer games, and small- to medium-sized application programs. Examples of small- to medium-sized application programs include programs to compute an individual's income tax, perform budgeting for a home or small business, and compute payroll for a small business.

Do We Need a Mainframe?

Bauer Industries (BI) is a manufacturer and wholesale distributor of jewelry making machinery and components. The company has 200 employees and approximately 2,500 products. The current inventory of automated systems—small for a company of its size—includes order entry, inventory control, and general accounting functions (e.g., accounts payable, accounts receivable, payroll, and financial reporting). These systems currently run on an older Digital Equipment Corporation minicomputer (a VAX 6000). Dedicated character-based video display terminals are used to interact with application programs.

BI has grown rapidly over the last decade, which has strained the company's computer hardware to its maximum capability. Currently, the VAX 6000 is configured with the maximum possible disk space and memory. The order-entry system fully uses memory and CPU capacity during daytime hours. The result is a rapid rise in response time (from a few seconds to as much as a minute or more) for the 25 order-entry clerks.

BI management realizes the seriousness of the situation, but they are uncertain how to solve it. There are additional factors complicating the decision, including:

► A planned fourfold increase in business over the next 10 years.
► A planned move to new facilities within the next year (the site/building has already been acquired).
► A desire on the part of some members of management to substantially increase and modernize automated support of operating and management functions.

Management has identified three viable options to address the current problem:

1. Purchase another VAX 6000 system (used, because they are no longer produced) and partition the application software among the old and new systems.

2. Upgrade to a small mainframe class DEC system (e.g., an AlphaServer 8000 series machine).

3. Develop a modern client/server based architecture consisting of microcomputers, a medium-sized server (e.g., an AlphaServer 4000 series computer), and a LAN.

The first option is least disruptive to the status quo but has two obvious disadvantages. The first is the use of outdated hardware—parts for VAX 6000s are only available from abandoned or broken machines. The second problem is data overlap among application programs. The current operating and application software will not support access to data stored on one machine by applications executing on another machine. Thus, data will have to be redundantly stored on both machines and periodically synchronized to reflect updates made by different application programs. Estimated two-year costs are $75,000 for hardware and $100,000 for labor.

The second option allows much of the current hardware investment to be reused (including disk drives and video display terminals). It also guarantees enough hardware capacity to operate current software at many times the current transaction volume. However, the new hardware will not support the old operating system. The effects of an operating system change on the application software are not fully understood. The new operating system was designed to support applications executing under the old operating system. But BI management has heard of and read about many problems involved in migrating application software. Estimated two-year costs are $350,000 for hardware and $150,000 for labor.

The last option is clearly the most expensive and the biggest departure from current operating methods. Migration problems are essentially identical to the second option. Existing video display terminals or microcomputers executing video display terminal simulation software (e.g., TELNET) could be used. This option appears to hold the most promise for implementing a new generation of application software. But BI has no experience and little expertise in building or maintaining client/server application software. Estimated two-year costs are $150,000 for hardware and $400,000 for labor (including hiring or contracting three new IS staff for two years).

Questions

1. What problems would you anticipate within two to five years if Option 1 is implemented? Option 2? Option 3?

2. Management has historically opted for the lowest-cost solutions (in "hard dollars"). How can the higher costs of Options 2 and 3 be justified?

3. Which alternative would you recommend, and why?

The term workstation often implies something more than a typical microcomputer. This is especially true of single-user computers in scientific and engineering organizations. Processing tasks in these environments demand more computer hardware capability than do typical business and home processing tasks. Examples of such tasks include complex mathematical computation, computer aided design, and the composition and display of high-resolution video (graphic) images. These tasks generally require more processor speed, more storage capabilities (in both quantity and speed), and higher-quality video display devices than are provided with an ordinary microcomputer. Particularly with respect to processor speed and storage capabilities, the power of a workstation is often similar to that of a minicomputer. However, the overall design of a workstation is still targeted toward a single-user operating environment.

Recently, a new type of microcomputer and/or workstation called a **network computer** has emerged. The primary distinguishing feature of a network computer is a lack of directly attached secondary storage capacity. Network computers rely on one or more network servers to supply their operating system and application software. "Ordinary" workstations load their operating software from secondary storage when they first start (i.e., on **boot up**). Network computers are configured to establish a connection to a server when they are powered on, and they boot up from the servers secondary storage.

Network computers provide some key management benefits compared to ordinary workstations. First, they are somewhat cheaper than workstations because of their lack of secondary storage. The cost reduction depends on the amount and speed of secondary storage not purchased (typically, between a few hundred and one thousand dollars). Second, it is far easier to update their software configuration. Any number of network computers can be reconfigured simply by updating the boot up and/or application software stored on a server. This is far simpler than updating the secondary storage content of many individual workstations.

A **minicomputer** is designed to provide information processing for multiple users and to execute many application programs simultaneously. Typically, they are designed to allow up to 100 users to interact with the computer system simultaneously. Supporting multiple users and programs requires fairly extensive capabilities for processor speed, storage, and I/O. It also requires more sophisticated system software than is typically found on microcomputers.

A **mainframe** computer system can handle the information processing needs of a large number of users and applications. These machines are capable of interacting with hundreds of users at a time and of executing hundreds of programs at a time. A mainframe might include hundreds of users entering customer orders, several programs generating periodic reports, various users querying the contents of a large corporate database, and an operator making backup copies of disk files—all simultaneously! The key hardware capability needed to support such a processing environment is the ability to quickly move large quantities of data from one place to another (e.g., secondary storage to primary storage or network I/O devices). A relatively fast CPU and large amounts of primary and secondary storage are also required, but the computer system is optimized primarily for data movement.

A **supercomputer** is intended primarily for one purpose—to perform large amounts of mathematical computation as quickly as possible. These machines are used for the most demanding of computational applications, including simulation, three-dimensional modeling, weather prediction, and computer animation. All of these tasks require an extremely large number of complex calculations and thus demand a CPU with the highest possible computational speed. Storage and communications requirements are also extremely high but are secondary to speed of the processing component(s). Typically, supercomputers are implemented using the very latest (and expensive) in computer technology.

The term **server** can be applied to computers as small as microcomputers and as large as supercomputers. The term does not imply any minimum set of hardware capabilities; rather, it suggests a specific mode of use. Specifically, it implies a computer system that manages one or more resources (e.g., shared file systems, databases, Web sites, printers, and high-speed CPUs) and allows access to those resources by users over a local (LAN) or wide area network (WAN).

The required hardware capabilities depend on the resources managed and the number of simultaneous users of those resources. For example, a microcomputer may be more than sufficient for low-intensity resource sharing on a LAN (e.g., sharing a file system and two printers among a dozen users). More demanding resources (e.g., large databases) accessed by thousands of users might require a mainframe computer.

Table 2-2 summarizes the configuration and capabilities of each class of computer system. Each class is represented by a typical model available in 1997. The performance and other specifications of all computer classes are in a constant state of flux. Rapid advancements in computer technology lead to rapid improvements in computer capability as well as a redefinition of both the classes themselves and the expected capabilities of computers within that class. There is also a technology pecking order. That is, the "latest and greatest" (and most expensive) hardware and software technology usually appears first in supercomputers. As experience is gained and costs decrease, these technologies start moving downward through the less powerful and costly computer classes.

TABLE 2-2	Class	Typical Product	Typical Specifications	Approximate Cost	Processor Speed
This table compares representative models of each class of computer and their typical costs and performance capabilities.	Micro-computer	IBM 300 PL 5/200	32 million main memory cells 4.2 billion disk storage cells Single CPU Single user	$2,000	25 million instructions per second
	Workstation	IBM IntelliStation 6/300	128 million main memory cells 9 billion disk storage cells 2 CPUs High-performance video I/O Single user	$10,000	100 million instructions per second
	Mini-computer	IBM RS/6000 S70	512 million main memory cells 26 billion disk storage cells 4 CPUs up to 256 interactive users	$100,000	250 million instructions per second
	Mainframe	IBM S/390 R65	24 billion main memory cells 100 billion disk storage locations 6 CPUs 32 high-capacity I/O channels Up to 1,000 users	$1,000,000	350 million instructions per second
	Super-computer	IBM RS/6000 SP	16 billion main memory cells 100 billion disk storage cells No interactive users 64 CPUs	$2,000,000	4 billion floating point operations per second

DEC Alpha-Based Computer Systems

Digital Equipment Corporation (DEC) first introduced the Alpha microprocessor in 1992. The processor was designed to be fast and usable in a wide variety of computers. DEC intended to build their next generation of computer systems (including everything from PCs to mainframes) using the Alpha. Their current product offerings provide ample evidence of both the power of the Alpha processor and the company's ability to design and build many different classes of Alpha-based computer systems.

The current generation of the Alpha processor is the 21164, which is a 64-bit processor with clock rates ranging from 266–600 MHz. The processor contains a small memory cache on the chip and can be combined with a larger memory cache (up to 4 megabytes) in a multiple-chip module. The 21164PC processor version is designed for full-motion video playback and editing. It has on-chip support for encoding and decoding compressed motion video.

Typical configurations of Alpha-based computer systems in the microcomputer, workstation, minicomputer, and mainframe classes are summarized in Table 2-3. The XL series of personal workstations fill the microcomputer portion of the product line. These are designed to run the Microsoft Windows NT Workstation operating system and function as relatively high-powered desktop computers. The AlphaStation series of desktop computers is designed to fill the workstation portion of the product line. These systems are designed for demanding computational processing tasks and can be purchased in configurations customized for computer-aided design, 3D modeling, and motion video production. AlphaStation systems can be configured with high-performance graphic displays and up to 512 megabytes of memory.

The AlphaServer series of computer systems is a large group of machines spanning the range from low-end LAN servers to mainframe class systems. The 4000 and 4100 model series cover the normal range of minicomputer capabilities. The systems include one to four CPUs and may include up to 4 gigabytes of memory. I/O bandwidth is as high as 1 gigabyte/second, and a wide range of I/O devices can be configured.

TABLE 2-3	Model	Cost	Processor	Memory Capacity	Data Transfer Rate	Disk Capacity
Typical base computer system configurations for DEC computer systems based on the Alpha microprocessor (1997)	Alpha XL 266	$6,000	One 266-MHz CPU	32 megabytes	133 megabytes per second	1 gigabyte
	AlphaStation 500–400	$22,500	One 400-MHz CPU	64 megabytes	266 megabytes per second	2 gigabytes
	AlphaServer 4000–5/400	$50,000	One 400-MHz CPU	512 megabytes	1 gigabyte per second	16 gigabytes
	AlphaServer 8400–5/330	$580,000	Dual 330-MHz CPU	2 gigabytes	2 gigabytes per second	64 gigabytes

The AlphaServer 8200 and 8400 have performance specifications that place them clearly in the mainframe class. They may include up to 14 CPUs and 28 gigabytes of memory. I/O bandwidth ranges from 1 to 2 gigabytes/second. Over 100 separate I/O devices may be incorporated into the system. Secondary storage capacities range into the terabytes. Clearly, these systems are designed to move large amounts of data.

Supercomputer performance levels are not beyond the range of Alpha-based computers. No one model fits in the supercomputer class, but up to eight AlphaServers can be combined in a processing array. The CPUs of individual machines can share primary storage and communicate at high speed using advanced data transfer and switching technology. A multisystem configuration with a total of 64 CPUs can execute over 25 billion computations per second.

PERFORMANCE MEASUREMENT AND DESIGN FACTORS

The overall performance of a computer system depends on the power and speed of its various components, the speed of communication among the components, and the degree to which the components are "well matched" in speed and capacity. Because so many factors influence system performance, there are many ways to measure and tune (or tailor) performance to the needs of an individual user or set of applications. An engineer tends to measure computer capability with respect to the specific capability of particular hardware components (or groups of components). From a user perspective, though, raw hardware capacity represents only a potential capability. Actual capacity is realized in the form of useful work performed by the hardware and software.

Workload Performance

When looking at a computer system as a whole, a user is concerned primarily with the amount of work that it can perform. The primary problem with this perspective is the difficulty inherent in precise definitions and measurements of "work," which vary widely from one application and user to another. A further complication is that larger computer systems must respond to different work demands of many users and applications simultaneously. For example, the definition of work to a secretary may involve creating and editing letters and memos whereas the definition of work to a manager may involve complex retrievals of data and the analysis thereof. Each kind of work places different demands on the computer system's components. To measure a computer system's capability, it is necessary to define the term **work** precisely and to define a specific time reference.

The definition of work varies not only with the type of processing but also with the nature of human-computer interaction (or lack thereof) during normal processing. There are two broad classes of human-computer interaction with application programs. **Batch processing** describes a situation in which a user does not normally interact with an application program once execution has begun. **Interactive processing** (or **online processing**) describes a situation in which a user interacts directly with an executing application program. A typical IS contains application programs of both types.

For example, consider an IS that supports mail order sales. Online processing in this environment occurs through telephone operators who speak directly with customers while simultaneously interacting with an order entry/query application program. Orders are entered as they are received and then stored for use in later processing steps. One of these later processing steps is the generation of purchase orders to suppliers. An application program examines the products ordered during the recent past (by examining all stored orders). Based on this examination, predictions are made of goods needed to satisfy future orders and are used to generate purchase orders to suppliers automatically. Such an application program might execute automatically once per hour or day and not require any direct user interaction. All of its data inputs have been previously stored by the order entry/query program.

There are several units of work in the preceding example. One unit of work is the acceptance, editing, and storage of a single customer order. Another unit of work is a query of the status of a single order. The commonly used term for such units of work is a **transaction**. By establishing a time reference, the work capacity of the computer system can be stated in terms of transactions processed per time period (e.g., orders per hour). This type of work capacity measurement is normally called **throughput**. Defining a unit of work for the batch purchase order generation program is more difficult. It could be measured in terms of its inputs (customer orders), outputs (purchase orders), or simply as a single unit of work (execution of the entire program). In the latter case, throughput would be measured in terms of programs executed per unit of time (e.g., one hour).

When considering computer systems running a mixed variety of application programs, the measurement of total throughput becomes more difficult because allowance must be made for multiple types (and thus definitions) of work. For instance, a computer might be capable of executing 500 transactions and 10 programs per hour or provide a range of possible combinations (e.g., 1,000 transactions and 0 programs to 0 transactions and 45 programs).

Another way of measuring the work capacity of a computer system is to examine the interval necessary to satisfy a single processing request (e.g., a query of the status of a single order) by a single user. This measurement is usually described by a number called **response time**. Unfortunately, this measure is also difficult to determine since it depends on the number of processing requests that the computer is attempting to service at any one time. Response time will typically be fast (a low number) when few processing requests are being serviced and slow (a high number) when many processing requests are being serviced. Thus, response time cannot be accurately stated with a single number. Instead, it is expressed either as a formula or as a statistical distribution.

A typical specification of response time shows a range and an average (e.g., best: 0.5 seconds per transaction, worst: 10 seconds per transaction, average: 2 seconds per transaction). Because this performance measurement varies with total processing demand, it can only be determined by knowing the demand likely to be placed on the computer system. In other words, it isn't possible to state an average response time unless the average demand for processing is known. In the previous example, the demand for online processing would be expressed in terms of the number of new orders and order query transactions per unit of time.

Computer System Performance

Because most computer systems are designed to be general purpose (i.e., to support a wide variety of application programs), it is not possible for computer vendors to specify the types of workload measurements defined earlier. Instead, performance measurements are provided for various types of machine actions (e.g., executing an instruction, accessing main memory, reading 1,000 data items from a magnetic disk, etc.). It is the user's responsibility to determine how these various measurements of hardware capacity can be translated into measurements of throughput and response time for user transactions and application programs. Generally, performance measurements are given for each individual component (or subsystem) of the computer, the communication capacity among the subcomponents, and a general measurement that attempts to summarize all of these performance measurements.

The primary performance consideration for a CPU is the speed at which an instruction can be executed (or the time required to execute a single instruction), which is generally stated in terms of **millions of instructions per second (MIPS)**. Unfortunately, not all instructions take the same time to execute. Certain instructions execute quickly because of their inherent simplicity (e.g., addition of whole numbers). Instructions that implement more complex functions (e.g., division of fractional numbers) execute more slowly. Another common measure of CPU performance is **millions of floating point operations per second (MFLOPS)**. The term **floating point** refers to the internal representation of real numbers (numbers with both whole and fractional components) within a CPU. This performance measure is commonly given for scientific and engineering workstations and for supercomputers since it is expected that applications running on these machines will primarily perform floating point computations.

The capacity of computer storage subsystems is generally measured in **bytes**. A byte is a unit of data capable of holding all or part of a single data item (e.g., a single text character or one-fourth of a numeric value). Larger units include **kilobytes** (thousands of bytes), **megabytes** (millions of bytes), **gigabytes** (billions of bytes) and **terabytes** (trillions of bytes). Typically, secondary storage capacity is measured in megabytes or gigabytes. Primary storage capacity is usually measured in megabytes.

The speed at which data can be read from or written to a storage or input/output device depends on the **access time** (speed) of that device and the capacity and speed of the communication channel to/from the device. Access time is normally measured in **milliseconds** (thousandths of a second), **microseconds** (millionths of a second), or **nanoseconds** (billionths of a second). Access time is generally the same for both reading and writing unless otherwise stated. Access time for some devices (e.g., disk drives) may vary from one request to another, depending on various factors (e.g., the location of the requested data on the disk). In that case, access time is normally stated as an average value.

The capacity of a communication channel for a storage or I/O device may be less than the device's access time. In this case, the speed of the device is effectively reduced to the speed of the communication channel. A useful analogy is the movement of water within a plumbing system. Water may be supplied to a house by pipes with a large transfer capacity (e.g., dozens of gallons per minute). A water-consuming device (e.g., a showerhead) may be designed to operate optimally with a specific flow rate (e.g., three gallons per minute). But if the pipes connecting the water main to the device cannot meet the demand (e.g., because they are too small or are clogged), then the device will not operate at its full capacity.

In a computer, the system bus is the primary "plumbing system." There also may be several other "pipes" and intermediate devices between the supplier of data (e.g., a disk) and the consumer of data (e.g., a CPU). The sending device, receiving device, and entire set of devices forming the "pipe" between them are collectively referred to as a **communication channel**. The ability of a communication channel to support data movement is normally expressed as a **data transfer rate**, which states the amount of data that can be moved over the channel in a specified interval of time (e.g., 10 megabytes per second).

A computer system's overall performance is a complex combination of CPU speed, access times of the storage and I/O devices, and data transfer rates of the system bus and other communication channels. Because different applications make different demands on these various subsystems, any overall performance measure for the computer system is valid only for a particular instance or class of application program.

For example, an online order entry application will normally utilize the secondary storage and I/O subsystems extensively while using the CPU rather sparingly. A numerical simulation program will heavily rely on primary storage and the CPU while using the other subsystems infrequently. The overall performance of a computer system with a fast CPU and memory and slow secondary storage will be very different, depending on which of these applications is run on the system. An overall measure of system performance is often expressed in terms of MIPS. Unfortunately, assumptions about the application are rarely stated with this number.

Cost/Performance Relationships

In 1952, the computer scientist H.A. Grosch asserted that computing power (as measured by CPU MIPS) is proportional to the square of the cost of hardware. According to Grosch, large, powerful computers always will be more cost-efficient than smaller ones. The mathematical formula that describes this relationship is called **Grosch's law**. For many years, computer system managers pointed to this law as justification for investments in ever-larger central computers.

In the years since Grosch's law was first asserted, several major changes have occurred in computer technology, including the emergence of distinct classes of computers, expanded abilities to change the configuration and capabilities of a given computer, an increase in software costs relative to hardware costs, a vast expansion in the size of computer databases, the widespread adoption of graphical user interfaces, and the widespread use of computer networks. These changes collectively have invalidated Grosch's law.

In Grosch's time, there was only one class of computer—the mainframe. Purchasing larger machines within a single class (e.g., the mainframe class) tends to offer more MIPS for the money, and thus Grosch's law appears to hold. But if all classes of computers are considered as a group, Grosch's law no longer holds. The cost of CPU power actually increases on a per unit basis as computer class increases.

Examining computer power based only on CPU MIPS is at best simplistic and at worst highly misleading. Today's computers are expected to do much more than just crunch numbers. The ability to search large databases and to cope with the variety of ways in which humans access and assimilate information are as important as pure number-crunching power. Thus, MIPS are only a part of a computer system's real power. This fact is best seen when comparing minicomputers and mainframes. Those classes typically differ little in raw processing power, but they do differ substantially in their ability to store, retrieve, and move large amounts of data. The implication of this is that computing power is a complex variable. The most cost-effective processing power is generally obtained in lower classes of computers. But the data storage and communication requirements of an application program may necessitate a larger machine.

Comparisons of cost-efficiency are even more difficult when networked computers are considered as an option. In Grosch's time, the hardware and software to support networking of computers did not exist, but it is now possible to tie together multiple machines and make them operate as a single machine. A dramatic demonstration of this was made by Sandia National Laboratories in the early 1990s when 1,000 microcomputers were networked and shown to mimic the processing power of a supercomputer on certain applications. The design of many modern supercomputers is, in fact, based on the parallel use of large numbers of microprocessors. Many organizations have opted to construct networked systems of microcomputers and minicomputers instead of a single large mainframe because of the inherent cost efficiencies.

Within a given class or model of computer system, cost is affected by the power or capacity of the various subsystems. Often, a deliberate choice can be made to cost-effectively tailor the performance of the system to a particular application or class of applications. For example, in a transaction processing application, the system might be configured with fast secondary storage and I/O subsystems while sacrificing CPU performance (especially as measured in MFLOPS). The same system might be cost-effectively tailored to a simulation application by providing a large amount of high-speed main memory, a CPU with a high MFLOPS rating, and a high-capacity communication channel between them. Secondary storage and I/O performance would be far less important, and thus less costly alternatives could be used.

Such "tuning" of computer systems by substituting subsystems of varying power and cost is common, and modern computers are designed with this substitution in mind. This fact has led to a significant blurring of the lines among the various classes of computers—particularly for microcomputers, workstations, and low-end minicomputers and for high-end minicomputers and low-end mainframes. Although a microcomputer was earlier defined as a single-user machine, high-powered versions of these machines can often be used for multi-user applications simply by upgrading the capacity of a few of the subsystems and incorporating more powerful systems software. Similar possibilities exist for upgrading the performance of a minicomputer to that of a mainframe.

Another difficulty in applying Grosch's law today is the reality of costly systems and applications software. The power of a computer system as expressed in MIPS and data transfer rates is only a potential as far as the end-user is concerned. The realization of this power requires systems and applications software, but the development of such software requires a substantial commitment of labor and its associated costs. This fact, coupled with the ever-decreasing cost of computer hardware, has made the cost of computer hardware an increasingly insignificant factor in an IS's total cost. Thus, although hardware costs should be managed effectively, a user or manager is far more likely to maximize the cost-effectiveness of an entire IS by concentrating on its software components.

SUMMARY

A computer is an automated device for performing computational tasks. It accepts input data from the external world, performs one or more computations on the data, and returns results to the external world. Early computers had extremely limited capabilities and were implemented with mechanical devices. Modern computers have more extensive computational capabilities and are implemented with a combination of electrical and optical devices. The advantages of electrical and optical implementation include speed, accuracy, and reliability.

A computer system contains a general purpose processor with specific capabilities and components. Capabilities include simple arithmetic functions, numeric comparisons, and the movement of data among storage locations and I/O devices. A command to the computer to perform one of these basic functions on one or more data items is called an instruction. Complex manipulations of data can be implemented as a sequence of instructions called a program. The actions of a computer system are determined by the program and can be changed by using a different program.

A computer processor alternates continuously between the instruction cycle and the execution cycle. During the instruction cycle, the control unit fetches an instruction and its related data from storage. During the execution cycle, the ALU or control unit executes the instruction. Computation and comparison instructions are executed by the ALU and data movement instructions are executed by the control unit. The next instruction cycle fetches the next instruction and its data. This process continues until the last instruction in the program has been executed.

A computer system consists of a central processing unit (CPU), primary storage, secondary storage, and input-output (I/O) devices. The CPU contains the ALU, control unit, and a small set of temporary storage locations called registers. Registers hold both data and instructions for use by the other CPU components. The ALU performs arithmetic and comparison functions. The control unit is responsible for managing the movement of instructions and data and for directing the ALU.

Primary storage consists of a large number of relatively fast storage locations. Primary storage holds programs and data that are currently in use by the CPU. Instructions and data are moved continuously between primary storage and the CPU during program execution. Primary storage is generally implemented by electrical devices called random access memory (RAM). These devices provide direct access and high speed at relatively high cost.

Secondary storage consists of one or more devices with high capacity and relatively low cost. Access speed is normally much slower than for primary storage. Secondary storage holds programs and data that are not currently in use by the CPU. I/O devices are the means by which the CPU communicates with the external world. They may allow communication with users or with other computers. I/O devices vary in speed and communication medium.

A computer system may be classified as a microcomputer, minicomputer, mainframe, or supercomputer. Microcomputers are designed for use by a single user. Minicomputers and mainframes are designed to support many programs and users simultaneously. Mainframe computers are designed for large amounts of data access and movement. Supercomputers are designed to perform large amounts of numeric computation quickly.

A computer system's performance may be measured in terms of its ability to perform basic computing functions or in terms of its ability to perform useful work (i.e., workload). Performance measures for basic computing functions include millions of instructions per second (MIPS) and millions of floating point operations per second (MFLOPS). Performance measures for workload include throughput and response time. Although both types of performance measures are directly related, the relationship is difficult to specify precisely.

Selecting the minimal-cost computer system for a given workload is a complex problem. With earlier generations of computers, economical computing solutions were generally achieved by sharing large general purpose computers. With modern computer systems, economical solutions are generally achieved by utilizing the smallest computer system applicable to a particular type of computing task. Economic solutions may also be obtained by customizing a computer system to a particular type or set of application programs.

access time
algorithm
arithmetic logic unit (ALU)
batch processing
boot up
branching
bytes
central processing unit (CPU)
communication channel
computer
computer system
control unit
data transfer rate
direct access
executing
execution cycle
fetch
fetch cycle
floating point
formulaic problem
general purpose processor
gigabyte
Grosch's law
input-output (I/O) unit
instruction
instruction cycle
interactive processing
kilobytes
logic instruction
main memory
mainframe
mechanics

megabyte
memory
microcomputer
microsecond
millions of floating point operations per second (MFLOPS)
millions of instructions per second (MIPS)
millisecond
minicomputer
nanosecond
network computer
online processing
personal computer (PC)
primary storage
processor
program
random access
random access memory (RAM)
register
response time
secondary storage
server
special purpose processor
supercomputer
system bus
terabyte
throughput
transaction
volatility
von Neumann machine
work
workstation

Vocabulary Exercises

1. A(n) _____ is moved from primary storage to the CPU during the _____ cycle.

2. The speed of a primary or secondary storage device is generally stated in terms of its _MHz_.

3. The capacity of secondary storage is generally stated in _MB . GB_.

4. A device that uses a general purpose processor, stored programs, and operates by alternating instruction and execution cycles is called a(n) _Comp_.

5. The speed and capacity of a communication channel is generally stated in terms of its _Access Time_

6. A(n) _main frame_ generally supports more simultaneous users than a(n) _mini comp_. Both are designed to support more than one user.

7. A(n) _Registers_ is a storage location implemented within the CPU.

8. The processing speed of a supercomputer is generally measured in terms of _floating point operation per sec_. The processing speed of other classes of computers is generally measured in terms of _instruction/sec_.

9. _Batch processing_ processing describes a mode of application program execution that does not require input directly from a user.

10. A problem-solving procedure that requires the execution of one or more comparison instructions is called a(n) _algorithm_.

11. The term _secondary_ storage describes any storage device that holds its data content indefinitely.

12. The _work capacity_ of a computer system may be measured in terms of programs per time interval and/or transactions per time interval.

13. A(n) _server_ is a computer that manages shared resources and allows access to them through a network.

14. The _response time_ of an online transaction processing application increases as processing load increases.

15. When powered on, a(n) _network_ establishes a connection with a server that enables it to _____.

16. A program that solves a(n) _to emulate problem_ requires no branching instructions.

17. The primary components of a CPU are the _ALU_, _control unit_, and a set of _Registers_.

18. Primary storage may also be called _main memory_ and is generally implemented using _devices_.

19. A set of instructions that is executed to solve a specific problem is called a(n) _program_.

20. A(n) _super computers_ is typically implemented with the latest and most expensive technology.

21. The _system bus_ is the "plumbing" that connects all computer system components.

Review Questions

1. What similarities exist among mechanical, electrical, and optical methods of computation?

2. What shortcomings of mechanical computation were addressed by the introduction of electrical computing devices?

3. What shortcomings of electronic computation were (will be) addressed by the introduction of optical computing devices?

4. What is a CPU? What are its primary components?

5. What are registers? What is/are their function(s)?

6. What is main memory? In what way(s) does it differ from registers?

7. Explain the operations of the instruction cycle. Explain the operations of the execution cycle.

8. What is a von Neumann machine?

9. What are the differences between primary and secondary storage?

10. How does a workstation differ from a microcomputer?

11. How does a supercomputer differ from a mainframe computer?

12. What class(es) of computer system(s) are normally used to implement a server?

13. In what ways (by what measures) is the performance of computer system components stated? What problems are encountered when using these measures to acquire computer hardware for specific user needs?

14. What is throughput? What is response time? How are they measured?

15. What are the components of a communication channel? How is its performance and/or capacity measured?

16. What is Grosch's law? Does it hold today? Why or why not?

17. Why have mainframe computers become a small segment of the computer market that they once dominated?

18. How may a computer system be "tuned" to a particular application?

Problems and Exercises

1. Obtain a catalog or access the Web site of a major distributor of microcomputer equipment such as Insight (www.insight.com) or Computer Discount Warehouse (www.cdw.com). Assume that you have a budget of $3,000 to purchase workstation hardware (exclusive of peripheral devices such as a scanner or printer). Select or configure a system that provides optimal performance for the following types of users:

   ► A home user who uses a word processor (e.g., Microsoft Word), a home accounting package (e.g., Quicken and/or TurboTax), and children's games.

   ► An accountant who uses typical office software (e.g., Microsoft Office), statistical software (e.g., SYSTAT), and who regularly downloads large amounts of financial data from a corporate server for statistical and financial modeling.

   ► An architect who uses typical office software and computer-aided design software.

   Pay particular attention to the adequacy of CPU power, disk space, and I/O capabilities.

2. Assume that you have been asked to recommend a computer system to use as a server for a 250-person office. The server must provide file access (approximately 10 gigabytes), access to several shared printers, access to the Internet (news, email, and Web site services), and database services (database size is approximately 5 gigabytes) to a variety of application programs. Assume further that Microsoft products (e.g., Windows NT Server and Microsoft SQL Server) will be the primary software used on the server.

   Obtain product information from IBM (www.ibm.com), Digital Equipment Corporation (www.digital.com), and Dell Computer Corporation (www.dell.com). Determine which of these products best meet the server requirements.

1. The measurement of computer performance is a difficult task because of the complexity of hardware and assumptions about application program behavior. As discussed in the text, many computer system vendors state system performance in terms of millions of instructions per second (MIPS). However, this measure can be misleading, partly because it does not account for differences among applications. Common differences among applications include the mix of CPU instructions executed, amount of secondary storage access, and amount and type of I/O device access. Three alternative measures of computer system performance are the Norton Utilities System Information (SI) index (www.symantec.com), the Byte Magazine BYTEmark test suite (www.byte.com), and the LINPACK benchmarks (www.netlib.org). Investigate each of these performance measures, and identify the class(es) of computer and type(s) of applications for which the benchmarks are intended.

2. A substantial depth and breadth of technical skills is needed to perform thorough and unbiased testing of computer hardware and software. Since most organizations do not have such resources "in house," they contract for testing services with organizations that specialize in testing. National Software Testing Laboratories (www.nstl.com) is a well-known provider of testing services. Investigate their services and products to determine which might be potentially useful to organizations for purposes of selecting computer hardware, system software, and application software.

Chapter 3

Computer Software

Chapter Goals

▶ Describe the translation process from user requests to machine instructions that satisfy those requests.

▶ Define the roles and functions of application and system software.

▶ Define the role of a programming language.

▶ Describe the mechanisms by which programs are developed, translated, and executed.

▶ Describe the functions and components of an operating system.

▶ Describe the economic role of system software.

THE ROLE OF SOFTWARE

Think of the process of asking a computer to perform useful work as a translation process from one level of detail to another. The need or idea that motivates a request for computer processing is generally stated at an abstract level. The computer actions, or execution of instructions, necessary to satisfy a request are specific and primitive.

The primary role of software is to provide a means to translate generally stated needs into a sequence of computer instructions that will produce a result that satisfies those needs. Because the instructions that a computer can execute are primitive, the number of instructions that must be executed to satisfy a need is large and the translation process is complex. The individual steps in this translation process are illustrated in Figure 3-1 along with the division of responsibility between the user, software, and hardware.

Figure 3-1

The division of responsibility for translation steps between the user, software, and computer hardware

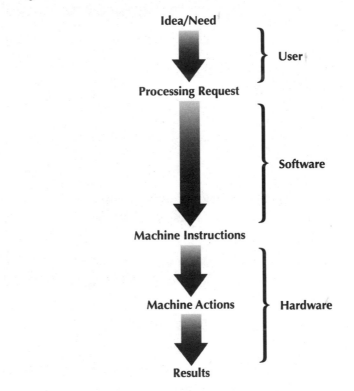

From Ideas/Needs to Machine Instructions

Imagine that you want someone to perform a complex task such as cooking a meal. How would you go about asking for this? If the person were an adult and an experienced cook, the request could be stated relatively simply. Once you described the meal you wanted, the cook would know what tasks to perform and how. You would have communicated a request for a complex task with relatively few words and left it to the person receiving the request to translate your request into the series of individual actions necessary to satisfy the request. The translation steps for this example are represented in Figure 3-2.

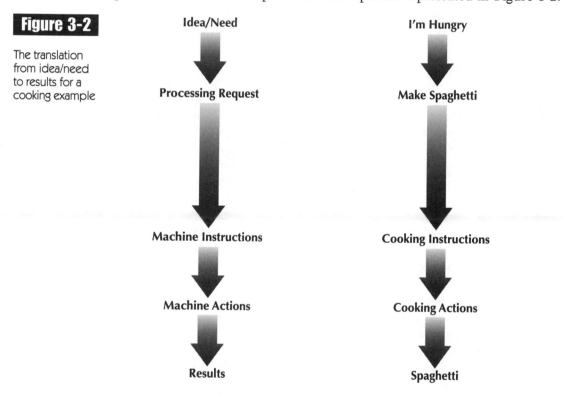

Figure 3-2

The translation from idea/need to results for a cooking example

Idea/Need → Processing Request → Machine Instructions → Machine Actions → Results

I'm Hungry → Make Spaghetti → Cooking Instructions → Cooking Actions → Spaghetti

Now imagine making the same request of a six-year-old child, who may know what a plate of spaghetti is but has no idea how to prepare it. However, the child does possess a set of primitive skills that could be combined for that purpose. If you wanted, you could instruct the child in a detailed manner about the individual steps necessary to prepare the meal. For instance, you could first tell the child to get the large metal pot from the cabinet on the left, put it in the sink, fill it two-thirds full with water, put the pot on the left burner of the stove, turn the left knob on the front of the stove to the position marked Medium, and so forth. If the child were old enough to read, you could write down the detailed instructions as a recipe.

Although the meal will eventually be "prepared" in either scenario, the distribution of intelligence and effort required to bring about the result differs substantially. In the first scenario, the request is stated simply and the cook is responsible for determining the appropriate detailed instructions and actions. Thus, the intelligence needed to prepare the meal is embedded within the cook. In the second scenario, the requestor cannot assume that the child has a high level of intelligence with respect to the task at hand. Therefore, the child receives detailed instructions to perform specific actions within his or her knowledge and ability. The knowledge of how to perform the task is thus embedded within the requestor.

Programs

The child's basic abilities in the preceding example are analogous to the processing capabilities of a computer system. The child can perform many simple (primitive) tasks but has little knowledge of how to combine them to respond to a complex demand. Similarly, a computer has a set of primitive capabilities but no direct knowledge of how to combine primitive actions to perform a complex task. The computer must be instructed in a detailed and specific manner, either directly by a user or by a previously defined set of instructions (i.e., a program).

Much as the child will eventually memorize the procedure for making spaghetti, the computer can "remember" a procedure by storing the corresponding program. This program can then be recalled as needed, much as a person can recall a recipe. The recipe is analogous to an **application program**, or **application software**, which is a stored set of instructions for responding to a specific request. Examples of application programs in a system devoted to payroll processing include programs to print checks, enter new employee information, produce annual tax reports, and so forth.

Other programs can be created and stored for more general purposes. In the cooking example earlier, the child needed instructions in the mechanics of boiling water in which to cook the spaghetti. Boiling water, however, is not a procedure specific to the task of cooking spaghetti; rather, it is a frequently used component of many other cooking tasks. A recipe for spaghetti will not provide detailed instructions for boiling water; instead, the recipe will simply state that a certain quantity of water should be brought to boil and assume that the user knows how to do that.

The procedure for boiling water is analogous to a **utility program**, which is a set of instructions for performing a relatively basic task that is a necessary component of many application-oriented tasks. Examples of utility programs include programs to print text files, log onto a multiuser computer, copy a file from one disk to another, and so forth. Each of these tasks is basic (although it may still require thousands of CPU instructions to complete) and is performed in conjunction with many different applications.

ECONOMICS OF SOFTWARE DEVELOPMENT

The process of developing application software is lengthy, complex, and costly. When a user wants a computer to perform a specific task, a program (set of machine instructions) must be developed for that purpose. This development process follows a system development life cycle (SDLC) as defined in Chapter 1. The development process consumes a substantial amount of human effort for survey, analysis, design, and programming. Thus, an investment of resources is required before the user can receive any benefit from the computer.

The benefits of this investment are realized by repetitive use of the programs developed. If the costs of operating the computer program are less than the costs of performing that same function manually, then the initial investment of resources will eventually be repaid through savings. If there is not a sufficient difference in operating costs, then the development of software is not worth undertaking.

To make the use of computers economically feasible, it is necessary to minimize the costs of developing software. There are numerous ways to approach this basic problem, including the following:

► Reuse software as much as possible.

► Minimize labor inputs to the software development process.

Most of the software that is commonly called **system software** is designed to further one or both of these approaches.

Software reuse is addressed primarily through the development of utility programs, which are designed to perform functions that are frequently used and/or needed by many different application programs. For example, most application programs must access permanent data on secondary storage devices. Therefore, it is economically advantageous to provide a set of utility programs for that purpose. These programs perform functions such as creating and deleting files, reading and writing data to/from files, initializing storage devices, and so forth.

The availability of utility programs allows the developers of application software to avoid "reinventing the wheel" each time they develop a new system. Thus, an application programmer can concentrate on developing software only for the functions that are unique to a particular application and can use the utility programs as necessary. Because application programmers no longer must write programs for these tasks, the amount of time and labor needed to develop application software is decreased.

Figure 3-3 illustrates the interaction between the user, application software, utility programs within system software, and computer hardware. Notice the similarity between this diagram and the translation model depicted in Figure 3-1. User input and output is limited to communication directly with the application program. Application programs, in turn, communicate with system software to request basic services (e.g., opening a file, reading data from a file, etc.). System software translates a service request into a sequence of machine instructions, passes those instructions to the hardware for execution, and receives the results of that execution.

Figure 3-3

The interactions between the user, application software, system software, and computer hardware

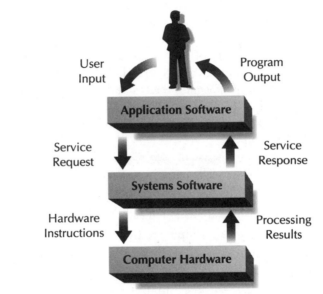

Figure 3-3 also illustrates what is commonly called a layered approach to software, in which application software is layered above system software which is, in turn, layered above the hardware. A key advantage of the layered approach to software is that users and application programmers do not need to know the technical details of computer hardware. Instead, they interact with hardware through a set of standardized service requests (or commands). Those requests are then translated into the actual hardware instructions needed to satisfy the requests. Knowledge of the machine's physical details and basic processing tasks is thus embedded within system software and hidden from the user and application programmers. This advantage is commonly referred to as **machine independence**, or **hardware independence**.

PROGRAMMING LANGUAGES

Another method of reducing the labor input for application software development is to minimize the number of instructions that a programmer must develop to perform a given task. Even simple programs (application tasks) often require hundreds of thousands of machine instructions. To a large degree, the time and effort required to develop software is proportional to the number of instructions that a programmer must write. Thus, minimizing the number of these instructions reduces the labor input to the software development process.

A **programming language** is a language for instructing a computer. A complete program in such a language is often referred to as **code**, and the programmer is sometimes referred to as a **coder**. The most primitive type of programming language is the language that the computer understands directly—that is, the instructions that can be executed by the CPU. This is commonly referred to as **machine language**. Machine language, the earliest form of programming, is also the most tedious and time-consuming because it requires programmers to determine the exact set of primitive machine actions (instructions) necessary to perform a complex task and to write down each instruction in a form understandable to the computer.

Machine Language Programming

Most of a computer's work consists of the execution of relatively simple computational functions as well as the movement of data between various computer system components. The capabilities of the computer system—referred to as the computer's **instruction set**—are defined by (and limited to) the set of data manipulation and movement functions of which the CPU is capable.

The instruction sets of different computers vary widely. In general, the size of the instruction set (number of different functions that can be performed) increases as the class of the computer system increases. Thus, mainframes and supercomputers usually have significantly larger instruction sets than microcomputers. But regardless of the class of the computer, the types of instructions tend to be similar and primitive.

The processing functions of a typical CPU can be classified into the following three categories:

► **Computation:** Perform a mathematical calculation using two numbers as input and store the result.

► **Comparison:** Compare the values of two stored numbers and determine if they are equivalent.

► **Data Movement:** Move an individual item of data from one storage location to another storage location.

A **simple machine instruction** performs a single function that can be classified as one of these types. A CPU also may provide **complex machine instructions** that combine two or more processing functions. For example, an instruction may be provided to perform data movement only if two stored values are not equal to one another. Such an instruction would be a combination of a simple comparison and a simple data movement. However, even if complex instructions are provided, many instructions are still necessary to perform any significantly complex task.

Each instruction is known to the CPU by a unique number called the **instruction code**, **operation code**, or **op code**. The execution of an instruction requires that the instruction code and related data be moved into CPU registers from primary storage. Both the instructions and data are stored in main memory and are accessed by their addresses (locations). Machine language instructions consist of the binary representation of the op code and data inputs. Thus, a machine language program is little more than an extremely long sequence of numbers, some of which represent instructions and some of which represent data items or the main memory addresses of data items or other instructions. This sequence of numbers is generally called **executable code** (or an **executable**) because it is code (instructions) that is ready to be executed.

To write a program in machine language, a programmer must know all of the available instructions by their instruction codes. In addition, the programmer must know where each of these instructions and all of the data items are located in memory. All program statements consist of numbers representing instruction codes, data items, and memory locations. A typical program consists of many thousands of these statements. Developing this mass of numbers is tedious and complex. This complexity and its resultant costs (i.e., human programming time) make machine language programming a very inefficient method of implementing software. To address this inefficiency, a new class of programming languages was developed called **assembly language**.

Assembly Language Programming

The primary difference between assembly language and machine language is the way in which instructions and memory locations are written. Machine language requires that these be specified by number or address, whereas assembly language allows the programmer to use names composed of alphabetic characters. For example, an instruction to add two numbers might be known to the CPU as instruction code 4. In assembly language, this instruction is given a character name such as ADD.

An add instruction might require the addresses in memory of the two numbers to be added. In machine language, the programmer must know where these numbers are stored in memory and must supply the memory addresses to the CPU with the instruction. In assembly language, the programmer can specify names for these storage locations (e.g., N1 and N2) and use these names instead of the addresses. Thus, the assembly language equivalent of an instruction such as 04 0020 0024 might be ADD N1 N2. Figure 3-4 shows sample assembly language code for the Intel Pentium processor family.

A portion of an assembly language program to compute string length

```
main_loop:
    mov   eax, dword ptr [ecx]  ;  read 4 bytes
    mov   edx, 7efefeffh
    add   edx, eax
    xor   eax, -1
    xor   eax, edx
    add   ecx, 4
test eax, 81010100h
je  short main_loop
; found zero byte in the loop
mov  eax, [ecx - 4]
test al, al                 ; is it byte 0
je  short byte_0
test ah,ah                  ; is it byte 1
je  short byte_1
test eax, 00ff0000h    ; is it byte 2
je  short byte_2
test eax, 0ff000000h   ; is it byte 3
je  short byte_3
jne  short main_loop        ; taken if bits 24-30 are clear
                            ; and bit 31 is set
byte_3:
    lea   eax, [ecx - 1]
    mov   ecx, string
    sub   eax, ecx
    ret
byte_2:
    lea   eax, [ecx - 2]
    mov   ecx, string
    sub   eax, ecx
    ret
byte_1:
    lea   eax, [ecx - 3]
    mov   ecx, string
    sub   eax, ecx
    ret
byte_0:
    lea   eax, [ecx - 4]
    mov   ecx, string
    sub   eax, ecx
    ret

strlen endp

end
```

The use of assembly language requires a means of translating it into machine language. This translation is performed by a program called an **assembler**. The assembler reads each assembly language instruction and creates an equivalent machine language instruction. After all of the assembly language instructions have been translated, the entire set of corresponding machine instructions can then be loaded into memory and executed by the CPU.

Although the use of assembly language still requires the programmer to specify each individual CPU instruction, it provides a more convenient way in which to write those instructions. The human memory is more efficient at recalling and manipulating character names than long strings of numbers. Thus, the programmer's effort required to develop an assembly language program is substantially less than that required to develop an equivalent program in machine language. The development of assembly languages led to dramatically improved programmer productivity.

High-Level Programming Languages

Further increases in programmer productivity required a fundamentally different approach to programming—one that eliminated the need to specify each individual CPU instruction. This was accomplished by the development of **high-level programming languages**, which allow the programmer to specify a processing action requiring many CPU instructions in a relatively small number of statements.

Examples of such languages include FORTRAN, COBOL, C, C++, and Visual Basic, among others. Figure 3-5 shows portions of two programs in COBOL and C. Although the program fragments differ substantially, they accomplish the same purpose. Depending on the capabilities of the CPU, a machine language program to perform this same function would require hundreds or thousands of instructions. An assembly language program would require a similar number of instructions.

Figure 3-5

COBOL Example

Sample program
fragments in the
COBOL and C
high-level
programming
languages

```
DETAIL-PARAGRAPH.
     READ CARD-FILE AT END GO TO END-PARAGRAPH.
     MOVE CORRESPONDING CARD-IN TO LINE-OUT.
     ADD CURRENT-MONTH-SALES IN CARD-IN, YEAR-TO-DATE-SALES IN
        CARD-IN GIVING TOTAL-SALES-OUT IN LINE-OUT.
     WRITE LINE-OUT BEFORE ADVANCING 2 LINES AT EOP PERFORM
        HEADER-PARAGRAPH.
```

C Example

```c
detail_procedure()
{
     int file_status, current_line, id_num;
     char *name;
     float current_sales, ytd_sales, total_sales;

     current_line=64;
     file_status = fscanf(card_file, "%d  %s  %f  %f"  &id_num,
         &name, &current_sales, &ytd_sales);
     while (file_status != EOF) {
         if (current_line > 63 ) {
             header_procedure();
             current_line=3;
         }
         file_status  =  fscanf(card_file, "%d  %s  %f %f,"
             &id_num, &name, &current_sales, &ytd_sales);
         total_sales = current_sales + ytd_sales;
         printf("%4d  %30s  %6.2f  %6.2f  %7.2f\n\n", id_num,
             name, current_sales, ytd_sales, total_sales);
         current_line++; current_line++;
     }
     end_procedure();
}
```

Much as assembly language programs are translated into machine instructions by an assembler, a high-level language program is translated into machine instructions by a **compiler**. In essence, the compiler translates each statement of the high-level language into an equivalent set of machine instructions. The compiler is also responsible for keeping track of instruction and data locations in memory. The translation process is depicted in Figure 3-6 on the next page. The statements of the assembly or high-level language program are called **source code**.

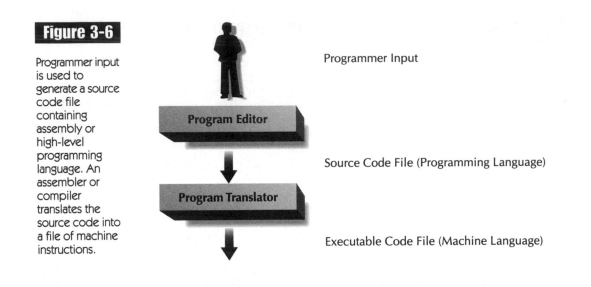

Figure 3-6

Programmer input is used to generate a source code file containing assembly or high-level programming language. An assembler or compiler translates the source code into a file of machine instructions.

Programmer Input

Program Editor

Source Code File (Programming Language)

Program Translator

Executable Code File (Machine Language)

TYPES OF SYSTEM SOFTWARE

The previous sections discussed system software in the contexts of translation and software development. This section provides a more pragmatic classification of system software into the following categories:

► Application development software.
► Application tools.
► Operating software.

The functions of system software illustrated in Figure 3-3 fall primarily within the category of operating software. The first two categories cannot be called either application or system software using the definitions given thus far; neither are they directed toward a specific need nor do they provide an interface to hardware services during the execution of an application program. They do, however, provide general purpose capabilities, and these distinctions will be clarified in the following sections.

Application Development Software

Application development software includes any program used to develop application programs. In the past, this was a relatively small class of software that included programs such as assemblers, compilers, and text editors. Because of advances in computer hardware, software, and the practice of application development, a much wider range of programs now fills this class. Application development software can be further classified into the following categories:

▶ Program translators.
▶ Program development tools.
▶ System development tools.
▶ Data manipulation tools.
▶ I/O tools.

Program translators include two types of software already discussed: assemblers and compilers. Both of these translate programming language statements into instructions that can be executed by the operating system and/or the hardware. Another type of software that fills this role is an **interpreter**, which differs from a compiler or an assembler primarily in when and how program statements are translated. Assemblers and compilers both translate an entire program into executable code that can be loaded into memory and run as a unit. An interpreter, on the other hand, translates one program statement at a time and executes it immediately before translating the next statement. It is, in essence, a program designed to execute other programs one statement at a time.

Program development tools include a wide range of programs to aid programmers in developing application programs. Examples of these include text editors, program verifiers, and debuggers. **Text editors** are basic tools that allow a programmer to create files containing assembly or high-level language programs. **Program verifiers** (also called **code checkers**) are programmers' aids that help evaluate program correctness and quality. These programs read an application program (source code) as input and check for such items as syntax errors, logic errors, and other actual or potential errors. Similar functions are often embedded within compilers and interpreters, especially those for the more recent high-level languages.

Debuggers are also tools for verifying program correctness. They differ from program verifiers in that they are used after a program has been compiled rather than before; that is, they verify executable code instead of source code. Generally, they allow a programmer to simulate the execution of a compiled program and to check for errors or problems during the simulated execution. They allow programmers to examine the contents of variables, restart the program in various places, and so forth.

System development tools, a relatively new class of application development software, are designed to support the earlier phases of the SDLC (survey, analysis, and design) and the development of groups of application programs comprising an entire IS. They address the complexity of developing an entire system of application programs, which is substantially greater than the complexity of developing a single program. System development tools are often called **Computer Assisted Software Engineering (CASE) tools**.

Data manipulation tools are designed to extend the basic capabilities of programming languages and operating systems to manipulate data stored on secondary storage devices. These are commonly called **database management systems (DBMSs)**. The need for these tools has developed in response to ever-larger and more complex ISs, which require complex methods of storing, retrieving, defining, and redefining data as well as methods of sharing data among competing programs and users. While operating systems and high-level programming languages provide some support for these functions, DBMSs are designed to provide a much higher level of support. These tools can be considered a form of system software because of their role in supporting application software and the interface service they provide to secondary storage.

I/O tools are similar to data manipulation tools in that they extend the basic capabilities of operating systems and programming languages. Complex I/O has become commonplace in application programs today. Examples include the use of full-screen menus, windows, and high-resolution graphics. These tools can be considered system software because they provide I/O support for application programs and an interface to I/O hardware.

Application Tools

Application tools include two broad types of software: those tools designed to automate repetitive clerical or technical tasks and those designed to take the place of application programs. Examples of the first type include word processors (e.g., WordPerfect) and automated drawing packages (e.g., Adobe Illustrator and AutoCAD). Rather than address a specific output or processing result, they address a particular type of processing.

Tools designed to take the place of (or augment) application programs allow users to perform processing tasks without the need to have an application program developed specifically for their processing requirements. Examples of these tools include spreadsheet processors (e.g., Microsoft Excel) and some DBMSs (e.g., Microsoft Access). These tools either allow a user to perform information processing without an application program or provide the ability to generate simple application programs.

Some application tools, which can be classified as **application generators**, provide an integrated set of system software including a high-level programming language (compiler and/or interpreter), DBMS, and advanced I/O tools. An integrated set of application development tools is often called a **fourth generation language (4GL)**, or **fourth generation development environment**. Microsoft Visual Basic is a good example of a fourth generation development environment.

Operating Systems

The final class of system software, operating software, encompasses a wide range of functions, including the following:

▶ Process control.

▶ File control.

▶ Secondary storage control.

▶ I/O device control.

▶ User control.

The common term for a set of software that performs all of these functions is an **operating system**. Examples of commonly available operating systems include Windows and UNIX.

Operating System Model and Functions All of the functions listed earlier provide an interface between computer hardware and the user or application programs. The primary components of an operating system are illustrated in Figure 3-7. Notice that this diagram (and the operating system) is organized according to the principle of software layers. Input from a user or an application program must travel through the various layers to reach the hardware, and results from the hardware must travel back to the user through those same layers.

Figure 3-7

Operating system layers (shaded) and their relationship to the user, application software, and computer hardware

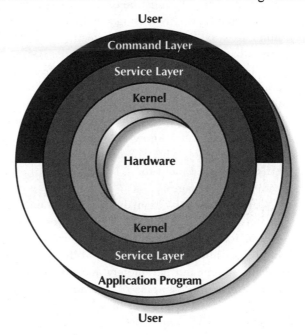

The **kernel** is the only operating system layer that interacts directly with the hardware, which provides a measure of machine independence within the operating system. In theory, an operating system can be altered to interact with a different set of computer hardware by making changes only to the kernel. This layering also insulates users of the application-level and command-level interfaces from direct knowledge of hardware specifics, thus simplifying these interfaces considerably. The name "kernel" is derived from the fact that this is the innermost layer of the operating system.

The **service layer**, also called the system service interface, is the layer that most closely resembles the system software description illustrated in Figure 3-3. This layer accepts service requests from application programs and/or the command layer and translates them into detailed instructions to the kernel. Processing results, if any, are passed back to the program that requested the service.

The **command layer** is the only portion of the operating system with which the user directly interacts. Because it is the outermost layer, it is sometimes referred to as the **shell**. The command layer responds to commands given directly by the user or by a previously stored list (or file) of commands.

Command Layer The command layer is the user's interface to the operating system. The user can directly request system services through a textually or graphically based command interpreter. The command interpreter decides which service calls must be executed to process the request and passes input to the service layer. Processing results received are displayed to the user in some human-readable form. The command layer may be implemented with a textual, forms-based, or graphical user interface (GUI).

Textual interfaces accept user input from the keyboard. The set of commands and their syntax requirements are referred to as a **command language** or, sometimes, as a **job control language** (JCL). Command languages tend to be difficult to learn and use. They are similar to programming languages in that the user (or programmer) must know the syntax and semantics of the language in order to communicate. The MS-DOS command interpreter (COMMAND.COM), IBM MVS JCL, and the UNIX Bourne Shell command set are all examples of command languages.

Alternatives to command processors include form-based (full-screen) interaction and graphically based interfaces. Form-based interfaces provide full-screen forms or prompts to the user for operating system commands. These interfaces alleviate some of the difficulties in remembering command syntax because command parameters and options are explicitly requested (prompted) from the user. Examples of operating systems that use form-based command interfaces include portions of VMS and certain IBM interactive operating systems such as TSO and SPF.

Modern operating systems generally interact with the user by a graphical user interface (GUI), which reduces or eliminates the need for users to learn a command language, or JCL. These interfaces allow users to execute many operating system commands by manipulating graphical images on a video display using a pointing device (e.g., a mouse). For example, files may be represented by graphical images of file folders, and moving a file might be accomplished by moving the associated folder image. The visual metaphor for commands and functions provides an easy-to-learn interface. Examples of these types of interfaces include those of the Macintosh and Windows operating systems.

Regardless of the type of interface, the functions of the command layer are similar. They provide an interface that allows a user to perform common operating functions without writing specific application programs. Examples of these common operating functions include loading and executing application programs, manipulating files, and initializing (formatting) secondary storage devices.

Service Layer Service calls implement commonly used functions in process control, file control, and I/O control. Examples of process control service calls include loading a **process** (program or program segment) into memory, starting its execution, and terminating it. Examples of file control service calls include opening, closing, creating, deleting, and renaming files as well as reading and writing data to and from files. Examples of I/O control service calls include initializing printers and display devices, sending characters to a printer, sending characters or graphic images to a video display, and reading characters from a keyboard.

The number of system service requests (or **service calls**) provided by the service layer is generally large. It also depends on the capabilities of the operating system and of the hardware that it controls. In general, more operating system capabilities and/or more powerful and complex hardware require more service calls. This relationship also applies to the amount of code needed to implement the kernel.

As an example of this relationship, consider the difference between the MS-DOS, Windows, and UNIX operating systems. MS-DOS was designed originally to operate the IBM personal computer (PC), which was a relatively simple machine. In addition, MS-DOS was designed to provide a minimal level of capability (e.g., no multiuser or graphical I/O capabilities). As such, the number of service calls available in MS-DOS is relatively small (less than 100) as is the size (amount of memory consumed by the kernel).

Windows was designed to operate higher-powered microcomputers. Partly due to the more powerful hardware, Windows also was intended to provide more capabilities than MS-DOS (e.g., window interfaces, multiple program execution, etc.). The additional power and complexity of the hardware and the more advanced capabilities of the operating system lead to a larger number of system calls. The kernel is also substantially larger; as a result, a computer needs more memory to use Windows than it does to use MS-DOS.

The UNIX operating system was designed initially to operate minicomputers and later augmented to operate computers ranging from workstations to supercomputers. Primarily because of the wide range and power of hardware that it is designed to operate, the UNIX operating system has a large number of system calls. This number would be even larger if the user-oriented I/O capabilities were extended to be similar to those of Windows. The kernel is larger than Windows, and a computer using UNIX with a graphically oriented command layer requires more primary storage.

Kernel The components of the operating system kernel are depicted in Figure 3-8. The kernel, which is primarily responsible for allocation and direct control of hardware devices, provides a set of interface programs for each hardware device in the computer system. These interface programs are commonly called **device drivers**. Examples include device drivers for input keyboards, video display devices, disk drives, tape drives, and printers.

Figure 3-8

Resource allocation and device driver components of the operating system kernel

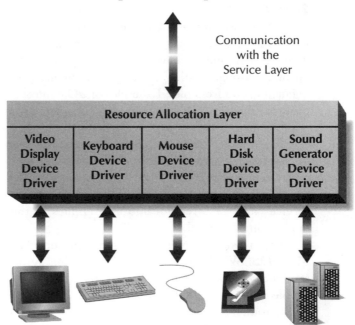

Communication with the Service Layer

The service layer uses the device drivers in much the same way that application programs use the utility programs within the service layer. That is, the service layer makes specific requests to kernel device drivers for access to hardware devices. There are many different service layer functions that use each device driver. For example, there may be several dozen different service functions for input and output to files. All of these will use the same device driver to perform required disk operations.

The advantages of this organization of service layer and kernel functions are similar to the advantages of the organization of application and service layer functions. The kernel provides a set of capabilities that are reused by different elements of the service layer. If these capabilities were not available, they would be redundantly implemented within the service layer (e.g., each file-manipulation service program would contain redundant code for control of disk hardware).

The use of kernel device drivers also provides a degree of modularity and hardware independence. As new hardware devices are added to the computer system, new or revised device drivers can be added, which allows the operating system to incorporate capabilities to control new hardware with a minimum amount of disruption to existing code and capabilities. In addition, existing capabilities can be modified without affecting the service layer. For example, a change in type of hard drive can be incorporated into the operating system by modifying the hard drive device driver. The service layer (and all layers above it) should be unaffected by this change.

Resource Allocation One of the most important and least visible operating system functions is resource allocation—a function normally implemented within the kernel. Resources to be allocated include CPU time, access to secondary storage devices, access to I/O devices, and access to memory. Resource allocation is a direct consequence of program execution and an indirect consequence of a program's requests to the service layer. The execution of a program requires the allocation of CPU time. Service calls generally require access to hardware resources for their satisfaction (e.g., access to secondary storage devices for file manipulation).

Resource allocation is relatively straightforward if the operating system supports the execution of only one program at a time (e.g., MS-DOS). The currently executing program is simply allocated whatever resources it requests. A request by an executing program through a service call is processed immediately. The only complexities consist in determining if the resources requested are available and checking for error conditions. For example, when a program requests a file to be opened, the operating system must determine if the file exists and monitor the actual operation for errors such as the inability to read the file from disk.

The complexities of resource allocation rise substantially as the number of users, programs, and hardware resources increase. Under such conditions, the kernel cannot immediately provide any resources a program or user requests. It must first determine whether or not those resources are currently being used by another user or process and must decide if and when to satisfy the service request. Therefore, the kernel must balance the competing demands of users and processes for access to the CPU and other hardware resources. This complexity reaches its zenith in modern mainframe computers where hundreds of users and thousands of programs may be simultaneously in contention for access to the CPU, memory, communication channels and devices, and multiple secondary storage devices.

PARALLEL HARDWARE/SOFTWARE DEVELOPMENT

Current system software bears little resemblance to that used in the past. In fact, system software as we know it today did not exist until many years after computers came into use. Programming languages other than machine and assembly language did not appear until the late 1950s. Operating systems also were not developed until that time, and their capabilities were extremely limited compared to modern operating systems.

Why is this so? Are we simply better at developing system software today than we were 50 years ago? The answer to these questions follows from two basic facts of computer hardware and software:

► System software requires hardware resources.

► The cost per unit of computing power has decreased at nearly an exponential rate.

These facts underlie a shifting trade-off between the costs of developing and supporting application programs and the cost of computer hardware resources.

When computers were first introduced, they were extremely expensive to purchase and operate. Because of this, computer resources were only expended on "high value" applications and only where the alternatives to their use were more expensive. For example, consider the operations of a typical accounts payable department of a large firm in the 1950s. Such an operation typically would employ hundreds of clerks to pay the firm's bills and maintain manual accounting records.

Although the labor costs associated with such an operation were high, they were usually lower than the cost of computer resources and software development necessary to automate accounts payable processing tasks. As the cost of computer power decreased, the trade-off in cost between automated and manual processing shifted. Repetitive and high-volume processing was now cheaper if automated. Thus, computer resources were substituted for manual labor in common transaction and information-processing tasks. This trend has continued to the present with an ever-greater substitution of computer resources for labor.

Computer resources also may be substituted for labor in the development of application programs. In the early years of computers, application development was an almost entirely manual process. Computer resources were too expensive to allow them to be used to directly support application development; thus, tools to support the development or execution of application programs did not exist.

As the cost of computer resources decreased, the economic trade-off shifted. Tedious programming in machine language and extensive manual checks of code were more expensive than application development using advanced tools. Thus began a trend toward increased automated support of application development, which took three parallel paths through the development of:

► Application development tools.
► Operating system service functions.
► Operating software.

Application Development Tools

The use of application development tools began with the introduction of programming languages and their associated tools. Assembly language and assemblers appeared first. These were soon followed by high-level languages and their associated interpreters and compilers. Text editors, debuggers, code verifiers, CASE tools, and a host of other software tools are also examples of this trend. All of these tools reduce the labor input to programming by substituting computer resources.

The use of modern application tools is a continuation of this trend. Tools such as word processors and spreadsheet programs minimize application development labor to nearly zero. The number and power of such tools continue to grow and will likely do so for the foreseeable future.

Service Layer Functions

A parallel trend in supporting application programs occurred in the development of service layer functions. Early programmers had to specify every primitive machine action. This included basic functions that occurred in every application program. Examples of the functions include access to I/O devices, input and output to secondary storage, file management, and memory management. Programmers often rewrote these same functions repeatedly in different application programs.

The trend toward using service layer functions started with these basic functions. A set of utility programs were provided within the operating system; application programmers used these utilities rather than writing equivalent functions for each application program, which resulted in reduced labor input for application development. However, there are some hidden costs to this approach. Consider, for example, the machine instructions comprising a program to read and write data from a file. These instructions are contained within one or more programs within the service layer. Because they are used for many different types of files and/or storage devices, they must be general in nature; that is, they must allow for variations in file content, file size, and other factors.

If the same function were implemented within an application program, the instructions could be more specifically oriented toward the file content and size for that application. This specificity would likely lead to a more efficient program (i.e., reduced use of computer hardware resources). This represents a basic trade-off in software. General purpose utility programs tend to be less efficient than programs written for a specialized function. By their nature, utility programs and service-layer functions must be general in nature.

When computer resources were relatively expensive, this difference in efficiency was critical. At that time, it was more cost-effective to implement specialized functions within application programs as compared to utilizing general purpose system software. The extra cost in application development (e.g., programmer labor) was more than offset by the reduced operating costs of the application program because of increased efficiency. As computer hardware became cheaper (and programming labor more expensive), this situation reversed and general purpose system software became widespread.

This trend—driven both by cheaper hardware resources and by increased expectations for the functionality of application software—continues to the present day. Examples of system software that addresses both of these issues include DBMSs and service layer support for graphical interfaces. Both of these types of software provide a set of utility programs that can be used by application programs. Both also address advanced requirements for application program functionality (e.g., the management of large databases and user friendly program interfaces). As computer technology advances and cost decreases, the development of such tools will continue.

Operating Software

It is difficult for present-day computer users to imagine operating a computer without the aid of software. But, as mentioned earlier, such software did not exist when computers were introduced. As an example, consider the loading of a program into memory and its subsequent execution. A PC user running Windows simply double-clicks an icon or selects a menu item, and the system software handles program loading and execution.

In the early 1950s, this conceptually simple task was a much more complex matter. A deck of punched cards containing the program instructions was first loaded into a card reader; then a set of switches on the computer were manually positioned. These switches instructed the machine to execute a small hardware-based program. This program consisted of a small loop of instructions that sequentially read each card and loaded the corresponding instruction into sequential memory locations. After all the cards were read, the machine stopped and displayed a status code through a set of lights.

Next, the computer operator loaded a deck of cards containing the input data into the card reader. Switches on the machine were then set to indicate the location in memory of the first program instruction. Another switch would then be set to instruct the computer to load and execute that first instruction. Then the application program would "take control" of the machine. When the last program instruction was executed, the machine would stop and display another status code. If an error occurred during program loading or execution, the machine stopped and a status code was displayed. The operator had to consult a reference manual to interpret the meaning of the status code and take corrective action.

These procedures were used partly because a typical computer did not contain enough memory to hold both operating and application software. Also, manual computer operations were cheaper than diverting hardware resources from application to operating software. Thus, a human operator, in concert with a few hardware-based operating programs, served as the operating system. As hardware became cheaper and more powerful, operating software was developed to automate such mundane tasks. This trend continues to the present day with the development of sophisticated operating software that is powerful and easy to use. As a result, the job of "computer operator" has virtually disappeared.

As computers became more powerful, multiple application programs could execute simultaneously on a single computer system. However, multiple-program execution required a new type of operating software with extensive capabilities for resource allocation and management. The complexity and function of this software was further increased with the introduction of online interactive applications and computer networks.

The resource management functions of a complex operating system consume a substantial amount of computer hardware resources. These resources do not directly support application programs. However, they do make resource sharing possible. A major goal in designing such operating systems is to minimize the amount of resources consumed by management functions.

Future Trends

Figure 3-9 (on the next page) shows the change in cost components of ISs over time. The early years of ISs are characterized by high hardware cost relative to the cost of application software. The balance of hardware and software cost shifts over time because of the declining cost of hardware and the emergence of system software. Typically, hardware is the cheapest component of current ISs. System and application software are now nearly equal components of total cost.

Figure 3-9

The change over time in the relative cost of hardware, application software, and system software in a typical information system

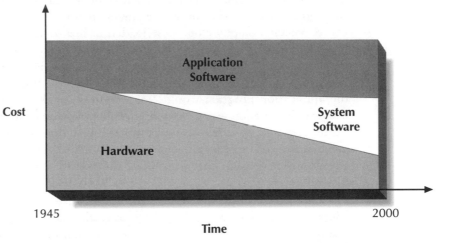

The use of system software will continue to increase as computer hardware becomes cheaper and more powerful. There is constant push and pull between hardware advances and software advances. Cheaper hardware allows a greater substitution of hardware for labor in the development and execution of application programs. Advanced hardware functions alter user expectations of application programs. These eventually result in increased functionality of system software for application development, support, and computer operations.

Consider graphical interfaces as a modern example. Windows, icons, and high-resolution graphics did not suddenly appear in the 1980s. They were under development at XEROX in the early 1970s. Why, then, did they take so long to come into use? Because they consume huge amounts of computer resources. Displaying a window or graphic image requires the movement of a large amount of data from main memory to the graphic display device. Moving or resizing a window requires a great deal of I/O and a large number of CPU instructions.

Until recently the cost of these resources was prohibitive. In the past, it was far more cost-efficient to train users in the obscure syntax and semantics of a command language than to pay for the extra computing resources needed to implement a graphical interface. Current hardware capabilities and costs make graphical interfaces cost-effective, which creates a demand for them in application and operating software.

Because the demand for such interfaces occurs across many application programs, a demand for system software to support such interfaces is created.

This demand has been answered by the development of system services for graphical and window functions (e.g., those embedded in the Macintosh and Windows operating systems). It also requires an advance in operating software to control the new hardware resources and to make using them easy and transparent.

Mathematicians, physicists, computer scientists, and other researchers ponder new possibilities of automated computation (even automated thinking). Engineers give physical reality to these ideas by building devices based on them. System and application programmers then ponder what to do with those devices. As time passes, manufacturing experience accumulates, and the devices become cheaper and more powerful. Users demand the benefits of this hardware which, in turn, causes programmers to look for new ways to exploit the additional power. Such has been the relationship between hardware and software advances to date and such is it likely to be for the foreseeable future.

Technology Focus

Intel CPUs and Microsoft Operating Systems

Intel and Microsoft have jointly defined the power, look, and feel of an IBM compatible PC since the first IBM machine was introduced in 1981. Since IBM lost control of the IBM PC "standard" in the mid-1980s, Intel and Microsoft have taken on increasingly larger roles in defining standards. Intel has pushed the hardware standard into new territory through its newer microprocessors and has also developed and championed enhancements to the original PC technology suite, including the PCI system bus and plug and play hardware compatibility.

Microsoft has taken an increasingly active role in defining the look and feel of software that runs on the IBM PC platform. Early versions of its MS-DOS operating system did little to support the user interface of application programs. Thus, application program developers were left to develop their own styles of user interface. As a result, there was a wide variation among user interfaces on the same machine.

Microsoft brought some consistency to the look and feel of application programs with the introduction of Windows. This operating system provided service layer support for GUI features and kernel-level support for advanced user-interface hardware. Over the years, Microsoft has tightened interface standards to the point that application programs must now pass a rigorous set of tests (performed by Microsoft) to receive certification as "Windows compatible." These tests address both the user interface and the ways in which the application interacts with system software and hardware.

Figure 3-10 shows a timeline of the release of various Intel CPUs and Microsoft operating system versions. The timeline shows a clear pattern in the timing of operating system releases and new CPUs. Major advances in operating systems generally have lagged the introduction of the last CPU by three years and led the introduction of the next CPU by a shorter time span.

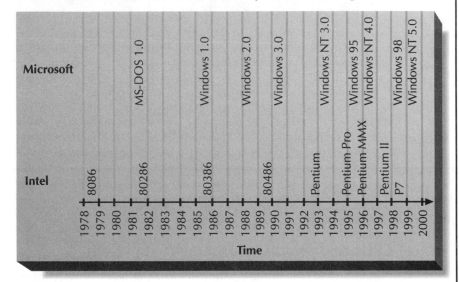

Figure 3-10

The timing of new CPU introductions by Intel and new operating system introductions by Microsoft

Since the mid-1980s, Microsoft and Intel have provided one another advance information about their developmental technologies. Given that relationship, one might expect operating system releases to be less than three years distant from CPU release or coincident with the next CPU release, but that expectation ignores the labor-intensive nature of software development. It also ignores the dependency of operating system capabilities on hardware other than the CPU (e.g., video displays, network communication, primary storage, and secondary storage). To a great extent, these hardware technologies also lag Intel's CPU advances.

The introduction of Windows is a good example of the dependency of software capability on hardware technology. Windows was simply too complex an operating system to run on the original IBM PC platform (using the Intel 8086). Processor power, communication capacity, and the video display capabilities of the platform were all too limited. The 80286 supplied the needed processor power. Several enhancements to the PC platform (e.g., a wider/faster system bus and the EGA video display standard) provided the rest of the required hardware capabilities.

Although the capabilities of the 80286 and the enhanced PC platform (called the AT) were sufficient to run Windows, they weren't necessarily optimal. Users who switched to an AT-compatible computer running Windows found that their Windows application programs frequently ran slower than the MS-DOS versions on their old hardware. Many of these users quickly switched to computers based on the 80386 and newer platform enhancements (e.g., VGA displays and 32-bit communication channels).

The 80386 made several important advances over the 80286 besides an increase in raw processor power. It replaced earlier approaches to memory addressing that had effectively limited application programs to no more than 640 Kbytes of memory and also included hardware support for the execution of multiple programs simultaneously. The 80386 simplified the partitioning of primary storage among programs and provided mechanisms to protect those programs from interfering with one another and the implementation of operating capabilities such as task switching and virtual memory management. These capabilities were incorporated with little change into the 80486 and Pentium family of processors.

The significant number of advances in the 80386 was a lot for Microsoft to handle all at once. Windows 3.0 was the first operating system to take advantage of them. It adopted the newer memory addressing and management methods and thus freed application programs to use more memory; it allowed easy switching from one active application program to another; and it also allowed certain programs to execute "in the background" during idle periods of execution by other programs.

Windows 95 did a much more complete job of utilizing all of the advances in the 80386. Application programs are much better protected from one another than under Windows 3.0. Task switching also was improved as are the management of memory and other hardware devices. Although Windows 95 did not require any specific capabilities of CPUs beyond the 80386, it does benefit greatly from their increased raw processing power and thus it fueled a rapid move by users of 80386- and 80486-based computers to Pentium-based computers.

The introduction of Windows NT required the advances in task and memory management of the 80386 and the raw processing power improvements of the 80486 and Pentium. Windows NT implemented much more sophisticated versions of the memory and hardware management capabilities found in Windows 3.x and Windows 95. Also, it was designed to be portable to hardware platforms other than IBM compatible PCs. This portability came at the price of a greatly increased need for raw processing power. The 80486 microprocessor provided barely enough power for that purpose.

The push and pull relationship of hardware power and software capability is well-demonstrated by Intel and Microsoft product advances. Hardware is the driving force—to a great extent, CPU makers can safely assume that their product improvements will quickly prove useful, even if most users don't need them immediately. This is not to say that CPU makers don't actively seek out and implement technologies that will be useful to software developers. It is simply that they know that those technologies won't be widely used until software developers have had time to assimilate them. Thus, software that fully uses significant hardware advances typically lags a generation behind the hardware.

Software makers can generally assume that raw processing power increases will arrive at regular intervals with no increase (or a decrease in cost). They can develop software that strains under the limitations of current hardware, confident that hardware manufacturers will soon eliminate those limitations. To the extent that software increases the consumption of memory and processor cycles, it fuels users' drive to acquire new and more powerful machines. The rapid arrival of more processing horsepower also gives software some breathing room to incorporate fundamental hardware advances at a more leisurely pace. Competition among software developers, though, guarantees continual and rapid assimilation of fundamentally new hardware technologies.

SUMMARY

The role of software is to translate user processing requests into machine instructions that will satisfy the request when executed. This translation is highly complex due to the wide disparity in detail and content between a user processing request and a machine language program. The two primary types of software are application software and system software. Application software consists of programs that satisfy specific user processing needs. Examples of these programs include payroll calculation, accounts payable, and report generation programs. System software consists of utility programs that satisfy a class of processing needs. These programs are designed to be general in nature and may be used many times by many different application programs and/or users.

A program consists of instructions that can be directly executed by the CPU. Such programs are called machine language programs. Writing machine language programs is a tedious, labor-intensive activity. Alternative languages for writing programs include assembly language and high-level programming languages. Assembly language programs are similar to machine language programs except that instructions and data may be written using names composed of characters instead of numbers. This difference makes assembly language programs easier to write. High-level language programs are designed to allow a programmer to write a small number of statements that correspond to many machine instructions. These languages require an interpreter or compiler to translate high-level language statements into machine instructions. They improve programmer productivity by allowing complex processing to be stated in a compact form.

System software can be classified into application development tools, application tools, and operating software. Application development tools are programs used to create (or help create) other programs. These include compilers, interpreters, CASE tools, and other programs commonly used by programmers. Application tools are programs that allow a user to perform certain types of processing without an application program. They are designed to address a specific type of processing such as document preparation, drawing, and data management. Operating software (or an operating system) exists to manage hardware resources and to perform functions that are common to many application programs. These functions include process, file, I/O, and user control.

An operating system is composed of the command layer, service layer, and kernel. The command layer, the user interface to the operating system, is the means by which the user controls computer hardware and executes other programs. Functions such as file manipulation, program execution, and secondary storage management are performed using command layer facilities. Typically, the user interacts with the command layer through a graphical user interface (GUI). Alternate command mechanisms include forms, menus, and command (or job control) languages.

The service layer consists of a set of utility programs that are used by application programs and the command layer. These programs implement commonly used functions such as input and output to files and I/O devices. Access to hardware by users and/or application programs is directed through the service layer, which provides a measure of independence between application software and physical hardware. This feature is commonly called hardware or machine independence.

The kernel is responsible for the control of hardware devices and the management of resources. Direct manipulation of the CPU, primary and secondary storage, and I/O devices is implemented within the kernel. The kernel contains a set of device drivers for each hardware device that serve as interface and control programs. The kernel also allocates hardware resources to application programs and users, which is a complex and relatively invisible process from the user's point of view.

The development of system software has been motivated by advances in hardware technology. These advances have resulted in lower-cost and more powerful computers. System software serves several economic purposes. Application tools and application development tools reduce the cost of meeting users' processing needs. Service layer functions also reduce this cost by reducing the amount of code that application programmers must write. Command layer functions automate a computer's operation and allow it to be shared by many users and programs. Each of these uses represent a substitution of relatively cheap computer hardware for relatively expensive labor.

application generator
application program
application software
application tool
assembler
assembly language
Computer Assisted Software
  Engineering (CASE) tool
code
code checker
coder
command language
command layer
compiler
complex machine instruction
data manipulation tool
database management system
  (DBMS)

debugger
device driver
executable
executable code
fourth generation development
  environment
fourth generation language
  (4GL)
hardware independence
high-level programming language
instruction code
instruction set
interpreter
job control language (JCL)
kernel
machine independence
machine language
op code

operating system
operation code
process
program development tool
program translator
program verifier
programming language
service call
service layer
shell
simple machine instruction
source code
system development tool
system software
text editor
utility program

Vocabulary Exercises

1. The translation of a high-level language into machine instructions is performed by a(n) _____ or a(n) _____. *compiler or assembler*

2. A(n) _____ extends the ability of an operating system or programming language to manipulate data stored on secondary storage devices. *DBMS's*

3. *instruction code* _____ must be translated into *binary code* _____ before it is ready for execution by the CPU.

4. A(n) *command* _____ is a processing request made by an application program to the operating system service layer.

5. The term _____ describes a programming language in which one statement may be translated into many CPU instructions.

6. Application software makes _____ to the operating system _____ to perform functions such as file manipulation and printer control.

7. _____ differs from machine language primarily in the use of short alphabetic names for instructions and data.

8. A(n) _____ performs a single computation, comparison, or data movement operation. A(n) _____ combines the function of two or more _____.

9. A(n) _____ translates assembly language programs into machine language programs. *assembler*

10. _____ is an alternative name for the operating system command layer. *command language*

11. The set of op codes recognized and processed by a CPU is called the CPU's _____.

12. _____ or _____ allows computer system hardware to be changed without changing application or service layer programs.

13. A component of the kernel that controls access and interface to a single hardware device is called a(n) _____. *device driver*

14. A(n) _____ is a tool used by a programmer to find errors in compiled code. *debugger*

15. Word processors, drawing programs, and spreadsheet packages are all examples of _____. *application tools*

16. Resource allocation and direct hardware control is the responsibility of the _____. *Kernel*

17. The language that a user uses to direct the actions of an operating system is called a(n) _____ or a(n) _____. *Command Language or Job Control language*

18. _____ software is general purpose. _____ software is specialized to a specific user need. *utility* *application*

19. The _____ is the user interface to an operating system. *service layer*

20. A(n) _____ is a numeric code that directs the CPU to perform one primitive processing function. *machine instruction*

21. A(n) _____ supports early phases of the SDLC including survery, analysis, and design. *system development tool*

Review Questions

1. What is the instruction set of a computer system?

2. What are the three basic types of instructions?

3. How are user requests for processing translated into machine instructions?

4. What characteristics differentiate application software from system software?

5. In what ways does system software make the development of application software easier?

6. In what ways does system software make application software more portable?

7. How does assembly language programming differ from machine language programming? Which is easier and why?

8. Describe the functions of an assembler and a compiler. Which is simpler and why?

9. What are application tools? In what way(s) do they differ from application development tools?

10. What are the primary components of an operating system?

11. What is the service layer? With what does it interact?

12. What is the kernel? What functions does it perform?

13. Why has the development of system software paralleled the development of computer hardware?

Research Problem

The Windows NT operating system is unusual in that it incorporates support for several different sets of service calls. These include service calls supported by an earlier version of Windows, the POSIX standard, and MS-DOS. In addition, these various service layer definitions can interact with several different CPU instruction sets including the Intel 80X86 series and DEC Alpha microprocessors. Such capabilities require a more complex scheme of operating system layers than is described in this chapter. What operating system layers and interactions between layers are used by Windows NT to support multiple service layer definitions and CPU instruction sets?

Chapter 4

Data Representation

DATA IN MOTION AND DATA AT REST

Humans can understand and manipulate data represented in a variety of forms. For example, we can interpret numeric data represented symbolically as Arabic numerals (e.g., 8714), Roman numerals (e.g., XVII), and simple lines or tick marks on paper (e.g., ||| to represent the value 3). Words of a natural language can be represented symbolically using a pictorial character representation (e.g., 电脑), an alphabetic character representation (e.g., "computer" and "ЛПНРШАФЕТ"), or as a sound wave (i.e., a word spoken in a particular language). Humans can also extract data content from visual images (e.g., still and motion pictures) and from our senses of taste, smell, and touch. The rich variety of data representations that can be recognized and used as input to the brain are an integral part of its considerable processing power and flexibility.

The previous examples are all external symbolic representations of data. To be manipulated by the brain, external data representations must be converted to an appropriate internal representation and transported to the brain's "processing circuitry." The exact processes of data conversion, internal transport, and internal processing in the human brain are poorly understood. Much of this activity is known to be electrically based, but other theories have postulated additional representation and communication mechanisms (e.g., data represented as vibrations transmitted through microtubules). All of our sensory organs act as a means of converting various data inputs (e.g., sights, smells, tastes, sounds, and pressures) into electrical impulses that are transported through our nervous systems to the brain.

Any data and information processor—whether organically, mechanically, electrically, or optically implemented—must be capable of the following data-related functions:

► Access to externally and internally stored data.
► Transport of data among external storage, internal storage, and internal processing components.

If external and internal data stores are not implemented using a technology directly compatible with that of the internal processing components, then various data conversion capabilities (e.g., from the printed word to corresponding neural impulses) are also required.

Physics and Mathematics of Automated Data Processing

The current generation of automated computers implement processing components with electrical switches. Data manipulation (e.g., addition, and multiplication) is accomplished by combining switches into processing circuits and by transporting numeric data represented as electrical impulses through those processing circuits. The circuits physically transform data inputs (e.g., the numbers on the left side of an equation) into data outputs (e.g., the answer on the right side of the equation).

Recall from your high school and/or college science classes that the terms "physics" and "mathematics" are often uttered in the same phrase. The laws of physics are described by mathematical equations, and this applies equally to the physics of mechanical objects, electricity, and light. The physical implementation of computer circuitry exploits this fundamental relationship between mathematics and physics. If the behavior of a device is based on a well-defined mathematically described law of physics then, in theory, that device can be used to implement a processor to perform the equivalent mathematical function. This thesis underlies the implementation and function of automated computation devices ranging from mechanical clocks (utilizing the mathematical ratios of various gears) to electrical microprocessors (utilizing the mathematics of electrical voltage and resistance).

Implementing computers based on the precise relationship between mathematics and physics has limits. It confines information processing functions to those based on mathematical functions (e.g., addition and equality comparison) using numerical data inputs and generating numerical outputs. Non-numeric data is difficult or impossible to represent and process. Familiar human ideas and concepts such as mother, friend, love, and hate have no direct or obvious numerical representation. Because there is no direct numeric representation, there is no way of directly implementing processing of those "data items" using traditional processing circuitry.

Humans are still studying how such data might be converted to numeric form as a necessary conversion step for automated processing. Many researchers are currently investigating the implementation of non-numeric processing devices. But the vast majority of computers currently in use are designed to perform only numerical computing functions. Some simple non-numeric data can be processed (e.g., characters and tiny components of pictures) — but only by first converting it to numeric form.

COMPUTER DATA REPRESENTATION AND TRANSPORT

Because the internal processing components of all modern computers are implemented as electrical devices, the internal representation of data is also electrically based. External data representation and data communication among computers is also electrically based for reasons of efficiency and accuracy. Thus, a key issue in the implementation and operation of computers is the means of representing data in electronic form.

Carrier Waves and Signals

Electrical current is the flow of electrons from one place to another. Given sufficient initial energy and an appropriate path over which to move, electrons will flow from place to place using the atoms of conductive material as stepping stones from start to finish. In nature, this process can be haphazard (e.g., lightning). Within an electrical device, this process is tightly controlled, and the conductive materials (e.g., wires and switches) are organized to implement specific data-processing functions.

The flow of electrons through a conductor is sometimes described as a "wave." Imagine each electron as the crest of a wave in the ocean and the crashing of ocean waves on the beach as the arrival of electrons at their destination. Other physical phenomena such as sound and light can also be described as waves. The importance of describing electrical current as a wave lies in:

1. Utilizing waves as a means of transport.

2. Manipulating and detecting patterns within waves.

Representation of data within a wave can be accomplished by intentionally manipulating patterns within or characteristics of the wave(s). The function of the wave as a transport mechanism allows data embedded within it to move from place to place.

As a simple example, consider ripples in a pool of water as waves and a potential data-carrying mechanism. Long and short distances between the ripples (see Figure 4-1) could be used to represent letters in essentially the same manner that Morse code (see Table 4-1) represents letters as long and short duration sounds ("dots" and "dashes"). A person at one end of the pool could intentionally embed letters in the ripples by precisely timing contact of a stick with the water surface. A person at the other end of the pool could extract data content from the ripples by interpreting them based on the sender's timing rules. Data thus would be embedded in a physical characteristic of the wave by the sender, transported from sender to receiver by the wave, and interpreted and understood by the receiver.

Figure 4-1

Distances between wave peaks used to represent the "dots" and "dashes" of Morse code

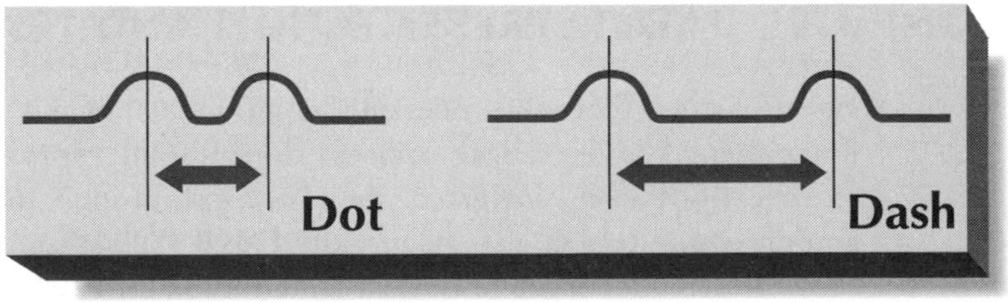

Data Representation

TABLE 4-1	A	•–	N	–•	0	–––––
	B	–•••	O	–––	1	•––––
Morse code: an encoding and decoding scheme for printed letters, numerals, and punctuation symbols. The • symbol represents a short duration sound and the – symbol represents a long duration sound.	C	–•–•	P	•––•	2	••–––
	D	–••	Q	––•–	3	•••––
	E	•	R	•–•	4	••••–
	F	••–•	S	•••	5	•••••
	G	––•	T	–	6	–••••
	H	••••	U	••–	7	––•••
	I	••	V	•••–	8	–––••
	J	•–––	W	•––	9	––––•
	K	–•–	X	–••–	?	••––••
	L	•–••	Y	–•––	.	•–•–•–
	M	––	Z	––••	,	––••––

The preceding example illustrates many of the key components of data representation and communication. Data is embedded within a wave (two waves in this example) by intentionally manipulating some physical aspect of the waves (e.g., time between their peaks). The wave is called a **carrier wave** because it transports (or carries) data from one place to another.

To convey data properly, both sender and receiver must agree to a common method for representing data within the wave. This is accomplished by defining a table of correspondences between physical wave characteristics and data values (e.g., letters on the left and distances between wave peaks on the right). The table describes both a method of **encoding** the data within the carrier wave (e.g., reading from the left column to the right) and **decoding** the carrier wave to extract the data (e.g., reading from the right column to the left).

Other rules are generally required in addition to encoding and decoding rules. In the previous example, sender and receiver must also agree on the spacing between "dots" or "dashes," the spacing between "words," and how large a wave should be to be interpreted as meaningful data (instead of a random wave generated by wind or a leaf falling on the water). The complete set of rules including the encoding/decoding method is called a **communication protocol** and is more fully discussed in Chapter 8.

A **signal** is a specific data transmission event or group of events representing either an entire data item or a well-defined component of a data item. In the previous example, a single "dot" or "dash" is represented by two wave peaks and the distance between them. Thus, the combination of the two peaks and the distance between them represent one signal or message. Other methods of data encoding might represent data using only a single wave peak or some other wave characteristic.

Digital Signals

Digital signals are those that can contain one of a finite number of possible values. A more precise term is **discrete signal** (i.e., a signal with a discrete or countable number of possible values). The coding of dots and dashes into waves is a digital signaling method because there are two possible values ("dot" or "dash") that the signal can contain. That coding is also an example of a **binary signal** (the term **binary** is defined as "having two distinct parts" or "having a numerical base of 2"). Other possible discrete signaling methods include trinary (3 values), quadrary (4 values), and so forth.

Note that the distance between wave peaks is not inherently binary. That distance could be any value between a slightly more than 0 number and a very large number (i.e., it is an inherently continuous variable). In the Morse code scheme with waves, we must choose two of the infinite number of possible distances—one to represent a dot and the other to represent a dash. In essence, we choose to ignore all of the other possible values of distance between wave peaks.

A more flexible approach is to allow groups or ranges of wave parameter values (e.g., distance) to represent a single data value. Any continuously variable parameter of a carrier wave can be used to carry a digital signal by dividing its range of possible values into a discrete number of intervals. For example, the range of frequency between 20 Hz and 20,000 Hz (the range of human hearing for sound waves)[1] could be divided into two intervals—less than 10,000 Hz and greater than or equal to 10,000 Hz. The dividing line between the two ranges is called a **threshold**.

A digital interpretation of a parameter value such as frequency is defined by assigning a meaning to each range of values (e.g., "dot" or "zero" for the lower range and "dash" or "one" for the upper range) as shown in Figure 4-2. Interpretation of the signal's data content is made by comparing the observed value of the parameter to the threshold. Thus, the parameter value can be any of several possibilities and still convey the same meaning or data (e.g., 0; 5,000; and 9,999.9 are all less than 10,000 and are, therefore, all interpreted as a "dot" or "zero").

---

[1] The frequency of sound is measured in cycles per second or hertz (abbreviated as Hz). For example, the note A, above middle C on a piano, corresponds to a pitch of 440 cycles per second (440 Hz).

Data Representation

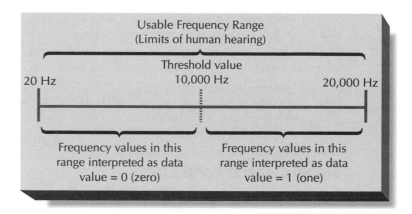

Figure 4-2

A binary signaling method using sound wave frequency. Frequency values below 10,000 Hz are interpreted as 0, and values above 10,000 Hz are interpreted as 1.

Usable Frequency Range
(Limits of human hearing)

Threshold value

20 Hz

10,000 Hz

20,000 Hz

Frequency values in this range interpreted as data value = 0 (zero)

Frequency values in this range interpreted as data value = 1 (one)

Analog Signals

Analog signals use the full range of possible values of a carrier wave parameter to encode data values. The parameter value is either equivalent to a data value or can be converted to a data value by a simple mathematical function. For example, assume that data is to be encoded as an analog signal using the frequency of a carrier wave. The numeric value 100 would be encoded by setting the wave frequency to 100 Hz. The number 9,999.9 would be encoded by setting the wave frequency to 9,999.9 Hz. Data values anywhere within the range of frequencies that can be manipulated could be encoded within the signal.

Analog signals are continuous in nature. That is, they can represent any data value within a continuum (or range) of values. The number of values that can be encoded is theoretically infinite. For example, even if the range of frequency variation is limited to between 20 Hz and 20,000 Hz there are still an infinite number of possible pitches. The frequency of a given sound could be 440 Hz, 441 Hz, or an infinite number of possibilities in between, such as 440.1, 440.001, 440.00001, and so forth. The number of possible frequency values within this range is limited only by the ability of the sender to generate them, the transport mechanism to carry them, and the receiver to distinguish among them.

Electricity can be used as a carrier wave for either analog or digital signals. Variations in voltage are the most typically used method of encoding data within the wave. For example, we could send the number one as an analog signal by sending one volt of electricity through a wire. To transmit a binary signal, an appropriate threshold must be defined: for example, voltages of less than 10 to mean 0 and voltages of greater than or equal to 10 to mean 1. We could then encode the number one within the electrical current by generating a voltage greater than or equal to 10 volts.

To send multiple messages across an electrical medium, a timing convention must be used. That is, an interval of time must be defined during which a single value of the signal will be present on the wire. For example, we could adopt one second as the time interval and send first 1 and then 0 across the wire (as a digital signal) by sending 20 volts across the wire for one second and then immediately lowering the voltage to 0 for the next second. As with all other aspects of the communication protocol, both sender and receiver must agree on the timing convention.

Signal Capacity and Errors

Analog and digital signals each have strengths and weakness relative to one another. The most important of these are message-carrying capacity and susceptibility to error. In general, analog signals can carry a greater amount of information than digital signals within a fixed time interval. Higher data carrying capacity results from the large number of possible messages that can be coded within an analog signal during one time period.

For example, assume the use of an analog signaling method using electrical voltage with a message duration of one second. Assume further that the range of voltage is 0 to 127 volts and that sender and receiver can accurately distinguish one-volt differences in voltage. Thus, there are 128 different messages that can be transmitted during any one transmission event. This number would be much larger (theoretically approaching infinity) if sender and receiver were capable of distinguishing relatively small differences in voltage.

Compare the previous example to a binary electrical signal (e.g., less than 64 volts is 0 and greater than 64 volts is 1). During any single transmission event, the number of possible data values sent is only two (either a 0 or a 1). Thus, the message-carrying capacity of the analog signal is 64 times greater (128 possible values with analog divided by two possible values with binary) than the binary signal in this example.

To transmit more than two possible values with binary coding, adjacent binary signals must be combined to form a single message. For example, we could send seven different binary signals in succession and combine them to form one message. It is thus possible to transmit 128 different messages ($2^7 = 128$ possible combinations of seven binary signals) during seven consecutive time periods. But an analog signal could have communicated any of 128 different values in each time period.

Although analog signals have a significant advantage in message-carrying capacity, they are at a distinct disadvantage in their susceptibility to transmission error. If the mechanisms for encoding, transmitting, and decoding electrical analog signals were perfect, then this would not be a problem. But errors are always possible; for example, electrical signals are subject to noise and disruption because of electrical and magnetic disturbances; the noise on a telephone during a thunderstorm is an example of such a disturbance. Voltage, amperage, and other parameters of an electrical carrier wave can be altered by such interference.

Consider the following hypothetical scenario. Assume that your bank's computer communicates with its automated teller machines by analog electrical signals. You are in the process of withdrawing $100. The value $100 is sent from the automated teller to the bank's computer by a signal of 100 millivolts. After the computer checks your balance and decides that you have enough money, it sends an analog signal of 100 millivolts back to the automated teller as a confirmation. During the transmission of this signal, a bolt of lightning strikes near the wire carrying the signal. The nearby lightning strike induces current in the wire of five volts (5,000 millivolts). When the signal reaches the automated teller, it dutifully dispenses $5,000 dollars to you in crisp $10 and $20 bills.

Susceptibility of electrical signals to noise and interference can never be eliminated. Normally, computer equipment is heavily shielded to prevent noise from interfering but external communications may not be protected so well. In addition, errors can be introduced by the resistance present in internal wiring or by the magnetic fields generated within the device.

A digital electrical signal is not nearly as susceptible to noise and interference. Consider the previous example with digital encoding used instead of analog encoding. If we use the previous protocol of 0 volts for 0, 20 volts for 1, and a threshold of 10 volts, we can tolerate almost 10 volts of noise without misinterpretation of a signal. If a 0 was being sent at the instant the lightning struck, the automated teller would still interpret the signal as a 0 because the five volts induced by the lightning is still below the threshold value of 10. If a 1 was being sent, then the lightning would raise the voltage from 20 to 25 volts, which is still above the threshold. Resistance in the wire or other factors degrading voltage would not be a problem as long as the strength of the signal did not decrease below 10 volts. The margin for transmission error is illustrated in Figure 4-3.

Figure 4-3

Margin of transmission error (voltage drop or surge) allowed before the data value encoded within a binary signal is altered

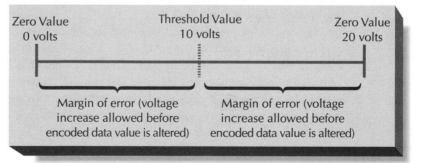

The accuracy desired in computer computation and communication requires the use of digital signals. The potential for error inherent in analog signaling is simply too high. Thus, every general purpose computer available today employs digital signals for representing data. This is true for internal processing components (e.g., the CPU) and most storage devices (e.g., hard drives and random access memory). It is also the rule for all data communication among internal computer system components and most data communication among computer systems.

Analog signals are, however, sometimes used when message carrying capacity is critical (e.g., for high-speed communication by narrow band radio waves). Analog devices also remain important and useful in handling inputs and outputs within digital systems. To connect these two technologies, a conversion step is required. The transformation of continuous, analog signals into discrete, digital data is called **analog to digital (A/D) conversion**. Similarly, digital signals must be converted to analog signals (**digital to analog**, or **D/A conversion**).

BINARY DATA REPRESENTATION

Most computers use a specific method of digital signaling—binary signaling. There are two primary reasons for the use of binary signaling and data representation in modern computers:

▶ Binary signals provide maximal protection against data-transmission errors.

▶ Binary signals are processed by two-state electrical devices that are relatively easy to design and fabricate.

Although digital computers are binary devices, there is nothing mandatory about the choice of two-state design. It is possible to create 4-state, 6-state, or even 8- or 10-state computing devices. In fact, many nonbinary digital devices have been developed in the course of the evolution of the computer industry. However, given current technology, binary signals and processing devices represent the most cost-efficient trade-off among signal and processing capacity, accuracy, reliability, and cost.

Binary codes are also well-suited to computer processing because they correspond directly with values in boolean logic. This form of logic is named for Nineteenth Century mathematician George Boole. He developed methods of reasoning and logical proof using sequences of statements that could be evaluated only as True or False. Similarly, a computer can perform logical comparisons of two binary data values to determine whether it is True or False that a presented value is greater than, equal to, less than, less than or equal to, not equal to, or greater than or equal to another value. As discussed in Chapter 2, this primitive logical capability is the means by which a computer can implement intelligent behavior.

The use of binary signals within a computer does not imply that the only two numbers that it can represent and process are 0 and 1. Both computers and humans can combine multiple digits for the purpose of representing and manipulating numbers larger than the base of their respective numbering systems. Both decimal and binary notation are forms of **positional number systems**. In a positional number system, numeric values are represented as patterns of symbols, that is, strings of numeric digits. The symbol used for each digit and its position within the string determine its value. The sum of the values of all positions in the string equals the value of the string.

For example, in the decimal number system, the number 5,689 is interpreted as follows:

$$(5 \times 1000) + (6 \times 100) + (8 \times 10) + 9 =$$
$$5,000 + 600 + 80 + 9 = 5,689$$

The same series of operations can be represented in columnar form or with positions of the same value aligned in columns:

$$
\begin{array}{r}
5,000 \\
600 \\
80 \\
+\ 9 \\
\hline
5,689
\end{array}
$$

For whole numbers, values are accumulated from right to left. In the example, the digit 9 is in the first position; 8 is in the second position; 6 in the third; and 5 in the fourth.

The maximum value, or weight, of each position is a multiple of the weight of the position to its right. In the decimal number system, the second position is 10 times that of the "ones," or first, position. The third is 10 times the second, or 100. The fourth is 10 times the third, or 1,000, and so on. In the binary numbering system, each position is two times the previous position. Thus, position values for whole numbers are 1, 2, 4, 8, and so forth. The multiplier that describes the difference between one position and the next is the **base** of the numbering system. Another term for base is **radix**. The base, or radix, of the decimal number system is 10. The radix of the binary numbering system is 2.

For fractional parts of real numbers, values are accumulated from left to right, beginning with the digit immediately to the right of the (.) notation, or **radix point**. In the decimal number system, the radix point usually is referred to as the decimal point, for example:

$$5,689.368$$

The fractional portion of this real number is (.368). The digits are interpreted as follows:

$$(3 \times .1) + (6 \times .01) + (8 \times .001) =$$
$$.3 + 0.06 + 0.008 = 0.368$$

Note that, proceeding rightward from the radix point, the weight of each position is a fraction of the position to its left. In the decimal number system, the first position to the right of the decimal point represents tenths (10^{-1}), the second position represents hundredths (10^{-2}), the third thousandths (10^{-3}), and so forth. In the binary numbering system, the first position to the right of the decimal point represents halves (2^{-1}), the second position represents quarters (2^{-2}), the third eighths (2^{-3}), and so forth. As with whole numbers, each fractional position has a weight that is 10 (or 2) times that of the position to its right. Table 4-2 shows a comparison between decimal and binary representations of the numbers 1 through 10.

TABLE 4-2

The equivalent notations for the values 0 through 10 under the binary and digital numbering systems

	Binary System (Base 2)					Decimal System (Base 10)			
Place	2^3	2^2	2^1	2^0		10^3	10^2	10^1	10^0
Values	8	4	2	1		1000	100	10	1
	0	0	0	0	=	0	0	0	0
	0	0	0	1	=	0	0	0	1
	0	0	1	0	=	0	0	0	2
	0	0	1	1	=	0	0	0	3
	0	1	0	0	=	0	0	0	4
	0	1	0	1	=	0	0	0	5
	0	1	1	0	=	0	0	0	6
	0	1	1	1	=	0	0	0	7
	1	0	0	0	=	0	0	0	8
	1	0	0	1	=	0	0	0	9
	1	0	1	0	=	0	0	1	0

Besides determining the numeric value of groups of digits, these positional relationships also affect the number of digits required to represent a given numeric value. The number of digits required to represent a value increases as the base of the numbering system decreases. Thus, values that can be represented relatively compactly in decimal notation may require lengthy sequences of binary digits. Table 4-3 summarizes the number of binary digits required to represent various decimal values.

TABLE 4-3	Number of Bits (*n*)	Number of Values (2ⁿ)	Numeric Range (Decimal)
Binary data representations for bit strings ranging from 1- through 16-bit positions	1	2	0..1
	2	4	0..3
	3	8	0..7
	4	16	0..15
	5	32	0..31
	6	64	0..63
	7	128	0..127
	8	256	0..255
	9	512	0..511
	10	1024	0..1023
	11	2048	0..2047
	12	4096	0..4095
	13	8192	0..8191
	14	16384	0..16383
	15	32768	0..32767
	16	65536	0..65535

Converting a number in binary notation to its decimal equivalent can be done by multiplying the value of each position by the decimal weight of that position and then summing the results. An example is shown in Figure 4-4.

Procedure for computing the decimal equivalent of a binary number

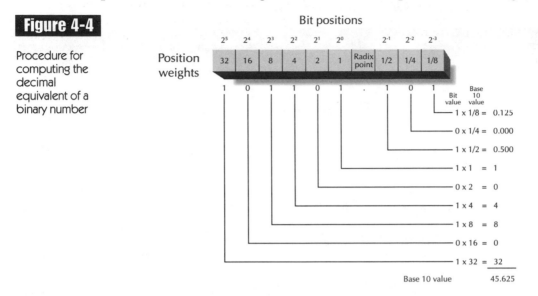

In computer terminology, each digit of a binary number is called a **bit**. Each bit occupies a unique **bit position**. A meaningful sequence of binary digits, or group of bits, is called a **bit string**. The leftmost digit, which has the greatest weight, is called the **most significant digit**, or **high order bit**. Conversely, the rightmost digit is the **least significant digit**, or **low order bit**. A string of eight bits is called a **byte**. A byte is generally the smallest unit of data that can be read or written to/from a storage device.

Addition operations involving positive binary numbers can be performed by applying the following four mathematical rules:

$$0 + 0 = 0$$
$$1 + 0 = 1$$
$$0 + 1 = 1$$
$$1 + 1 = 10$$

Addition of two binary, real numbers is done by first aligning the respective radix points:

$$101101.101$$
$$+\ 10100.0010$$

Data Representation

At each position, an addition operation is performed, the result of which is determined by applying the appropriate addition rule. Values at each position, or in each column, are added, starting with the rightmost, or least significant, digit. Columnar addition proceeds just as it does for decimal numbers. If the result exceeds the weight assigned to that position, the excess value must be "carried" to the next column, or position. The excess value then is added to the values at the next position. The result of adding the two numbers above is:

```
    111      1
    101101.101
 +   10100.0010
    1000001.1100
```

Note that the result is the same as that obtained by adding the values in base 10 notation:

	Binary		Real fractions		Real decimals
	101101.101	+	45 ⅝	+	45.625
+	10100.0010	+	20 ⅛	+	20.125
	1000001.1100	+	65 ¾	+	65.750

The representation of data as binary numbers poses problems for programmers. Binary numbers require many, potentially error-producing digits. To minimize these errors, alternate number systems are sometimes used in assembly language programming. These number systems include hexadecimal and octal. Hexadecimal is the most commonly encountered method.

Hexadecimal Notation

There are not enough numeric digits (Arabic numerals) to represent 16 different values. Thus, the digits of the hexadecimal numbering system include several English letters to represent the upper values (see Table 4-4 on the next page). The primary advantage of **hexadecimal notation** as compared to binary notation is its compactness. Numeric values expressed in binary notation require four times as many digits as those expressed in hexadecimal notation. For example, the content of a byte of storage requires eight binary digits (e.g., 11110000) but only two hexadecimal digits (e.g., F0). The more compact representation greatly reduces programmer error.

TABLE 4-4	Base 16 Digit	Decimal Value	Base 16 Digit	Decimal Value
The digits of the hexadecimal numbering system and their corresponding decimal (base 10) values	0	0	8	8
	1	1	9	9
	2	2	A	10
	3	3	B	11
	4	4	C	12
	5	5	D	13
	6	6	E	14
	7	7	F	15

Hexadecimal numbers are often used to designate memory addresses. An area that contains 64 K memory cells can hold 65,536 bytes (64 × 1,024 bytes/K). Thus, the range of possible memory addresses must be 0 to 65,535 (decimal). In binary notation, 16 bits will be required to cover this range:

00000000 00000000 – 11111111 11111111

A much more concise and understandable representation is to code the addresses in four-bit groups, using one hexadecimal digit for each group. In this scheme of notation, the full range of addresses within 64 K of memory is 0000 to FFFF hex. This is a widely accepted numbering scheme for memory addresses.

To differentiate binary and hexadecimal numbers in printed text, it is necessary to indicate the base of the written number. In mathematics, it is common to denote the base of a number with a subscript. Thus, 1001_2 would indicate that 1001 should be interpreted as a binary number and 6044_{16} would indicate that 6044 should be interpreted as a hexadecimal number.

When writing about computers, it is more common to designate the base of a number by affixing a letter to the end of the number. Thus, 1001B would indicate a binary number, and 6044H would indicate a hexadecimal number. No letter is normally used to indicate a decimal (base 10) number. Some programing languages also use the prefix 0x to indicate a binary number (e.g., 0x1001 is equivalent to 1001_2).

Unfortunately, these conventions are not observed consistently. Often, it will be left to the reader to guess at the base by the content of the number or the context in which it appears. Bit strings, for example, generally are assumed to be expressed in binary, and memory addresses are assumed to be expressed in hexadecimal. The content also can guide the reader. Any numeral other than a 0 or a 1 indicates that the number cannot be binary. Similarly, the use of letters A through F indicates that the contents are expressed in hexadecimal.

Octal Notation

In some assembly language applications, it is convenient to work with groups of three bits, or 000 to 111 (binary). An alternate method of coding these groups is base 8 or **octal notation**. Octal digits range from 0 to 7, which correspond exactly with the minimum and maximum values that can be represented in three bits. For binary values that extend beyond three bits, decimal 8 is octal 10, decimal 9 is octal 11, and so on. The only assembly languages that commonly use octal notation are those designed for use with IBM mainframe computer systems (specifically, those compatible with IBM 360/370 processor architecture).

GOALS OF COMPUTER DATA REPRESENTATION

Although all modern computers represent data with binary digits they do not necessarily represent larger numeric values in exactly the format described in previous sections (i.e., in ordinary binary format). Any representation format for numeric data represents a trade-off between several factors, including compactness, ease of manipulation, accuracy and range, and standardization.

As with many computer design decisions, alternatives that perform well in one dimension or category often perform poorly in others. Thus, for example, a particular data format with a high degree of accuracy and a large range of representable values is usually difficult (and therefore expensive) to manipulate and not very compact.

Compactness

In general, compactness of representation and the range of representable values are inversely related. That is, large ranges require more digits (larger bit strings) to represent values within the range. A large number of digits also has implications for the implementation of processing and storage devices. Large data representation formats require relatively large amounts of storage (e.g., disk and RAM). Processing devices for large data formats are also large and complex. These characteristics translate directly into costs of those devices; that is, larger data formats require more expensive processing and storage devices.

Ease of Manipulation

When discussing computer processing, the term "manipulation" refers to executing processor instructions (e.g., addition, subtraction, and equality comparison). The term "ease" equates to machine efficiency. The efficiency of a processor is dependent on its complexity (i.e., the number of its primitive components and the complexity of the wiring that binds them together). Efficient processor circuits perform their function relatively quickly due to a relatively small number of components (short distance) through which electricity must travel. More complex devices require correspondingly longer times to perform their functions.

Data representation formats vary in their ability to allow efficient implementation of processor circuits. As a simple example of this phenomenon, consider the difference between humans manipulating fractions versus decimal numbers. Most people can perform computation and comparison with decimal numbers far more easily than they can with fractions. The decimal-representation format is more efficiently processed because it uses fewer "brain circuits" and operates more quickly than the fractional-representation format.

Unfortunately, there are trade-offs among the efficiency of different types of processing functions and various data-representation formats. A data-representation format that is highly efficient for one type of processing function (e.g., addition) may not be the optimal choice for efficient implementation of another processing function (e.g., division). Consider the difference between ordinary decimal representation and logarithms. Representation of decimal numbers as logarithms may increase their length and complexity (i.e., more complex storage), but it makes certain types of processing operations (e.g., multiplication and division) much easier to implement[2].

The desire of computer users for fast computation at an affordable price forces computer designers to tailor data-representation formats to increase the efficiency of certain processing functions. This design task may require trade-offs among various processing functions as well as among design factors other than ease of manipulation.

Accuracy

Although concise coding minimizes the complexity and cost of hardware elements, it may do so at the expense of accurate data representation. The accuracy (or precision) of representation increases with the number of data bits that are used. As discussed previously, the number of data bits also limits the largest (and smallest) number that can be directly represented within the machine.

It is possible for routine calculations to generate quantities that are either too large or too small to be contained within the finite circuitry of a machine (i.e., within a fixed number of bits). For example, the fraction ⅓ cannot be

---

[2] Multiplication of two numbers can be done by adding their logarithms. Division of two numbers may be accomplished by subtracting their logarithms.

represented within a fixed number of bits because it is a nonterminating fractional quantity (i.e., 0.333333333 with an infinite number of 3s). In such cases, the quantities must be manipulated and stored as approximations. Each approximation introduces a degree of error. If approximate results are used as inputs for other computation operations, errors can be compounded. In some cases, the extent of error can be significant. Thus, it is possible for a computer program to be without apparent logical flaws and yet produce incorrect results.

If all data types were represented in the most concise form possible, the approximations required in some instances would introduce unacceptable margins of error. If, instead, a large number of bits were allocated to each data value, machine efficiency and performance would be sacrificed and hardware cost would be substantially increased. The most favorable trade-offs can be achieved by using an optimum coding method for each type of data and/or for each type of transformation that must be performed. This is the main reason for the variety of binary coding formats around which processing circuitry must be designed.

Standardization

Data must be communicated among devices in a single computer system and among computer systems. To ensure correct and efficient data transmission, it is desirable to code data in a format suitable to a wide variety of devices and computer systems. This is especially true with the alphanumeric data that forms the bulk of human-readable data communication. For this reason, various standards organizations have proposed data encoding methods for communication between computer system components. Adherence to such standards provides computer users with the flexibility to configure systems of "mixed" equipment with minimal problems in data communication.

CPU Data Types

The CPUs of most modern computers can represent and process five basic (or primitive) data types:

► Integer numeric.
► Real numeric.
► Character.
► Boolean.
► Memory address.

Some CPUs include additional variants of the types for specialized application software. The representation format for each data type is designed to represent a balance among the various design factors described in "Ease of Manipulation."

INTEGER DATA

An **integer** is a whole number, or a value that does not have a fractional part. Thus, the values 2, 3, 9, and 12,964 are integers. Arithmetic operations on integers, or **integer arithmetic**, have some specific characteristics. If addition, subtraction, or multiplication are performed entirely with integers, the result will be an integer. In the case of dividing one integer by another, it is possible for fractional values to be generated:

$$18 \div 4 = 4, \text{remainder } 2$$
$$= 4\,\tfrac{3}{4}$$
$$= 4\,\tfrac{1}{2}$$

If integer arithmetic is being used, the fractional portion (the remainder of 2 in this case) is discarded. The result (4) is an approximation.

Note that all of the examples of numeric data presented thus far have contained positive numbers. However, many application programs need to manipulate negative numbers as well. Representation of the sign of a number is straightforward since the sign has only two possible values. That is, a sign can be represented by a single bit with one value corresponding to + and the other to − . The choice of which bit value (0 or 1) corresponds to which sign (+ or −) is essentially arbitrary. The term **sign magnitude notation** refers to any storage format for numeric data that uses one of the bit positions to store the sign of the value.

Integers that are stored with a sign bit are sometimes called **signed integers**. Integers stored without a sign bit are called **unsigned integers** and are always assumed to represent positive values. A question arises as to whether the value 0 is positive or negative. In most numeric storage formats, zero is assumed to be positive. The term **non-negative** applies to all of the positive numbers and zero. Thus, the term **negative** refers only to negative values and does not include 0.

The sign bit is normally the high order bit in a numeric storage format. In most commonly used storage formats, the sign bit is 1 for a negative number and 0 for a non-negative number. It is important to note that a sign bit occupies a bit position that would otherwise be available to store part of the data value. Thus, the use of sign-magnitude notation reduces the largest positive value that can be stored in any given number of bit positions. For example, the largest positive value that can be stored in an 8-bit unsigned binary format is 255 (2^8-1). If 1 bit is used for the sign, then the largest positive value that can be stored is 127 (2^7-1).

Offsetting the reduction in maximum positive value is the ability to store negative numbers. With unsigned numbers, the lowest value that can be stored is always 0. With signed numbers, the lowest value that can be stored is the negative of the highest value that can be stored (e.g., –127 for 8-bit signed binary). Thus, the choice between signed and unsigned formats determines the largest and smallest values that can be represented.

Excess Notation

Another coding method that can be used to represent signed values is **excess notation**. Under this method, a fixed number of bit positions is used. For example, bit strings might be restricted to four positions. Thus, values could range from 0000 to 1111. Within this range, the lowest value with a 1 as its left-most digit—1000—is designated to represent 0. In ascending order, bit strings that are greater than 1000 are used to represent positive values. As shown in Table 4-5, all non-negative values have a 1 as the high-order bit, and negative values have a 0 in this position. Thus, in descending order, bit strings that are less than 1000 are used to represent negative values. This is the reverse of sign magnitude notation.

TABLE 4-5	Bit String	Decimal Value	
A specific example of excess notation using four bits called excess eight	1111	7	
	1110	6	
	1101	5	
	1100	4	Non-negative values
	1011	3	
	1010	2	
	1001	1	
	1000	0	
	0111	-1	
	0110	-2	
	0101	-3	
	0100	-4	Negative values
	0011	-5	
	0010	-6	
	0001	-7	
	0000	-8	

A system of excess notation that uses four bit positions is called excess eight, because the value for zero is binary 1000, which otherwise would be the equivalent of decimal 8. Because this value is used as a starting point for counting, the binary representations of all values exceed their actual values by decimal 8, or binary 1000. Similarly, a 5-bit excess notation scheme, would use 10000 as the zero value, which would be the equivalent of decimal 16. Thus, excess notation in 5 bits is called excess 16. Excess notation is not commonly used for representing integers. However, it is commonly used as a part of the coding method for real numbers, as described in "Character Data."

Two's Complement Notation

In the binary number system, the complement of 0 is 1, and the complement of 1 is 0. The complement of a bit string is formed by substituting 0 for all values of 1, and a 1 for each 0. This transformation is the basis of **two's complement** notation. Non-negative values are rendered in ordinary binary notation. For example, a two's complement representation of 7_{10} using 4 bits is 0111.

Unlike sign-magnitude notation, a negative value is not represented by appending a sign bit to the binary equivalent of a positive number. Instead, negative values are represented as follows:

complement of positive value + 1 = negative representation

Thus, in 4 bit, two's complement representation, decimal -7_{10} would be[3]:

$$(0111) + 0001 =$$
$$1000 \ \ + 0001 =$$
$$1001 = -7_{10}$$

In this example, the positive binary value for 7_{10} is converted to its complement, 1,000. Then, a binary value of 1 is added to the least significant (or rightmost) digit. The result, 1001, is the two's complement equivalent of, -7_{10}.

The use of two's complement notation is awkward for most humans. Recall that one of the goals of data representation is the efficient implementation of processor circuitry. Two's complement notation is highly compatible with digital electronic circuitry for reasons that include:

▶ The leftmost bit, though part of the data value, may be interpreted as a sign bit.

▶ A fixed number of bit positions is used, requiring a minimum number of electronic circuits.

---

[3] Parentheses are a common mathematical notation for the complement of a value. Thus, if A is a numeric value, (A) is a mathematical notation representing the comlement of A.

► Subtraction may be performed as addition of a negative value, which simplifies processing circuitry.

► The number of logic circuits required to perform addition is reduced to two.

For these reasons, virtually all CPUs in use today represent and manipulate signed integers using two's complement format.

Range and Overflow

Two's complement, like most numeric data formats, uses a fixed number of bit positions. In most modern CPUs, either 32 or 64 bits are used to store a two's complement value (we'll assume 32 bits for the remainder of this text unless otherwise specified). Because the high-order bit is a sign bit, the **numeric range** of a two's complement value is $-(2^{n-1})$ to $(2^{n-1}-1)$, where n is the number of bits used to store the value.

Two's complement and most other numeric storage formats always use the same number of bits to store a value regardless of its magnitude (i.e., they are **fixed width formats**). Thus, a small value such as 1 occupies 32 bits even though, in theory, only one bit is required. The "unneeded" bits are padded with leading zeros as in the example above. A fixed number of bits are always used because that restriction allows processor and data communication circuitry to be more efficiently implemented. Although humans can easily deal with numeric values of varying lengths, computer circuitry is not nearly as flexible. The additional complexity required to handle variable length data formats in electronic processing and storage devices results in unacceptably slow performance.

The use of fixed width data storage formats introduces the possibility that an application program may use a numeric value that is too large (or small) to represent. For 32-bit two's complement format, no number with an absolute value larger than $2^{31}-1$ can be represented and stored (i.e., the numeric range is $-2{,}147{,}483{,}646_{10}$ to $2{,}147{,}483{,}647_{10}$). Unfortunately, it is possible for a CPU to generate (or attempt to generate) larger numbers in the course of normal instruction execution. For example, an addition operation with $2^{31}-1$ and any other positive number as the operands will generate a result that is too big to be stored in 32 bits. This condition is referred to as **overflow** and is treated as an error by the CPU. Note that overflow is not simply a matter of a number being too big in value. It is a matter of a number being too big in **absolute value**. That is, an overflow condition can also be generated as a result of subtraction (e.g., $-[2^{31}] - 1$). Overflow occurs when the result of a computation contains too many bits to fit into the required storage format.

As with most other aspects of computer design, the length of a storage format represents a design trade-off. Large formats (many bits) reduce the chance of overflow by increasing the maximum absolute value that can be represented. But many of these bits are "wasted" when smaller values are stored. If the bits were free, then there would be no trade-off. But bits cost money to store (e.g., additional RAM and secondary storage), and they also cost money to process due to the more complex processing circuitry required to handle large data formats. Thus, the size of data formats represents a trade-off among usable range, the chance of overflow during program execution, and the complexity, cost, and speed of processing and storage devices.

To avoid overflow and to increase accuracy, some computers and programming language implementations define additional numeric data types called **double precision** data types. In double precision representation, two fixed length units of storage are combined to hold each value. In the case of whole numbers, double precision representations sometimes are called **long integers**.

Overflow also can be prevented through careful programming. If the programmer anticipates that overflow is possible, the units of measure for program variables can be made larger. For example, if calculations were being performed in centimeters, the values might be represented instead in meters. Converting to different units of measure, as well as performing some types of calculations with precision, may not be practical if data values are restricted to whole numbers. Numbers with fractional parts, or real numbers, may have to be introduced. Manipulating such quantities within computers requires a different scheme of data representation.

REAL NUMBERS

In mathematics, real numbers may contain both whole and fractional components. In the decimal number system, the fractional portion is indicated by digits appearing to the right of the radix point. In the equation:

$$18.0 \div 4.0 = 4.5$$

the result is a real number containing the fractional component (.5) that is interpreted as $5/10$. The equivalent equation in binary notation is:

$$10010 \div 100 = 100.1$$

In program source code, real numbers usually are written in decimal notation.

Storage and representation of real numbers requires a mechanism to separate the whole and fractional components of the value (i.e., the equivalent of a written radix point). A simple way to accomplish this is to simply declare a storage format in which a portion of the bit string holds the whole portion, and the remainder of the bit string holds the fractional portion. Figure 4-5 illustrates such a format with the addition of a sign bit.

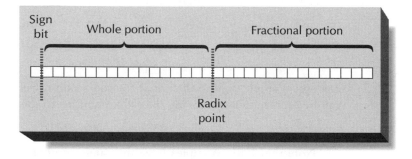

Figure 4-5

A 32-bit storage format for real numbers using a fixed radix point

Sign bit — Whole portion — Fractional portion — Radix point

The format illustrated in Figure 4-5 is structurally simple due to the fixed location of the radix point; this provides an advantage during the performance of computations. Humans simplify many computational operations on real numbers by creating columns of values and "lining up" the radix points of those values (see some of the examples near the beginning of this chapter). A similar approach can be taken to the design of electronic processing circuitry resulting in highly efficient (small and fast) processor modules.

Unfortunately, there is a trade-off of this simplicity against the numeric range of the storage format. Although the sample format uses 32 bits, its numeric range is substantially less than 32-bit two's complement. The reduction arises from the fact that only 16 bits are allocated to the whole portion of the value. Thus, the largest possible whole value is 2^{16}-1, or 65535. The remaining bits are used to store the fractional portion of the value which, by definition, can never be greater than or equal to 1.

The numeric range of the format could be increased by allocating a larger number of bits to the whole portion. But if the total size of the format is fixed at 32 bits, then the reallocation would result in a reduction in the number of bits used to store the fractional portion of the value. The reallocation of bits from the fractional portion to the whole portion results in a reduction in the precision or accuracy of fractional quantities stored. The nature of fixed width formats combined with fixed position radix points forces a trade-off between numeric range and fractional precision. For many application programs, this trade-off is too severe unless a very large number (e.g., 128 or more) of bit positions are used. Although such large formats provide satisfactory range and precision, they do so at the expense of large amounts of data storage and substantially increased complexity of processor circuitry.

Floating Point Notation

Humans deal with the trade-off between range and complexity by abandoning the concept of a fixed radix point. When we want to represent extremely small (precise) values, we move the radix point far to the left. For example, the value:

$$0.0000000013526473$$

has only a single digit to the left of the radix point. Similarly, very large values can be represented by moving the radix point far to the right, as in:

$$1352647300000000.0$$

Note that both examples have the same number of digits but have traded precision of the fractional portion for maximum range of the whole portion (or vice versa) by allowing the radix point to "float" left or right. Values can be either very large or very small (precise), but not both at the same time.

Humans tend to commit large numbers of representational and computation errors when manipulating long strings of digits. Thus, numbers such as those above are often represented in a more compact format called "scientific notation." In scientific notation, the two numbers above are represented as $13,526,473 \times 10^{-16}$ and $13,526,473 \times 10^8$. Note that the base of the number system (i.e., 10) is used as part of the multiplier. The exponent attached to the base can be interpreted as the number and direction of "positional moves" of the radix point as illustrated in Figure 4-6. Negative exponents indicate movement to the left and positive exponents indicate movement to the right.

Figure 4-6

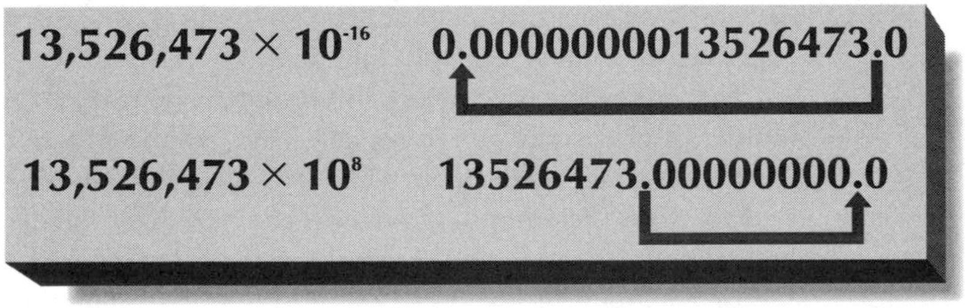

Conversion of scientific notation to decimal notation. The exponent is interpreted as the number of positions to shift the radix point.

$$13,526,473 \times 10^{-16} \qquad 0.0000000013526473.0$$

$$13,526,473 \times 10^8 \qquad 13526473.00000000.0$$

Real numbers are represented within computers in **floating point notation**. Floating point notation is similar to scientific notation except that 2 (rather than 10) is the base. A real value is derived from a floating point bit string according to the following formula:

$$\text{value} = \text{mantissa} \times 2^{\text{exponent}}$$

The **mantissa** holds the bits that are interpreted to derive the digits of the real number. By convention, the mantissa is assumed to be preceded by a radix point. The exponent field indicates the position of the radix point.

Many CPU-specific implementations of floating point notation are possible. Differences among them might include the length and coding formats of the mantissa and exponent and the location of the radix point within the mantissa. Although two's complement can be used to code either or both fields, other coding formats might represent better design trade-offs. The large number of possible floating point design choices resulted in very little compatibility in floating point format among CPUs prior to the 1980s. The primary problem that resulted from this situation was a lack of transportability of floating point data among different computers.

The **Institute of Electrical and Electronics Engineers (IEEE)** addressed this problem by defining a number of standard formats for the representation of floating point data. These formats have been widely adopted by computer manufacturers. IEEE Standards 754 and 854 define 32- and 64-bit floating point coding formats. The 32-bit format is described in detail next. Both formats have been widely adopted by microprocessor manufacturers and are implemented in most modern computers. For the remainder of this chapter, all references to floating point coding formats will refer to the IEEE 32-bit format unless otherwise specified.

The IEEE 32-bit floating point format is illustrated in Figure 4-7. The leading sign bit applies to the mantissa, not the exponent that immediately follows. The exponent is coded in excess-128 notation (this implies that its first bit is a sign bit). The 23-bit mantissa is coded as an ordinary binary number. It is assumed to be preceded by a binary 1 and the radix point. This extends the precision of the mantissa to 24 bits, although only 23 are actually stored.

Figure 4-7

IEEE 754 standard 32-bit floating point format

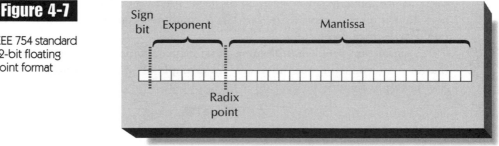

Range, Overflow, and Underflow

The number of bits in a floating point string and the formats of the mantissa and exponent fields impose limitations on the range of values that can be represented. The number of digits in the mantissa corresponds with the number of significant (i.e., nonzero) digits in the largest and smallest values that can be represented. The number of digits in the exponent determines the number of possible bit positions to the right or to the left of the radix point.

Using the number of bits assigned to each field, the largest absolute value of a floating point value appears to be:

$$1.11111111111111111111111 \times 2^{11111111}$$

However, exponents containing all zeros and ones are used to represent special data values. Thus, the usable exponent range is reduced and the decimal range for the entire floating point value is approximately 10^{-45} to 10^{38}.

As with integers, some computations may generate results with an absolute value too large to be stored within the number of available bits (i.e., less than 10^{-45} or greater than 10^{38}). As with integers, this condition is called overflow. Data items with large absolute values are represented by large positive exponents. Thus, overflow occurs when a large positive exponent will not fit within the bits allocated to store it.

Floating point representation is also subject to another error condition called **underflow**. Underflow occurs when a processing result is too small in absolute value to be represented. Very small numbers are represented by negative exponents. Thus, underflow occurs when a negative exponent is too large in absolute value to fit within the bits allocated to store it. Underflow of a floating point value is, thus, a result of overflow in the exponent.

Precision and Truncation

Recall that scientific notation (and, thus, floating point notation) trades range for accuracy. Accuracy is reduced as the number of digits available to store the mantissa is reduced. The 23-bit mantissa used in floating point format represents approximately seven decimal digits of precision. However, many numbers that might be represented within a computer contain more than seven nonzero decimal digits.

A significant problem in floating point representation and computations is that of nonterminating real numbers. An example of such a real number is the decimal equivalent of the fraction ⅓:

$$⅓ = 0.33333333 \ldots$$

In mathematics, the number of digits to the right of the decimal points in this example is assumed to be infinite. However, only a limited number of mantissa digits are available. Thus, a nonterminating number must be truncated to fit within the bits allocated to the mantissa.

Truncation is a representational error, but it is not generally a significant problem for single data values. The lost digits are relatively insignificant compared to the large (significant) value that is stored. Problems can arise when truncated values are used as input to computations. The error introduced by truncation can be magnified as truncated values are used to generate inaccurate computational results. The error resulting from a long series of computations that starts with truncated inputs can be relatively large.

An added difficulty is that more values have nonterminating representations in the binary system than in the decimal system. For example, the fraction ⅒ is nonterminating in binary notation. The representation of this value in floating point notation would be a truncated value. Such problems, however, usually can be avoided through careful programming. As a rule of thumb, programmers reserve floating point calculations for quantities that may vary continuously over wide ranges. Measurements that are made by scientific instruments, for example, might be handled in floating point notation.

Integer representation and calculation do not suffer from any potential errors except overflow. Thus, programmers typically avoid using real numbers when possible. One common example is financial applications. Because dollars and cents usually are shown as decimals in source code and in printouts, novice programmers sometimes assume that floating point representations are appropriate for financial calculations in binary. However, even if overflow and underflow are avoided, rounding errors due to nonterminating representations of tenths will persist. Cumulative errors will mount as approximations are input to subsequent calculations. For this reason, careful programmers use integer arithmetic for accounting and financial applications. In such applications, using integer representations is equivalent to expressing all values in pennies. Then, a decimal point is inserted when the result is converted to decimal notation for output.

Processing Complexity

The difficulty that most people encounter when learning to use scientific and/or floating point notation is understandable. The formats are relatively complex notational formats—far more complex than integer representation methods. This complexity exists for electronic processing circuitry as well. Although floating point formats are optimized for processing efficiency, they still require relatively complex processing devices. The simpler two's complement format used for integers requires much less complex computational circuitry.

The difference in complexity of processing circuitry for two's complement and floating point numbers translates directly to a speed difference when performing calculations. The magnitude of the difference depends on a number of factors, including the computational function being performed and the exact details of the processing circuitry. As a general rule, simple computational functions (e.g., addition and subtraction) take at least twice as long for real numbers as for integers. The difference is even greater for more complex functions (e.g., division).

For this reason and for reasons of accuracy, careful programmers never use real numbers (floating point representation) if integers will suffice. This is particularly important for data items that are frequently updated by computational operations (e.g., an array element incremented within a loop).

CHARACTER DATA

Much of the data that a computer is expected to process is written text. English and many other written languages use alphabetic letters, numerals, punctuation marks, and a variety of other special purpose symbols (e.g., $ and &) in their written form. Each of these symbols is called a **character**. A sequence of characters that forms a meaningful word, phrase, or other useful grouping of text is called a **string**. In most programming languages, single characters are surrounded by single quotation marks (e.g., 'C'), and strings are surrounded by double quotation marks (e.g., "Computer").

Character data cannot be directly represented or processed within a computer since computers are designed to process only digital (i.e., numeric) data. Character data can be represented indirectly by defining a table of correspondences between individual characters and specific numeric values. For example, the integer values 0–9 could be used to represent the numerals (characters) '0'–'9'. The uppercase (i.e., capital) letters 'A'–'Z' could be represented as the integer values 10–36. The lowercase letters 'a'–'z' could be represented as the integer values 37–63. The integer value 64 could be used for '$' and so forth.

Such a table of correspondences between one set of values and another is a simple coding method. It shares many important characteristics of the other coding methods discussed earlier in this chapter (e.g., binary signals using electrical voltage), including:

► A coding table must be defined and agreed to by all users.

► The values used to represent the data must be capable of being encoded, transmitted, and decoded.

► The choice of a specific coding method represents a trade-off among compactness, ease of manipulation, accuracy, range, and standardization.

The following sections describe some of the more common coding methods for character data.

BCD and EBCDIC

Binary coded decimal (BCD) is a character coding method used by early IBM mainframe computers. Characters are encoded as strings of six bits. Because only six bits are used, the method can represent only 64 (2^6) different symbols. The character set includes alphabetic letters in uppercase form only. Also included are the numerals 0–9 and some punctuation marks. Table 4-6 shows a portion of the coding table.

TABLE 4-6	Symbol	ASCII	EBCDIC	BCD
A partial listing of the ASCII, EBCDIC, and BCD binary codes for numerals, uppercase letters, and lowercase letters	0	0110000	11110000	000000
	1	0110001	11110001	000001
	2	0110010	11110010	000010
	3	0110011	11110011	000011
	4	0110100	11110100	000100
	5	0110101	11110101	000101
	6	0110110	11110110	000110
	7	0110111	11110111	000111
	8	0111000	11111000	001000
	9	0111001	11111001	001001
	A	1000001	11000001	010001
	B	1000010	11000010	010010
	C	1000011	11000011	010011
	a	1100001	10000001	
	b	1100010	10000010	
	c	1100011	10000011	

BCD's relatively compact form was an advantage in the era of limited memory and slow processors. Small bit strings require less memory and simpler processing circuits. BCD was also designed with another processing efficiency in mind. The representation of the symbols '0'–'9' is intentionally the same as the internal representation of the integer values 0–9. This simplifies or eliminates conversion problems between numeric and character data formats (e.g., when calculating the amount of a payment and then printing it on the face of a check).

The limited range inherent in BCD's six-bit format became a less acceptable trade-off for small memory and simple processing requirements as memory and processors became faster and less expensive. It is also a proprietary standard that was never adopted by many computer manufacturers other than IBM. BCD is obsolete as a general purpose code for data processing and transmission.

IBM extended the capabilities of BCD by expanding its length to eight bits and defining additional characters within the expanded table. The coding method is called **Extended Binary Coded Decimal Interchange Code (EBCDIC)**. This coding method is still used internally in all IBM mainframe computers and their binary compatible counterparts.

ASCII

The **American Standard Code for Information Interchange (ASCII)** is a coding method that has been adopted in the United States and that is used widely in data communication and within many computers and peripheral devices. A subset of ASCII codes are shown in Tables 4-6 and 4-7. The international equivalent of this character set, which is used widely to transmit text in languages other than English, is **International Alphabet Number 5 (IA5)**. This standard is promulgated by the **International Standards Organization (ISO)**.

TABLE 4-7	Decimal code	Control character	Description
Control codes from the ASCII character set	000	NUL	Null
	001	SOH	Start of Heading
	002	STX	Start of Text
	003	ETX	End of Text
	004	EOT	End of Transmission
	005	ENQ	Enquiry
	006	ACK	Acknowledge
	007	BEL	Bell
	008	BS	Backspace
	009	HT	Horizontal Tabulation
	010	LF	Line Feed
	011	VT	Vertical Tabulation
	012	FF	Form Feed
	013	CR	Carriage Return
	014	SO	Shift Out
	015	SI	Shift In
	016	DLE	Data Link Escape
	017	DC1	Device Control 1
	018	DC2	Device Control 2
	019	DC3	Device Control 3
	020	DC4	Device Control 4
	021	NAK	Negative Acknowledge
	022	SYN	Synchronous Idle

Data Representation

TABLE 4-7 continued	Decimal code	Control character	Description
	023	ETB	End of Transmission Block
	024	CAN	Cancel
	025	EM	End of Medium
	026	SUB	Substitute
	027	ESC	Escape
	028	FS	File Separator
	029	GS	Group Separator
	030	RS	Record Separator
	031	US	Unit Separator
	127	DEL	Delete

ASCII was designed to achieve two primary goals:

► Define a vendor-independent standard for character coding.

► Support data communication between computers and peripheral devices.

At the time ASCII was defined, IBM, still controlled most the world market for computers. For various technical and political reasons, EBCDIC was not chosen as a standard and the code that was developed is incompatible with EBCDIC. For purposes of backward compatibility, IBM has continued to use EBCDIC in its mainframe computers. Virtually all other computers (including IBM's non-mainframe computers) use ASCII.

Two key characteristics of ASCII arose from its intended use for data communication among computers and peripheral devices:

► A seven-bit format.

► The inclusion of device control codes within the coding table.

The use of a seven-bit format was based on the commonly used byte size of eight bits and the common use of an error detection method known as parity checking. Parity checking and other error detection and correction methods are fully discussed in Chapter 8. For this discussion, the only important characteristic of parity checking is that it requires one bit per character communicated to be used as a check digit. Thus, one bit out of eight is not used as part of the data value, which leaves only seven bits for data representation.

The standard version of ASCII used for data communication is sometimes called ASCII-7 to emphasize its 7-bit format. This coding table has 128 (2^7) defined characters. However, computers that use 8-bit bytes are capable of representing 256 (2^8) different characters. In most computers, the ASCII-7 characters are included within an eight-bit character coding table as the first (or lower) 128 table entries. The additional (or upper) 128 entries are defined by the computer manufacturer and are typically used for graphic characters (e.g., line drawing characters) and/or multinational characters (e.g., á, ñ, Ö, and Ω). This encoding method is sometimes called ASCII-8. The term is a misnomer as it implies that the entire table (all 256 entries) is standardized. In fact, only the first 128 entries are defined by the ASCII standard.

Device Control

When text is printed or displayed on an output device, it is often formatted in a particular way. For example, a customer record may be displayed on a video display terminal in a manner similar to a printed form. Text output to a printer is normally formatted into lines and paragraphs as it is in this textbook. Certain text may be highlighted when printed or displayed (e.g., using underlining, a bold font, or a reversal of background and foreground color).

Data communication methods in use at the time ASCII was defined were based on the transmission of one character at a time over a single data communication wire (i.e., serial transmission, as described in Chapter 8). Because only a single wire is used, formatting commands must be sent in the same transmission stream as the text to be printed or displayed. The textual data and formatting commands are interspersed with formatting commands immediately before or after the text that they modify.

ASCII defines a number of device control codes (see Table 4-7) that can be used for this purpose. Among the simpler codes are carriage return (move the print head or cursor to the beginning of a text line), line feed (move the print head or display cursor down one line), and bell (ring a bell or generate a short sound such as a beep). In ASCII, each of these functions is assigned a numeric code and a short character name (e.g., CR for carriage return, LF for line feed, and BEL for bell).

Not all of the device control codes are used for formatting. Some are used for controlling various aspects of the data-communication process. For example, ACK is sent to acknowledge correct receipt of data, and NAK is sent to indicate that an error has been detected. Some control codes have defined names, but their function was left for future definition. ASCII defines a total of 32 device control codes, and they occupy the first 32 table entries (numbered 0–31).

The inclusion of these control codes in ASCII was a substantial benefit to users and computer equipment manufacturers. It allowed users the flexibility of using peripheral devices from many different manufacturers. To the extent computer manufacturers and software developers agreed to abide by the ASCII standard, systems of "mixed" equipment could be configured without compatibility problems. It also standardized the transfer of textual data between computer systems and users.

Software and Hardware Support

Because characters are usually represented as integers, there is little or no need for the CPU to provide special support for character processing. Instructions that move and copy integers behave the same, whether the content being manipulated is a "True" numeric value or an ASCII encoded character. Similarly, an equality (or inequality) comparison instruction that works for "True" integers also works for integers that represent characters. That is, an equality comparison of two integers representing ASCII codes produces the same result that a human produces when comparing the printed characters.

The results of nonequality comparisons are less straightforward. The assignment of numeric codes to characters follows a specific order called a **collating sequence**. A greater-than comparison for two characters (e.g., 'a' greater than 'z') returns a result based on the numeric comparison of the corresponding ASCII codes. If the character set has a well-accepted order and if the coding method follows the order, then less-than and greater-than comparisons produce expected results.

Note that the use of numeric values to represent characters can produce some unexpected or unplanned results. For example, the collating sequence of letters and numerals in ASCII follows accepted norms (e.g., the standard alphabetic order for letters and numeric order for numerals). But because uppercase and lowercase letters are represented by different codes, the result of an equality comparison between uppercase and lowercase versions of the same letters is false (i.e., 'a' ≠ 'A'). Also, punctuation symbols have a specific order in the collating sequence though there is no underlying ordinal relationship in common use.

The only potential need for CPU support of a separate character data type is to allow the CPU to detect attempts by a program to perform "illegal" operations on a character. For example, it is acceptable to allow an integer add instruction to be executed with two integers in it. However, it makes no sense to execute that instruction with two ASCII codes as input even if their internal coding format is the same as for integers. Thus, the CPU can act as an error detection device if characters are treated as a separate data type. However, characters are seldom treated as a distinct data type by the CPU because most CPU designers seek to simplify their designs as much as possible. Thus, for the vast majority of CPUs, characters are represented using an integer data coding format, and software (e.g., the programmer and/or the operating system) is responsible for keeping track of which integers are "true" numeric values and which are ASCII codes.

ASCII Limitations

Although the designers of ASCII left some room for growth (i.e., the undefined control codes), they could not foresee the long lifetime (approaching 40 years) the code would enjoy or the revolutions in I/O device technologies that would occur during that lifetime. The ASCII designers never envisioned common characteristics of I/O today such as color, bitmapped graphics, and selectable fonts. Unfortunately, ASCII does not have sufficient range to allow the definition of a set of control codes large enough to account for the full range of formatting and display capabilities in modern I/O devices.

ASCII is also an English-specific coding method (the word "American" is the first word of its full name), which is not surprising given that the United States accounted for most computer use and virtually all computer production in the world at the time ASCII was defined. As such, the coding method has a heavy bias toward Western languages in general and American English in particular. This has become a significant issue as computers have proliferated beyond the United States.

Recall that seven-bit ASCII has only 127 table entries, 32 of which are used for device control. Thus, only 95 printable characters can be represented within ASCII. This is enough for a usable subset of the characters commonly employed in American English text. But that subset does not include any modified Roman characters (e.g., á and ç) or those from other alphabets (e.g., δ, ㄅ, Я, and ε). The International Standards Organization (ISO) has partly addressed the problem by defining a number of 256 entry tables. One of these, called **Latin-1**, contains the ASCII-7 characters in the first 128 table entries and most of the characters used by Western European languages in the upper 128 table entries. The upper 128 entries are sometimes called the **multinational characters**. But the number of available character codes in a 256-entry table is still much too small to represent the full range of the world's printable characters.

Further complicating matters is the fact that some printed languages are not based on characters as they are understood in most Western languages. Chinese, Japanese, and Korean compose written text with thousands of ideographs (pictorial representations of words or concepts). Ideographs are, in turn, composed of individual graphic elements (sometimes called strokes) that number in the thousands. Other written languages present similar, though less severe, problems.

ASCII was clearly never designed to handle the full range of written human language. Despite this limitation, it has persisted as a standard for decades due to the dominance of the United States as an economic power and as a designer and manufacturer of computer equipment. A substantial expansion (or replacement) of ASCII is required to enlarge the standardized character set.

Unicode

The Unicode Consortium was founded in 1991 to develop a multilingual character encoding standard encompassing all of the world's written languages. The original members were Apple Computer Corporation and Xerox Corporation, but many computer companies soon joined. The effort was seen as a necessary step in the evolution of software that would allow software and data to cross international boundaries. The Unicode 1.0 standard has been adopted by the International Standards Organization as a subset of ISO Standard 10646.

Like ASCII, **Unicode** is a coding table that assigns non-negative integers to represent individual printable characters. The ISO Latin-1 standard (which includes ASCII-7) is incorporated into Unicode as the first 256 table entries. Thus, ASCII is a subset of Unicode. The primary difference between ASCII and Unicode lies in the number of table entries. Unicode assumes the availability of 16 bits to represent a non-negative integer. Thus, there are 65,536 table entries numbered 0 through 65,535.

The additional table entries are used primarily to represent characters, strokes, and ideograms of languages other than English and its Western European siblings. It includes a large number of other alphabets (e.g., Arabic, Cyrillic, and Hebrew) and thousands of Chinese, Japanese, and Korean ideograms and characters. Some extensions to the ASCII device control codes are provided also. As currently defined, Unicode can represent written text from all modern languages. Approximately 29,000 codes are currently defined. An additional 29,000 are held in reserve for future expansion, and 6,000 user or software-defined entries are provided to allow customization.

Unicode is widely supported in modern software including the latest versions of most operating systems and word processors. Because few CPUs directly implement character data types, there is no real need to upgrade processor hardware to support Unicode. Most modern CPUs use 32 bits to represent an integer so the existing practice of internally coding characters and integers can be continued.

The primary impact on hardware is in storage and I/O devices. Because the size of the numeric code is double that of ASCII, pure text files are twice as large. This appears to be a problem at first glance, but the impact of that problem is largely mitigated by the use of custom word-processing file formats and the ever-declining cost of secondary storage. Most text is not stored as pure ASCII. Instead, it is stored in a format that intermixes text and formatting commands that apply to the text. Because formatting commands also occupy file space, the file size increase experienced when switching from ASCII to Unicode is generally far less than 100%.

Devices designed for character I/O have traditionally used ASCII by default and have implemented vendor-specific methods (or used older ISO standards) to process character sets other than Latin-1. The typical method is to maintain an internal set of alternate coding tables, in which each table contains a different alphabet or character set. Device-specific commands are used to switch from one table to another. Unicode unifies and standardizes the content of these various tables and thus offers a standardized method for processing international character sets. That standard is being widely adopted by I/O device manufacturers. Backward compatibility with ASCII is not a problem because Unicode includes ASCII.

BOOLEAN DATA

The **boolean** data type has only two data values: True and False. Most people do not ordinarily think of boolean values as data items. But the primitive nature of CPU processing requires the ability to store and manipulate boolean values. Recall that processing circuitry acts to physically transform input signals into output signals. If the input signals represent numbers and the processor is performing a computational function (e.g., addition), then the output signal represents the numerical result. When the processing function is a comparison operation (e.g., greater than or equal to), then the output signal represents the boolean result of True or False. This result is stored in a register as is any other processing result and may be used by other instructions as input (e.g., a conditional branch or GOTO).

The boolean data type is potentially the most concise in coding format since only a single bit is required (e.g., binary 1 can represent True, and 0 can represent False). However, most CPU designers seek to minimize the number of different coding formats in use so as to simplify processor design and implementation. Thus, CPUs generally use the coding format for integers (usually two's complement) to also represent boolean values. When coded in this manner, the integer value zero corresponds to False and all nonzero values correspond to True.

To conserve storage, programmers will sometimes "pack" many boolean values into a single integer by using each individual bit to represent a separate boolean value. Although this conserves memory, it generally slows program execution because of the complicated instructions required to extract and interpret individual bits from an integer data item.

MEMORY ADDRESSES

As described in Chapter 2, primary storage is a series of contiguous bytes of storage. In conceptual memory models, it is typical to consider each memory byte as having a unique identifying number. The identifying numbers (or addresses) are assumed to start with 0 and continue in sequence (with no gaps) until the last byte of memory is addressed.

In some CPUs, this conceptual view is also the physically implemented method of identifying memory locations. That is, memory bytes are identified by a series of non-negative integers. This approach to assigning memory addresses is called a **flat memory model**. In CPUs that use a flat memory model, it is logical (and typical) to use two's complement as the coding format for memory addresses. Thus, memory addresses are stored in the same internal format as integers. The primary advantage of this approach is a minimization of the number of different data types and, thus, the complexity of processor circuitry.

For a variety of reasons, some CPU designers and implementers choose not to use simple integers as memory addresses. In most current cases, this decision is an effort to maintain backward compatibility with earlier generations of CPUs that didn't use a flat memory model (e.g., the Intel Pentium CPU maintains backward compatibility with the 8088 CPU used in the first IBM PC). Earlier generations of processors typically used an alternate approach to memory addressing called a **segmented memory model**. In a segmented memory model, primary storage is divided into a series of equal sized segments (e.g., 64 kilobytes). Segments are identified by sequential non-negative integers. Each byte is also identified by a sequentially assigned non-negative integer. Thus, each byte of memory has a two-part address. The first part identifies the segment, and the second part identifies the byte within the page.

Segmented memory models are discussed in more detail in Chapter 6. This discussion is concerned only with the fact that a memory address in a segmented memory model consists of two different numbers. Therefore, the data coding method for single integers cannot be used. Instead, a specific coding format for memory addresses must be defined and implemented. Thus, CPUs with segmented memory models have an extra data type (coding format) for storing memory addresses.

Intel 80X86 Memory Address Format

Intel microprocessors have been used in PCs since the first IBM PC was sold in 1981. That computer used an Intel processor code named the 8088. Subsequent generations of Intel processors (the 8086, 80286, 80386, 80486, and Pentium family of processors) have all maintained backward compatibility with the 8088. That is, each subsequent processor can execute machine language programs designed for the 8088. Backward compatibility has been considered an essential ingredient in the success and proliferation of PCs based on the Intel 80X86 family of CPUs.

Early Intel microprocessors were designed and implemented with speed of processor operation as a primary goal. Because a general purpose CPU uses memory addresses in nearly every instruction execution (e.g., to fetch the instruction and its data inputs), it is important to make memory access as efficient as possible. The complexity of processor circuitry rises with the number of bits that must be processed simultaneously. Thus, large coding formats for memory addresses and other data types require complicated and, therefore, relatively slow processing circuitry. Intel designers sought an approach to minimize the complexity of processor circuitry associated with memory addresses. They also needed to balance the desire for efficient memory processing with the need to maintain a relatively large range of possible memory addresses.

Intel designers made two significant design decisions for the 8088 to achieve this balance. These were the adoption of a 20-bit size for memory addresses and the use of a segmented memory model. The 20-bit address format limited usable (addressable) memory to 1 MB (2^{20} addressable bytes). This did not seem like a significant limitation at the time because few computers (including most mainframes) had that much memory at the time. The 20-bit format is divided into two parts: a 4-bit segment identifier and a 16-bit segment offset. The 16-bit segment offset identifies a specific byte within a 64 (2^{16}) KB memory segment. Because four bits are used to represent the segment, there are 16 (2^4) possible segments.

Intel designers anticipated that most programs would fit within a single 64 KB memory segment. Further, they knew that 16-bit memory addresses could be processed much more efficiently than larger addresses. Thus, they defined two types of address processing functions—those that used the four-bit segment portion of the address and those that ignored it. Memory could be accessed rapidly when the processor ignored the segment identifier. When the segment identifier had to be used, memory accesses were slowed considerably. Intel designers assumed that most memory accesses would not require the segment identifier.

Both the 64 KB segment size and the 1 MB total memory limit soon became significant constraints. Although early programs for the IBM PC were generally less than 64 KB in size, later versions of those same programs usually surpassed that limit. Because the processor design imposed a performance penalty when the segment identifier was used, these larger programs ran more slowly than their earlier versions. In addition, both the operating system and the computer hardware were becoming more complex, and, thus, consuming more memory resources. As a result, computer designers, operating systems designers, and users chafed under the 1-MB limit.

Intel designers first addressed these constraints in the 80286. They increased the segment identifier to eight bits, thus increasing total addressable memory to 16 MB. The 80386 went a step further by providing two methods of addressing. The first was a method compatible with that of the 8088 (called real mode). The second was an entirely new method based on 32-bit memory addresses (called protected mode). Protected mode addressing largely eliminated the performance penalty for programs larger than 64 KB and provided other benefits as well. Both addressing methods have continued in the 80486 and Pentium family of microprocessors.

Although the older design limits were effectively bypassed by protected mode, the need to also implement real mode has proved to be a problem in later microprocessors. The implementation of two different addressing modes has increased the complexity of those processors. This, in turn, has made it more difficult to implement fast versions of the processors. Intel has struggled to match the performance of other CPUs that implement only a single memory addressing method.

This discussion and the earlier explanation of ASCII illustrate a few of the pitfalls for computer designers in choosing and implementing data coding methods and formats. They also illustrate a fundamental characteristic of any design: it always represents a trade-off. For Intel memory address format and processing, the trade-off was made among processor cost, speed of program execution, and the memory requirements of typical PC software. The trade-off may have been optimal (or, at least, reasonable), given the state of all these variables in 1981. But neither time nor design parameters stand still; design decisions have a limited life, which can be very short for computer technology. Attempts to maintain backward compatibility with previous design decisions can be difficult, expensive, and may limit future design choices.

DATA STRUCTURES

The previous sections outlined the various basic data types that a computer can manipulate directly. However, it would be difficult to develop useful programs of any kind if these were the only possibilities for data representation. Most application programs require that individual data elements be combined to form useful aggregations of data. A simple and common example is that of character strings. Although few CPUs define instructions that directly manipulate strings, most application programs must manipulate them. Application development is simplified if strings can be defined and manipulated (e.g., read, written, and compared) as a single unit instead of as the individual characters of which they're composed.

A **data structure** is a set of basic data elements organized for some type of common processing. Data structures are defined and manipulated within software. Computer hardware cannot manipulate them directly but must deal with them in terms of their basic components (e.g., integers, floating point numbers, single characters, etc.). Thus, software must provide the means for translating operations on data structures into equivalent sets of machine instructions operating on individual basic data elements.

The complexity of data structures is limited only by the imagination and skill of systems and application programmers. As a practical matter, however, certain data structures are useful in a wide variety of situations and are thus commonly implemented. Examples of such data structures include character strings (or arrays), records, and files. Because of their frequent use and a desire to implement them efficiently, support for the representation and manipulation of such data structures is commonly provided in system software. For example, the operating system normally provides service routines for the input and output of records to a file.

Other types of data structures may or may not be supported within system software. Examples of these include numeric arrays, indexed files, and complex database structures. Indexed files are supported in some, but not all, operating systems. Normally, numeric arrays are supported within a programming language, but not within the operating system. Database structures are usually supported by a database management system (DBMS) (separate from the operating system). Most programming languages (and thus their compilers and interpreters) support the direct manipulation of strings. Many data structures are commonly supported through a library of reusable program modules or objects provided with an operating system.

Pointers and Addresses

Whether implemented within system or application software, virtually all data structures make extensive use of **pointers** and **addresses**. A pointer is a data element containing the address of another data element. An address is the location of some data element within a storage device. The storage device may be memory, a disk drive, or some other device.

Addresses vary in content and representation depending on the storage device being addressed. Typically, secondary storage devices are organized as a sequence of data blocks. A block is a group of bytes that is read or written as a unit by the device. For example, diskette drives on PCs often read and write data in 512-byte blocks. For storage organized as a sequence of blocks, an address can be represented as an integer containing the sequential position of the block. Integers also can be used to represent the address of an individual byte within a block.

As described earlier, memory addresses may be complex if a segmented memory model is used. In this discussion, we assume that a flat memory model is used and, thus, that non-negative integers may be used as memory addresses.

Arrays and Lists

Many sets of data can be considered as lists, or sets of related data values. In mathematics, a list is considered unordered. That is, there is no concept of the first, second, or last element of a list. When writing software, a programmer usually prefers to impose some ordering on the list. For example, a list of the days of the week might be ordered sequentially starting with Monday.

An **array** is an ordered list in which each element can be referenced by an index to its position. An example of an array for the first five letters of the alphabet appears in Figure 4-8 on the next page. Note that the index values are numbered starting at zero, as is common (although not universal) practice in computer programming. Although the index values appear in the diagram, they are not stored. Instead, they are inferred from the location of the data value within the storage allocated to the array. Within a high-level programming language, individual array elements normally are referenced by the name of the array and the index value. For example, the third letter of the alphabet stored in an array might be referenced as:

alphabet[2]

where alphabet is the name of the array and two is the index value (numbered from zero).

Figure 4-8

Array elements in contiguous storage locations

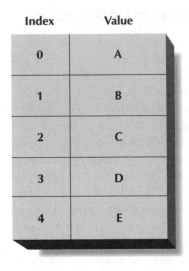

Figure 4-9 shows a character array (string) stored in contiguous (sequential) memory locations. In this example, each character of the name "John Doe" is stored within a single byte of memory and the characters are ordered in sequential byte locations starting at byte 1000. An equivalent organization might be used to store the name on a secondary storage device. Access to an individual element of the array can be achieved by using the starting address of the array and the index of the element. For example, if we wish to retrieve the third character in the array, we can compute its address as the sum of the address of the first element plus the index (assuming index values start at zero).

Figure 4-9

Character array stored in contiguous (sequential) memory locations

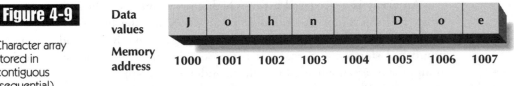

The use of contiguous storage locations (especially in secondary storage devices) complicates the allocation of those locations. If an array must be enlarged in order to add more data, the locations at the end of the existing array may already be allocated to other purposes. Therefore, contiguous allocation generally is used only for fixed-length arrays.

For a variety of reasons, it may be desirable to store individual elements of the array in widely dispersed storage locations. One common reason is to make it easier to expand or shrink an array. In such a situation, each element must be stored along with a pointer to the next element of the array. An example is depicted in Figure 4-10. The letters of the name "John Doe" are stored in noncontiguous locations, and each letter is followed by the address of the next letter in the array. In the figure, the addresses are depicted as arrows instead of numeric addresses.

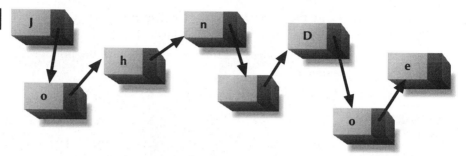

Note that the use of pointers increases the amount of storage needed for the array. In addition, it complicates the task of locating individual array elements. Instead of a simple address computation as earlier, references to array elements must be resolved by following the chain of pointers starting with the first array element. This can be highly inefficient if the number of array elements is large.

The type of data structure used in Figure 4-10 is called a **linked list**. Linked lists can be used for many different things besides the storage of arrays. Figure 4-11 shows a generic example of a **singly linked list**. The term is derived from the use of one pointer (or link) per list element.

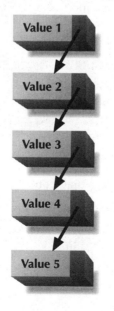

Although access to individual elements in a linked list may be slow, the addition or deletion of elements is relatively fast. The procedure to add a new element is as follows:

1. Allocate storage for the new element.

2. Copy the pointer from the element preceding the new element into the pointer field of the new element.

3. Write the address of the new element into the pointer field of the preceding element.

An example of the procedure is depicted in Figure 4-12.

Figure 4-12

Insertion of a new element into a singly linked list

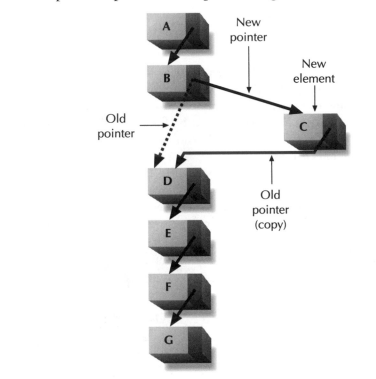

In contrast, insertion of an element into a list stored in contiguous memory locations can be very time-consuming. The procedure is as follows:

1. Allocate a new storage location to the end of the list.

2. For each element past the insertion point, copy the element value to the next storage location.

3. Write the new element value in the storage location at the insertion point.

The insertion procedure is depicted in Figure 4-13. For insertion near the beginning of the list, efficiency is degraded by the large number of copy operations required.

Figure 4-13

Procedure for inserting an element into an array stored in contiguous memory locations

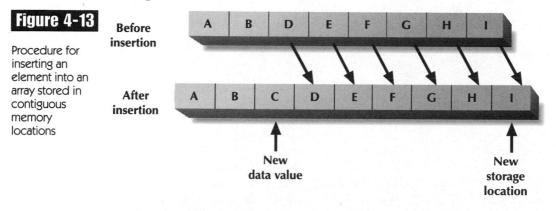

Figure 4-14 depicts a more complicated form of linked list called a **doubly linked list**. As the term implies, elements of a doubly linked list have two pointers each. One pointer goes to the next element in the list and the other one points to the previous element in the list. The primary advantage of this storage method is that lists may be traversed in either direction with equal efficiency. The primary disadvantage is that more pointers must be updated each time an element is inserted into or deleted from the list. In addition, more storage locations are required by the extra set of pointers.

Figure 4-14

A doubly linked list

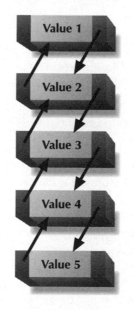

Records and Files

A **record** is a data structure composed of other data structures or basic data types. Records are commonly used as a unit of input and output to files as well as for grouping related data components together into a single named unit. As an example, consider the following items of data:

Account-Number   Street-Address
First-Name   City
Last-Name   State
Middle-Initial   Zip-Code

This set of data might be the contents of the data structure for a customer record, as shown in Figure 4-15. Each component of the record is either a basic data element (e.g., Middle-Initial) or another data structure (e.g., a character array for Street-Address). To speed input and output, records are normally stored in contiguous storage locations, which restricts array components of the record to be of fixed length.

Figure 4-15

A data structure for a record

A sequence of records on secondary storage is called a file. A sequence of records stored within main memory is usually called a table, although the structure of the storage may be similar. Files may be organized in many different ways, but they are most commonly sequential and indexed.

In a sequential file, the records are stored in contiguous storage locations. As with arrays stored in contiguous storage, accessing a specific record is relatively simple. The address of the nth record in a file can be computed as:

$$\text{address-of-first-record} + ([n\text{-}1] \times \text{record-size})$$

Thus, if the first byte of the first record is at address 1 and the size of a record is 200 bytes, the address of the fourth record is 601.

Sequential files suffer the same problems as contiguous arrays when inserting and deleting records. A copy procedure similar to that depicted in Figure 4-13 must be executed to add a record to the file. The efficiency of this procedure is even less for files than for arrays because of the relatively large size of the data (records) that must be copied.

One method of solving this problem is to use linked lists. With files, the data elements of a linked list are entire records instead of basic data elements. The methods for searching, record insertion, and record deletion are essentially the same as previously described.

Another method of organizing files involves the use of an **index**. An index is an array of pointers to records, which can be ordered in any sequence the user desires. For example, a file of customer records might be ordered by ascending account number, as shown in Figure 4-16.

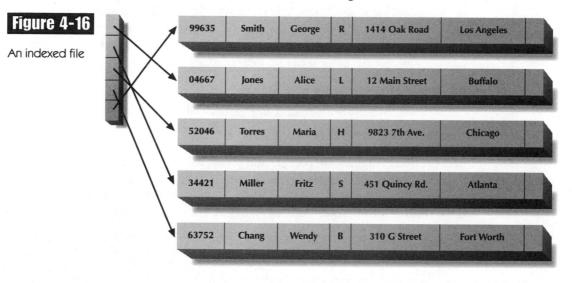

Figure 4-16

An indexed file

The advantage of using an index lies in the efficiency of record insertion, deletion, and retrieval. When a record is added to a file, storage is allocated for the record and the data placed inside. The index is then updated by inserting the address of the new record. The index update is accomplished by the same procedure as an array update. Because the array contains only pointers, it is small and thus relatively fast to update.

SUMMARY

A computer can manipulate data only if it is encoded in pulses of electricity representing numbers. Thus, all data must be encoded numerically and converted into pulses of electricity for storage, transmission, or processing.

Data can be encoded in electricity using analog or digital signaling methods. Analog signals offer superior message-carrying capacity at the expense of high susceptibility to noise and interference. Digital signals offer superior reliability and are thus exclusively used within modern digital computers.

The electrical representation of numbers within digital computers is based on the binary number system. This method is used because it corresponds with clearly identifiable states of electron flow. Devices that transmit, store, and process binary signals are reliable and inexpensive.

The various data types that may be represented within a computer include integer, real, character, and boolean data. Integers are whole numbers. Real data consists of numbers that may contain fractional components. Characters are made up of printable symbols such as the letters of the alphabet. Characters may also include special control codes for controlling the behavior of I/O devices. Boolean data elements may only have the values True or False.

There are a variety of methods (or coding formats) by which data may be encoded as binary numbers. These methods vary in ease of manipulation, conciseness, accuracy, and standardization among devices and computers. Different coding methods represent different trade-offs between these characteristics and the cost of computer hardware.

Numeric data may be encoded in binary form using sign magnitude, excess, or two's complement notation. Two's complement notation is the preferred method of encoding integer values because it simplifies processing circuitry within the CPU. Fractional numbers require a specialized representation method in which the whole and fractional portions of the number are stored separately.

Characters are encoded as binary numbers by assigning a unique number to represent each character. Various coding schemes exist for characters, including BCD, EBCDIC, ASCII, and Unicode. Each of these has its own relative advantages. ASCII is the most widely accepted coding method and is a standard method of data communication between computers and peripheral I/O devices.

Memory addresses are often represented as a distinct data type by the CPU. This is particularly true of CPUs that implement a segmented memory model. Some CPUs that implement a flat memory model use a separate data type for memory addresses. Others use the same coding format as for signed or unsigned integers.

Data structures are structured sets of related data items. They are useful in processing and transmitting related groups of data. Within a computer, data structures are stored as sequences or groups of primitive data items (i.e., integers, characters, etc.). Commonly used data structures include arrays, linked lists, records, and files.

Key Terms

absolute value
address
American Standard Code for
    Information Interchange
    (ASCII)
analog signal
analog to digital (A/D)
    conversionarray
array
base
binary
~~binary coded decimal (BCD)~~
binary signal
bit
bit position
bit string
boolean
byte
carrier wave
character
collating sequence
communication protocol
data structure
decoding
digital signal
digital to analog (D/A)
    conversion

discrete signal
double precision
doubly linked list
encoding
excess notation
Extended Binary Coded
    Decimal Interchange Code
    (EBCDIC)
fixed width format
flat memory model
floating point notation
hexadecimal notation
high order bit
index
Institute of Electrical and
    Electronics Engineers
    (IEEE)
integer
integer arithmetic
International Alphabet
    Number 5 (IA5)
International Standards
    Organization (ISO)
Latin-1
least significant digit
linked list
long integer

low order bit
mantissa
most significant digit
multinational character
negative
non-negative
numeric range
octal notation
overflow
pointer
positional number system
radix
radix point
record
segmented memory model
sign magnitude notation
signal
signed integer
singly linked list
string
threshold
truncation
two's complement
underflow
Unicode
unsigned integer

1. An element in a(n) _____ contains pointers to both the next and previous list elements.

2. _____ notation encodes a real number as a mantissa multiplied by a power (exponent) of two.

3. A(n) _____ is an integer stored in double the normal number of bit positions.

4. Two's complement notation represents _____ values as ordinary binary numbers.

5. _____ signals can represent a large range of values but are highly susceptible to transmission error.

6. Increasing the size (number of bits) of a numeric representation format increases the _____ of values that can be represented.

7. Assembly language programs for IBM mainframes commonly use _____ notation to represent numeric values. Assembly language programs for other computers commonly use _____ notation to represent numeric values and memory addresses.

8. A value is _____ within a(n) _____ so that it can be transported from one place to another.

9. A(n) _____ is a data item composed of multiple primitive data items.

10. In IBM mainframe computers, characters are coded as numbers according to the _____ coding scheme. In most other computers, the _____ character coding scheme is used.

11. A(n) _____ is the address of another data item or structure.

12. Binary, trinary, and quadrary signals are all examples of _____ or _____ signals.

13. In a positional numbering system, the _____ separates digits representing whole number quantities from digits representing fractional quantities.

14. A(n) _____ is an array of characters.

15. Most Intel CPUs implement a(n) _____ where each memory address is represented by two integers.

16. A set of data items that can be accessed in a specified order using a set of pointers is called a(n) _____.

17. _____ signals are widely used within computer systems because of their low susceptibility to transmission error.

18. A(n) _____ contains eight _____.

19. A(n) _____ list stores one pointer with each list element.

20. The result of adding, subtracting, or multiplying two _____ may result in overflow, but never _____ or_____.

21. A(n) _____ is a sequence of primitive data elements stored in sequential storage locations.

22. A(n) _____ is a group of data elements that usually describe a single entity or event.

23. A(n) _____ data item can only contain the values True or False.

24. A(n) _____ is an array of data items, each of which contains a key value and a pointer to another data item.

25. A(n) _____ signal can carry one of an infinite number of messages during one unit of time.

26. Many computers implement _____ numeric data types to provide greater accuracy and prevent overflow and underflow.

27. Unlike ASCII and EBCDIC, _____ is a 16-bit character coding table.

28. The _____ is the bit of lowest magnitude within a byte or bit string.

29. The value of a binary signal parameter is compared to a(n) _____ value to determine whether it represents a 0 or a 1.

30. _____ occurs when the result of an arithmetic operation exceeds the number of bits available to store it.

31. Within a CPU, _____ arithmetic is generally simpler to implement than _____ arithmetic because of a simpler data-coding scheme and data-manipulation circuitry.

32. Under the _____, memory addresses consist of a single integer.

33. A(n) _____ is a means of carrying a message between two points.

34. A(n) _____ can carry only one of two possible messages during one unit of time.

35. The _____ has defined a character coding table called _____, which combines the ASCII-7 coding table with an additional 128 Western European multinational characters.

36. Portability of floating point data from one computer to another is guaranteed if each computer's CPU represents real numbers using _____ standard floating point notation.

37. The ordering of characters within a coding table is called the table _____.

Review Questions

1. How does a digital signal differ from an analog signal? What are their comparative advantages and disadvantages for data representation and communication?

2. What is the binary representation of the decimal number 10? What is the octal representation? What is the hexadecimal representation?

3. Why is binary data representation and signaling the preferred method of computer hardware implementation?

4. What is sign magnitude notation? What is excess notation? What is two's complement notation? What criteria determine which notation is used by computer hardware?

5. What is overflow? What is underflow? How may the probability of their occurrence be minimized?

6. How or why are real numbers more difficult to represent and process than integers?

7. Why might a programmer choose to represent a data item in IEEE 64-bit floating point format instead of IEEE 32-bit floating point format? What additional costs are incurred at run time (when the application program executes) as a result of using the 64- instead of the 32-bit format?

8. What are the differences between ASCII and EBCDIC?

9. What basic data types can usually be represented and processed by a CPU?

10. Define a data structure, and list several common types. What software layer(s) support their definition and manipulation?

11. What is an address? What is a pointer? What is their purpose?

12. How is an array stored in main memory? How is a linked list stored in main memory? What are their comparative advantages and disadvantages? Give an example of data that would be best stored as an array. Give an example of data that would be best stored as a linked list.

1. Develop an algorithm or program to implement the following function:

   insert_in_linked_list (element,after_pointer)

   The parameter 'element' is a data item to be added to a linked list. The parameter 'after_pointer' is the address of the element after which the new element will be inserted.

2. Develop an algorithm or program to implement the following function:

   insert_in_array (element,position)

   The parameter 'element' is a data item to be added to the array. The parameter 'position' is the array index at which the new element will be inserted. Be sure to account for elements that must be "moved over" to make room for the new element.

3. Assume that Type XYZ copper wire can carry an electrical voltage between 0 and +12 volts (including exactly 0 and 12 volts). The voltage can be carried for a distance of 500 feet with a signal loss of no more than 0.01 volts. Assume further that sending and receiving devices can vary or detect voltage changes in Type XYZ wire up to 10,000,000 times per second. How many bits per second can be successfully transmitted over short distances (less than 500 feet) if binary signals are used? How many bits per second (best case) can be successfully transmitted over short distances if analog signals are used?

4. Consider the following binary value:

   1000 0000 0010 0110 0000 0110 1101 1001

   What number (base 10) is represented if the value is assumed to represent a number stored in two's complement notation? Excess notation? IEEE floating point (32-bit) notation?

5. What is the numeric range of a 16-bit two's complement value? A 16-bit excess notation value? A 16-bit unsigned binary value?

Research Problems

1. Choose a commonly used microprocessor such as the Intel Pentium (www.intel.com), the DEC Alpha (www.digital.com), or the IBM/Motorola PowerPC (www.mot.com). What data types are supported? How many bits are used to store each data type? How is each data type internally represented?

2. Most application tools (e.g., word processors) are marketed internationally. To minimize development costs, software producers develop a single version of the program that can be configured for many different languages (e.g., menu items and error message text). Investigate a commonly used development environment for application tools (e.g., Microsoft Developer Studio, www.microsoft.com). What tools and techniques are supported for developing multilingual programs (e.g., Unicode data types and string tables)?

3. Object-oriented programming has been widely adopted due to its inherent ability to allow code reuse. Many object libraries provided with application development software (e.g., the Microsoft Foundation Classes, www.microsoft.com) provide extensive support for complex data structures, including various types of linked lists. Investigate one of these libraries and the data structure objects provided. What types of data are recommended for use with each data structure object?

Data Representation

Chapter 5

Processor Technology and Architecture

Chapter Goals

▶ Describe the instruction set of a typical CPU.

▶ Describe the key design features of a CPU including instruction format, word size, and clock rate.

▶ Describe the function of general and special purpose registers.

▶ Describe CISC and RISC CPUs.

▶ Describe the principles and limitations of semiconductor-based microprocessors.

PROCESSORS AND INSTRUCTION SETS

The most primitive command that software can direct a processor to perform is called an **instruction**. An instruction is a command to the CPU to perform one of its primitive processing functions on specifically identified data inputs. As stored in memory, an instruction consists of a contiguous sequence of bits. This bit string is logically divided into a number of components. The first group of bits represents the unique binary number of the instruction, commonly called its **op code**. Subsequent groups of bits hold the input values for the instruction called **operands**. The content of an operand can represent a data item (e.g., an integer value) or the location of a data item (e.g., a memory address, a register address, or the address of a secondary storage or I/O device).

At the implementation level, an instruction directs the CPU to route electrical signals representing data input(s) through a predefined set of processing circuits that implement the desired function. Data inputs are accessed from storage (or extracted directly from the operands) and temporarily stored in one or more registers. For computation and logic functions, the arithmetic logic unit (ALU) is directed to access these registers and send the corresponding electrical signals through the appropriate processing circuitry. This circuitry physically transforms the input signals into output signals that represent the processing result. Then the result is stored in a register prior to movement to a storage device or I/O device or for use as input to another instruction.

The collection of instructions that a CPU is designed to process is called the CPU's **instruction set**. Instruction sets vary widely among CPUs in the following ways:

► Size of the instruction set (number of instructions supported).

► Size of an individual instruction and of its op code and operands.

► Set of data types supported (used as input and/or generated as output) by the instruction set.

► Number and complexity of processing functions implemented as individual instructions.

Variations in these design parameters reflect differences in design philosophies, processor fabrication technology, and intended use (e.g., class of computer system and type of application software). Cost of the CPU and the speed at which it operates depend on these design parameters.

The full range of processing functions expected of a modern computer can be implemented with approximately 12 separate instructions. Such an instruction set can directly support all computational, comparison, data movement, and branching functions for integer and boolean data types. Processing functions for real numbers can be implemented with appropriate combinations of integer instructions on the component parts of the real number. In fact, the earliest CPUs and microprocessors used this relatively small set of instructions to implement the full range of computer processing power. The following sections describe the operations performed by this minimal instruction set.

Data Movement

Data movement instructions copy data bits from one location within the computer to another. The following instructions are supported:

► Load

► Store

► Move

Before executing a data transformation instruction, the processor must fetch data inputs from specific locations in memory to temporary storage areas (registers) within the CPU. Data transfers from main memory into the processor's internal registers are called **load** operations. In performing output functions, the processor must cause data to be sent from a register to a specified location in main memory. The writing of data from a processor register back to a location in main memory is called a **store** operation.

To support processing functions, intermediate results often must be transferred among registers within the CPU. Transfers among registers are referred to as **move** operations; the term move may also be used to describe transfers between memory locations. However, movement from one memory location to another also may be implemented as a combination of a load operation and a store operation. This implementation method is required in a CPU that does not implement a direct memory-to-memory move instruction.

In all three operations, the essential task to be performed is copying a data value, bit by bit, from one storage location to another. Data bits in such copying operations are not moved physically. The bit value in a storage location at the point of origin is tested, and an electronic signal of the same value is sent to the same position in a storage location at the destination. This signal causes the storage location at the destination to be set to the same value. When the copying operation is completed, both sending and receiving locations hold a copy of the same data byte(s). The term move is a misnomer because the data content of the source location is unchanged.

Load and store instructions may be used to implement input and output functions. I/O devices and their communication channels can be configured to communicate with the CPU through specific memory addresses. Thus, an input device "fills" the content of a memory location and that content can be "read" by the CPU by loading the content of that memory address into a register. Similarly, data can be "written" by the CPU by storing the content of a register to a memory address. The output device must be configured to continually monitor the content of that memory address and to read and display it each time it changes. Communication between the CPU and a peripheral device via memory addresses is sometimes called **memory mapped I/O.**

Data Transformations

In data transformation instructions, data content is modified according to a specific arithmetic or logical rule. The most primitive rules are taken from boolean logic and include only five transformations:

► AND
► OR
► XOR
► NOT
► ADD

These functions are sufficient to implement all possible numeric comparisons and addition. An additional transformation called SHIFT is normally provided as a more efficient basis for implementing multiplication and division. Each of these transformations is described fully in the following sections.

AND

In boolean logic, the **AND** operation is a specific method of combining two boolean (True or False) data values. The result of executing an AND instruction is True if both of the data inputs are True. Thus, if a 0 bit represents False, and a 1 bit represents True, the following processing results are generated with single-bit data inputs:

$$0 \text{ AND } 0 = 0$$
$$1 \text{ AND } 0 = 0$$
$$0 \text{ AND } 1 = 0$$
$$1 \text{ AND } 1 = 1$$

The effect of a bitwise AND operation on two separate bit strings is shown in the following example:

```
          10001011
AND       11101100
          10001000
```

For each bit position, the result string contains a 1 only if both inputs contained a 1.

OR

The inverse of the AND transformation is the **OR** transformation. This transformation also may be called an **inclusive OR** to distinguish it from another boolean transformation. If the OR transformation is applied, both operands must be 0 to yield a result of 0. Any other combination of values produces a result of 1, that is:

$$0 \text{ OR } 0 = 0$$
$$1 \text{ OR } 0 = 1$$
$$0 \text{ OR } 1 = 1$$
$$1 \text{ OR } 1 = 1$$

Exclusive OR

An **exclusive OR (XOR)** transformation compares two bits and produces a result of 1 only if exactly one of the operand bits has a value of 1. The full set of processing results is:

$$0 \text{ XOR } 0 = 0$$
$$1 \text{ XOR } 0 = 1$$
$$0 \text{ XOR } 1 = 1$$
$$1 \text{ XOR } 1 = 0$$

A common use of the XOR operation is to generate the complement of a bit string. To produce this result, each bit position in the second operand must contain a value of 1. For example:

```
      01010010
XOR 11111111
      10101101
```

This transformation is necessary as a first step in performing subtraction or division by complementary arithmetic.

NOT

In boolean logic, the **NOT** transformation turns the boolean value True into False and the boolean value False into True. If 0 represents False and 1 represents True, then possible processing results are:

$$\text{NOT } 0 = 1$$
$$\text{NOT } 1 = 0$$

The exact implementation of a NOT instruction for bit strings depends on the coding format for boolean values. In general, performing a bitwise NOT

---

of the value True will not yield the value False because False is represented by 0 and True is represented by any nonzero integer. Because such a True value can contain some 0 bits, applying the NOT transformation to each bit will not guarantee that all bits in the result are 0.

ADD

As its name implies, the **ADD** operation accepts two numeric values as input and produces their arithmetic sum as the processing result. The ADD operation differs from other boolean operations because its operation extends across bit positions. That is, adding together two bits in one position may produce a result (a carryover) that must be applied to the next position.

Another key difference between the ADD operation and boolean operations is that its implementation depends on the data input's coding format. If multiple numeric data types (using different coding formats) are supported by the CPU, then a separate implementation of the ADD operation is required for each type. In essence, multiple instructions must be defined, one for each type of possible data input.

SHIFT

The effect of a **SHIFT** instruction is shown in Figure 5-1. In Figure 5-1(a), the value 01101011 occupies an 8-bit storage location. A shift may be implemented to the right or left, and the number of positions shifted may be greater than one. Typically, a second operand is used to hold an integer value that indicates the number of bit positions by which the value will be shifted. Positive or negative values of this operand may be used to indicate right or left shifting.

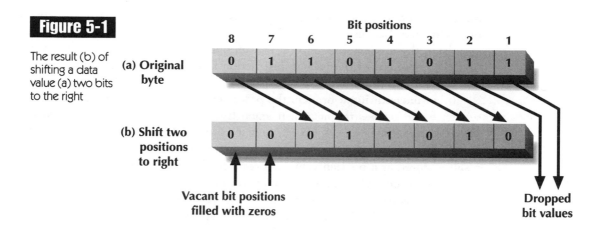

Figure 5-1

The result (b) of shifting a data value (a) two bits to the right

Bit positions

(a) Original byte

(b) Shift two positions to right

Vacant bit positions filled with zeros

Dropped bit values

Figure 5-1(b) shows the result of shifting the value two positions to the right. The resulting value is 00011010. In this case, the values in the two least significant positions of the original string have been dropped. This is an example of a **logical shift** operation—an operation that includes all of the bit positions in the shift.

Arithmetic shift operations also may be used to perform arithmetic operations. In a positional number system of base n, each position represents n times the value of the position to the right. Thus, in a binary number system, shifting a digit to the left has the effect of multiplying its value by two. Similarly, shifting a digit to the right is the same as dividing by two. Arithmetic shifts differ from logical shifts because they must preserve the original sign bit. Thus, arithmetic shift operations must shift contents "around" the sign bit, leaving its content unchanged.

Sequence Control

Sequence control operations, which include the following, alter the flow of instruction execution in a program:

► Unconditional branch.
► Conditional branch.
► Halt.

Branch

A **branch**, (or **jump**), operation causes the processor to depart from the normal sequence of a program. Recall that the control unit fetches the next instruction from memory at the conclusion of each execution cycle. A branch instruction requires an operand referencing an address in memory where the next instruction is located. Because loading of the instruction is controlled by the content of a register, the implementation of a branch operation is actually a move operation. The operand of the branch instruction is simply loaded into the instruction pointer.

In an **unconditional branch**, the processor always departs from the normal sequence and fetches the next instruction from a different address. In a **conditional branch**, the branch will occur only if a specified condition is met. Typically, the condition (e.g., the equivalence of two numeric variables) is evaluated, and the result is stored as a boolean value in a register. The conditional branch operation checks the content of that register and performs the branch only if the value contained there is true. In some architectures, the register containing the comparison result is predetermined, and in others it may be specified as an operand.

Halt

A **halt** operation suspends the normal flow of instruction execution in the current program. In some CPUs, it causes the CPU to cease all operations whereas in others, it causes a branch to be performed to a predetermined memory address. A portion of the operating system is typically present at this address and thus control is effectively transferred to the operating system and terminates the previously executing program.

Complex Processing Operations

Complex processing operations may be performed by combining the simpler operations described thus far. For example, subtraction may be implemented as complementary addition. That is, the operation $A - B$ may be implemented as $A + (-B)$. As described in Chapter 4, a negative two's complement value can be derived from its positive counterpart by taking the complement of the positive value and adding 1. The XOR operation can be used to derive the complement of a value as the result of an XOR operation applied to the value and a string of binary 1 digits.

Thus, for example, the complement of 0111, represented as a two's complement value, can be derived as:

$$XOR(0111,1111) + 0001 = 1000 + 0001 = 1001$$

This result can then be added to implement a subtraction operation. For example, the result of subtracting 0111 from 1000 can be calculated as:

$$ADD(ADD(XOR(0111,1111),0001),1000)$$
$$ADD(ADD(1000,0001),1000)$$
$$ADD(1001,1000)$$
$$10001$$

Because four-bit values are used, the result of 10001 is truncated from the left, which results in a value of 0001.

Comparison operations can be implemented in much the same way as subtractions. Generally, the purpose of a comparison operation is to generate an output that is interpreted as a boolean (True or False) value. Typically, an integer value of 0 is interpreted as False and any nonzero value is interpreted as True. The comparison $A \neq B$ may be implemented by generating the complement of B and adding it to A. If the two numbers are equal, the result of the addition will be a string of zeros (interpreted as False). An equality comparison can be implemented simply by negating the boolean result of an inequality comparison.

Greater than and less than comparisons also can be implemented as subtraction (complementary addition) followed by extraction of the sign bit. For the condition $A<B$, the subtraction of B from A will generate a negative result if the condition is True. In two's complement notation, a negative value will always have a one in the leftmost position (i.e., the sign bit). The shift operation can be used to extract the sign bit from the remainder of the value. For example, the two's complement value 10000111 is a negative number. The sign bit can be extracted by shifting the value seven bits to the right, which results in the string 00000001. The result of the shift can be interpreted as a boolean value (nonzero, or True, in this case).

For example, the comparison:

$$0111 < 0011$$

can be evaluated as:

$$SHIFT(ADD(0111,ADD(XOR(0011,1111),0001)),0011)$$
$$SHIFT(ADD(0111,ADD(1100,0001)),0011)$$
$$SHIFT(ADD(0111,1101),0011)$$
$$SHIFT(0100,0011)$$
$$0000$$

The result is 0 and is interpreted as the boolean value False.

Short Programming Example[1]

Consider the following high-level programming language program:

IF (BALANCE < 100) THEN
    BALANCE = BALANCE − 5
ENDIF

       Such a computation might be used in a program that applies a monthly service charge to checking or savings accounts that have a balance below a certain minimum. A program that implements this computation using only the previously defined low-level CPU instructions is shown in Figure 5-2. Table 5-1 shows the register contents after each instruction is executed when the account balance is $64. Eight-bit two's complement is the assumed coding format for all numeric data.

Figure 5-2				
A simple program using primitive CPU instructions	1	LOAD	M1, R1	' load BALANCE
	2	LOAD	M2, R2	' load minimum balance 100_{10}
	3	LOAD	M3, R3	' load service charge 5_{10}
	4	LOAD	M4, R4	' load constant all 1 bits
	5	LOAD	M5, R5	' load constant 1_{10}
	6	LOAD	M6, R6	' load constant 7_{10}
	7	XOR	R2, R4, R2	' start < comparison
	8	ADD	R2, R5, R2	'
	9	ADD	R1, R2, R0	'
	10	SHIFT	R0, R6	' end < comparison
	11	XOR	R0, R4, R0	' invert comparison result
	12	BRANCH	R0, 14	' branch if comparison false
	13	XOR	R3, R4, R3	' start negation of service chg
	14	ADD	R3, R5, R3	' end negation of service chg
	15	ADD	R1, R3, R0	' subtract svce chg from balance
	16	STORE	R0, M1	' store new balance
	17	HALT		' terminate program

[1] You may wish to review the discussion of two's complement notation in Chapter 4 before studying these examples.

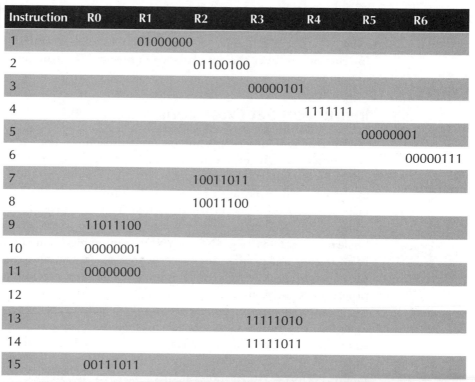

Instruction	R0	R1	R2	R3	R4	R5	R6
1		01000000					
2			01100100				
3				00000101			
4					1111111		
5						00000001	
6							00000111
7			10011011				
8			10011100				
9	11011100						
10	00000001						
11	00000000						
12							
13				11111010			
14				11111011			
15	00111011						

TABLE 5-1 Register contents after the execution of each instruction in Figure 5-2 when account balance (content of M1) is $64

Instructions 1 through 6 load the account balance, minimum balance, service charge, and needed binary constants from memory locations M1 through M6. A less than comparison is implemented in instructions 7 through 10. The right side of the comparison is converted to a negative value by XORing it with a string of 1s (instruction 7) and adding 1 to the result (instruction 8). This value is then added to the account balance (instruction 9), and the result is shifted 7 places to the right to extract the sign bit (instruction 10). At this point, register R0 holds the boolean result of the less than comparison.

For this example, all branch statements are assumed to be conditional on the content of a register. The branch is taken if the register holds a boolean True value and otherwise is ignored. Because we wish to branch around the code that implements the service charge if the account balance is above the minimum, we must invert the boolean result of the condition prior to branching. Instruction 11 inverts the sign bit stored in R0 by XORing it against a string of ones. The conditional branch is then executed. Because the original sign bit was 1, the inverted value is 0. Thus, the branch is ignored and processing continues with instruction 13.

Instructions 13 through 15 subtract the service charge (stored in register R3) from the account balance. Instructions 13 and 14 convert the positive value to a negative value, and instruction 15 adds it to the account balance. Instruction 16 saves the new balance in memory. Instruction 17 halts program execution.

Instruction Set Extensions

The instructions described thus far are sufficient to implement a general purpose processor. All of the more complex functions normally associated with computer processing can be implemented by combining those primitive building blocks. However, most processors will provide a substantially larger set of instructions. Additional instructions that might be provided include advanced computation functions (e.g., multiplication and division), combinations of boolean operations (e.g., NOT–AND), and single-bit manipulation functions (e.g., testing a sign bit). Such instructions are sometimes referred to as **complex instructions** because they represent complex combinations of primitive processing operations.

The decision to implement (or not implement) such instructions represents a trade-off between processor complexity and programming simplicity. For example, consider the three-step process used to implement subtraction described in "Complex Processing Operations." Given that subtraction is a commonly performed operation, a CPU designer might decide to implement it directly as a single instruction. Thus, a compiler or human programmer would be provided with the convenience of a single instruction. The CPU design would be more complex because of the additional processor circuitry required to implement the instruction.

Complex instructions also represent a trade-off between processor complexity and speed of program execution. Many multistep operations are executed faster if they are fully implemented within hardware than if they are implemented within software through the execution of multiple primitive instructions. The overhead of fetching extra instructions is bypassed when all processing steps are implemented as a single instruction. Other efficiencies may also be realized by "hardwiring" the steps together. However, these efficiencies have limits, as described later in this chapter.

Generally, extensions to the instruction set are required when new data types are added. For example, most CPUs provide separate instructions for adding, subtracting, multiplying, and dividing integers. If floating point numbers are also supported, then an additional set of instructions is required to implement those computation operations with floating point inputs. The provision of double-precision integers and floating point numbers also requires additional instructions designed for those data types. It is not unusual to see up to six different ADD instructions in an instruction set (e.g., to support integers and real numbers in single and double precision as well as signed and unsigned "short" integers).

Some instruction sets also provide instructions that combine data transformations with data movement. The primitive instructions described and used earlier all used registers for both data input and data output. In some CPUs, data transformation instructions allow one or more of the operands to be a memory address. In essence, this combines data movement (i.e., load or store) with data transformation.

INSTRUCTION FORMAT

Recall that an instruction consists of an op code (instruction number) and 0 or more operands (representing data values or storage locations). An **instruction format** is simply a template that determines the length, number, and position of the individual components of an instruction. CPUs differ in the instruction format(s) that they implement. Instruction format depends on a number of factors, including the number of instructions, the data types used as operands, and the length and coding format of each type of operand. The term instruction format erroneously implies that each CPU uses a single format. Because instructions vary in the number and type of operands they require, a CPU generally uses multiple instruction formats. These formats represent various combinations of operand types.

Figure 5-3 shows an instruction format consisting of an op code and three operands with a total length of 20 bits. Such a format is typical of instructions that use register inputs and outputs. Eight bits are allocated to the op code. Most CPUs represent the op code as an unsigned binary number. Thus, an 8-bit op code allows for 256 possible instructions numbered 0–255. Each operand is four bits in length. If these are unsigned integers representing registers, then a total of 16 registers can be referenced.

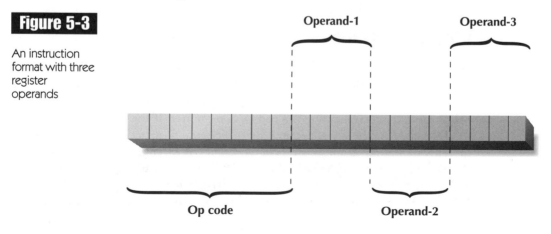

Figure 5-3

An instruction format with three register operands

Figure 5-4 shows another instruction format that could be used for load and store instructions. The format contains three op codes and is 32 bits in length. The first two op codes store register numbers. The first contains the data value to be moved (for a store instruction) or to hold the value being moved from memory (for a load instruction). The second register contains a partial memory address.

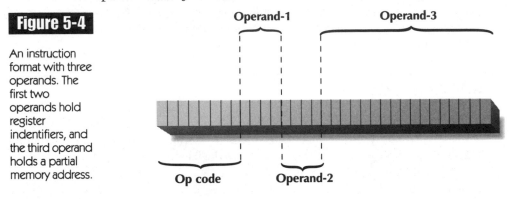

Recall from Chapter 4 that some CPUs use a segmented memory addressing scheme in which a memory address consists of two parts: a memory segment identifier and a segment offset. For segmented memory schemes, the second operand holds the segment identifier and the third operand holds the segment offset. Even if a flat memory model is used, a two-part addressing mechanism may still be used. For a flat 32-bit address, a register is used to hold the 16 high-order bits and the 16 low-order bits are held in an operand. In either case, the control unit uses both the contents of the register identified in the second operand and the content of the third operand to construct a "complete" memory address.

Instruction Length

Instruction formats may be either fixed or variable in length. **Fixed length instructions** and instruction components simplify the instruction fetching process implemented within the control unit. Recall that the control unit is responsible for incrementing the instruction pointer. If the instruction format is of fixed length, then the amount by which the instruction pointer must be incremented after each fetch is a constant (the length of an instruction).

Although it would appear that a CPU using the two instruction formats shown in Figures 5-3 and 5-4 is using variable length instructions (since the formats aren't of equal length), this is not necessarily the case. To turn a set of instruction formats into a set of fixed length instructions, one simply "pads" the shorter instruction formats with trailing 0 bits. Thus, for example, the format in Figure 5-3 could be padded with 12 zero bits to increase its length to match the format in Figure 5-4. The extra bits are simply ignored by the CPU when processing an instruction of that format.

If instruction length is variable (i.e., a **variable length instruction**), then the amount by which the instruction pointer is incremented is the length of the current instruction. Thus, the CPU must check the op code of the instruction last fetched to determine the proper incrementation value. Conceptually, one might think of a table with op-code values on the left and format lengths on the right. The control unit would look up the op code in the table and extract the appropriate value to add to the instruction pointer. Although this method is conceptually simple, most CPUs use more efficient means of determining instruction lengths based on the op code (e.g., testing a single bit or pair of bits).

Variable length instructions also complicate the instruction fetching operation because the number of bytes to be fetched is not known in advance. One method of addressing this uncertainty is to always fetch the number of bytes in the longest instruction format, but this can result in a great deal of unnecessary memory access. Another possible method is to fetch the first few bytes (e.g., the number of bytes in the shortest format). Because the op code is always stored first, such a fetch will always retrieve it. It can then be used, as before, to determine the length of the instruction and, thus, the number of additional bytes to fetch, if any. In either case, extra computer resources are used in the fetch operation that would not have been used if the instruction length were fixed (and thus known before the fetch).

Although fixed length instructions and fields simplify control unit functions, they do so at the expense of efficient use of memory. Some instructions have no operands (e.g., a HALT) and those with operands can have either one, two, or three. If fixed length instructions are used, then the instruction length must be the length of the longest instruction (e.g., an instruction with two or three large operands). Smaller instructions stored in memory must be padded with empty bit positions to extend their length to the maximum. This is an inefficient use of memory due to the amount of space wasted by the "padding." Programs are larger than necessary and require more memory and time to load into memory. Execution also may be slowed by the movement of empty instruction fields to and from the CPU. The trade-off between these inefficiencies and the simplification of control unit circuitry for instruction fetching is a fundamental CPU design decision.

Reduced Instruction Set Computing

Reduced instruction set computing (RISC) is a relatively recent philosophy of processor and computer system design. The primary architectural feature of a RISC processor is the absence of some (but not all) complex instructions in the instruction set. In particular, instructions that combine data transformation and data movement operations are usually avoided. For example, a non-RISC processor might provide a data transformation instruction of the form:

Transform(Address1,Address2,Address3)

where Address1 and Address2 are the memory addresses of data inputs, and Address3 is the memory address to which the result will be stored. Depending on the processor implementation, this instruction might require as little as one or as many as four execution cycles to complete. In a RISC processor, transformation operations always use register inputs and outputs. Thus, the single complex instruction shown earlier would require four separate RISC instructions:

Load(Address1,R1)
Load(Address2,R2)
Transform(R1,R2,R3)
Store(R3,Address3)

Each of these instructions would be independently loaded and executed, thus consuming at least four execution cycles.

Although the lack of many complex instructions is the primary distinguishing feature of a RISC processor, other differences are also typical. These include the use of fixed length instructions, relatively short instruction length, and a relatively large number of general purpose registers. To fully describe these differences, RISC must be contrasted with its opposite design philosophy—**complex instruction set computing (CISC)**.

Complex instruction sets were developed because early computers had extremely limited memory and processing power. Memory, in particular, was very expensive and many computers had barely enough to hold an entire application program. One response to the lack of memory and processing power in early computers was to implement complex instructions. As discussed in "Instruction Set Extensions," complex instructions can generally be executed more quickly than combinations of simpler instructions thanks to savings in overhead (fetching of multiple instructions) and design efficiencies that can be achieved by "hardwiring" the individual processing components of a complex instruction. Memory also is saved because of the representation of complex functions as a single instruction instead of two, three, or more instructions.

For example, assume that a floating point addition operation, implemented as a single complex instruction, can be executed in one processor cycle. Assume further that an equivalent operation can be performed with five integer math instructions, each of which requires one processor cycle. In this example, a direct implementation of the (complex) floating point instruction saves four processor cycles each time it is executed.

If complex instructions are so beneficial why, then, would anyone want to eliminate them? The most important reason is that the previously described benefits of complex instructions are subject to the economic law of diminishing returns. Each complex instruction provides benefits if measured in isolation, but as more and more complex instructions are added, the instruction set as a whole becomes bloated. This, in turn, leads to costs that affect the entire instruction set.

A bloated instruction set is problematic in two ways. First, it complicates the job of the control unit. There are more possible instructions to decode and (usually) more instruction formats to manage. Also, a large set of complex instructions often goes hand in hand with variable length instruction formats, which, in turn, adds complexity to the job of fetching and decoding instructions. That complexity then translates to reduced speed of fetching and decoding due to the extra work and overhead for the control unit. Because every instruction (whether simple or complex) must be fetched before it is executed, a performance penalty is applied to every instruction (even the simple ones).

The second problem arising from a bloated instruction set is that of microprocessor size. As discussed later in "Speed," speed improvements in microprocessors are primarily achieved by shrinking them. The simpler the processor, the simpler the task of shrinking it and the more reliably the smaller versions can be fabricated. But CISC processors are much more complex than RISC processors and are thus more difficult to shrink and reliably fabricate. A RISC processor is deliberately designed to trade the size and complexity of the instruction set for raw speed in fetching and executing instructions. RISC design follows a "less is more" strategy and only extends the instruction set when the benefits are very high.

The primary disadvantage inherent in RISC processing is extra memory requirements for program storage and execution. Since many complex instructions are not present, programs must use multiple simple instructions in their place. These occupy more space than their complex equivalents and, thus, increase program memory requirements. Although this is a relative disadvantage for RISC, it does not translate to a significant dollar amount. The price of primary storage has fallen rapidly, thus making RISC's disadvantage relatively inexpensive.

RISC processors are also relatively inefficient at executing programs that do many of the functions for which complex instructions are designed. Although this is a potential disadvantage, detailed studies of typical program behavior have shown that many complex instructions are used infrequently. Further, "typical" programs spend most of their time executing relatively primitive instructions (e.g., load, store, addition, and comparison). In most cases, then, the speed advantage of complex instructions is not realized frequently enough to make up for the performance penalty applied to every instruction because of a more complex instruction set.

The choice of CISC versus RISC is an excellent example of the multiple factors that a computer design must balance. CISC and RISC aren't fundamentally different strategies. They are simply different points along a design continuum representing different trade-offs among design factors. The economics and technology of computer use, processor fabrication, and memory led to CISC as an optimal trade-off in the early days of computing. As the economics and technological capabilities shifted, so did the optimal trade-off to the point that RISC is now a better choice. Twenty more years of shifting technology and economics may move the pendulum back toward CISC or toward a fundamentally different approach to CPU design and implementation.

CLOCK RATE

The **system clock** is a digital circuit that generates timing pulses (signals) and transmits them to other devices within the computer. The clock is generally implemented as an entirely separate device (i.e., it is not a part of the CPU) with a dedicated communication line that is monitored by all devices within the computer system. The clock signal is the "heartbeat" of the computer. All actions, especially the instruction and execution cycles of the CPU, are timed according to this clock. Other devices such as primary and secondary storage devices are also timed by the clock signal. The shared clock signal serves as one means of coordinating the activity of all devices within a computer system.

The frequency at which the system clock generates timing pulses is called the **clock rate** of the system. Each "tick" of the clock is referred to as a **clock cycle**. CPU and computer system clock rates are expressed in **hertz** (abbreviated Hz). One hertz corresponds to one clock cycle per second. Modern CPUs and computer systems typically use clocks that generate millions of timing pulses per second. The frequency of these clocks is measured in **megahertz** (MHz) meaning millions of cycles per second.

CPU instruction and execution cycles usually represent some fraction of the clock rate. Thus, the clock rate is a significant measure of processor performance.

It is also, however, a frequently misused measure of processor and computer system performance. Two ways in which the clock rate is often misused are:

► Equating CPU clock rate with CPU MIPS rate.

► Assuming that computer system performance is equivalent to that of the CPU.

The amount of time required to fetch and decode an instruction is generally the same for all instructions in the CPU instruction set. The same cannot be said for executing those instructions. However, instructions that implement relatively simple processing functions (e.g., boolean logical functions, equality comparison, and integer addition) are executed relatively quickly whereas instructions that implement more complex functions (e.g., multiplication and division) require longer execution times.

The clock rate of a CPU generally corresponds to the time required to fetch and execute the simplest instruction(s) in the instruction set. For example, assume that the CPU clock rate is 100 MHz and that the bitwise OR instruction is the simplest, fastest instruction. The time required to fetch and execute that instruction can be computed as the inverse of the clock rate:

time = 1/clock rate = 1/100,000,000 = 0.00000001 seconds = 10 nanoseconds

The inverse of the clock rate is sometimes called the processor's **cycle time**.

More complex instructions typically require more than one CPU clock cycles to complete. The fetch portion of these additional clock cycles is essentially ignored. A new instruction fetch does not begin until the previous instruction has completed execution. Typical numbers of clock cycles required to complete complex instructions are two to four for complex integer computations (i.e., multiplication and division) and three to ten for floating point computations.

Programs contain a mix of instructions, some of which are very simple and some more complex. Only the simplest instructions will execute within a single CPU clock cycle, whereas the rest will require multiple cycles. Thus, the number of instructions executed in a given time interval depends on the mix of simple and complex instructions. Assume, for example, that a program executes 100 million instructions. Further assume that 50% of these are simple instructions requiring a single clock cycle, and 50% are complex instructions requiring an average of three clock cycles. Thus, the average program instruction requires two clock cycles ([1 × 0.50]+(3 × 0.50]). Therefore, the MIPS rate of the CPU is 50% of the clock rate when executing this program.

Note that the previous calculation assumes that nothing hinders the CPU in fetching and executing instructions. Unfortunately, the CPU relies heavily on other computer system devices to keep it supplied with instructions and data. These devices are typically slower than the CPU and are, therefore, a potential source of delay. For example, main memory is usually two to ten times slower than the processor. That is, the time required to complete a main memory read or write operation is typically two to ten CPU clock cycles. Secondary storage accesses are thousands or millions of times slower than the CPU. Thus, accesses by the CPU to these storage devices may cause the CPU to sit idle, waiting for the devices to deliver the requested data or instructions. Each clock cycle that the CPU spends waiting for a slower device is called a **wait state**.

Unfortunately, a typical CPU spends much of its time in wait states. Depending on the CPU, computer system, and program being executed, the CPU might spend up to 90% of its clock cycles in wait states. No instructions are being executed during these wait states. Thus, the effective MIPS rate of a computer system can be as little as 10% of the effective MIPS rate of the CPU because of delays imposed by other devices within the computer system. Combining the two worst case scenarios described in this chapter for a CPU with a 200-MHz clock rate produces an effective computer system MIPS rate of 10 MIPS (200 ÷ 2 × 0.10). Various improvement methods are discussed in Chapter 7.

CPU REGISTERS

An important variable in CPU design is the number of registers, or temporary holding areas, that can be used by programs to store intermediate processing results. The more registers within the processor, the greater its capacity to hold intermediate results. In general, register contents can be accessed without any wait states. Thus, the use of registers (instead of main memory) to store intermediate processing results increases the speed of program execution. If the need to store intermediate results exceeds the capacity of the available registers, then memory must be used to hold the intermediate results. Program execution speed will decrease due to delays (wait states) when loading and storing intermediate results to or from memory.

Each register is referenced by a unique register number. The maximum value is a constraint imposed by the number of registers that can be referenced. For example, if four bits are used to specify register numbers, the range of identifiers would be from 0000_2 to 1111_2, for a total of 16 registers. For purposes of notation, register numbers, like memory addresses, usually are expressed in hexadecimal form. Thus, for a processor that contains 16 registers, the register numbers would range, from 0 (zero) to F.

General Purpose Registers

Registers can be divided into two classes: general purpose and special purpose. Although general purpose registers can be used in a variety of ways by a program, typically, they hold intermediate results or data values that will be used

Business Focus

RISC or CISC for the Desktop?

With the notable exception of Intel, most CPU developers have abandoned CISC in favor of RISC. Computer system manufacturers that heavily or exclusively utilize RISC CPUs include IBM, SUN Microsystems, and Digital Equipment Corporation. The majority of computer systems manufactured by these companies since the late 1980s use a RISC CPU. The only CISC-based computers produced by these companies are those aimed directly at the IBM PC-compatible market (i.e., compatible with the Intel 8086 used in the original IBM PC).

Most application software for the desktop market is sold in binary form. That is, source code instructions have already been translated into executable code (binary CPU instructions). The application program thus directly depends on the CPU instruction set for which it was compiled. Any new CPU that executes an "old" binary application program must implement a compatible instruction set for the program (sometimes referred to as **binary compatibility**). Newer Intel CPUs have implemented binary compatibility by preserving the instruction set of older Intel processors. This allows users to upgrade CPUs and computers while preserving their investment in application software.

A CPU that does not implement an older instruction set must provide a means of emulating its function to run binary software designed for that older instruction set. Emulation is accomplished by translating CPU instructions for the older processor into equivalent instructions for the new processor. This translation is performed during the execution of the program on an instruction-by-instruction basis. Instruction set emulation may be performed by special purpose hardware, operating system software, or a combination of the two. Several Pentium CPU clones (e.g., those manufactured by AMD and Cyrix) combine a RISC processor with hardware emulation to translate Intel 80X86 instructions into "native" RISC instructions.

Preserving older instruction sets within newer CPUs provides maximal backward compatibility, but it does so at the expense of CPU complexity. In theory, the added complexity becomes a drag on performance and limits the ability of the CPU manufacturer to substantially increase clock rate. Intel processors have thus far appeared immune to this fate. Although Intel clock rates lag behind many RISC designs, they are not so different as to cause a significant performance difference. Intel has kept pace by investing huge sums of money into processor design and world class fabrication facilities. In essence, Intel has held RISC designs at bay by staying one step ahead in their fabrication processes. The huge profits realized from the IBM PC-compatible boom have made this possible.

RISC processors generally operate at higher clock rates than their CISC counterparts do. The increasing gap in maximum clock rate currently approaches a 2:1 difference, which is apparent to a user when executing software designed for the RISC processor. However, much of the difference is lost when executing older software because of the cost of emulating older instruction sets. Thus, a newer RISC processor provides little or no performance improvement if it must emulate an older CISC instruction set. Its advantage in raw clock speed is whittled away by the extra cost of instruction set emulation. The lack of a clear performance advantage has left desktop computer purchasers with little reason to switch to newer RISC CPUs from "tried and true" Intel CPUs.

The latest generation of operating systems have begun to change this scenario. Newer operating systems such as Windows NT do not allow application programs to directly issue CPU instructions. Instead, CPU instructions are issued to the operating system kernel which then passes them on to the hardware. The newer operating systems define a generic (or virtual) CPU environment for application programs. CPU instructions issued by application programs are targeted to this virtual CPU. The operating system kernel translates virtual CPU instructions into instructions specific to the "real" CPU. This approach incurs the same costs of translation and emulation that RISC CPUs incur when emulating CISC instructions. But it also provides important benefits including improved security, improved reliability, operating system and application program portability, and an improved ability to efficiently manage all hardware resources.

The new order being ushered in by newer operating systems and RISC competitors to Intel has complicated the task of acquiring PC-compatible desktop computers. Until recently, there was no real choice in CPUs. Customers simply purchased the fastest and most current Intel CPU that they could afford. But now there is a choice of Intel CPUs, RISC-based Intel clones, and "true" RISC CPUs with high clock rates. How does a purchaser decide which is best?

Questions

1. A family member asks you to recommend a replacement for his aging PC-compatible system. He has a large inventory of software, including games, entertainment, and personal productivity programs. Most of this software is less than three years old, but some of it is as much as ten years old. He has heard from a friend that substantial cost savings can be achieved by *not* buying a computer system with an Intel processor. What do you recommend to him?

2. You are a system administrator for a small (50-employee) building contractor business. Your user base consists of clerical, managerial, and technical personnel (including architects and draftsmen). The company has just obtained a large capital infusion in preparation for a substantial business modernization and expansion. You have been asked to prepare a proposal to replace all of the company's desktop computers over the next two years. Should you seriously consider alternatives to Intel-based PC-compatible systems?

frequently (e.g., loop counters or array indices). By holding such data in registers rather than memory, program execution speed can be increased.

Unfortunately, registers are very expensive in addition to being fast. The cost of memory locations is substantially cheaper and, thus, a trade-off between processor cost and program execution speed exists. There is also a diminishing rate of return for additional registers beyond a certain point. Although a CPU with eight general purpose registers may provide substantial improvements in program execution speed as compared to a CPU with four registers, the same amount of speed difference is unlikely when increasing the number from 8 to 12.

In order for the potential speed increase of additional registers to be realized, they must be used effectively, yet there are only a limited number of intermediate results or frequently used data items in any given process or program. Thus, a primary CPU design decision is the optimal trade-off between the number of general purpose registers, the extent to which those registers will be used by a typical process, and the cost of implementing those registers. Typical numbers of general purpose registers are 8–16 for CISC processors and 12–32 for RISC processors.

Special Purpose Registers

Every processor makes use of a number of special purpose registers. The content and use of each register is specified as part of the CPU design. The implementation of processor circuitry and the instruction set are integrally connected with these registers; several of the more important ones are:

▶ Instruction register.
▶ Instruction pointer.
▶ Program status word.

The **instruction register** is used by the control unit to hold an instruction just loaded from memory. Once loaded, the instruction is **decoded** and then executed. The instruction register thus serves as the data input to the decoding process and the circuitry of the control unit is designed accordingly. Decoding refers to the separation of the instruction into its op code and operands, the movement of data (e.g., loading data into a register from a memory address in one of the operands), and the generation of control signals to the ALU for instruction execution.

The **instruction pointer** may also be called the **program counter**. Recall that the CPU constantly alternates between the instruction (fetch and decode) and execution (data movement and/or transformation) cycles. At the end of each execution cycle, the control unit starts the next instruction cycle by retrieving the next instruction from memory. The instruction retrieved is the instruction at the memory address in the instruction pointer. Because a von Neumann processor assumes sequential execution of program instructions, the instruction pointer is incremented by the control unit either during or immediately after the instruction cycle.

The CPU will deviate from sequential execution only if a branch instruction is executed. A branch instruction is implemented by overwriting the value of the instruction pointer with the address of the instruction to which the branch is directed. An unconditional branch is thus implemented as a copy from one of the operands in the instruction register (containing the branch address) to the instruction pointer.

The **program status word (PSW)** contains numerous individual bit fields that indicate the current status of program execution. The content, format, and use of these bit fields varies considerably from one processor to another. In general, they have two primary uses—to direct the execution of a conditional branch instruction and to indicate actual or potential error conditions.

A single bit within the PSW may be used to store the result of a comparison instruction. Recall that a conditional branch is a branch instruction that is only executed if a specified condition (numerical comparison) is True. The True or False nature of the condition is determined by the ALU, and the result may be stored as a bit within the PSW[2]. The control unit can then check this bit to determine whether or not to overwrite the contents of the instruction pointer to implement the branch.

Other bits within the PSW may be used to represent status conditions resulting from instruction execution by the ALU. Conditions such as overflow, underflow, or an attempt to perform an undefined operation (e.g., divide by zero) are represented in status bits within the PSW. These bits are tested by the control unit at the conclusion of the execution cycle to determine if an error has occurred. They also can be tested by software (e.g., an operating system) to determine appropriate error message display and corrective measures.

WORD SIZE

A **word** is a unit of data that consists of a fixed number of bytes or bits. A word can be loosely defined as the amount of data that a CPU "processes" at one time. Depending on the CPU, this processing may include arithmetic, logic, fetch, store, and copy operations. Word size will normally match the size of CPU registers because these generally serve as input and output to/from

---

[2] Recall that the example program in Figure 5-2 implemented comparison using the XOR, ADD, and SHIFT instructions. The result was stored in a general purpose register because no compare instructions were assumed to exist. In many CPUs with directly implemented compare instructions, the boolean result is automatically stored in the PSW.

instructions. Word size is a fundamental CPU design decision with implications for most of the other components of a computer system.

In general, a CPU with a large word size will be capable of performing a given amount of work faster than a CPU with a small word size. Consider, for example, the addition or comparison of two 64-bit (double precision) integers. A processor with a word size of 64 bits will be able to perform addition or comparison by executing a single instruction because the registers that hold the operands as well as the circuitry within the ALU are designed to process 64-bit values simultaneously.

Now consider the same operation upon the same data performed by a processor with a 32-bit word size. Because the size of the operands exceeds the word size of the processor, the operands must be partitioned and the operation(s) carried out on the pieces. In a comparison operation, the processor must first compare the first 32 bits of the operands and then, as a separate execution cycle, compare the latter 32 bits. Inefficiency arises as a result of loading and storing multiple operands to and from memory. Inefficiency also arises from executing multiple instructions to accomplish what is logically a single operation. Because of these inefficiencies, a 64-bit processor is usually more than twice as fast as a 32-bit processor when processing 64-bit data values. These inefficiencies are compounded as the complexity of the operation increases. Operations such as division and exponentiation are extremely complex to perform in a piecewise fashion, with resultant slowed program execution.

Processor word size also has implications for system bus design. In general, maximal processor performance is achieved when the width of the bus is at least as large as the processor word size. If the width is any smaller, then every load or store operation requires multiple transfer operations. For example, the movement of eight bytes of character data to contiguous memory locations will require two separate data movement operations on a 32-bit bus, even if the word size of the processor is 64 bits. Similarly, fetching a 64-bit instruction will require two separate transfers.

Processor word size also has implications for the physical implementation of memory. Although the storage capacity of memory is always measured in bytes, data movement between memory and the CPU is generally performed in "word-sized" units. For a 64-bit CPU, it is desirable to organize memory such that eight contiguous bytes can be accessed in a single memory cycle. Lesser capabilities will cause the CPU to incur wait states.

As with many other CPU design parameters, increases in word size are subject to the law of diminishing returns. The two primary issues of concern are the usefulness of the extra "bits" of processing and their cost. The key determinant of the usefulness of larger word size is the relationship between word size and the size of data items that the CPU normally processes.

Consider, for example, the early days of microprocessors. Typical microprocessors of the 1970s had an 8-bit word size. This limited the range of a two's complement integer within a single word to −128 to +127. Little useful processing can be performed on such a restricted set of numbers. Thus, to implement useful computer programs, it was necessary to use multiword storage for numeric data (e.g., two words for a "short" integer and four words for a "normal" 32-bit integer). These multiword formats, in turn, required the use of complex and inefficient programs to implement even the simplest of processing functions (e.g., addition and subtraction). Dozens of instructions were required to perform simple processing operations on pieces of data and to combine the piecewise results into a "complete" result.

As processor technology and fabrication improved, it became possible to increase word size. Improvements in fabrication were necessary because doubling word size requires at least double (typically, 2.5 to 3 times) the number of electrical components within the CPU. Sixteen-bit processors, which appeared in the late 1970s and early 1980s, were highly useful because they allowed a single instruction to execute with two 16-bit data inputs. However, there were still many data items that could benefit from larger word sizes. Microprocessor technology continued its rapid evolution. In the mid to late 1980s, 32-bit processors appeared and 64-bit processors emerged in the 1990s.

Beyond 64 bits, the benefits of increased word size start to diminish. Recall from Chapter 4 that integers are typically stored in 32-bit two's complement and that real numbers are typically stored in either 32- or 64-bit IEEE floating point representation. Of what use are the "extra" 32 bits of a 64-bit processor when combining two 32-bit data items? The answer is none at all. The "extra" 32 bits simply store zeros that are carried through the computational process. Those "extra" bits only provide benefit when 64-bit data items are being processed. But many programs (e.g., word processors and most financial applications) never manipulate 64-bit data items.

The added benefit of increased word size drops off sharply past 64 bits. Only a few types of application programs—numerical processing applications that need very high precision (e.g., navigation and numerical simulations of complex phenomena) and certain database and text processing applications (e.g., text searching with large comparison strings)—can effectively use the "additional" bits. For other programs, the increased computational power of the larger word size is never used.

The "waste" of extra word size wouldn't be a major concern if cost were not an issue. However, as stated earlier, doubling word size generally increases the number of CPU components by 2.5 to 3 times. This increase, in turn, increases the cost of the CPU. This is particularly problematic when the increased number of components is at or beyond the capability of current fabrication technology. Costs tend to rise at nonlinear rates when approaching limits of fabrication technology.

Intel Pentium Processor Family

The Pentium processor is a direct descendent of the Intel 8086 and 8088 processors used in the original IBM PC. It maintains backward compatibility with those processors in all areas including instruction set, data types, and memory addressing. The Pentium was introduced in 1993, and several enhancements have since been made to the processor. The first was the introduction of the Pentium Pro in 1995. The second was the addition of multimedia processing capabilities (called MMX technology by Intel) to the Pentium in 1996. The third was the introduction of the Pentium II, which combines the features of the Pentium Pro with MMX technology. This discussion will concentrate on features common to the entire Pentium family (i.e., those of the original Pentium processor) and the MMX technology extensions.

Data Types

The Pentium processor supports a rich variety of data types. For purposes of describing data types, a word is considered to be 16 bits in length (as it was in the 8086/8088). Data types are of four sizes including byte (8 bits), word (16 bits), doubleword (32 bits), and quadword (64 bits).

Integers may be signed or unsigned and may be of byte, word, or doubleword length. Thus, there are six different integer data types. Signed integers are stored in two's complement format, and unsigned integers are stored as ordinary binary numbers. Floating point data is stored in all three IEEE standard formats (32-bit, 64-bit, and 80-bit format). Floating point range is approximately $\pm 10^{38}$, $\pm 10^{308}$, and $\pm 10^{4932}$ for the three formats, respectively.

No special data types are provided for boolean or character data; instead, these are assumed to use appropriate integer data types. A bit field data type is defined consisting of up to 32 individually accessible bits. Two variable-length data types are also defined. A bit string is a sequence of 1 to 2^{32} bits. This data type is commonly used for large binary objects such as compressed audio or video files. A byte string consists of 1 to 2^{32} bytes. Elements of a byte string may be bytes (e.g., ASCII characters), words (e.g., Unicode characters), or doublewords.

Four 64-bit MMX data types are provided. Three of these are packed data types in which multiple integers are "packed" within a single 64-bit unit. The packed integers may be 8, 16, or 32 bits in length. The fourth data type is a quadword (64-bit) integer. The use of packed data types allows efficient parallel processing to be implemented. MMX instructions operate on each component of a packed data type in parallel. Thus, an MMX instruction can execute the same instruction on up to eight data inputs at once.

Three different memory address formats are provided. Two of these are segmented memory addresses that are backward compatible with earlier

80X86 processors. The third is a 32-bit flat memory model address compatible with the 80386 and later processors. Segmented memory addresses contain offsets with respect to predefined segment registers. Six different segment registers are available and are assumed to point to specific portions of the program's address space. Flat 32-bit addresses are the preferred format for newly developed programs.

Instruction Set

The Pentium instruction set contains 28 system instructions; 92 floating point instructions; 52 MMX instructions; and 164 integer, logical, and other instructions. This is one of the largest instruction sets to be found among modern microprocessors. As should be expected, such a large instruction set has many different instruction formats of varying length. Instructions may have from zero to three operands. Legal operands include words, registers, and 32-bit memory addresses.

A detailed discussion of the instruction set fills an entire volume (several hundred pages) of Intel documentation. The large number of instructions represents both the rich variety of supported data types and the desire to implement as many processing functions as possible in hardware. This clearly places the Pentium in the CISC camp of CPU design.

Word Size

There is no clear definition of word size for the Pentium. For backward compatibility (with the 8088 and 8086 microprocessors), a word is defined as 16 bits. However, all registers in the integer ALUs are 32 bits in length (the upper and lower 16 bits can be accessed separately for backward compatibility). The internal processing paths of the integer ALUs (there are two) are also 32 bits wide. The external bus interface is 64 bits wide, and the on-chip memory cache interface is 128 bits wide.

The Pentium has a separate floating point ALU with its own registers and internal processing paths. Registers within this ALU are 80 bits wide, as are the internal processing paths, which allows single instruction operations to be performed on the largest floating point data type. Although this is a powerful floating point processing capability, it is seldom used by most application programs.

Clock Rate

Although the original Pentium debuted at a clock rate of 60 MHz, clock rate for this processor eventually reached 200 MHz. The Pentium with MMX instructions debuted at 166 MHz; the Pentium Pro at 200 MHz; and the Pentium II at 266 MHz. Clock-rate increases are continually achieved through smaller fabrication processes and corresponding reductions in operating voltage. Some processors in the family rely on an external clock signal (typically provided by the PCI bus). PCI clock signal inputs are either 50, 60, or 66 MHz and the selected clock rate is multiplied (by 1.5, 2, 2.5, 3, 3.5, or 4) to derive the processor clock rate.

PROCESSOR IMPLEMENTATION

A modern CPU is implemented as a complex system of interconnected electrical switches. Early CPUs typically contained several hundred to a few thousand switches. Modern CPUs contain millions of switches. Various technologies are available to implement both the switches and the wiring among them. All of these technologies are subject to the similar properties and limitations of electricity and electrical devices.

Switches and Gates

The basic building blocks of computer processing circuits are electronic **gates**. Gates are, in turn, composed of one or more electrical switches wired together in specific ways. Gates perform their functions by applying one or more primitive transformations to individual binary signals (digits). Basic processing functions on binary digits are performed with the logical functions AND, OR, XOR, and NOT; thus, gates that implement them are sometimes called **logic gates**. Processing circuitry is composed primarily of groups of gates representing these basic transformations.

The simplest gates are those that implement the boolean functions NOT (Figure 5-5[a]), AND (Figure 5-5[b]), and OR (Figure 5-5[c], on the next page). The NOT operation is implemented by a gate called a signal inverter. This gate transforms a value of 0 into a 1 and vice versa. Logically, the gate may be used to represent a transformation of A to &A , or NOT A. An AND gate combines the separate inputs A and B to create AB, or A AND B; an OR gate generates the result A+B. Each of these gates can be implemented with a small number of electrical switches.

Processing circuits for more complex operations are constructed by combining the gates for more primitive operations. One example is the XOR gate. Although it is represented by a single symbol (Figure 5-5[d]) is actually constructed by combining NOT, AND, and OR gates. For inputs of A and B, this gate produces (A AND &B) OR (&A AND B), as shown in Figure 5-5(d). Another frequently used complex gate is the **NAND gate** (Figure 5-5[e]).

Figure 5-5

Electrical component symbols for a signal inverter or NOT gate (a), an AND gate (b), an OR gate (c), an XOR gate (d), and a NAND gate (e)

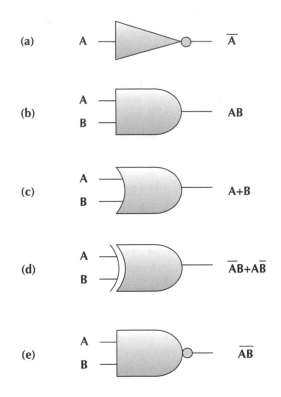

Figure 5-6 shows the implementation of a half adder and full adder processing circuit. An addition operation on full word data inputs is implemented by parallel adder circuits, one for each bit position. The least significant bit position is implemented by a half adder. A half adder circuit adds two inputs (A and B) representing single bits. The sum output represents the value of the corresponding bit position after addition. The carry-out signal represents the bit value "carried over" to the next bit position. The remaining bit positions are implemented with full adder circuits. A full adder is constructed of two half adders and an XOR gate. One half adder adds the bits for a given bit position. The other half adder adds the sum of the current bit position to the carry in signal from the previous bit position. The XOR gate combines the carry-out bits of the two half adders.

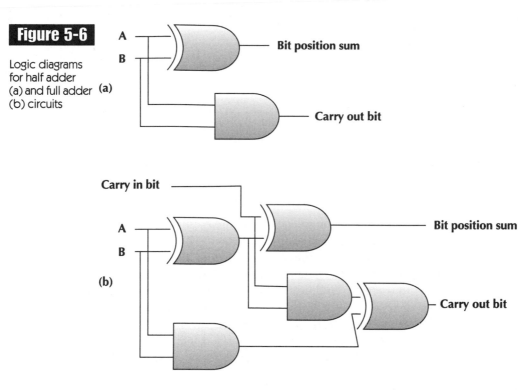

Figure 5-6

Logic diagrams for half adder (a) and full adder (b) circuits

A
B

(a)

Bit position sum

Carry out bit

Carry in bit

A
B

(b)

Bit position sum

Carry out bit

More complex processing functions require even more complicated arrangements of gates. The variety of processing functions implemented and the large number of bits that must be processed in parallel account for the extremely large number of gates (and switches) that comprise a CPU's processing circuitry.

Electrical Properties

Electrical wiring and switches can be implemented with a wide variety of materials. Some design and implementation challenges arise from the materials used whereas others are inherent in the nature of electricity and may be magnified or reduced by the materials used.

Electrical Conductors Switches and gates act as the means of transforming electrical signals representing binary zeros and ones. To do useful work, those signals must be carried from their original source, through the various devices comprising the circuit, and to their ultimate destination. A connection that allows electrons to travel from one place to another is termed a **wire**. Although the switches used in a CPU number in the millions, there are an even larger number of wires connecting those switches. The construction of the wires is as important as the construction of the switches in determining the CPU's speed and reliability.

Electrical current is the flow of electrons from one place (or device) to another. An electron requires an initial input of energy to "excite" it sufficiently to move. Once excited, electrons can move from place to place using the molecules of a substance as stepping stones. A single electron can only absorb so much energy. To carry more energy requires more electrons, so the power of electrical current is primarily a function of the number of electrons that are moving.

Elements and compounds vary in their ability to enable electron movement. The ability of an element or substance to enable electron flow is called **conductivity**. Substances that allow electrons to flow through them are called **conductors**. Conductivity is a matter of degree. A perfect conductor allows electrons to travel through it with no loss of energy. With less than perfect conductors, energy is "lost" as electrons pass through. For any conductor, this "loss" is proportional to the amount of the substance through which the electrons travel given a constant power input and temperature. If enough energy is "lost," the electron ceases to move and the flow of electrical current is halted.

Resistance   The loss of electrical power that occurs within a conductor is called **resistance**. The conductor is said to "resist" the flow of electrons through it. A perfect conductor has zero resistance. Unfortunately, all known substances have some degree of resistance. Conductors with relatively low resistance include some well-known (and valuable) metals such as silver, gold, and platinum. Fortunately, some cheaper materials can serve nearly as well.

The laws of physics tell us that energy is never really "lost" but merely converted from one form to another. Thus, resistance isn't really the loss of electrical energy. Instead, it is the conversion of that energy to another form. In most cases, the result of that conversion is heat. The heat generated due to resistance depends on the amount of electrical power being transmitted and the resistance of the conductor. Higher power and/or higher resistance increase the amount of heat generated.

Heat   Heat has two highly negative effects on electrical conductivity. The first is physical damage to the conductor. Why can't a wire the width of a human hair conduct 10,000 watts of electrical power for a mile? The answer is simply that the resistance in the wire generates so much heat that the wire melts and, thus, breaks. This is precisely the principle on which a fuse is based. To successfully transmit electrical power in this example, either a very low-resistance wire should be used, power input should be reduced, or the wire's diameter should be increased.

Processor Technology and Architecture

The second negative effect of heat is a change to the conductor's inherent resistance. The resistance of most materials increases as their temperature increases. In a closed environment, this phenomenon can raise the temperature of the conductor to destructive levels. The flow of power through even a mildly resistive conductor generates heat. This, in turn, increases the resistance of the conductor which then causes the generation of additional heat. If the conductor doesn't reach a destruction point, then a steady state is reached when the increased resistance virtually shuts down the flow of electricity. But with little or no electricity flowing, the attached devices are unable to perform their tasks.

To keep operating temperature and resistance low, some heat must be removed or dissipated from the wire. This can be accomplished in many ways. The simplest is to provide a cushion of moving air around the wire. Heat migrates from the surface of the wire (or its coating) to the air and is transported away. The use of fans and vent holes in computer equipment accomplishes this form of heat transfer and dissipation.

A **heat sink** may be employed to improve heat dissipation (see Figure 5-7). A heat sink is an object specifically designed to hold heat and to allow rapid dissipation of that heat through air or water movement. A heat sink is placed in direct physical contact with an electrical device (e.g., a microprocessor) to maximize heat transfer from the device. A heat sink exposes a large surface area to the moving air to allow more rapid dissipation. Dissipation can be given a boost by using denser heat absorbers (e.g., a liquid instead of air) and by cooling the heat absorbers before use (e.g., using liquid helium or nitrogen).

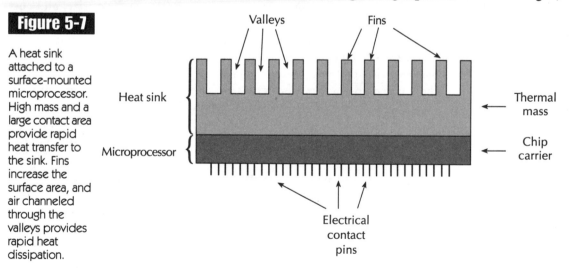

Figure 5-7

A heat sink attached to a surface-mounted microprocessor. High mass and a large contact area provide rapid heat transfer to the sink. Fins increase the surface area, and air channeled through the valleys provides rapid heat dissipation.

Speed To perform a processing function, electrons must move through all of the wires and switches comprising a processing circuit. Electrons move at the (constant) speed of light (approximately 180,000 miles per second). Thus, a fundamental relationship exists between the length of an electrical circuit and the speed with which the circuits processing function is performed. That is, the time required to perform the function is simply the quotient of the length of the circuit and the speed of light. For example, if a processing circuit is one inch long, then the circuit performs its function in:

$$\frac{1 \text{ inch}}{180{,}000 \text{ miles/second}} = \frac{1 \text{ inch}}{11{,}404{,}800{,}000 \text{ inches/second}}$$
$$= 0.00000000008768 \text{ seconds}$$
$$\approx 88 \text{ picoseconds}$$

Because circuit length is the only variable in the equation, the path to faster processing is clear. That is, faster processing is only achieved by reducing the length of the circuit. This, in turn, implies smaller (and thus narrower) switches as well as shorter wires to connect them. Miniaturization has, in fact, been the primary basis for CPU speed improvement since the first electrical computer.

Processor Fabrication

Construction of reliable and efficient data transmission and processing circuits requires a balance among power requirements, resistance, heat, size, and cost. The history of electrical computers represents an ever-changing balance among these factors. Improvements in construction materials and fabrication techniques have vastly increased the performance and reliability of processors, but the basic electrical design trade-offs persist.

The earliest computers were constructed with ordinary copper wire and vacuum tube switches. They were highly unreliable devices because of the large amounts of heat generated by the vacuum tubes. They were also large, typically filling an entire room with processing circuitry less powerful than that found today in a typical $25 calculator.

Transistors and Integrated Circuits In 1947, research engineers at Bell Laboratories discovered a class of materials called **semiconductors**. The electrical conducting properties of these materials vary in response to the electrical inputs applied. **Transistors** are made of semiconductor material that has been treated (or doped) with chemical impurities to enhance these semiconducting effects. Both silicon and germanium are basic elements with resistance characteristics that can be tailored through use of chemicals called dopants.

Figure 5-8 shows the components and current flow of a transistor. The emitter and the collector are composed of an N-type semiconductor with surplus electrons. Transistors and other semiconductors are called solid state electronic devices because a solid material, called the base, is used as the medium through which electrons are transferred from emitter to collector. The base is a P-type semiconductor, which is missing electrons. The resistance of the base is normally high, which prevents the flow of current from collector to emitter. When current is applied to the metal by the control line, the resistance of the base is dramatically lowered. Thus, the transistor functions as a switch controlled by the current (or lack thereof) present in the control line.

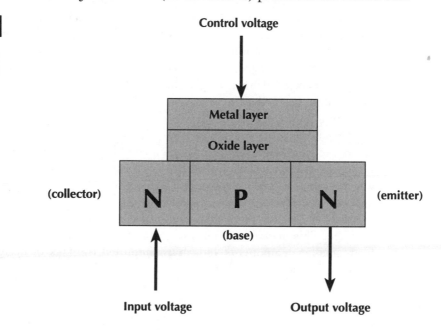

In the early 1960s, techniques were developed for fabricating miniature electronic circuits from multiple layers of metals, oxides, and semiconductor materials. These layers are built up on small, thin chips. Some of these devices are named metal oxide semiconductors (MOS), a term derived from the materials used in each of three layers on the chip. MOS technology made it possible to implement several transistors and their interconnections on a single chip to form an **integrated circuit (IC)**.

Integrated circuits represented advances in reduction of manufacturing cost, space utilization, processing performance, and reliability. Costs were reduced because many chips could be manufactured in a single sheet, or wafer. The wafer is cut apart, and chips are mounted on dual in-line packages (DIPs), or chip carriers with dual rows of pins for interconnection with larger circuits on printed circuit boards. Combining multiple gates on a single chip also reduced the manufacturing cost per gate and created a compact, modular package that simplified circuit design and manufacturing. Integrated circuits were more reliable than circuits built of separate (or discrete) components because the circuits were sealed within a single package.

Microchips and Microprocessors Further economies were realized as it became possible to put increasing numbers of gates on integrated circuits. The term **microchip** was coined to refer to this new class of electronic device. Early microchips housed 4, 10, and then 100 gates on a single chip. Later generations of devices that contained 100 to 1,000 gates came to be called medium-scale integration (MSI). At levels above 1,000 gates, a chip is said to implement large-scale integration (LSI). At levels above 10,000 gates, a chip is said to implement very large-scale integration (VLSI). With each increase in integration scale, the size of individual switches and wires decreased. As a result, processing speed increased. Current fabrication technology allows millions of gates to be implemented on a single chip. Terminology based on the words **scale integration** has fallen into disuse since the English language has run out of adjectives to precede those words.

One important application of the chip technology is to implement all of the gates, registers, and wiring for a complete CPU on a single microchip. Such a device is called a **microprocessor**. The first microprocessor (see Figure 5-9) was designed by Ted Hoff of Intel and introduced in 1971. Microprocessors ushered in a new era in computer design and manufacturing. The most important part of a computer system could now be produced or purchased as a single package. This negated the need for computer designers to construct processors from smaller components and, thus, simplified the task of designing computer systems. It also led to a degree of standardization as a few specific microprocessors became used widely. The revolution of IBM-compatible microcomputers would not have been possible without standardized microprocessors.

Figure 5-9

The Intel 4004
microprocessor
containing 2,300
transistors

Current Technology Capabilities and Limitations

Gordon Moore (a founder of Intel) made an observation during a speech in 1965 that has come to be known as **Moore's Law**. Moore observed that the rate of increase in transistor density on microchips had increased at a steady rate — roughly doubling every 18 to 24 months. He further observed that each doubling was achieved with no increase in unit cost. Moore's Law has proven insightful and surprisingly durable. Figure 5-10 shows that the increases in transistor density of the Intel microprocessors have followed Moore's Law, which has become a cornerstone of planning for computer power increases. Software developers and users have come to expect CPU performance to double every couple of years with no price increase.

Increases in transistor density and CPU for Intel microprocessors. These increases follow the trend predicted by Moore's Law.

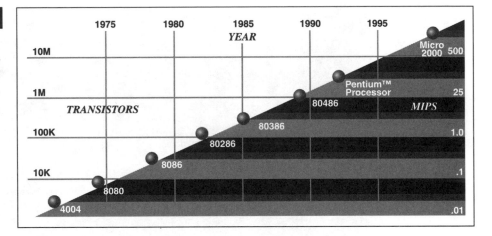

Arthur Rock (a venture capitalist) made a short addendum to Moore's Law. **Rock's Law** states that the cost of fabrication facilities for the latest generation of chips doubles every four years. This law has also proven durable. A fabrication facility using the latest production processes currently costs several billion dollars. Improvements in chip fabrication have been achieved by ever-more-expensive, exotic techniques. The added expense hasn't led to price increases because the demand for chips has increased tremendously, thus providing new sources of revenue to recover the additional fabrication costs.

Rock's Law was based on historical observation but it also has well-founded technological bases. There are limits to both the operational properties of semiconductors and the processes by which they are produced. Costs spiral upward, and problems multiply as these limits are approached. At some point, these costs can be expected to rise at faster rates than predicted by Rock's Law.

Current fabrication technology can squeeze millions of transistors onto a wafer of silicon less than one square centimeter (see Figure 5-11). The process starts with a wafer (a flat disc) of pure silicon. A thin layer of conductive or semiconductive material is spread over the wafer's surface. The material is then exposed to a laser or X-ray focused through a patterned map (similar to an overhead transparency). An etching chemical is then applied, which removes the exposed portion of the layer. The remaining portion is in a very specific pattern (or map) in correspondence to one layer of the microprocessor circuitry or wiring. Additional layers are added and then etched to gradually build up complete transistors and wires attaching them.

Figure 5-11

The Intel Pentium Pro microprocessor containing 5.5 million transistors

The wafer contains hundreds of microprocessors, which are cut apart, encased in individual carriers, and tested. It is not unusual for 70 to 80% of the microprocessors to fail initial testing. Those that survive initial tests are then subjected to further tests at successively higher clock rates. A microprocessor that fails to perform reliably at a higher clock rate (e.g., 300 MHz) is still usable (and rated for sale) at a lower clock rate (e.g., 200 MHz). Testing failures can result from many factors, including impurities in materials, contamination of layers as they are laid down on the wafer, errors in etching, and variations in any number of production-process parameters.

Increases in the number of circuits (packing density) on a chip have been achieved by two means. The first is shrinking the size of each transistor and the wires connecting them. This shrinkage is a result of etching processes that leave increasingly thinner lines of conductive and semiconductive material on the wafer. An etching process is described by the width of these lines measured in fractions of a micron (a micron is one millionth of a meter). Processes as low as 0.18 micron are currently in use. Improvements (reductions in size) in the etching processes are generally achieved every two to three years. Packing density increases are also accomplished by increasing the number of layers added to the wafer. In effect, additional layers put additional components one on top of another on the chip.

Several properties of electricity and electrical conductivity must be considered when increasing packing density. On the plus side, shrinking circuits shortens their length and, thus, increases processing speed. This, in turn, allows the microprocessor to operate at higher clock rates. The first IBM PC used an Intel 8088 microprocessor with a clock rate of 4.77 MHz. Current (1997) IBM-compatible microcomputers use microprocessors with clock rates as high as 300 MHz. Some RISC microprocessors have achieved clock rates above 500 MHz.

Increased packing density must compensate for some potential problems. One of these is susceptibility of the circuits to damage from excess voltage. As the size of the devices and wiring decreases, their ability to transmit electrical power is reduced. This requires a reduction in normal operating voltage (currently as low as two volts down from five volts a few years ago). It also makes the circuits far more susceptible to damage from voltage surges and from static electricity.

Electrical resistance is also a problem for two reasons. First, wires and devices must have uniform resistance properties throughout the chip. This is required due to the low voltages in use. A slight increase in resistance in one part of the chip may prevent sufficient electrical power from reaching another part of the chip. Uniform resistance requires very high and uniform purity of the conductive and semiconductive material used in the chip layers.

The second resistance related problem is heat. Because a chip is a sealed package, there is no way to circulate air (or any other substance) among its circuits. That heat must migrate through the chip carrier to be dissipated by a fan or heat sink. Heat generation per circuit is minimized by their small size and low-resistance materials, but each device and wire does generate heat. Higher clock rates cause even more heat to be generated as more electrons are forced through the circuits. Inadequate or barely adequate cooling can substantially shorten a microprocessor's operating life.

The speed increase in microprocessors to date is entirely due to the use of increasingly smaller fabrication processes. But further shrinkage will be difficult to achieve, partly because of the nature of the etching process. The width of wires and devices constructed by etching depends on the wavelength of the etching beam. Smaller wires require smaller-wavelength beams. But there are limits to the ability to generate and focus short wavelength lasers and X-rays. Another part of the problem lies in the semiconducting materials themselves. The width of chip wiring is rapidly approaching the molecular width of the materials from which the wires are constructed. A wire can't be less than one molecule wide. Thus, the molecular width of the wiring material is a lower bound on process shrinkage.

Various industry experts have estimated that a "wall" will be reached early in the Twenty-first Century. The cost of speed and packing density improvements will soar as the wall is approached. Subsequent improvements (beyond the wall) will require fundamentally different approaches to microprocessor design and fabrication. These approaches may be based on any number of technologies, including optics and quantum physics.

FUTURE TRENDS

Many different technologies for implementing processors—optical processors, electro-optical processors, and devices that operate the atomic and subatomic levels—are currently under study. None of these technologies has yet been proven commercially.

Optical Processing

Because electrons and photons both move at the speed of light, there is no obvious speed advantage to optical communication and computation. The advantages of optics are based on other differences, including energy requirements, susceptibility to external interference, and communication among computational subcomponents. The energy requirements of a computational device are directly related to the mass of its moving parts. Mechanical gears have a large mass compared to electrons; thus, the energy requirements of mechanical computation are much larger than for electrical computation. Similarly, a photon has much less mass than an electron and, thus, optical computation has lower energy requirements.

Moving electrons are subject to many sources of interference. Nearby magnetic fields can interrupt or re-direct their flow. Resistance in wires can sap their energy, thus reducing their ability to perform computational work. Light is far less susceptible to external interference and is thus a more reliable means of data communication. This difference explains much of the reason that voice and data communication is rapidly being converted from electronics to optics.

In an electrical computer, computation is performed by forcing electrons to flow through electrical switches. Early computers had from a few dozen to a few thousand of these electrical switches. Modern computers have millions of switches. For electricity to flow from one switch to another, an electrical conductor (e.g., a wire) must connect the two devices. Many of the switches have multiple interconnections. Thus, the use of millions of switches requires an even higher number of electrical interconnections. The task of reliably fabricating millions of switches and the interconnections among them is daunting. Current fabrication technology requires the investment of billions of dollars in plant and equipment to produce reliable, fast, and affordable electronic computation devices.

Theoretically, the use of light bypasses many of these interconnection and fabrication problems. This is because of several properties of light, including its low power requirements, immunity to many types of interference, and the ability of two light paths to cross without interfering with one another. Although fiber-optic wires are usually used to connect optical devices, it is possible to interconnect them without wires. If the sending and receiving devices are precisely aligned, then signals (light) sent between them can travel through free space. The ability to interconnect processing components without wires promises vast improvements in fabrication costs and reliability.

Commercially available (and affordable) optical computers are at least two decades from introduction. There is no clear leader among current technological approaches for fabricating optical processing circuits or for combining them to build an entire processor or computer. As long as semiconductors have room for improvement, there isn't sufficient economic incentive to vigorously develop optical technologies. However, there is economic incentive to pursue specialized processing applications in the areas of telecommunications and networking. Thus, practical optical processors will first appear as dedicated communication controllers and that technology will later evolve into full-fledged computers.

Electro-Optical Processing

A potential replacement for silicon in microchips is **gallium arsenide**. Because of its molecular structure, it is possible to fabricate much smaller electrical circuits with this substance than with silicon. Thus, chips with larger numbers of circuits can be manufactured within the same size constraints.

Processor Technology and Architecture

Although gallium arsenide is a promising material, it has yet to see wide application outside the realm of supercomputers. It is substantially more difficult to fabricate electrical components with gallium arsenide than with silicon, partly because gallium arsenide is more brittle. In addition, materials scientists and engineers have gained a substantial amount of experience in the production of silicon-based devices over the last 30 years. It will take some time before enough experience accumulates with gallium arsenide to reach the production yields and costs of silicon today.

Gallium arsenide has both electrical and optical properties. That is, it may also be used to implement devices that combine aspects of electrical and optical processing (i.e., opto-electrical devices). For example, gallium arsenide is capable of switching the flow of electricity based on an optical input. In essence, it may be used to implement a transistor that allows electrical current to flow when an optical input is applied. Thus, gallium arsenide devices can form a bridge between optical data communication and electrical processing.

Such transistors hold great promise as the interface between semiconductor processing circuits and optical communication channels. They are already in use for exactly that purpose in some massively parallel supercomputers. They are advantageous in that environment because dozens or hundreds of processors require gigabytes per second of data transfer capacity to keep them supplied with instructions and data. Such rates simply cannot be achieved with conventional electrical buses, but they can be accomplished with fiber-optic buses.

Electro-optical transistors also hold promise as the interface between semiconductor processors and purely optical memory and storage devices. Technologies such as holographic storage have the potential to deliver gigabytes of data in a matter of seconds. Optical communication channels will be required to provide processor access to all of that data.

SUMMARY

A computer system's CPU is composed of a control unit, arithmetic logic unit, and processor registers. A CPU's basic capabilities are defined by its instruction set. Other CPU design considerations include instruction format, word size, register use, bus capabilities, clock rate, memory organization, and memory access methods. The overall capability and power of a computer system depends on the interaction of all these architectural considerations.

The instruction set of a CPU contains data transformation, data movement, and sequence control instructions. The minimal set of data transformation instructions include, AND, OR, exclusive OR, ADD, shift, and rotate instructions. The first four operate on two inputs (operands) to produce a separate result. The latter two instructions modify the content of a single operand.

Data movement instructions include load, store, and move. A load instruction copies the contents of a memory location to a register. A store instruction copies the contents of a register to a memory location. A move instruction copies data from one register to another or from one memory location to another.

Sequence control instructions include unconditional branch (jump), conditional branch, and halt. A jump instruction causes the instruction sequence of the current program to be altered. A conditional branch alters instruction sequence only if a stated condition is satisfied. A halt instruction terminates a program and generally returns control to the operating system.

Extensions to the minimal instruction set are normally made. Some are required to support additional data types (e.g., floating point), some are provided as a programming convenience, and some are provided to reduce processing overhead.

The instruction format of a CPU refers to the size of an instruction and the number and location of parameters within the instruction. Instructions generally consist of an op code (instruction number) and zero, one, or two operands (data inputs). An instruction may be of fixed or variable length. Fixed length instructions contain fixed length components in predetermined positions. Variable length instructions allow instruction components to vary in size and/or position. Variable length instructions economize on memory used to store instructions at the expense of more complex decoding by the control unit.

CPU registers may be either general or special purpose. General purpose registers are used to store intermediate processing results. Their use improves program efficiency by avoiding accesses to memory by the CPU. Within practical limits, a higher number of general purpose registers implies a more powerful CPU.

A reduced instruction set computer (RISC) processor is a CPU with several specific architectural features. These features, which include fixed length instructions and the absence of complex instructions, allow the CPU circuitry to be substantially simplified as compared to a complex instruction set computer (CISC) processor. As a result of this simplicity, these circuits can be smaller and faster. Thus, the clock rate of a RISC processor is generally higher than that of a CISC processor. Although the absence of complex instructions makes some operations slower, the faster clock rate for simple instructions usually offsets this, resulting in a net performance gain for RISC processors.

There are a number of special purpose registers within a CPU, including the instruction register, instruction pointer, program status word, offset register, interrupt register, and stack pointer. The instruction register holds the current instruction (op code and operands) for decoding by the control unit. The instruction pointer holds the memory address of the next instruction to be fetched. Altering the content of the instruction pointer is the mechanism by which branch and halt instructions are implemented. The program status word is used to hold status and error codes.

The word size of a CPU is the number of bits processed by the CPU during a single execution cycle. It is also the size of processor registers and is usually equivalent to the number of bus data lines. Matching word size to the number of bus data lines ensures that the CPU can be efficiently supplied with instructions and data. Within limits, larger word size implies greater processing power and efficiency.

Modern CPUs are implemented as microprocessors using semiconductors. Millions of transistors and wires are fabricated on a single chip. The technology of these chips represents a careful balance of materials and fabrication technology within the limits of electrical conduction. Performance increases have been achieved by rapid miniaturization of microprocessor components, but the limits of this process are in sight. The next CPU technology will probably be based on optical technology, possibly in combination with improved semiconducting technology.

Key Terms

ADD
AND
arithmetic shift
binary compatibility
branch
clock cycle
clock rate
complex instruction set
    computing (CISC)
complex instructions
conditional branch
conductivity
conductor
cycle time
decoded
exclusive OR (XOR)
fixed length instructions
gallium arsenide
gates
halt
heat sink

hertz
inclusive OR
instruction
instruction format
instruction pointer
instruction register
instruction set
integrated circuit (IC)
jump
load
logic gates
logical shift
megahertz
memory mapped I/O
microchip
microprocessor
Moore's Law
move
NAND gate
NOT
op code

operands
OR
program counter
program status word (PSW)
reduced instruction set
    computing (RISC)
resistance
Rock's Law
scale integration
semiconductor
SHIFT
store
system clock
transistor
unconditional branch
variable length instruction
wait state
wire
word

1. The _____ time of a processor is 1 divided by the clock rate (in Hz).

2. A CPU typically implements multiple _____ to account for differences among instructions in the number and type of operands.

3. _____ generates heat within electrical devices.

4. _____ is a semiconducting material with optical properties.

5. A(n) _____ is an electrical switch built of semiconducting materials.

6. A left _____ instruction multiplies a numeric value by two.

7. A(n) _____ improves heat dissipation by providing a thermal mass and a large thermal transfer surface.

8. One _____ is one cycle per second.

9. Applying a(n) _____ OR transformation to input bit values 1 and 1 generates True. Applying a(n) _____ OR transformation to those same inputs generates False.

10. When first fetched from memory, an instruction is placed in the _____ and then _____ to extract its components.

11. The use of _____ instructions simplifies the process of fetching and _____ those instructions.

12. A(n) _____ is an electrical circuit that implements a boolean or other primitive processing function on single-bit inputs.

13. A microchip containing all of the components of a CPU is called a(n) _____

14. In a(n) _____ operation, bit pairs of 1/1, 1/0, and 0/1 produce a result bit of 1.

15. The address of the next instruction to be fetched by the CPU is held in the _____.

16. The contents of a memory location are copied to a register while performing a(n) _____ operation.

17. A(n) _____ or _____ contains multiple transistors or gates in a single sealed package.

18. A(n) _____ instruction alters the sequence of instruction execution. A(n) _____ instruction alters the sequence of instruction execution only if a specified condition is True.

19. A(n) _____ processor does not directly implement complex instructions.

20. The instruction cycle and execution cycles are a fraction of a processor's _____.

21. A(n) _____ operation copies data from one memory location to another.

22. The CPU incurs one or more _____ when it is idle pending the completion of an operation by another device within the computer system.

23. A(n) _____ is the number of bits processed by the CPU in a single instruction. It also describes the size of a single register.

24. In many CPUs, a register called the _____ stores condition codes, including those representing processing errors and the results of comparison operations.

25. The components of an instruction are its _____ and one or more _____.

26. Two 1-bit values generate a 1 result value when the _____ operation is applied. All other input pairs generate a 0 result value.

27. A(n) _____ CPU typically uses variable length instructions and has a large instruction set.

28. A(n) _____ operation transforms a 0 bit value to 1 and a 1 bit value to 0.

29. _____ predicts that transistor density and processor power will double every two years or less.

1. Describe the operation of the load, move, and store instructions. Why is the name "move" a misnomer?

2. What computations can be implemented using an arithmetic shift operation?

3. Why does the execution speed of an application program generally increase as the number of general purpose registers increases?

4. What are special purpose registers? Give three examples of special purpose registers and explain how each is used.

5. What are the advantages and disadvantages of fixed length instructions as compared to variable length instructions? Which type are generally used in a RISC processor? Which type are generally used in a CISC processor?

6. Define the term **word size**. What are the advantages and disadvantages of increasing word size?

7. What characteristics of the CPU and of primary storage should be "balanced" to obtain maximum system performance?

8. How does a RISC processor differ from a CISC processor? Is one processor type better than the other? Why or why not?

9. What factor(s) account for the dramatic improvements in microprocessor clock rates since their invention in the early 1970s?

10. What potential advantages are provided by optical processors as compared to electrical processors?

1. Develop a program consisting of primitive CPU instructions to implement the following procedure:

   ```
   Integer i, a;

   i=0;
   while (i<10) do
       a=i*2;
       i=i+1;
   endwhile
   ```

2. Assume that a microprocessor has a cycle time of four nanoseconds. What is the processor clock rate? Assume further that the instruction cycle is 40% of the processor cycle time. What memory access speed is required to implement load operations with zero wait states? What memory access speed is required to implement load operations with two wait states?

3. Processor R is a RISC processor with a 400-MHz clock rate. The average instruction requires two cycles to complete (assuming zero wait state memory accesses). Processor C is a CISC processor with a 250-MHz clock rate. The average simple instruction requires two cycles to complete (assuming zero wait state memory accesses). The average complex instruction requires four cycles to complete (assuming zero wait state memory accesses). Processor R cannot directly implement the complex processing instructions of Processor C. Executing an equivalent set of simple instructions requires an average of ten cycles to complete (assuming zero wait state memory accesses).

   Program S executes 90% simple instructions and 10% complex instructions. Program C executes 70% simple instructions and 30% complex instructions. Which processor will execute program S more quickly? Which processor will execute program C more quickly? At what percentage of complex instructions will the performance of the two processors be equal?

4. Assume the existence of a CPU with a 400-MHz clock rate. The instruction and execution cycles are each 50% of the clock cycle. The average instruction requires two nanoseconds to complete execution. Main memory access speed for a single instruction is eight nanoseconds on average. What is the expected average MIPS rate for this CPU?

1. Investigate the instruction set and architectural features of a modern RISC processor such as the Digital Equipment Corporation (www.digital.com) Alpha or Motorola/IBM (www.mot.com) PowerPC. In what ways does it differ from the architecture of the Intel Pentium processor family?

2. Several companies, including Cyrix (www.cyrix.com) and AMD (www.amd.com), produce microprocessors that execute CPU instructions designed for the Intel Pentium processor family. Investigate the current product offerings of Cyrix and AMD. Do they offer "true" Pentium compatibility? Is their performance comparable to that of Intel processors?

Chapter 6

Data Storage Technology

Chapter Goals

▶ Describe the characteristics of primary and secondary storage.

▶ Describe the devices used to implement primary storage.

▶ Describe the mechanisms by which the CPU accesses primary storage.

▶ Describe magnetic devices used to implement secondary storage.

▶ Describe optical devices used to implement secondary storage.

STORAGE ALTERNATIVES AND TRADE-OFFS

A typical computer system uses multiple storage devices (see Figure 6-1), which are needed to support the immediate execution of programs as well as provide long-term storage to support future program execution. **Primary storage** is the collection of devices used to support immediate storage needs. **Secondary storage** is the collection of devices used to support long-term storage needs.

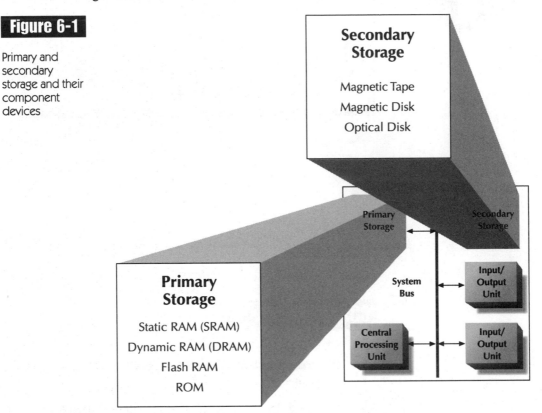

Figure 6-1

Primary and secondary storage and their component devices

A storage device consists of a **read/write mechanism** and a **storage medium** (or multiple storage media). Generally, a **device controller** provides an interface between the storage device and the system bus. (Device controllers are discussed in detail in Chapter 7.) In some devices, the read/write mechanism and storage medium are implemented as a single unit that uses similar technology for both storage and access. For example, some types of random access memory (RAM) use transistors to implement both the read/write mechanism and the storage medium. In other devices, the read/write mechanism and the storage medium are distinct components with fundamentally different implementations. For example, tape storage implements the read/write mechanism as an electromechanical device and uses a storage medium composed of polymers and metal oxides.

In some devices (e.g., tape and diskette), the storage medium can be removed from the device (i.e., a **removable media** storage device), which allows physical transport of data and programs and the creation of backup copies. It also permits multiple media to be used with a single read/write mechanism, thus effectively increasing the device's storage capacity. Storage media that are removed from a storage device are said to be **off-line**. Storage media that are contained within a storage device and ready for immediate use are said to be **online**.

Primary and secondary storage represent trade-offs among storage-device characteristics. Different devices are appropriate for different storage purposes because no single storage device is "best" in all characteristics. The primary characteristics that distinguish storage devices are:

► Speed.
► Volatility.
► Access method.
► Portability.
► Cost and capacity.

Each of these characteristics is discussed in detail next.

Speed

Speed is the most important differentiating characteristic between primary and secondary storage. Primary storage extends the limited capacity of CPU registers. The CPU continually moves data and instructions between registers and primary storage. CPU instruction and execution cycles are extremely short (generally less than 10 nanoseconds). If primary storage is implemented with devices that cannot be read or written within a single CPU cycle, then the CPU must wait for the read/write operation to complete. Any CPU cycle spent waiting for a storage device to complete a read/write operation is called a **wait state**.

Wait states reduce the performance of the CPU and the entire computer system. For example, if every fetch operation incurs one wait state, then half of the CPU's available execution cycles are not used. This effectively reduces CPU performance by 50% (e.g., a 300-MHz processor will run at an effective rate of 150 MHz). To avoid this, all or part of primary storage is implemented with the fastest available storage devices. With current technology primary storage speed is typically greater than secondary storage speed by a factor of 10^5 or more (tens of nanoseconds versus hundreds of milliseconds).

Speed is also an important issue for secondary storage. Many information system applications require access to large databases to support ongoing processing (e.g., a customer order query system). Program response time in such systems directly depends on the speed of secondary storage access. Secondary storage access speed also affects response time in other ways. For example, starting program execution generally requires reading the corresponding executable code from secondary storage and copying it to primary storage. Thus, the time delay between a user request for program execution and the first prompt for user input directly depends on the speed of both primary and secondary storage.

The speed of a storage device generally is expressed in terms of its **access time**: that is, the time required by the device to completely execute one read or write operation. Access time is assumed to be the same for both reading and writing unless otherwise stated. For some storage devices, the nature of access to specific locations is such that access time varies by location. For such devices (e.g., disks), access time is expressed as an average (i.e., **average access time**). Access times of primary storage devices are expressed in nanoseconds (billionths of a second), whereas access times for secondary storage devices are expressed in milliseconds (thousandths of a second).

By itself, access time is an incomplete measure of data access speed. The unit of data transfer to/from the storage device also must be specified to provide a complete picture of access speed. Various units of data transfer can be stated, and these units vary from one storage device to another. The unit of data transfer for primary storage devices is usually a word. Depending on the CPU, a word can represent 2, 4, 8, or more bytes (4 and 8 bytes are the most common word sizes). Data transfer unit size is sometimes used as an adjective as in the term "32-bit memory."

Secondary storage devices read or write data in units larger than words. The word **block** is a generic term for such units. The size of a block is normally stated in bytes. Unfortunately, block size varies widely from one storage device to another and can vary even for the same storage device (e.g., tape backup units). Thus the term block alone has little significance and requires a specific definition of **block size**.

The term **sector** is frequently used to describe the unit of data transfer to/from a magnetic or optical disk. As with blocks, sector size is stated in bytes and can vary from one device to another. A 512-byte block size is a widely used (but not universal) standard unit of data transfer for magnetic disks.

A storage device's **data transfer rate** is computed by dividing one by the access time (expressed in seconds) and multiplying the result by the unit of data transfer (expressed in bytes). Thus, for example, the data transfer rate for a primary storage device with 60-nanosecond access time and a 32-bit word data transfer unit can be computed as:

$$\frac{1 \text{ second}}{60 \text{ nanoseconds}} \times 32 \text{ bits} = \frac{1}{0.00000006} \times 4 \text{ bytes} = 66,666,667 \text{ bytes/second}$$

The same formula can be used to compute an average data transfer rate using an average access time as the input variable.

Volatility

A storage device is said to be **nonvolatile** if its content can be held stable over long periods. A storage device is said to be **volatile** if its content cannot be held for long periods of time. Primary storage devices are generally volatile, whereas secondary storage devices are generally nonvolatile.

The terms volatile and nonvolatile imply absolute permanence or lack thereof. In fact, though, volatility is a matter of degree and stated conditions. RAM, for example, is nonvolatile as long as external power is continually supplied, but it is generally considered a volatile storage device because continuous power cannot be guaranteed under normal operating conditions. Magnetic tape and disk generally are considered to be a nonvolatile storage media, but data held on magnetic media is often lost after a few years because of natural magnetic decay and other factors.

Access Method

The physical structure of storage devices and media determine the ways in which data can be accessed. There are three broad classes of recognized access:

▶ Serial access.
▶ Random access.
▶ Parallel access.

Often, these methods are combined within a single storage device or group of storage devices.

Serial Access

A **serial access** storage device stores and retrieves data items in a linear (sequential) order. Magnetic tape is the only widely used form of serial access storage. Data is written to a linear storage medium in a specific order. Once written, data can be read back only in that same order. For example, to play the last song on a cassette or DAT audio tape requires playing (or fast-forwarding past) all of the songs that precede it. Computer data storage on tapes has the same access characteristics. Thus, if there are n data items stored on the device, then access to the nth item is accomplished by first accessing the preceding $n-1$ items in linear order.

Access time for a serial access device is not a constant; rather, it depends on the current storage position and on the position of the data item to be accessed. If both positions are known, then access time can be computed as the difference between the current and target positions multiplied by the amount of time required to move from one position to the next.

Serial access devices are not used for frequently accessed data because of the inherent inefficiency of their access method. Even when data is accessed in the same order it was written, serial access devices generally are much slower than other forms of storage. Thus, the primary role for serial access storage is as a backup to other forms of secondary storage.

Random Access

Random access devices are not restricted to any specific order when accessing data; rather, they can directly access any desired data item stored on the device. Random access can also be called **direct access**. All primary storage devices and all forms of disk storage are random access devices.

Access time for a random access storage device may or may not be a constant. It is a constant for most primary storage devices. It is not a constant for disk storage due to the physical (spatial) relationship among the read/write mechanism, storage media, and locations on the media in which specific data items are stored. This issue is fully discussed in the "Magnetic Disk" section.

Parallel Access

A **parallel access** device can simultaneously access multiple storage locations. Although RAM is generally considered a random access storage device (random access is part of its name), it is also a parallel access device. The confusion arises from differences in the definition of the unit of data access. If the unit of data access is considered to be a bit, then access is parallel. That is, the construction of RAM circuitry is such that individual bits are accessed simultaneously. Individual bytes within a word also can be accessed in parallel.

Parallel access can be implemented by subdividing data items and storing the component pieces on multiple storage media. If these multiple storage media can be read and written simultaneously, parallel access exists. Generally, parallel access implies the use of multiple read/write mechanisms, one for each storage medium or device. In turn, this suggests an increased device cost as compared to devices incapable of parallel access.

Data Storage Technology

Portability

Secondary storage is portable if the storage medium (or the entire storage device) can be removed from one computer system and installed on another. For example, compact disc and tape storage are considered portable because the storage media are removable and easily transported. External disk drives are portable if they can be added or deleted easily from a computer system and if they can be easily transported without damaging the device or its data content.

Access speed is often sacrificed when implementing removable storage media. High-speed access in many storage devices requires tight control of environmental factors. For example, low access times for magnetic disks are achieved, in part, by sealed enclosures for the storage media (thus minimizing or eliminating dust and air-density variations). The very nature of removable media places the media contact with a larger variety of environmental factors than are encountered within a sealed device. Thus, access speed typically is reduced compared to nonremovable (or fixed) storage media.

Cost and Capacity

A storage device's speed, volatility, access method, and portability are all directly related to its cost. An increase in speed, permanence, or portability generally comes at increased cost, if all other factors are constant. Cost also increases as access method moves from serial to random to parallel. Primary storage is expensive compared to secondary storage because of its high speed and combination of parallel and random access. Nonvolatile primary storage devices are more expensive than volatile primary storage devices.

Often, capacity is considered to be a distinguishing characteristic among primary and secondary storage devices, but considering capacity in this manner is misleading. Capacity, in fact, represents a compromise between cost and other device characteristics. If cost were not a consideration, then most users would opt for nonvolatile RAM instead of disk drives to implement secondary storage. But few users can afford gigabytes of nonvolatile RAM. The widespread use and typically high capacities of disk storage reflect its relatively low cost and not an inherent storage-capacity advantage.

Memory-Storage Hierarchy

A typical computer system utilizes a variety of primary and secondary storage devices. A small amount of high speed RAM is usually used within the CPU. Slower RAM is usually used to implement the bulk of primary storage and multiple secondary storage devices are generally used. One or more magnetic disk drives are typically complemented by an optical disk drive and at least one form of removable magnetic storage.

The range of storage devices used within a single computer system forms a memory-storage hierarchy, as illustrated in Figure 6-2. Cost and access speed generally decrease further down the hierarchy. Because of lower cost, capacity tends to increase as one moves down the hierarchy. A computer system designer or purchaser attempts to find an optimal mix of cost and performance for a particular purpose.

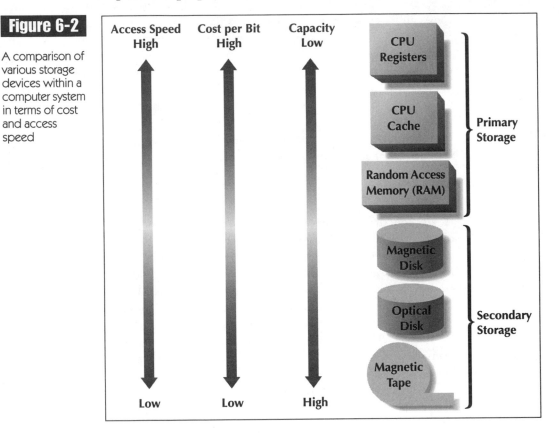

Figure 6-2

A comparison of various storage devices within a computer system in terms of cost and access speed

Hardware and software technology can be used to increase performance by judicious substitution of one form of storage for another. The exact mechanisms to accomplish this are varied and complex. For example, memory capacity can be increased by supplementing primary storage with secondary storage using a technique known as virtual memory management. Memory access speed can be improved by using relatively small amounts of very high-speed RAM using a technique known as caching. A similar technique can be used to improve access speed to disk storage devices. These and other performance-enhancing storage substitutions are discussed in later chapters.

PRIMARY STORAGE DEVICES

The critical performance characteristics of primary storage devices are their access speed and the number of bits that can be accessed in a single read or write operation. Because they must supply data and instructions to the CPU, primary storage devices must closely match both its speed and word size. Thus, the evolution of memory devices has closely mirrored the evolution of processing devices. Technologies applied to the construction of processors have been applied simultaneously to the construction of primary storage devices.

Storing Electrical Signals

As discussed in Chapters 4 and 5, the internal representation of data within the CPU is based on digital electrical signals. Because this is also the basis of data transmission among all devices attached to the system bus, any storage device must accept electrical signals as input (i.e., for a write operation) and generate electrical signals as output (i.e., for a read operation).

Electrical power can be stored directly by several different electrical components including batteries and capacitors. Unfortunately, there is a trade-off between the speed at which electrical current can be stored/retrieved and the speed of the storage/retrieval process. Batteries are slow to accept and/or regenerate an electrical current, and they lose their ability to accept charge with repeated use. Capacitors can charge or discharge with much greater speed than batteries. However, capacitors lose their charge at a fairly rapid rate and, thus, require a fresh injection of electrical current at regular intervals (e.g., hundreds or thousands of times per second).

Electrical power can be indirectly stored by many methods, all of which use electrical current to generate some other energy form. This alternate energy form is then stored directly or indirectly (i.e., a write operation). A read operation is performed by accessing the stored energy and using it to regenerate the original electrical signal. For example, a magnetic field can be generated with an electrical current. The strength of the magnetic field can induce a permanent magnetic charge in a nearby metallic compound. The stored magnetic charge then can be used to generate an electrical signal equivalent to the one used to create the original magnetic charge. Magnetic polarity (either positive or negative) can be used to represent the values zero and one.

Early computers used this method to store bits of data in rings of ferrous material (iron and iron compounds). These rings were embedded in a two-dimensional mesh of wires. The rings were called cores and the technology is referred to as **core memory**. Signals sent along the wires induced magnetic charges in the metallic rings. The polarity of the charge depends on the direction in which current flows through the wires. Thus, the value of a bit is set depending on the direction of the current flow.

Modern computers use memory implemented with semiconductors. Thus, memory devices are constructed and fabricated using technology similar to that of microprocessors. Basic types of memory that are built from semiconductor microchips include random access memory (RAM) and read only memory (ROM). There are many variations on the implementation of each memory type.

Random Access Memory

Random access memory (RAM) is a generic term describing primary storage devices with the following characteristics:

▶ Microchip implementation using semiconductors.

▶ Ability to read and write with equal speed.

▶ Direct access to stored bytes, words, or larger data units.

The semiconductor implementation of RAM is similar to that of microprocessors. Electrical components for bit storage are combined with the circuitry needed to read and write individual bits in parallel. Large quantities of bits can be stored on a single chip.

It might be assumed that microprocessor clock rates are well matched to RAM access speeds, given their similar technical implementations, but unfortunately, this is not the case. The process of reading and writing many bits in parallel is complex and requires a large number of electrical circuits. Additional complexity is added by the need to allow multiple devices to access memory over multiple communication channels (e.g., direct access by a video display controller). The implementation of RAM as a separate component (microchip or set of microchips) also slows the transfer of data to/from the CPU. Thus, RAM cannot match its access speed to the clock rate of a typical microprocessor unless both devices are implemented within the same microchip.

There are two primary types of RAM—static RAM and dynamic RAM—and several variations on each. **Static RAM (SRAM)**, implemented entirely with transistors, uses a flip-flop circuit (see Figure 6-3) as the basic unit of storage. Each flip-flop circuit requires at least two transistors to store a single bit and two or four additional transistors to allow that bit to respond to read and write commands. The transistors of the flip-flop circuit form an electrical switch that "remembers" its last position. One position represents zero, and the other represents one. Unlike mechanical switches, flip-flop circuits require a continuous supply of electrical power to maintain their "positions." Thus, SRAM is volatile unless a continuous supply of power can be guaranteed.

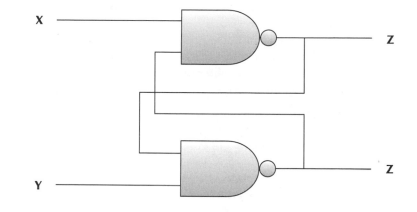

Figure 6-3

A flip-flop circuit composed of two NAND gates: the basic component of static RAM and CPU registers

Dynamic RAM (DRAM) is implemented using transistors and capacitors. (The capacitors are considered to be the "dynamic" element.) The required circuitry is less complex than that of SRAM; however, because capacitors quickly lose their charge, they require a fresh infusion of power thousands of times per second. The process of supplying this power is called a **refresh operation**. DRAM chips include the circuitry to perform refresh operations and perform that operation automatically. Each refresh operation is called a **refresh cycle**. Unfortunately, a DRAM chip cannot perform a refresh operation simultaneously with a read or write operation.

The more complex circuitry of SRAM is more expensive to implement than the simpler circuitry of DRAM: The current cost difference is approximately 10:1. Although simpler circuitry usually translates to faster speed, DRAM is slower than SRAM because of its required refresh cycles. Thus, it is very difficult to implement DRAM with less than 50-nanosecond access time using current technology. Currently, SRAM is implemented with access time as low as 10 nanoseconds. There is an approximately 5:1 difference in access speed between DRAM and SRAM.

Neither type of RAM can match current microprocessor clock rates. Current clock rates range from 150 MHz to 600 MHz. These clock rates translate to zero wait state access times for primary storage as follows:

$$\frac{1}{150 \text{ MHz}} = \frac{1}{150,000,000} = 0.00000000667 = 6.67 \text{ nanoseconds}$$

$$\frac{1}{600 \text{ MHz}} = \frac{1}{600,000,000} = 0.00000000167 = 1.67 \text{ nanoseconds}$$

The fastest DRAM is approximately 10 to 50 times slower than modern microprocessors, and SRAM is 2 to 10 times slower. In both cases, the performance gap can be improved if the RAM circuitry is implemented on the same microchip as the CPU. For SRAM, this generally eliminates the performance gap. Processor registers and on-processor memory caches are, in fact, implemented using flip-flop circuits.

The performance gap among microprocessors and semiconductor memories has led to a proliferation of technologies designed to minimize the gap. Several design approaches have been taken, including:

► Read-ahead memory access.

► Synchronous read operations.

► On-chip caching.

Fast page mode (FPM) and **extended data out (EDO)** DRAM are two types of read-ahead DRAM. When FPM DRAM receives a read request, it satisfies that request and also activates a portion of the circuitry necessary to read the next sequential word of data. If the next read request is for the next word, the request can be satisfied more quickly. EDO DRAM goes a step further by accessing the next word and storing it temporarily near the output circuits. The term **extended** is used because the availability of the prefetched data is extended by the use of temporary storage. FPM was the standard implementation for DRAM during the late 1980s and early 1990s. EDO supplanted FPM as the standard DRAM implementation during the mid-1990s.

Synchronous DRAM is a read-ahead RAM that uses the same clock pulse as the CPU. It starts a memory fetch with each clock cycle and assumes a sequential series of memory accesses. Sequential read access is improved because processing of read requests generally starts—and sometimes is completed—before the CPU requests the next word of data.

Cached DRAM, which sometimes is called enhanced DRAM or EDRAM, uses a small amount of SRAM (e.g., 256 bytes) on each DRAM device. When a read request is received for a word, that word and several words that follow are retrieved. The "extra" words are stored in the SRAM cache in anticipation of future sequential read requests. If those requests are made, the needed data is already waiting in the SRAM cache and can be more quickly retrieved than from the original DRAM location. Other uses of caching are discussed in Chapter 7.

A relatively new RAM technology is **ferroelectric RAM**. As its name implies, ferroelectric RAM uses iron to store individual bits of data in much the same manner as old-fashioned core memory. The read/write can be implemented in the same manner as conventional DRAM or SRAM. Ferroelectric RAM provides no access speed improvements over other technologies. Like core memory, however, its content is not lost during power interruptions. Thus, it seems to be a technology capable of truly bridging the gap between primary and secondary storage devices.

Read Only Memory

Microchips within which data are embedded permanently or semi-permanently are called **read only memory (ROM)**, which means ROM is a type of non-volatile memory. A common application of ROM chips is to store programs such as computer system boot subroutines (i.e., system BIOS). These instructions can be loaded at high speed into main memory from ROM for execution by the processor. Instructions that reside in ROM are called **firmware**.

There are several different types of ROMs, and ROM technology has changed greatly over the last two decades. The earliest ROMs were manufactured with bits implemented as circuits that always generated the same bit value. In essence, the program or data stored in the ROM was embedded within the design of the chip. Because it is expensive to generate new chip designs, this method was cost-effective only if tens of thousands of identical chips were needed.

A more cost-effective approach was the **programmable read only memory (PROM)**, a type of chip manufactured as a "blank slate" containing all zeros. The circuitry is designed so that the 0 bit values can be "burned out," thus enabling another set of circuitry representing a 1. The bit-writing process requires repeated applications of relatively high voltage. The process is destructive and, therefore, irreversible. PROM technology allowed millions of identical chips to be produced, thus considerably lowering manufacturing cost. The trade-off was the relatively long time required to write the content of an entire chip.

An **erasable programmable read only (EPROM)** memory chip is functionally identical to a PROM but uses a nondestructive and reversible means of writing bits. On an EPROM microchip, a tiny glass window is mounted above the chip on the outer packaging. Through this window, the chip can be exposed to a strong ultraviolet light that, after a period of minutes, erases data content. Once erased, the EPROM can be reprogrammed in another writing operation. After this step, the glass window is covered with a paper label, or sticker, to prevent stray light from striking the chip and altering the data. EPROM technology allowed reuse of chips. However, production costs of all forms of ROM eventually became very low (i.e., pennies or less per chip). Thus, it became cheaper to buy new blank chips than to reprocess used chips.

The next step in ROM evolution was the **electronically erasable programmable read only memory (EEPROM)**. EEPROMs can be programmed, erased, and reprogrammed by signals sent from an external control source (e.g., a CPU). As with PROM and EPROM technology, the process of writing to the chips is slow and requires relatively high voltages.

A newer form of EEPROM called **flash memory** (sometimes erroneously called **flash RAM**) requires much less time to complete a write operation (although still much longer than a read operation). It is competitive with DRAM in storage density (capacity) and read performance. Unfortunately, flash memory tends to wear out after 100,000 or more write operations and thus cannot be considered a direct competitor to RAM for implementing primary storage. Flash memory is used for large blocks of data that aren't updated frequently (e.g., computer system BIOS) and in some consumer products such as digital cameras.

Memory Packaging

The packaging of memory circuits has closely paralleled the packaging of processor circuits in CPUs. In recent years, memory packaging technology has surpassed microprocessor technology because of the need for tens or hundreds of megabytes of memory in modern computer systems. Minimizing access time and increasing total capacity of primary storage requires that memory circuits be packaged densely in small components that can be located as close to the CPU as possible.

Early microchip implementations of RAM and ROM used dual in-line packages (DIPs) as a memory package (see Figure 6-4). Improvements in fabrication and miniaturization have allowed increased numbers of bits to be implemented within each chip. Current generations of memory chips pack as much as 64 MB into a single DIP.

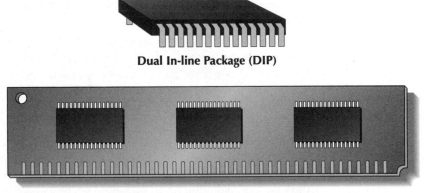

Figure 6-4

A dual in-line package (DIP) chip and a single in-line memory module (SIMM). Individual DIP chips are mounted within a SIMM to create a more modular and compact primary storage device.

Dual In-line Package (DIP)

Single In-line Memory Module (SIMM)

The installation of DIPs in a printed circuit board is a tedious, precise operation. In addition, single DIPs mounted on the board surface occupy a relatively large percentage of the total surface area. To address both problems, memory manufacturers adopted the **single in-line memory module (SIMM)** as a standard RAM package (see Figure 6-4). Each SIMM incorporates multiple DIP memory chips on a tiny printed circuit board. The edge of the circuit board has a row of electrical contacts, and the entire package is designed to lock into a SIMM slot on a computer motherboard. The **double in-line memory module (DIMM)**, a newer packaging standard, is essentially a double-sided SIMM with memory DIPs and electrical contacts on both sides of the module.

The trend in RAM packaging is toward ever-smaller and denser packages, which shorten circuitry pathways within the memory devices. However, relatively long inter-chip and inter-module connections are still required to connect memory to the CPU. These relatively long connections have become the predominant hindrance to faster memory access speeds.

Current microprocessors implement a small amount of primary storage within the chip. As fabrication improvements allow higher-circuit densities, on-chip memory capacity of microprocessors is expected to grow. The logical extension of this trend is the implementation of a full CPU and all of its primary storage (as SRAM) on a single chip. This would eliminate the gap that currently exists between microprocessor clock rates and memory access speed.

CPU MEMORY ACCESS

Although main memory is not currently implemented as a part of the CPU, the need for the CPU to load instructions and data from memory and to store processing results requires a close coordination between both devices. Specifically, the physical organization of memory, the organization of programs and data within memory, and the method(s) of referencing specific memory locations are critical design issues for both primary storage devices and processors.

Physical Memory Organization

The main memory of any computer can be regarded as a sequence of contiguous, or adjacent, memory cells, as shown in Figure 6-5. Addresses of these memory locations are assigned sequentially so that available addresses proceed from zero (low memory) to the maximum available address (high memory). By convention, memory capacity is expressed in units of either 1,024 bytes (1 kilobyte, Kbyte, or KB) or $1,024^2$ bytes (1 megabyte, Mbyte, or MB). Thus, a microcomputer with 64 MB of main memory has 67,108,864 ($64 \times 1,024^2$) storage locations in memory.

Figure 6-5

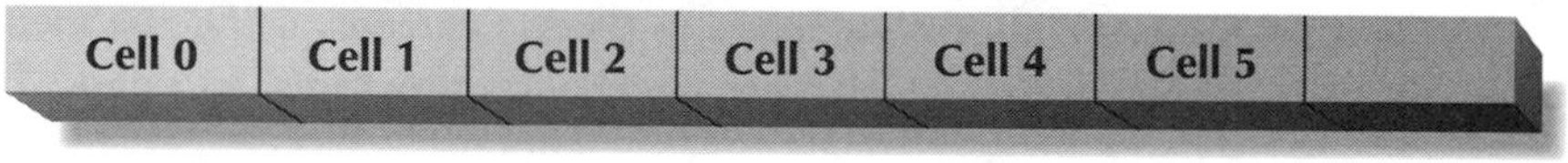

| Cell 0 | Cell 1 | Cell 2 | Cell 3 | Cell 4 | Cell 5 | |

A depiction of the sequential physical organization of memory cells

Data values and instructions generally require multiple bytes of storage. For example, an integer typically requires 4 consecutive bytes (32 bits) of storage. When written in a program, the bits (and bytes) of a numeric value typically are ordered from highest-position weight on the left to lowest-position weight on the right. Thus, if a 32-bit integer is written as:

00110010 11010011 01001100 01101001

it is assumed that the left position weight is the largest (2^{31}) and the right position is the smallest (2^0). When considered as a sequence of bytes, the leftmost byte is called the **most significant byte**, and the rightmost byte is called the **least significant byte**.

In many CPU and memory architectures, the least significant byte of a data item is placed at the lower memory address. For example, the 32-bit word $02468ACE_{16}$ would be stored in memory with CE at the lowest memory address, 8A at the next address, 46 at the next address, and 02 at the highest address. The term **endian** is used to describe storage sequence. **Big endian** storage describes the storage of the most significant data byte at the lowest memory address. **Little endian** storage describes the storage of the least significant byte at the lowest memory address. Little endian storage is sometimes called "back words" storage because the value is stored in the reverse of the order that a person would normally expect.

The number of bits used to define memory addresses determines the maximum number of locations that can be referenced. This potential capacity is called the **addressable memory** of the processor. For example, in a computer that uses 24-bit addresses, there are 16 MB of addressable memory. By contrast, the computer's **physical memory** corresponds with the number of memory cells that exist in a given hardware configuration. Thus, a computer's physical memory capacity can be less than or equal to the size of its addressable memory. Addressable memory limits should be the same for both the CPU and the system bus; that is, the number of bits used to represent an address within the CPU should correspond to the number of lines used to carry an address value on the system bus.

Memory Allocation and Addressing

Memory allocation refers to the assignment of specific memory addresses to elements of system software, application programs, and data. The memory-allocation scheme for a relatively simple computer is shown in Figure 6-6. Such a computer supports a single user running a single application program. For visual clarity, the total available memory space is usually represented as an array of addresses from bottom to top, as shown, rather than as a long (left to right) continuous string of memory cells. The lowest memory addresses are allocated to system software.

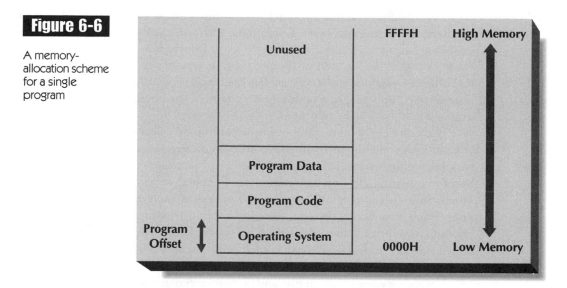

Figure 6-6

A memory-allocation scheme for a single program

Application programs are written as though their instructions started at memory address zero. This approach assumes that the program's instructions will be loaded into memory in sequence, starting at the lowest memory location and proceeding up through higher addresses, until the entire program has been loaded. Loading is performed from a secondary storage device either locally or through a network connection. The utility program that performs this function is called a **loader**.

Under the memory allocation shown in Figure 6-6, however, the application program cannot be loaded starting at address zero; rather, the next available memory cell follows the highest address occupied by system software. The number of memory cells that must be skipped before the program can be loaded is the **offset** by which the program's starting address must be adjusted. Under the memory allocation scheme shown in the diagram, calculating the correct addresses for program instructions is relatively straightforward. The offset must be added to any memory addresses explicitly stated within the program (e.g., addresses used in load, store, and branch instructions).

One way of implementing an offset is to account for it when a program is created. Although program compilation proceeds under the assumption that the program will start at address zero, the instruction addresses could be changed after compilation and before the program is loaded. This process, called **relocating** a program, is performed by a system software program called a **relocator**. Essentially, the relocator searches through the program looking for any explicitly stated memory addresses and adds the offset value to each such address. During program execution, the CPU is said to be using **absolute addressing** because the memory references within the program refer to actual physical memory locations.

The advantage to this approach is that the address computations (addition of offset) are performed once. Computing addresses at the time the program is executed can add complexity and inefficiency to program execution. The disadvantage is that the location of the program in memory is still fixed. Recall that the offset is essentially the size of the operating system, and, should that size change (e.g., updating to a new, larger version of the operating system), all programs would have to be relocated.

The more common method of implementing an offset is to perform address calculations while the program is running. Typically, these calculations are performed within the control unit because it is responsible for memory accesses and for updating the instruction pointer. To speed execution, the offset can be stored in a special purpose register and added to each explicit memory reference. Such a method is referred to as **indirect addressing**, and the register that contains the offset value is called an **offset register**. This method also can be called **relative addressing** because each program address is determined relative to the content of the offset register.

Microprocessors that implement a segmented memory model often have multiple offset registers, and each offset register is intended to point to a different memory segment. The offset registers are called **segment registers**. The use of multiple offset registers is advantageous when the size of memory segments is relatively small. For example, the Intel 8088 (used in the original IBM PC) used 64-Kbyte segments and four segment registers. Thus, a program could address up to 256 Kbytes of memory at any one time by storing the offset value for a different segment in each segment register (see Figure 6-7). Later generations of Intel microprocessors have implemented a flat memory model but also have maintained the segmented model (and the four segment registers) for backward compatibility.

Memory addressing and memory-allocation schemes become much more complex in architectures that permit multiple programs to be held in memory. Figure 6-8 shows system software and two programs stored in memory simultaneously. Note that the starting address of each program is different; thus, the offset value for each program is also different (3000H for program 1 and 9000H for program 2).

Figure 6-7

The segmented memory architecture of the Intel 8086 microprocessor. Primary storage is divided into 16 segments of 64 Kbytes each. Up to four segment registers (named CS, DS, SS, and ES) can be used by a program. A reference to a specific memory address consists of a segment register and a 16-bit offset value (e.g.,CS:1F40)

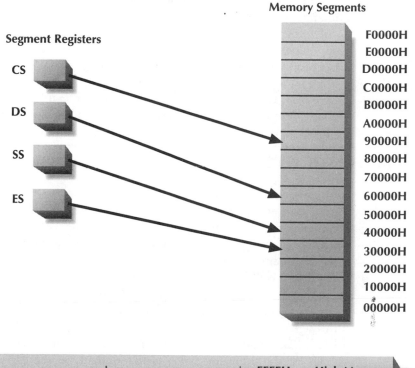

Segment Registers

Memory Segments

CS

DS

SS

ES

F0000H
E0000H
D0000H
C0000H
B0000H
A0000H
90000H
80000H
70000H
60000H
50000H
40000H
30000H
20000H
10000H
00000H

Figure 6-8

Memory allocation for multiple programs

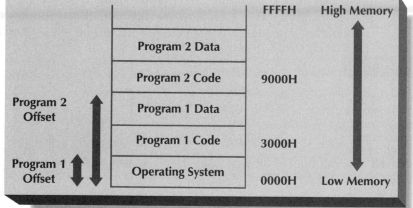

	FFFFH	High Memory
Program 2 Data		
Program 2 Code	9000H	
Program 1 Data		
Program 1 Code	3000H	
Operating System	0000H	Low Memory

Program 2 Offset

Program 1 Offset

In this case, relative address calculations must use the offset applicable to the process currently being executed, which implies either the existence of multiple offset registers or the use of a table of offset values in memory. The latter approach is more common because it is less likely to restrict the maximum number of processes in memory and is less costly. As CPU attention shifts from one process to another, the offset value for that process is loaded into the offset register from the table in memory. Programs also can be moved in memory, if necessary, and the appropriate entry in the offset table updated to reflect the move.

MAGNETIC STORAGE

Magnetic storage devices are based on the duality of magnetism and electricity. That is, electrical current can generate a magnetic field, and a magnetic field can generate electricity. A magnetic storage device requires a means of converting electricity into magnetism, capturing the magnetic charge, and later using the magnetic charge to generate an electrical current. Such a device also requires a well-defined method of differentiating stored zeros and ones.

Figure 6-9 illustrates the principles of writing to a magnetic storage device. A wire is coiled around a metallic **read/write head**. As electrical current passes through the wire, a magnetic field is generated from the gap in the read/write head. The polarity of the field (i.e., the position of the positive and negative poles of the magnetic field) is determined by the direction of current flow through the wire. Reversing the direction of current flow reverses the polarity, and each possible polarity corresponds to a bit value (0 or 1).

Figure 6-9

The principles behind magnetic data storage: An electrical current is passed through wire coiled around a read/write head; a magnetic field is generated and a magnetic charge is captured on the surface of a recording medium.

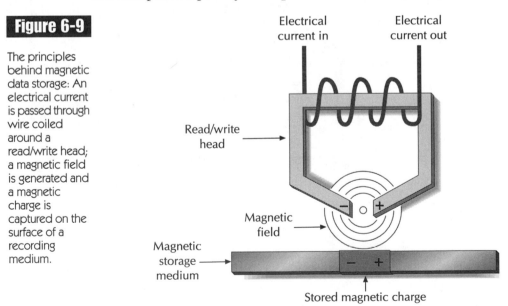

Electrical current in

Electrical current out

Read/write head

Magnetic field

Magnetic storage medium

Stored magnetic charge

A magnetic storage medium is placed next to the gap and the portion of the medium surface closest to the magnetic field is permanently charged. The polarity of the stored charge is identical to that of the generated magnetic field. The strength of the stored charge is directly proportional to the strength of the magnetic field. The strength of the magnetic field is determined by several factors including the number of coils in the wire, the strength of the electrical current, and the mass of the metallic read/write head. The storage medium must be constructed of (or coated with) a substance capable of accepting and storing a magnetic charge. Metals or metallic compounds are used for this purpose.

A read operation uses the same components and reverses the write process. A portion of the storage medium with a stored charge is placed near the gap of the read/write head. The stored charge radiates a small magnetic field. When the magnetic field comes in contact with the read/write head, a small electrical current is induced. Electrical current also is induced in the wire coiled around the read/write head. The direction of current flow depends on the polarity of the stored charge. Electrical switches at either end of the coiled wire detect the direction of current flow and, thus, sense either a 0- or 1-bit value.

Magnetic storage devices must all compensate for some undesirable characteristics of magnetism and several other design factors. These include:

▶ Magnetic decay.
▶ Magnetic leakage.
▶ Minimum threshold current for read operations.
▶ Coercivity of the storage medium.
▶ Long-term integrity of the storage medium.

A magnetic storage device must carefully balance all of these factors to achieve cost-effective, reliable storage.

Magnetic Decay and Leakage

Magnetically charged particles lose their charge over time in a process of **magnetic decay** that is relatively constant and proportional to the power of the charge. The electrical switches used in a read operation require a minimum (or threshold) level of magnetic charge for successful operation. Over a sufficient length of time, magnetic decay causes the amount of stored charge to fall below the threshold and, thus, data to become unreadable. To address this problem, data bits must be written with sufficient charge to compensate for decay. Thus, a write operation typically charges a bit area of the media at a substantially higher level than is required for proper reading. As long as the charge does not decay below the threshold for reliable reading, the data will remain recoverable.

The strength of individual stored bits also can decrease due to **magnetic leakage** from adjacent bits. This phenomenon can be clearly heard in audio recordings made on analog magnetic tape[1]. Any charged area continuously generates a magnetic field that can affect nearby bit areas. If the polarity of adjacent bits is opposite, then their magnetic fields will tend to cancel out the charge of both areas. Thus, the strength of both charges will fall below the threshold of readability.

Storage Density

Different materials have different capacities to hold magnetic charge. The relative capacity to accept magnetic charge is called **coercivity**. In general, metals and metallic compounds offer the highest coercivity at reasonable cost. For any specific material, the total amount of charge that can be held is directly proportional to the mass of material. In magnetic recording media, this mass is a function of the surface area in which a bit is stored, the thickness of the medium or its coating, and the density of the chargeable material within the coating. Larger, thicker, and/or denser areas can hold higher amounts of charge because of their higher mass of coercible material.

Most users want to store as much data as possible on the media. The simplest way to do this is to reduce the amount of surface area used to store a single bit value, thus increasing the total number of bits that can be stored on the media. For example, the storage density of a two-dimensional medium can be quadrupled by halving the length and width of each bit area (see Figure 6-10). The amount of surface area allocated to a bit is referred to as the **recording density**. Recording density is expressed (measured) in bits, bytes, or tracks per inch (abbreviated as bpi, Bpi, and tpi, respectively).

[1]Try the following experiment. Find an analog cassette tape that was recorded at least a year ago. Start the playback in the blank section preceding a musical selection with the volume turned up fairly high. You will hear the song "start" at a very low volume a few moments before it actually "starts" at normal volume. This effect is from the leakage of magnetic charge between portions of the analog tape that are wound on one another within the cassette.

Figure 6-10

Storage density is a function of the length and width of an individual bit area (a). Density can be quadrupled by halving the length and width of bit areas (b).

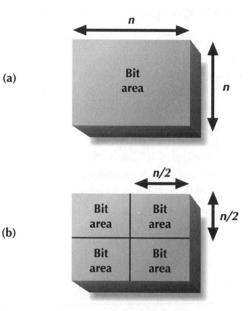

In some magnetic devices, recording density is fixed, but in others (e.g., tape) it is variable. But since reductions in area per bit reduce chargeable material per bit, the problems of magnetic decay and leakage are increased as recording density is increased. Designers and users of magnetic media and devices must find a balance between high recording density and the reliability of the media, especially over relatively long periods of time.

Media Integrity

The integrity of a magnetic storage medium depends on the nature of its construction and the environmental factors to which it is subjected. Some magnetic storage media (e.g., tape and floppy disk) use thin coatings of coercible material layered over a plastic or other substrate. Age and environmental stress can loosen the bond between the coating and substrate, thereby causing the coating to wear away. Physical stress on the media (e.g., fast forwarding and rewinding tape) can accelerate the process, as can extremes of temperature and humidity.

Loss of coercible coating represents a loss of strength in stored magnetic charges. As with magnetic leakage and decay, this loss becomes a problem when the remaining charge falls below the threshold of readability. To extend the life of magnetic media, they must normally be protected from physical abuse and from extremes of temperature and humidity. Removable magnetic media are sometimes stored in climate-controlled vaults to ensure long-term media integrity.

Magnetic Tape

Magnetic tapes are ribbons of plastic with surface coatings in which particles of metal oxide are suspended. Tapes are mounted in a tape drive for reading and writing. The **tape drive** contains motors to wind and unwind the tape as well as a fixed-position read/write head and related circuitry.

Tapes and tape drives come in a variety of shapes and sizes. Older tape drives (e.g., Figure 6-11) used wide tapes (0.25-, 0.5-, or 1-inch) wound on large (e.g., 8- or 10-inch) open reels. A tape reel was mounted on one side of the tape drive, the tape was physically placed through the read/write mechanism, and the end of the tape was fastened to an empty tape reel on the other side. A motor turned the empty reel, which caused the tape to pull itself through the read/write mechanism.

Figure 6-11

A depiction of components and configuration of an open-reel tape drive

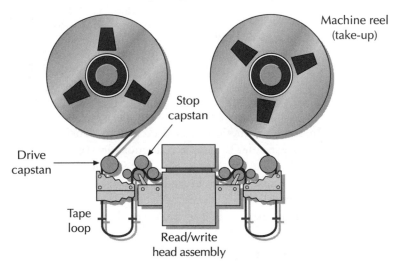

Modern tape drives have abandoned open reels in favor of tapes permanently mounted in plastic cassettes or cartridges (see Figure 6-12). The cassettes have small doors that remain closed until the cassette is inserted into the tape drive, and the read/write mechanism is then automatically positioned around the exposed tape. Thus, the tape is better protected from the environment than its open reel predecessors. Tape widths have shrunk as technology improvements have allowed substantial increases in recording density, with a resulting small, self-contained tape package and relatively small tape drive.

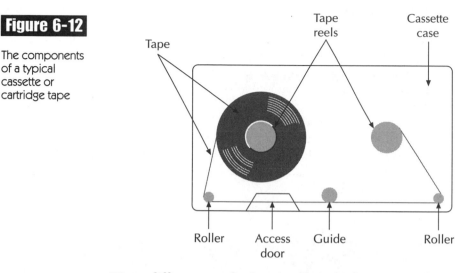

Figure 6-12

The components of a typical cassette or cartridge tape

Tape

Tape reels

Cassette case

Roller

Access door

Guide

Roller

Many different standards exist for cartridge sizes and data recording formats. 3M Corporation introduced the first widely used cartridge (approximately four by six inches using quarter-inch tape) in the 1970s. This cartridge is now referred to as the "data cartridge" and usually designated by the abbreviation DC. A smaller cartridge (3.25 by 2.5 inches) was introduced later and called the "minicartridge" (abbreviated MC). Data recording formats for these cartridges varied widely from one tape drive manufacturer to another until a standard set of formats was specified by the **Quarter Inch Committee (QIC)** in the early 1980s. The QIC continues to develop and publish new recording format standards. (See Table 6-1 for a partial listing of QIC recording format specifications.)

TABLE 6-1

A portion of the Quarter Inch Committee (QIC) tape format specifications

Format	Cartridge Size (inches)	Capacity (MB)	Tracks	Recording Density (bpi)	Tape Length (feet)
QIC-80	4×6	80	28	14,700	205
QIC-120	3.25×2.5	125	15	10,000	600
QIC-525	3.25×2.5	525	26	20,000	1,000
QIC-2100	3.25×2.5	2,100	30	50,800	875
QIC-5GB	3.25×2.5	5,000	44	62,182	1,200

QIC recording standards use a parallel track method of recording. The length of the tape is divided into a number of parallel tracks, with space between each track and between tracks and the edge of the tape (see Figure 6-13[a] on the next page). Bits are recorded linearly along each track, and all tracks are read/written in parallel. QIC format tapes, which use from 9 to 144 tracks, also vary in recording density. For tapes, recording density is measured in bits per inch (bpi) along each track. QIC recording densities range from 8,000 to 76,000 bpi. Tape length varies from 200 to 1,200 feet, and tape capacity (uncompressed) ranges up to 25 gigabytes.

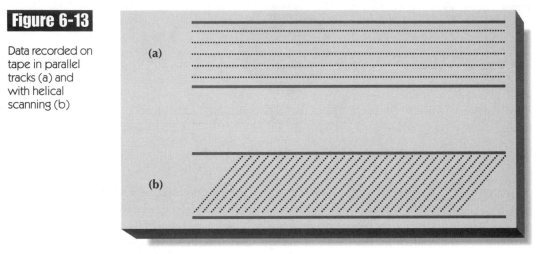

A newer set of tape-recording methods have come into use in the last decade. These methods are based on technology similar to that used in **digital audio tape (DAT)** drives and 8-millimeter (mm) video cassette recorders. These recording formats are based on a method called helical scanning (see Figure 6-13[b]). Helical scanning reads and writes data to/from tape by rotating the read/write head at an angle to the tape and moving from tape edge to tape edge. This method requires much more complex read/write mechanisms that employ multiple heads. However, it packs data much more densely onto the tape, which results in substantial improvements in recording density and data storage capacity.

All tape drives based on DAT record data using a standard method and format developed and licensed by Sony Corporation, which guarantees partial compatibility among tape drives. Tape capacity ranges up to 12 gigabytes with a recording density of over 3 megabits per inch. Tape drives based on 8-mm video follow no industry standard for recording method or format. Tape capacity is similar to DAT recording, but recording time is two to three times faster.

Tapes were once commonly used as data input and output devices for executing programs. Data records were stored on tape in a sorted order and processed by the program in that order. Access speed for tape drives is much slower than for optical and magnetic disk, which is too slow to support the current data transfer requirements of an executing program. Thus, tapes are now used primarily as a media for keeping backup and archival copies of programs and data files that are normally stored on disk.

Tapes are subject to all of the problems of magnetic decay and leakage discussed in the "Magnetic Decay and Leakage" section. They compound the problem of leakage by winding the tape upon itself, and thus, leakage can occur from adjacent bit positions on the same area of the tape as well as from the layer of tape wound above or below on the reel. Tapes are also susceptible to problems arising from stretching, friction, and temperature variations. As a tape is wound and unwound, the base layer of plastic tends to stretch in length; if enough stretching occurs, the distance between bit positions can be altered, which makes individual bits difficult or impossible to read.

In some tape drives, the read/write head is in direct contact with the tape as it passes, which causes friction as the tape moves. Chargeable material is scraped off of the tape each time it is read or written, which results in a lower capacity to hold charge and, therefore, information. This problem is accelerated by dirt on the tape or on the read/write head. Some modern high-speed tape drives position the read/write head just above the surface of the tape to avoid these problems.

Magnetic Disk

Magnetic disk media are flat, circular **platters** with metallic coatings that are rotated beneath read/write heads (see Figure 6-14 on the next page). Multiple platters can be used if they are mounted on and rotated with a common spindle. Data is normally recorded on both sides of a platter. A **track** is one concentric circle of a platter, or the surface area that passes under a read/write head when its position is fixed. A sector is a fractional portion of a track (see Figure 6-15 on the next page). The number of sectors per track ranges from eight for low-density floppy disks to several dozen for high-capacity hard drives. A single sector (or block) usually holds 512 bytes of data and is the normal unit of data transfer to/from the drive. Within a sector, data bits are stored sequentially, or serially. The read/write head reads or writes bits one at a time as the sector passes underneath.

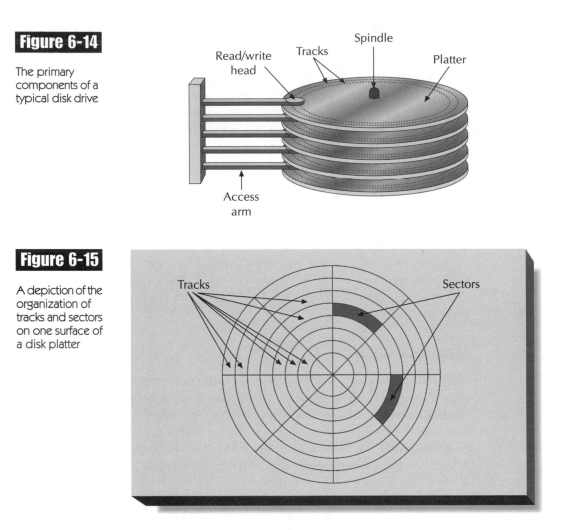

Figure 6-14

The primary components of a typical disk drive

Read/write head   Tracks   Spindle   Platter

Access arm

Figure 6-15

A depiction of the organization of tracks and sectors on one surface of a disk platter

Tracks   Sectors

There is a read/write head for each side of each platter, and read/write heads are mounted on the end of an **access arm**. Access arms are of equal length and are attached to a positioning servo. The servo moves the access arms so that the read/write heads are positioned anywhere between the outermost track (closest to the outer edge of the platters) and the innermost track (closest to the center of the disk). Read/write heads do not typically make contact with the recording surface of a platter; rather, the head floats above the platter on a thin cushion of air as the disk spins.

The term **hard disk** refers to magnetic disk media with rigid metal bases (or substrates). Typical platter size for hard disks is between three and five inches in diameter. Multiple platters are used and they are spun at relatively high speeds (up to 10,000 revolutions per minute). Drive capacity depends on the number of platters, size of the platters, and storage density. Typical hard disk drive capacities range from 1 to 20 gigabytes. Multiple hard disk drives often are enclosed in a single storage cabinet and referred to as a **drive array**. The term can also be used for arrays of optical disk drives.

Data storage media called **floppy disks** use bases of flexible or rigid plastic material and are otherwise similar in function to hard disk drives. Another term for floppy disk is **diskette**. Modern implementations of floppy disk typically use plastic media ranging from 2.5 to 3.5 inches in diameter. Diskettes consist of a single platter with data recorded on both sides, which is permanently mounted in a plastic case that can be removed from a diskette drive. The case has an access door that is closed when the diskette is removed from the drive and opened automatically when the diskette is inserted into the drive. Diskettes are spun at lower speeds than hard disks. Diskette capacities range from less than 1 MB to as much as 5 MB.

Larger-capacity removable magnetic disks are externally similar to floppy disks. Examples include disks designed for the Iomega Jaz drive and Syquest SyJet drive. Platter size is typically 3.5 or 5.25 inches. The platter is rigid and has a thick dense coating of chargable material. Spin rates are higher than for floppies but not generally as high as for hard disk drives. Storage capacity ranges as high as 1 gigabyte.

When originally manufactured, all magnetic disks are blank. Although the location of tracks within a platter is fixed (by the position of read/write heads), the location of sectors within a track is not. One of the operations performed when formatting a disk is to fix the location of sectors within tracks, which is accomplished by writing synchronization data at the beginning of each sector. Synchronization information also can be written within a sector either during formatting or during subsequent write operations. The read/write circuitry uses synchronization information to compensate for minor variations in rotation speed and other factors that might disturb the precise timing needed for reliable reading and writing.

Read or write operations are made in response to a command that identifies the platter, track (within platter), and sector (within track). Controller circuitry switches activate the read/write head for the platter. The positioning servo moves the access arms so that they are positioned over the track. Then, the controller circuitry waits for the appropriate sector to rotate underneath the read/write head. As the sector starts to rotate beneath the read/write head, the read or write operation is started, and the operation is completed just as the last portion of the sector passes the read/write head. The access arms are left in their final position until the next read or write command is received.

The access time of a disk drive depends on several factors, including:

► Time required to switch among read/write heads.

► Time required to position the read/write heads.

► Rotational delay.

The first of these factors is electrical, whereas the remaining two are primarily mechanical.

Disk drives share one set of read/write circuits among all of the read/write heads, and so the read/write circuitry must be electronically attached (switched) to the appropriate read/write head before a sector can be accessed. The time required to perform this switching is no more than a few nanoseconds and is referred to as **head-to-head switching time**. Switching occurs in sequence among the read/write heads, and multiple switching operations may be required to activate a desired head. For example, if there are 10 read/write heads, switching from the first head to the last requires nine separate switching operations (first to second, second to third, and so forth). Head-to-head switching time is the least significant factor that determines access time or delay.

The positioning servo that moves the access arms and read/write heads is a mechanical device and is essentially a bi-directional motor that can be started and stopped quickly. The access arms are attached to it so that they can be precisely positioned over tracks on the platters. The disk controller must keep track of the current position of the heads at all times. When receiving a new access command, the disk controller computes the difference between the current track and the desired track. It then activates the positioning servo motor for precisely the amount of time (and proper direction) needed to move the read/write heads to the desired track. The time required to move from one track to another is called **track-to-track seek time**. Track-to-track seek time, measured in milliseconds, is an average number because of variations in the time required to start, operate, and stop the positioning servo.

Once the read/write heads are positioned over the desired track, the controller waits for the desired sector to rotate beneath the heads. The controller keeps track of the current sector position by continually scanning the sector synchronization information written on the disk during formatting. When the appropriate sector is sensed, the read or write operation is begun. The time that the disk controller must wait for the proper sector to rotate beneath the heads is called **rotational delay**. Rotational delay heavily depends on the speed at which the platters are spun. High spin rates (e.g., 10,000 RPM) yield relatively low times for rotational delay.

Both track-to-track seek time and rotational delay can vary from one read or write operation to another. For example, if two consecutive read requests access adjacent sectors on the same track and platter, then rotational delay and track-to-track seek time for the second read operation will be zero. If the second read access is to the sector preceding the first, then rotational delay will be high (i.e., almost a full platter rotation). If one read request references a sector on the outermost track and the other requests references the innermost track, then track-to-track seek time will be high.

Whither the Floppy Disk?

Floppy disk drives have been a standard component of virtually every PC ever manufactured. They were also a commonly used medium for data archival and transport with minicomputers and some mainframe systems during the 1970s and early 1980s. Their attractiveness (particularly in 3.5-inch hard case format) is obvious. They offer cheap removable storage that can fit in a shirt pocket. Until relatively recently, floppy disks were the technology of choice for data transfer, data backup, and the distribution of both software and data. Data transport through floppy diskette is sometimes called SneakerNet because floppies were the primary means of data exchange before networks were widely implemented.

Table 6-2 shows specifications for the floppy disks used in IBM-compatible PCs since 1981. Storage capacity has increased by approximately 20 times during the 10-year period of 1981–1991. The 5.25-inch flexible format virtually disappeared and was replaced by the hard-shell 3.5-inch format common today. However, no format or capacity improvements are listed after 1991. Why, then, are there no improvements after that date? Has floppy diskette technology peaked out?

TABLE 6-2	Year	Capacity (Kbytes)	Platter Size (inches)	Number of Sides	Tracks per Side	Sectors per Side
A comparison of IBM PC floppy disk formats	1981	160	5.25	1	40	8
	1982	180	5.25	1	40	9
	1982	320	5.25	2	40	8
	1983	360	5.25	2	40	9
	1985	1,200	5.25	2	80	15
	1985	720	3.5	2	80	9
	1987	1,440	3.5	2	80	18
	1991	2,880	3.5	2	80	36

The primary reason that no standard format improvements have been adopted since 1991 is that IBM no longer defines the "standard" for IBM-compatible PCs. Intel and Microsoft have been the source of many new standards since the late 1980s, but they have been largely silent on the issue of improved floppy disk drives. Thus, there has been no authority willing and able to define any new "standard" formats.

Removable magnetic storage technology has continued to evolve since 1991. Iomega Corporation has been a leader in removable magnetic media technology since the mid-1980s, and that company is responsible for many of the advances since that time. Their Bernoulli Box products were commonly used as backup devices and "hard disk drive extenders" although they never became standard equipment on a PC. The Bernoulli Box products have evolved into two newer product lines: the Zip drive and the Jaz drive.

The Zip drive is a removable magnetic media drive that uses 100-MB 3.5-inch cartridges. Average access time is 29 milliseconds, rotation speed is just under 3,000 RPMs and sustained data transfer rate is 1 MB per minute or more. These performance statistics are approximately 10 times that of the fastest "standard" floppy drives.

The Jaz drive also uses 3.5-inch removable magnetic cartridges. It increases all of the statistics of the Zip drive (including the cost!). Average access time is 10 milliseconds for reading and 12 for writing, rotational speed is 5,400 RPM, and sustained data transfer rate is approximately 5 MB per second (Mbps). These specifications rival those of many magnetic hard disk drives.

The Jaz drive is too expensive to be considered the "next floppy drive standard" but the Zip drive is closer in price. The primary differences between the Zip drive and existing floppy drives are the lack of a standard drive controller interface—it uses either a SCSI interface or a parallel port interface—and its predominant manufacture as an external drive. Iomega has attempted to make the Zip drive the next floppy standard by producing them in large quantities (over five million in its first two years of production) and selling them cheaply ($140 as of 1997). Iomega has signed license agreements for drive manufacture and installation with many PC manufacturers. It also has entered production agreements with several magnetic media manufacturers to meet the growing demand for disk cartridges. Thus, the Zip drive appears to be the closest thing we'll ever have to a new floppy disk drive standard.

The amount of track-to-track seek time and rotational delay cannot be known in advance because the location of the next sector to be accessed can't be known in advance. Thus, access time for a disk drive is not a constant. Instead, it is usually stated using a variety of performance numbers including raw times for head-to-head switching and rotation speed. The most important performance numbers are **average access time**, **sequential access time**, and **sustained data transfer rate**.

Data Storage Technology

Average access time is computed by assuming that two consecutive accesses are made to random locations. Given a large number of random accesses, the expected value of head-to-head switching time will correspond to switching over half the number of recording surfaces. The expected value of track-to-track seek time will correspond to movement over half the tracks. The expected value of rotational delay will correspond to one-half of a platter rotation. Thus, if head-to-head switching time is five nanoseconds, there are 10 read/write heads, track-to-track seek time is five microseconds, there are 1,000 tracks on each recording surface, and the platters are rotated at 7,500 RPM, the average access delay is computed as:

$$\text{average access delay} = \text{HTH switch} + \text{TTT seek} + \text{rotational delay}$$
$$= \frac{(5 \text{ nanoseconds} \times 10)}{2} + \frac{(1{,}000 \times 5 \text{ microseconds})}{2} + \frac{(60 \text{ seconds} \div 2)}{7{,}500 \text{ RPM}}$$
$$= .000000025 + (500 \times .000005) + (.008 \div 2) \text{ seconds}$$
$$= .000000025 + .0025 + .004 \text{ seconds}$$
$$= 6.5 \text{ milliseconds}$$

Average access time is the sum of average access delay and the time required to read a single sector. The time required to read a single sector depends entirely on the rotational speed of the disk and the number of sectors per track. Thus, if the above figures are assumed and there are 24 sectors per track, the average access time is computed as:

$$\text{average access time} = \text{average access delay} + \frac{(60 \text{ seconds} \div 24)}{7{,}500 \text{ RPM}}$$
$$= .0065 + (.008 \div 24)$$
$$= 6.83 \text{ milliseconds}$$

Sequential access time is the time required to read the second of two adjacent sectors on the same track and platter. Head-to-head switching time, track to track seek time, and rotational delay are all zero for such an access. Thus, the only component of access time is the time required to read a single sector (i.e., the second half of the above equation, or 0.33 milliseconds in that example).

Because sequential access time is so much faster than average access time, it is generally preferred that related data be stored in sequential sectors. Thus, for example, access to a stored program on disk is relatively fast if the entire program is stored in sequential locations on the same track. If the program won't fit in a single track, then storage in equivalently positioned tracks on multiple platter surfaces is preferred.

If portions of the program are scattered around the disk in random sectors, then a great deal of time will be spent switching read/write heads, positioning the access arm, and waiting for the desired sectors to rotate beneath the heads. The performance improvement resulting from storing files in sequential locations is only realized if those files are accessed sequentially (e.g., as in loading a program into memory from disk). If access to portions of the file is random (e.g., a random access data file), then sequential storage will not provide significant, if any, performance improvements. Most operating systems provide a utility program that performs an operation called **disk defragmentation**. One of the primary components of this operation is the reorganization of disk content so that all of the portions of an individual file are stored in sequential sectors.

The final component of drive performance is the speed at which data can be moved through the disk controller circuitry to/from memory or the CPU. This circuitry is not a communication bottleneck for a single disk drive. A summary performance number that combines the physical aspects of data access with the electronic aspects of data movement is the disk drive data transfer rate. As with access times, there are generally two different numbers stated. Maximum data transfer rate is the fastest rate that the drive can support and assumes sequential access to sectors. Since mechanical delays are minimal with sequential accesses, this number is typically high (e.g., several Mbps). The maximum data transfer rate for the previously described performance figures (assuming no controller delays) is:

$$\frac{1}{0.00033} \text{ seconds} \times 512 \text{ bytes per sector} \approx 1.5 \text{ MB per second}$$

Average data transfer rate is computed with much less optimistic assumptions about data location. The most straightforward method is to use average access time. Thus, for the previous example, average data transfer rate is:

$$\frac{1}{0.00683} \text{ seconds} \times 512 \text{ bytes per sector} \approx 75 \text{ kilobytes per second}$$

However, because operating systems tend to allocate disk space to files sequentially, the average data transfer rate is much higher than the previous computation would imply. Table 6-3 provides various performance statistics for various hard disk drive models from Seagate Technology, Inc.

TABLE 6-3	Model	Platters	Cylinders	Average Seek Time (milliseconds)	Rotation Speed (RPM)	Average Access Time (milliseconds)
Various architectural and performance statistics for some Seagate hard disk drives	ST52520	2	4,970	11	5,400	11
	ST32151	4	4,176	10.5	5,411	9
	ST32272	2	6,311	9.8	7,200	8
	ST19101	8	6,526	9	10,033	8

OPTICAL MASS STORAGE DEVICES

Optical storage has come of age in the last decade. Its primary advantage as compared to other forms of storage is a relatively high storage density. Optical storage devices achieve storage densities at least 10 times greater than magnetic storage devices. Higher densities are achieved through the use of tightly focused lasers that can be reflected from a relatively small area of a recording medium. Magnetic fields cannot be focused to as small an area without overwriting surrounding bit positions.

Optical storage is not subject to problems of magnetic decay and leakage. Thus, in theory, it represents a permanent form of storage. Optical storage is also popular because of the availability of widely standardized and popular storage format for removable media. The compact disc (CD) format used for musical recordings also is supported by most optical storage devices. The cost to manufacture CDs is considerably less per bit than for other removable storage technologies (e.g., floppy disk). Thus, CD is a popular format for distributing thousands or millions of copies of software and large data sets used by many users (e.g., maps, census data, and telephone directories).

Optical storage devices do not store light directly; instead, they store bit values as variations in light reflection. The storage medium is a surface of highly reflective material. The read mechanism consists of a low-powered laser and a photoelectric cell. A laser is focused onto the recording medium at a specific angle. The photoelectric cell is positioned at a complementary angle so as to capture any light reflected back from the recording medium. Binary values are represented by low and high reflectivity of the recording surface. A highly reflective surface spot causes the photoelectric cell to generate a detectable electrical current. A poorly reflective surface spot does not reflect enough light to cause the photoelectric cell to fire. The current or lack thereof produced by the photoelectric cell is interpreted as a binary zero or one.

The implementation of a write operation varies widely from one type of optical storage device to another. An optimal surface material would allow a bit area to be rapidly changed from highly to poorly reflective and back again an unlimited number of times. The search for such a material and a reliable and efficient method of using it has proven problematic to developers. Some materials change quickly in one direction but not another (e.g., similar to the properties of many "automatic" sunglass materials). The ability of most materials to change reflectivity states degrades over time and/or with repeated use. The lack of a clearly dominant and economical material/technology is the reason for the wide variety of write technologies used in current devices.

All current optical storage devices use a disk storage format similar to that of magnetic disks. The only significant difference from magnetic disk format is the use of a single track spiraling outward from the center of the disk. Because the recording surface is a disk, there are many similarities to magnetic disk storage, including the use of a read/write mechanism mounted on an access arm, a spinning disk, and performance limitations due to rotational delay and movement of the access arm.

Although magnetic and optical devices both use disk technology, their performance is not comparable. Optical disk storage is slower for a number of reasons, including the use of removable media, the use of a single track spiral, and the inherent complexity of a write operation. The performance gap has narrowed considerably in recent years and is approximately 3:1 at present. Typically, magnetic hard disks have access times under 10 milliseconds; read access time for optical drives is greater than 25 milliseconds; and write access time is two or three times longer than read access time. The gap is expected to shrink as optical technology matures and magnetic technology reaches its physical limits.

Magnetic and optical storage are not currently direct competitors because of their performance difference and the difficulty in writing to optical media. Magnetic hard disk drives are still the cost-performance leader for implementing online general purpose secondary storage. Optical storage is favored for read-only storage with low performance requirements and/or extremely high-capacity requirements. The two technologies will compete more directly as optical access times improve and as optical write technology develops and matures.

Optical storage devices are currently available in a wide variety of storage formats and write technologies, including some of the more popular noted here:

- ► CD-ROM.
- ► WORM.
- ► Magneto-Optical.
- ► Phase-Change.

Each of these technologies is discussed next.

CD-ROM

CD-ROM (compact disk read only memory) is the optical media familiar to most people. Audio CDs, the most prolific example of this type of media, are called "read only" because data content is permanently embedded in the disk when the disk is manufactured. The (original) standard 120-mm disc holds approximately 600 MB of data. A standard format for directory and file organization (sometimes called the **high sierra format**) provides for compatibility of discs among CD-ROM readers.

The structure of a CD-ROM and the implementation of a read operation are depicted in Figure 6-16. The disc consists of a rigid plastic substrate, a thin reflective layer, and a top coating of clear plastic to protect the reflective layer. Bits are stored on a CD-ROM as **lands** and **pits**. Lands are flat, bit-sized areas in the reflective layer of a disc. When striking a land, a laser light will reflect back nearly all of its energy at a complementary angle. Laser light striking a pit will be scattered with only a small portion reflecting at a complementary angle. A photoelectric cell positioned along the complementary angle interprets the reflected laser light as a zero or one.

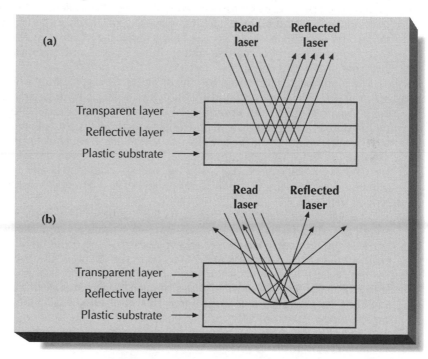

Figure 6-16

Optical disk read operations. Most light is reflected at a complementary angle from a land (a and b), and most light is scattered from a pit.

CD-ROM is a popular medium for distributing software and large databases. Its standardized format, high density, and cheap manufacturing cost make it well suited for this purpose. The physical disk format and bit encoding method were specified by Sony and Phillips in the late 1970s. Newer formats promise substantially increased storage density and, thus, capacity, but no format has yet emerged as a widely accepted replacement for the original Sony/Phillips standard.

Optical Phase-Change Disks

Phase-change optical technology is a promising candidate for the leading position in optical read/write technology and, as such, is a direct competitor to magneto-optical drives. The key to phase-change technology are materials that change state from noncrystalline (generally called amorphous) to crystalline and back again. The reflective characteristics of the materials differ in their amorphous and crystalline states, which allows a reflected laser to be interpreted as a zero or one.

The material currently used in optical phase disks is a compound of tellurium (Te), selenium (Se), and tin (Sn), which changes from an amorphous state to a crystalline state when heated to a precise temperature. The crystalline state is highly reflective. Changing the material back to an amorphous state requires heating the material to its melting point, which returns it to its amorphous state. The amorphous state is poorly reflective. The sharp contrast between reflective states is similar to that of lands and pits on a CD, which means the same read mechanism can be used. Storage density is also similar to that on a CD.

The laser power required by the write laser to heat the material is somewhat lower than for magneto-optical. Thus, write times are potentially faster because of the need for fewer write passes to heat up a bit area. The read/write head is also simpler and of lower mass. Theoretically, this should lead to faster track-to-track seek times, but the technology is still relatively immature. The primary shortcoming of the technology is that the recording medium "wears" after several hundred thousand uses. Current magneto-optical media "wear out" after millions of uses.

Matsushita (Panasonic) holds most of the patents on this technology and is the sole producer of phase-change optical drive mechanisms. Current products based on these drive mechanisms use 5.25- and 12-inch recording platters. The 5.25-inch drives also read conventional CDs, which allows them to do "double duty" as storage devices. However, the Sony/Phillips CD format introduces complexity that has thus far limited drive performance.

CDs are read at a **constant linear velocity**. That is, data bits must pass under the read/write head at a constant rate. Because the outer portions of the single spiral track contain more data than the inner portions, the speed of rotation is not constant. Rotational speed must be slowed when reading the outer portions of this disk to ensure a constant number of bits per second passing under the read head. As the read head is moved toward the center of the disk, the rotational speed must be increased to keep linear velocity constant. The need to vary the rotational speed based on disk (track) location complicates the drive motors and controller circuitry and acts as a hindrance to high access speeds.

Faster rotational speeds are easier to implement when rotational speed is a constant. Constant rotational speed is also called **constant angular velocity**. The Matsushita dual-function 5.25-inch drives switch between constant linear velocity (for CDs) and constant angular velocity (for phase change disks), but even when operating at constant angular velocity, their access time is relatively slow (165 ms. average) because of a long track-to-track seek time and slow rotational speed (2,026 RPM). This performance must improve substantially if phase-change drives are to effectively compete against more mature (and faster) magneto-optical drives.

WORM

A **WORM (write once/read many) disk** is the optical storage equivalent of a programmable ROM. Applications for this technology are similar to those for CD-ROM. They are used primarily by organizations that need to generate large quantities of data for distribution infrequently. Automobile parts catalogs, for instance, can be updated annually or quarterly. If the distribution volume is large (several thousand or more), then it is generally cheaper to have CD-ROMs manufactured. For lower volumes, WORM drives and media are more cost-effective.

A WORM disc is essentially a blank CD that can be written with two different methods. The first uses a high-powered laser to burn a pit into the surface, and the pit is sufficiently distorted to diffuse a large amount of light from the read laser. The second type of WORM drive uses a dye that changes color and reflectivity when a high-powered laser is applied. The dye is normally light and highly reflective but changes to a darker less reflective color when a sufficient amount of laser light is applied. Both processes are irreversible. The dye-based method is most commonly found in current WORM drives because of its lower power requirements and relatively faster write time.

Magneto-Optical

A **magneto-optical** drive uses a laser and reflected light to sense bit values. However, the variations in reflectivity of the disk surface are achieved by applying a magnetic charge to the bit area. Reading is based on the polarity of the reflected laser light, not its absolute amount. The magnetic charge shifts the polarity of the reflected laser light. The direction of the shift is determined by the polarity of the magnetic charge in a shift known as the Kerr effect.

The disk surface material resists magnetic charge at normal temperatures. At higher temperatures (approximately 150 degrees Celsius), the material can be charged by a magnetic field in much the same way as purely magnetic media. To bring the material to a chargeable temperature, a high-powered laser is focused on a bit position. A magnetic write head positioned below the media surface then generates the appropriate charge. Because the laser beam focuses on a single-bit position, only that bit position is raised to a sufficient temperature. Surrounding bits are unaffected by the magnetic field from the write head because they are cooler.

The write operation is relatively slow compared to the read operation, but it is considerably faster than with WORM drives. Multiple passes are required to heat the bit area sufficiently to accept the magnetic charge. Rewriting bit values is also slow and must be performed in two steps: the first to reorient a bit to its original (neutral) state and the second to write the new bit value. Magneto-optical media are relatively expensive and tend to wear out after a few million write operations.

Magneto-optical technology is complex and cumbersome to implement. High-capacity removable magnetic media offer lower cost and lower storage density. Alternative optical technologies (see the following "Business Focus" section) are maturing and may become more cost-effective alternatives for high-density storage. However, many companies are investing large sums of money in magneto-optical research and development. Storage density, media life, and access speed have all steadily improved while cost has dropped. Thus, the short-term prospects for magneto-optical technology appear bright, but longer-term prospects are uncertain.

Business Focus

A Modern SneakerNet?

Computer Publishing Company (CPC) is a medium-sized publisher of textbooks (high school, trade school, and college), software manuals, and computer-oriented "how to" books. The company currently has over 300 titles in active distribution and usually has between 50 and 100 titles under development.

CPC has approximately 50 full-time employees and occupies a single office building in Los Angeles. This workforce is much too small to do all of the work associated with the number of titles CPC manages. To fill the gap, CPC contracts out much of its development and production work (e.g., copy editing, content reviews, and illustrations). All printing and distribution is subcontracted to another large publisher.

Many of the copy editors, graphic artists, and other development and production personnel are independent contractors working out of their homes, and many others are employees of small companies with 10 or fewer employees. All of the authors and reviewers are independent contractors. Although many of the contractors are located in Southern California, others are scattered throughout North America. The authors and reviewers are also geographically dispersed over a wide area.

Development projects (both new and revised titles) require a great deal of communication among authors, editors, reviewers, product managers, and other production staff. A manuscript (or portions thereof) typically is shipped a dozen times or more during development. Until recently, all of this communication was performed by sending paper by overnight mail (e.g., Federal Express). Over the last few years, CPC has standardized the software used for development work. Microsoft Word, Adobe Illustrator, and a handful of other software products are used in most projects.

CPC would like to go one step further and transfer manuscripts and other information electronically. This would be a relatively easy task if CPC had a network with all of its workers directly attached, but the large number of contract employees and the wide variation in their computer configurations makes this impossible. An additional problem is the size of the files comprising a typical title (200 MB to 1 gigabyte). The company does maintain a Web site and e-mail links. Some data is currently moved by these means, but many of the authors, reviewers, and contract employees lack reliable and efficient connections to the Internet. Thus, CPC still relies heavily on overnight delivery of paper.

CPC is considering adopting a standard storage medium for electronic data exchange. Although this wouldn't eliminate the need for overnight delivery, it would eliminate the production and consumption of paper. It also would reduce overnight shipping costs because of reduced weight of the shipment. CPC has decided on the following requirements for their standard storage medium:

- ▶ A removable cartridge that is small, light, and environmentally "tough" (i.e., something that is easily and inexpensively mailed without damage to itself or the data).
- ▶ Reliable hardware and media (i.e., drives should last at least five years, and data should be 100% readable for at least five years).
- ▶ A long-lived and widely available format (i.e., a drive that is widely available now and a cartridge that is likely to be readable by drives purchased three to five years hence).
- ▶ Inexpensive drive (e.g., $500 maximum, $250 or less preferred) and media (e.g., 20¢ per megabyte maximum, 10¢ or less preferred).
- ▶ 100-MB minimum capacity per cartridge (e.g., 250 to 1,000 MB would be preferred).
- ▶ Must work with both IBM-compatible and MacIntosh-compatible PCs.

SUMMARY

The range of storage devices used within a single computer system forms a memory-storage hierarchy. Cost and access speed generally decrease as one moves down the hierarchy. Storage devices are required to support the immediate execution of programs (primary storage) and as a means of long-term storage to support future program execution (secondary storage). Primary and secondary storage represent different trade-offs among storage device characteristics, including speed, volatility, access method, portability, cost, and capacity.

Speed is the most important characteristic that differentiates primary and secondary storage. The speed of a storage device is generally expressed in terms of its access time. Access time alone is an incomplete measure of data access speed. The unit of data transfer to/from the storage device also must be specified to provide a complete picture of access speed. A storage device is said to be nonvolatile if its content can be held without loss over long periods of time. A storage device is said to be volatile if its content cannot be held for long periods of time.

The physical structure of storage devices and media determine the ways in which data can be accessed. A serial access storage device stores and retrieves data items in a linear (sequential) order. Random access devices are not restricted to any specific order when accessing data. A parallel access device can simultaneously access multiple storage locations. Portability of stored data is implemented through removability of the storage medium or of the entire storage device.

The critical performance characteristics of primary storage devices are their access speed and the number of bits that can be accessed in a single read or write operation. Modern computers use memory implemented with semiconductors. Basic types of memory that are built from semiconductor microchips include random access memory (RAM) and read only memory (ROM). RAM is a generic term describing primary storage devices implemented as microchips.

There are two primary types of RAM—static RAM and dynamic RAM—and several variations on each. Static RAM (SRAM) is implemented entirely with transistors, and dynamic RAM (DRAM) is implemented using transistors and capacitors. The more complex circuitry of SRAM is more expensive to implement but provides faster access times. Neither type of RAM can match current microprocessor clock rates.

Microchips within which data are embedded permanently or semi-permanently are called read only memory (ROM). The earliest ROMs were manufactured with bits implemented as circuits that always generated the same bit value. A more cost-effective approach was the programmable read only memory (PROM). An erasable programmable read only (EPROM) memory chip is identical in function to a PROM but uses a nondestructive and reversible means of writing bits. EEPROMs may be programmed, erased, and reprogrammed by signals sent from an external control source. Flash memory is the latest form of ROM.

The physical organization of memory, the organization of programs and data within memory, and the method(s) of referencing specific memory locations are critical design issues for both primary storage devices and processors. Programs are created as though they occupied contiguous memory locations starting at the first location (i.e., low memory). Programs are rarely physically located in low memory because of system software or other programs that may occupy those locations. Offset addressing is a mechanism to reconcile the difference between where a program "thinks" it is located and where it is actually located. The actual address of a program's first instruction is stored in an offset register, and this address is added to each address reference the program makes. This mechanism allows multiple programs to reside in memory and allows programs to be moved during execution.

Secondary storage is implemented using magnetic or optical storage devices. Magnetic storage devices store data bits as polarized metallic. Data bits are read and written from/to magnetic recording media by a read/write head. Types of magnetic media include magnetic tape and magnetic disk. Magnetic tapes are ribbons of acetate coated with a magnetically chargeable coating. Data is written to (or read from) tapes by passing it over the read/write head of a tape drive. Tapes, sequential access devices because data may be read or written in physical order on the tape surface, are stored on open reels or in enclosed cartridges.

Magnetic disks are platters coated with magnetically chargeable material. The platters are rotated within a disk drive and a read/write head access data at various locations on the platter(s). Magnetic disk drives are random access devices because the read/write head can be moved freely directly to any location on a disk platter. Hard disk drives are rigid platters coated with highly chargeable material. Floppy disks are flexible platters and can be removed from the disk drive. Hard disk drives offer faster access speeds and higher capacity than floppy disks.

Optical disks store data bits as variations in the ability to reflect light. An optical disk drive reads data bits by shining a laser beam onto a small disk location. High and low reflections of the laser are interpreted as ones and zeros. The cost per bit of optical storage generally is less than magnetic disks at the expense of slower access speed. Types of optical disks (and disk drives) include compact disk read only memory (CD-ROM), write once/read many (WORM) disks, magneto-optical drives, and phase change drives.

CD-ROMs are written with data during manufacture. Once written, the data cannot be altered. WORM drives are manufactured without data content, and they can be written to once by a special disk drive. Magneto-optical drives combine optical and magnetic storage technology. Reading is accomplished by optical means, whereas writing is accomplished by a combination of optical and magnetic means. Magneto-optical disks can be read and written multiple times. Phase-change storage media can change a state from non-crystalline (generally called amorphous) to crystalline and back again. The reflective characteristics of the materials are different in their amorphous and crystalline states. This difference allows a reflected laser to be interpreted as a zero or a one.

Key Terms

absolute addressing
access arm
access time
addressable memory
average access time
big endian
block
block size
coercivity
compact disk read only memory (CD-ROM)
constant angular velocity
constant linear velocity
core memory
data transfer rate
device controller
digital audio tape (DAT)
direct access
disk defragmentation
diskette
double in-line memory module (DIMM)
drive array
dynamic ram (DRAM)
electronically erasable programmable read only memory (EEPROM)
endian
erasable programmable read only (EPROM)
extended
extended data out (EDO)

fast page mode (FPM)
ferroelectric RAM
firmware
flash memory
flash RAM
floppy disk
hard disk
head-to-head switching time
high sierra format
indirect addressing
land
least significant byte
little endian
loader
magnetic decay
magnetic leakage
magnetic tape
magneto-optical
memory allocation
most significant byte
nonvolatile
off-line
offset
offset register
online
parallel access
physical memory
pit
platter
primary storage
programmable read only memory (PROM)

Quarter Inch Committee (QIC)
random access
random access memory (RAM)
read only memory (ROM)
read/write head
read/write mechanism
recording density
refresh cycle
refresh operation
relative addressing
relocating
relocator
removable media
rotational delay
secondary storage
sector
segment register
sequential access time
serial access
single in-line memory module (SIMM)
static ram (SRAM)
storage medium
sustained data transfer rate
tape drive
track
track-to-track seek time
volatile
wait state
write once/read many (WORM) disk

Vocabulary Exercises

1. _____ RAM requires frequent _____ to maintain its data content. *[handwritten: Dynamic] [handwritten: refresh cycles]*

2. The speed at which data can be moved to/from a storage device over a communication channel is described in terms of its _____ rate. *[handwritten: data transfere]*

3. An optical storage medium that is written only during manufacture is called a(n) _____ *[handwritten: CD-ROMs]*

4. The _____ of a hard disk drive are mounted at the end of a(n) _____. *[handwritten: Read/write head] [handwritten: access arm]*

5. Data that are stored on magnetic media for long periods of time may be lost because of _____ and _____. *[handwritten: Magnetic leakage]*

6. A(n) _____ stores data in magnetically charged areas on a rigid platter. *[handwritten: Hard disk drives] [handwritten: magnetic decay]*

7. _____ is read or written by moving it past a fixed read/write head. *[handwritten: Magnetic tape]*

8. The _____ endian storage format places the _____ byte of a word in the lowest memory address. The _____ endian storage format places the _____ byte of a word in the highest memory address. *[handwritten: little] [handwritten: least significal] [handwritten: big] [handwritten: most significal]*

9. The contents of most forms of RAM are _____, thus making them unsuitable for long-term data storage. *[handwritten: volatile]*

10. The CPU incurs a(n) _____ when it must access storage devices that are slower than its own cycle time.

11. Under relative addressing, a program's memory reference is added to the content of a(n) _____ to calculate the corresponding physical memory address.

12. The number of bits used to represent a memory address determine the amount of a CPU's _____.

13. _____ is typically stated in milliseconds for secondary storage devices and nanoseconds for primary storage devices. *[handwritten: Access time]*

14. Average access time for a disk drive is determined by its _____, _____, and _____. *[handwritten: ave access, rotational speed, delay] [handwritten: # of sectors per track]*

15. _____ is slower, cheaper, and less volatile than _____. *[handwritten: S.S] [handwritten: P.S]*

16. The access method for RAM is _____ or _____ if words are considered the unit of data access. The access method is _____ if bits are considered the unit of data access. *[handwritten: direct] [handwritten: Random] [handwritten: parallel]*

17. _____ is similar to core memory in its use of iron to store a bit value as a magnetic charge.

18. Under _____, program memory references correspond to physical memory locations. Under _____, the CPU must calculate the physical memory locations that correspond to a program memory reference.

19. A(n) _____ is a semiconductor storage device that can be erased only by using ultraviolet light.

20. A(n) _____ optical storage device/medium is manufactured blank and can be written to once.

21. A(n) _____ uses a combination of optical and magnetic storage technologies.

22. A(n) _____ is a series of sectors stored along one concentric circle on a disk _____.

23. The mechanical factors that limit disk access time are _____ and _____.

24. The data transfer rate of a secondary storage can be calculated as the product of its access time and _____.

25. CD-ROMs are read at a(n) _____ (variable RPMs). Hard disk drives are read at a(n) _____ (fixed RPMs).

26. Tape drives are _____ devices. _____ are random or direct access devices.

27. The areas of a CD-ROM that store single bit values are called _____ or *land*
    Pils, depending on whether the data content is a zero or one.

28. _____ drives use helical scanning to achieve high storage density.

29. Secondary storage devices generally are not attached directly to the system bus. Instead, they are linked to a(n) _____ which is, in turn, attached to the system bus.

30. Average access time can usually be improved by _____ the data content of a disk.

31. Modern PCs generally use memory packaged on small standardized circuit boards called _____ or _____.

32. The _____ of a magnetic or optical storage medium is the ratio of bits stored to unit of medium surface area.

33. A(n) _____ supports parallel disk access by storing fragments of a single data item on different storage devices.

34. For most disk drives, the unit of data access and transfer is a(n) _____ or _____.

35. The _____ defines a standard method for storing directories and files on a CD-ROM. *high Sierra format*

36. Software programs permanently stored in a(n) _____ are called _____.

37. Under _____, programs reference physical memory locations. Under _____ or _____, programs memory references do not refer directly to physical memory locations.

38. Many standard recording formats for cartridge tapes have been defined by the _____.

Review Questions

1. What factor(s) limit the speed of an electrically based processing device?

2. What is/are the difference(s) between static and dynamic RAM?

3. What improvements are offered by fast page RAM and extended data out (EDO) RAM as compared to "ordinary" dynamic RAM (DRAM)?

4. Why isn't flash RAM commonly used to implement primary storage?

5. Describe serial, random, and parallel access. What types of storage devices use each method?

6. What is direct addressing? What is relative addressing?

7. What are the costs and benefits of indirect addressing?

8. How is data stored and retrieved on a magnetic mass storage device?

9. Describe the factors that contribute to a disk drives average access time. Which of these factors are improved if spin rate is increased? Which are improved if recording density is increased?

10. What problems contribute to read/write errors on magnetic tapes? Are these problems also present with other magnetic storage media/devices?

11. Why is the recording density of optical disks higher than the recording density of magnetic disks? What factor(s) limit this recording density?

12. Describe the processes of reading and writing from/to a magneto-optical disk? How does the performance and storage density of these disks compare to purely magnetic and purely optical disks?

Problems and Exercises

1. Assume that a magnetic disk drive has the following characteristics:
   ► 10,000-RPM spin rate.
   ► 2-nanosecond head-to-head switching time.
   ► 3-microsecond average track-to-track seek time.
   ► 5 platters, 1,024 tracks/platter side (recorded on both sides), 50 sectors/track.
   a. What is the storage capacity of the drive?
   b. How long will it take to read the entire contents of the drive sequentially?
   c. What is the serial access time for the drive?
   d. What is the average (random) access time for the drive? (Assume movement over half the tracks and half the sectors within a track.)

2. Assume that a CPU has a clock rating of 400 MHz. Assume further that half of each clock cycle is used for fetching and the other half for execution.
   a. What access time is required of primary storage for the CPU to execute with zero wait states?
   b. How many wait states (per fetch) will the CPU incur if all of primary storage is implemented with 15-nanosecond static RAM (SRAM)?
   c. How many wait states (per fetch) will the CPU incur if all of primary storage is implemented with 60-nanosecond "ordinary" dynamic RAM (DRAM)?

Research Problems

1. Investigate a current magneto-optical disk drive product offering such as those produced by Olympus (www.olympusamerica.com). How does the cost per unit of storage compare to magnetic hard disk drives and phase change optical drives? How does the access time compare (for both reading and writing)? How does the recording density compare?

2. In current computers, primary storage is always implemented with semiconductor devices. However, some optical devices could conceivably be used for primary storage. Experimental devices have been developed that store data in an optically reactive surface or solid. In a write operation, one or more lasers are focused on a single point to change the reflectivity of that point. In a read operation, one or more lasers are focused on a point, and the reflection from or transparency through that point is used to indicate the data value stored there. What types of materials are used to construct such devices? How is random access (addressing) achieved? What is the access time of such devices? Is storage by this method nonvolatile?

7

System
Integration & Performance

▶ Describe the implementation of the system bus.

▶ Describe the configuration and control of I/O channels.

▶ Describe the use of interrupts to coordinate actions of the CPU with secondary storage and I/O devices.

▶ Describe the use of buffers, caches, and data compression to improve computer system performance.

SYSTEM BUS

The term **bus** describes a shared communication channel that connects two or more devices. The **system bus** is a set of parallel communication lines connecting the CPU with main memory and other system components, as shown in Figure 7-1. Each bus line can carry a single bit value during any **bus transfer** operation. Subsets of **bus lines** are dedicated to carry specific types of information. Computer system buses commonly use three subsets: the data bus, the address bus, and the control bus.

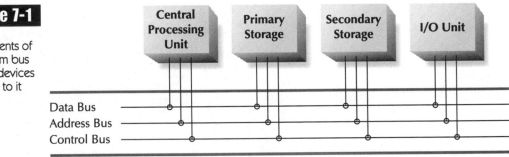

Figure 7-1

Components of the system bus and the devices attached to it

System Bus

As its name implies, the **data bus** is a set of lines dedicated to moving data from one computer-system component to another. The number of lines is typically the same as the CPU word size or a whole multiple of it. Thus, a computer system with a 32-bit CPU generally dedicates 32, 64, or 128 bus lines to carry data. The size (or width) of the data bus should always be at least equal to CPU word size, which helps prevent CPU wait states for bus data transfers. Data bus widths larger than CPU word size can provide performance improvements if the CPU can effectively use multiple word inputs.

The **address bus** is a set of lines dedicated to carrying the bits of a memory address. Commonly, 32-bits are allocated to this bus in modern computers. When a device attempts to transfer data to main memory, it simultaneously transmits the data item on the data bus and the address where that data item should be stored on the address bus. The address bus is unused for bus transfers in which main memory is not the target, or destination, device.

The **control bus** is a set of lines dedicated to carrying commands, command responses, status codes, and similar messages. Coordination of communication activity among computer-system devices is achieved by sending appropriate signals over the control bus. The format, content, and timing of messages sent across the control bus is an integral part of the **bus protocol**. Every device attached to the bus must understand and obey the rules of protocol. Thus, a bus is a combination of data transfer circuitry, a communication protocol, and the circuitry within the bus and the attached devices that implement that protocol.

Bus Protocol

As with other communication protocols, strict timing requirements are an integral part of the bus protocol. One line of the control bus is dedicated to carrying the **bus clock** pulse, which serves as a common timing reference for all attached devices. The frequency of the bus clock, measured in megahertz (MHz), may be equivalent to the CPU clock rate or to some well-defined fraction of it. Each clock pulse marks the start of a new opportunity to transmit data or a control message. The time interval from one clock pulse to the next is called a **bus cycle**. Each device continuously monitors the clock pulse and will only initiate a bus transfer at the start of a bus cycle.

Another important aspect of the bus protocol is the method by which attached devices initiate messages. If two devices attempt to send a message across the bus at the same time, the messages collide to produce noise. The bus protocol must regulate device access to the bus to prevent collisions. There are two broad approaches to this regulation—a master-slave approach and a multiple-master (i.e., peer-to-peer) approach.

In a von Neumann architecture, the CPU is assumed to be the focus of all computer activity. In that role, the CPU is also the **bus master**, and all other attached devices are **bus slaves**. This master-slave relationship has two important consequences: First, no device other than the CPU can access the bus except in response to an explicit instruction from the CPU; second, all data communication on the bus must be routed through the CPU.

The first consequence allows simplification of the bus protocol and bus interface circuitry. However, the second consequence can be a substantial limitation on overall system performance. The requirement that all data pass through the CPU can cause a communication bottleneck when large amounts of data need to be transferred from one device to another (e.g., from secondary storage to main memory). It is also a poor use of an expensive device (the CPU) to utilize it largely as a data conduit. The CPU is effectively prevented from executing computation and other instructions when it is actively managing data transfers among peripheral devices.

System buses that use the master-slave approach modify the von Neumann assumptions to allow direct data transfers between devices without routing those transfers through the CPU. This aspect of bus design is sometimes called **direct memory access (DMA)**, or a **multiple master bus**. The terms are not, however, equivalent. Direct memory access sometimes is used generically to refer to the capability of I/O and mass storage devices to move data directly to or from memory. It is also, however, a specific technology that can be used in conjunction with a bus that does not have true multiple-master capability. A device called a **DMA controller**, which provides a direct interface between the bus data lines and main memory, bypassing the CPU, is attached to the bus and to main memory.

Multiple-master capability implies that devices other than the CPU can assume control of the bus (i.e., act as a bus master). In this situation, peripheral devices must possess sufficient intelligence to act as a bus master. The bus itself must also possess intelligence to arbitrate among conflicting demands by devices to become a bus master. The intelligence is implemented in processing circuitry within the attached device or attached directly to the bus (i.e., a **bus arbitration unit**). The complexity of the bus protocol as well as the number of control and status signals increase substantially when multiple masters are allowed.

The benefit of the multiple-master approach to bus design is that data transfers among peripheral devices can be made without the assistance of the CPU, which allows the CPU to perform other functions while the transfer is taking place. It is, in effect, a form of parallel processing because the computer system is doing two things at once.

Bus Performance Factors

Two of the more important bus performance factors have been discussed already: the width of the data bus and the bus protocol. Another key performance feature is the **bus clock rate**. Each bus cycle represents an opportunity to transmit a message across the bus, and the data transfer rate of the bus is directly related to its clock rate. For example, a bus with a 33-MHz clock rate and a 32-bit data width has a theoretical data transfer rate of

$$33 \text{ MHz} \times 32 \text{ bits} = 33,000,000 \times 4 \text{ bytes} = 132 \text{ MBps}$$

As clock rate increases, so does the theoretical data transfer rate.

The effective data transfer rate of the bus is less than the theoretical maximum because of the overhead associated with implementing the bus protocol. Commands and command responses consume bus cycles. For example, data transfer from a disk drive to main memory usually occurs as a result of an explicit command from the CPU. Sending that command requires a bus cycle. In most bus protocols, that command is followed by an acknowledgment from the recipient and, later, by a confirmation that the command was carried out. Both signals consume a bus cycle, and each bus cycle used to send a command or response is a cycle that is unavailable to transmit data.

Theoretically, the easiest way to increase the bus data transfer rate is to increase the bus clock rate, but this method is limited. The primary limit is the physical length of the bus. By its nature, the bus must be relatively long because it connects many different devices within the computer system. Physical space is required to implement the connections between these devices and the bus. Although miniaturization of all computer-system components has dramatically reduced the length of a typical system bus, it is still large (e.g., 20 centimeters or more) in most computer systems. A bus cycle cannot be any shorter than the time required for an electrical signal to traverse the bus from end to end. Therefore, bus length imposes a theoretical maximum on bus data transfer rate.

In practice, the clock rate is set far below this theoretical maximum, which allows for some noise and interference and for reasonably priced peripheral devices. Consider the analogy of a modern highway. A highway could support traffic at 150 miles per hour and require all traffic to drive at that speed (i.e., a specified clock rate), but then we'd all be forced to drive cars capable of that speed. Although the 150-mph design limit might work under ideal conditions, it could be disastrous at night or in rain or snow.

LOGICAL AND PHYSICAL I/O

All secondary storage and I/O devices must be connected directly or indirectly to an **I/O port** on the system bus. Although the system bus provides for some variation in device control procedures (e.g., through the use of various command and status codes), the flow of data between the CPU and all devices is handled similarly. Yet there are major differences in many important characteristics of these devices, including capacity, data transfer rate, data coding methods, and many others.

The design of the CPU instruction set and the system bus would be much more complex if all possible differences between I/O and storage devices were considered. Physical-device details such as how a disk read/write head is positioned or how a certain color is displayed on a video display would require explicit processor instructions and/or bus control signals. Accounting for these complexities would increase the cost of the bus and processor as well as limit the types of secondary storage and I/O devices that could be included in a computer system. Only those devices that were "designed in" could be used.

The alternative (and commonly used approach) is to treat each device similarly: that is, provide a small set of generic capabilities for communication with and control of all devices. These generic commands must be limited and simple so as to streamline the CPU and the bus protocol. However, they also must provide sufficient power to allow processor control over a wide range of present and future storage and I/O devices.

The use of a generic set of commands and status signals for communication with a storage or I/O device is termed a **logical access**. In contrast, a **physical access** refers to the physical actions of the device that are carried out to meet the request (e.g., the physical actions required to read a sector of a disk or print a character on a printer).

Logical Access Commands

Generic I/O capabilities can be provided with two simple commands:

▶ read <device-id> <address>

▶ write <device-id> <address>

and three status signals:

▶ ready <device-id>

▶ busy <device-id>

▶ result <device-id> <condition-code>

The parameter <device-id> is a unique numeric code identifying one of the devices attached to the system bus. The parameter <address> is a number identifying a unique location within a device. The parameter <condition-code> is a number that indicates the result of processing. In its simplest form, it has only two values—one to indicate success (e.g., 1) and the other to indicate failure (e.g., 0). More elaborate coding schemes can be used to differentiate among various reasons for failure.

A processor executes I/O commands by placing a command code and device identifier on the bus control lines. An I/O or storage device detects these signals and responds by issuing a ready or busy status signal. When the device completes its processing, it places a result status code on the bus. The exact nature of data transfer varies from one computer system to another. In simple architectures, the CPU can use the data lines when a command is issued. This method is inefficient for large transfers because the CPU must transfer data in small increments by multiple commands. For example, storing 512 bytes of data over a bus with 32 data lines would require 128 separate write commands.

Modern computer architectures usually provide for block transfers (sometimes called burst-mode transfers). A write command issued by the CPU can include the starting address of the data in memory (i.e., using the bus address lines) as well as the number of bytes to transfer (i.e, using the bus data lines). Once the command is acknowledged, the CPU exercises no further control over the transfer. This method of I/O requires that a portion of main memory be dedicated to I/O for each individual I/O and storage device.

Logical Device Views

The CPU assumes that the storage locations of a secondary storage device are numbered sequentially starting at location zero and extending to the maximum capacity of the device. This organization is referred to as a **linear address space**. The physical organization of storage locations and the method of accessing those locations will vary among secondary storage devices. Disk devices are normally organized into blocks (or sectors) of 512 bytes. The <address> component of a read or write command identifies one of these blocks (see Figure 7-2).

Figure 7-2

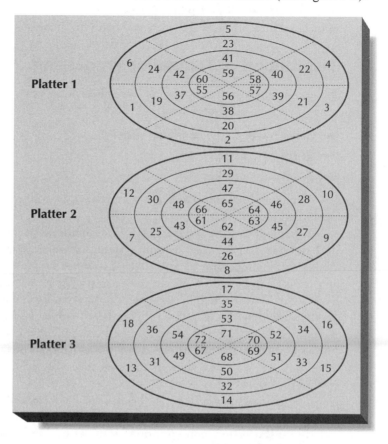

An assignment of logical sector numbers to physical disk locations. Note that the logical order minimizes head-to-head switching and track-to-track seeks for when assessing logical sectors in sequential order.

Translating logical disk addresses into physical addresses requires that the address parameter of a read or write command be converted into a specific disk location expressed in terms of a platter surface—or the read/write head dedicated to it—as well as track and cylinder. With tape drives, the translation between logical and physical addresses is more straightforward than for disk drives because the physical organization of blocks on a tape is linear. An access to a logical block is translated into commands to position the tape head over the appropriate portion of the tape before a read or write operation.

Some I/O devices communicate only in terms of sequential streams of characters (e.g., most ink-jet printers). With these devices, the concept of an address or location is irrelevant. Other I/O devices have storage locations in the traditional sense; For example, the individual character positions of a video display can be considered individual storage locations. With such devices, it is often desirable to write to individual display locations without disturbing data already being displayed in other locations (i.e., to overwrite only a portion of the display). This form of output requires that individual display locations be assigned addresses and that logical addresses in read and write commands be translated into physical (i.e., row and column) locations.

Distribution of Translation Processing

The translation between logical and physical accesses can occur in various places within a computer system, including:

► Software (e.g., the operating system).
► Within the device.
► A device controller.

Each of these approaches has relative advantages and disadvantages, as described next.

Knowledge of the physical organization and access methods of storage and I/O devices can be used within system software. For example, information about the organization of a disk (e.g., number of platters, tracks, and sectors per track) can be stored within the operating system. Service requests for disk access can be translated by the operating-system kernel into explicit commands to the disk drive to access a specific head, track, and sector. The commands necessary to instruct the device can be numerically encoded and communicated through the system bus in one or more transfers. Thus, all of the intelligence necessary to translate logical accesses into physical accesses would reside within the operating system.

The primary advantage to this approach is flexibility. Changes in device characteristics (e.g., replacing a smaller disk with a larger one) are incorporated into the system configuration by updating the corresponding kernel programs. The addition of new devices also is dealt with in this manner. Simple translation would require a relatively simple program whereas more complex translations would require correspondingly more complex programs. The primary disadvantage of a software-based approach is inefficiency. Recall that the motivation for using logical accesses is to offload complexity and work from the CPU. Although a software-based translation does simplify CPU and bus hardware, it still requires CPU resources to execute the translation programs. For this reason, complete logical/physical translation within software is used rarely.

Translation intelligence also can be embedded within a storage or I/O device. That is, each device implements both its normal physical actions and the processing actions necessary to perform logical to physical translation. The primary advantage to this approach is efficiency. Processing overhead associated with storage and I/O accesses is moved to the devices. Other than for logical command and bus control, no CPU cycles are consumed for I/O processing.

The primary disadvantage of device-based translation is cost and redundancy. Processing power must be included in every storage and I/O device. This cost is replicated over all similar devices in a computer system. Consider, for example, a mainframe computer system with 20 identical disk drives. Logical/physical translation within the operating system would use a single processor (the CPU) and store the translation program once. Translation within each disk drive would require 20 different, although relatively simple, processors, each with a separate copy of the translation program.

The most common approach to logical/physical translation is a distributed approach. Small portions of the translation process are implemented within system software and within device hardware. The majority of the translation process (and associated control procedures) are implemented within auxiliary hardware devices, as described next.

DEVICE CONTROLLERS

It is rare to connect storage or I/O devices directly to I/O ports. Instead, such devices are connected indirectly through **device controllers**, as shown in Figure 7-3. Device controllers perform several functions, including the following:

► Implement the bus interface and access protocols.
► Translate logical accesses into physical accesses.
► Allow several devices to share access to a single I/O port.

Figure 7-3

A depiction of the connection of mass storage and I/O devices using device controllers as interfaces to I/O ports

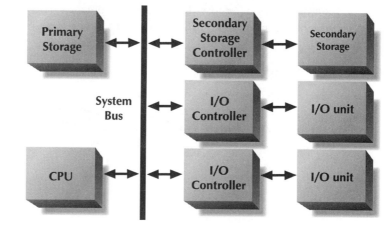

Because they are attached directly to I/O ports on the system bus, device controllers are responsible for implementing the bus protocol. The controller monitors the bus control lines for signals to attached devices. When detected, these signals must be translated into the appropriate commands to the storage or I/O device. Similarly, data and status signals from the device must be translated into the appropriate bus signals and placed on the bus. All protocols regarding timing, interrupts, signal encoding, handshaking, and so forth, are thus implemented by the controller.

The translation of signals between the bus and devices is a portion of the conversion between logical accesses and physical accesses. Device controllers must "know" the physical details of the attached devices and must issue specific instructions to the device. For example, a request for a certain sector of data from a disk must be converted into a command to read from a specific head, track, and sector. The CPU (or the program executing on it) sees the disk as a linear address space. The device controller provides the translation between that logical view of the disk drive and the physical platters, read/write heads, and sectors.

The number of physical I/O ports on the system bus is usually limited (16 is a typical maximum). Device controllers can serve as a mechanism for sharing I/O ports among several storage or I/O devices. A device controller may allocate bus access in either dedicated or multiplexed mode. In **dedicated mode**, all lines are used to transmit a burst of data between a single device and the CPU. In essence, the controller acts as a switch or selector, allocating access to the I/O port and translation facilities to only one device at a time. Such controllers are used for devices capable of high-speed data transmission such as disk drives.

In **multiplex mode**, transmissions involving multiple devices are interleaved, or handled concurrently. Such transmission is typically implemented using time-division multiplexing, as described in Chapter 8. The controller combines or separates the communication between several devices and the I/O port simultaneously. Such a controller is used with slow-speed devices such as modems and keyboards. This method efficiently allocates the relatively large communication bandwidth of an I/O port to many slower devices, none of which can individually match the port's data transfer rate.

Mainframe Channels

The use of device controllers greatly reduces the amount of CPU time dedicated to I/O and storage accesses. Within many mainframe computers, the concept of device control is taken one step further. That is, a dedicated, special purpose, computer is attached to each I/O port. This computer is called an **I/O channel**[1] or simply a **channel**.

The difference between an I/O channel and a device controller is not clear. It is a function of power and capability in several key areas, including:

► Number of devices that can be controlled.

► Variability in type and capability of attached devices.

► Maximum communication capacity.

A typical secondary storage controller can control up to eight devices (e.g., magnetic disk drives, optical disk drives, and magnetic tape drives). A typical I/O device controller can control up to 32 I/O devices (e.g., modems and terminals). In general, a device controller can only control devices of similar type and capacity. Its maximum communication capacity—with the CPU—is typically less than 100 MBps.

A channel dedicated to secondary storage will typically control up to a few dozen secondary storage devices. These devices may be of mixed types and capacities (e.g., four tape drives and 24 disk drives). A channel dedicated to online terminals might control as many as 256 devices. The maximum communication capacity of a channel is generally several hundred MBps.

Technology **Focus**

SCSI

The acronym SCSI (pronounced "scuzzy") stands for small computer system interface. It is an ANSI standard bus designed primarily for mass storage and I/O devices encompassing both physical and logical parameters. Although its name implies that it is used primarily in small computer systems, it is commonly applied in computer systems ranging from PCs to mainframes. SCSI exists in three primary forms—namely, SCSI-1, SCSI-2, and SCSI-3. SCSI-1 was standardized by ANSI in 1986 and SCSI-2 in 1994. SCSI-3 is currently in the final stage of ANSI approval. This description will concentrate primarily on SCSI-1.

---

[1] The term **channel** was originally coined by IBM to describe a specific component of its 300 series (and later 3000 series) mainframe computers. Due to the predominance of IBM mainframe computers, it has since gained the generic meaning described here. Vendors of other mainframe computer systems often use different terms (e.g., **peripheral processing unit** and **front-end processor**) to describe functionally similar components.

A SCSI bus implements both a low-level physical I/O protocol and a high-level logical device control protocol. The bus can connect up to eight devices, each of which is assigned a unique device identification number (0 through 7). The bus can be as long as 25 meters. Logically, the bus can be divided into two subsets—the control and data buses. The control bus is used to transmit control and status signals, whereas the data bus is used primarily to transmit data and may also be used to send device identification numbers. In SCSI-1, the data bus is eight bits wide; in SCSI-2, the data bus can be up to 32 bits wide.

A SCSI bus allows multiple devices to be bus masters, thus requiring a mechanism for arbitration when more than one device wants control of the bus. Theoretically, a SCSI bus could be used as the system bus for an entire computer system. Thus, all computer devices (including one or more CPUs) would be attached to the bus. In practice, the SCSI bus is used to connect and control mass storage devices. A SCSI controller translates signals between the system bus (or a CPU) and the SCSI bus. This architecture is shown in Figure 7-4.

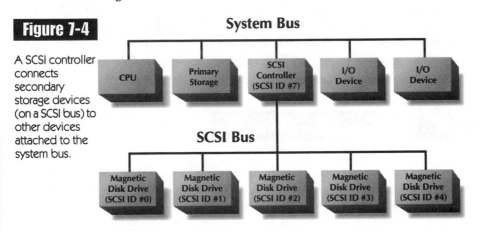

Figure 7-4

A SCSI controller connects secondary storage devices (on a SCSI bus) to other devices attached to the system bus.

A partial listing of the SCSI-1 control signals is shown in Table 7-1. To initiate a transaction, a device requests control of the bus by sending a busy signal and placing its identification number on the data bus. If multiple devices send the busy signal simultaneously, the device with the highest identification number becomes the bus master and is designated as the initiator. The initiator then selects a target device by sending a select signal and placing the identification number of the target on the data bus.

TABLE 7-1	Name	Description
A subset of the SCSI-1 control signals	Acknowledge	In combination with the Request signal, used to control asynchronous data transfer between devices.
	Attention	Instructs a target to place a message to the initiator on the bus.
	Busy	Sent continuously by the initiator and/or target while a bus transaction is in progress.
	Control/Data	Set by the target to indicate whether the data bus contains data or a control signal.
	Input/Output	Indicates the direction of data transfer on the bus (target to initiator or initiator to target).
	Message	Set by the target to inform the initiator that a message is being sent on the data bus.
	Request	In combination with the Acknowledge signal, used to control asynchronous data transfer between devices.
	Select	Used by the initiator to select a particular device as the target. The identification number of the target is simultaneously placed on the data bus.

Once the target device has been selected, the initiator sends it a data structure containing one or more commands. The ability to transfer multiple commands simultaneously is an important feature of the SCSI bus in that it allows a series of commands to be sent to a target with control of the bus released during the time the target processes those commands. The target queues the commands and executes them in sequence. This process avoids much of the control overhead associated with multiple commands. In addition, other devices can use the bus while the target is processing commands. Thus, interleaved execution of command sequences between multiple devices is possible.

A target can send several different status signals to an initiator, including:

- ► OK.
- ► Busy.
- ► Check condition.
- ► Condition met.
- ► Intermediate.

The OK signal is sent to indicate the successful receipt (and intent to satisfy) of a command or command sequence. A busy signal is sent if the target cannot process a request at the current time (e.g., because it is busy processing previously received commands). A check condition signal tells the initiator that an error has occurred (e.g., attempt to access a nonexistent storage location). The initiator must explicitly ask the target to send a message indicating the nature of the error. The condition met status signal is sent when a search command is executed successfully. (SCSI defines several different search commands including searches based on data content.) An intermediate signal is sent to indicate the successful completion of one command in a command sequence.

Data transfer between target and initiator can be implemented either synchronously or asynchronously. In a synchronous transfer, the target sends a request signal simultaneously with the data. The initiator simultaneously sends an acknowledge signal, but the target does not wait for that signal to begin sending the next data item. Synchronous data transfer thus uses the full capacity of the bus (i.e., 5 Mbps with SCSI-1) for data transfer. In asynchronous data transfer, the target sends a request signal and data simultaneously. The target doesn't send the next data item until the initiator explicitly acknowledges the receipt of the previous data item. Thus, asynchronous data transfer can proceed no faster than half the data transfer rate of the bus; every other cycle is used by the target to send an acknowledge signal.

The importance of the SCSI bus in modern computer systems follows from several of its characteristics, including:

- ► Standardized and nonproprietary definition.
- ► High data transfer rate.
- ► Multiple master capability.
- ► High-level (logical) data access commands.
- ► Multiple command execution.
- ► Interleaved command execution.
- ► High-level (logical) data access commands.

The standardized definition allows any hardware vendor to manufacture and distribute SCSI compatible devices. Data transfer rates can be as high as 40 MBps with SCSI-2, which approaches that of mainframe channels. Multiple-master capability provides for efficient bus control with a minimum of control overhead. Target devices can become initiators, or bus masters, in order to transfer data to a requestor. The ability to combine multiple commands and to interleave command sequences increases the likelihood that most of the bus capacity will be used to transfer data rather than control and status signals.

The access command set is rich and fairly high level. The standard hides many of the complexities of physical device organization from other computer-system components. For example, disk devices appear as a linear address space to the CPU and/or operating system. Bad storage blocks are detected within a SCSI disk drive, and the linear address space ignores them when processing requests. A wide variety of devices can be interfaced to the SCSI bus including magnetic and optical disk drives, conventional and digital tape drives, and data communication devices.

I/O PROCESSING

In most I/O operations, data transfer rates are much slower than the speed of the processor. The time interval between a request by the CPU for input and the moment that input is received can span many CPU cycles, which is due primarily to physical, often mechanical, limitations of secondary storage and I/O devices. If the CPU waits for the completion of the request, the CPU cycles that could have been devoted to instruction execution are "wasted." These wasted CPU cycles are referred to as **I/O wait states**.

To prevent such inefficient use of the processor, access to the CPU by peripheral devices is controlled by **interrupt** signals. In its simplest sense, an interrupt is a signal to the CPU that some event has occurred that requires processor action. It is a coded signal that is generated and sent over the control bus by a hardware device (e.g., an I/O device) to request access to the CPU or to trigger some processing action.

The implementation of interrupts requires both a means for recognizing and responding to the signal. Recognition is implemented by assigning a storage location for the interrupt signal and by checking this storage location at the conclusion of each execution cycle. Each type of interrupt is assigned a unique code and an interrupt is made by placing that code on the bus. The bus interface logic of the CPU continuously monitors the bus for such signals. When an interrupt is detected on the bus, a corresponding **interrupt code** (or number) is placed in the **interrupt register**.

At the conclusion of each execution cycle, the control unit checks the interrupt register for a nonzero value. If one is present, the CPU suspends execution of the current process (i.e., it does not fetch its next instruction) and proceeds to process the interrupt. When the interrupt has been processed, the interrupt register is reset to zero and execution of the suspended process resumes.

Communication with I/O devices through interrupts is only advantageous if the CPU has something else to do while it is waiting for an I/O-related interrupt to occur. If the CPU is executing only a single process or program, then I/O wait states are not avoided. However, if the CPU is sharing its processing cycles among many processes, then performance improvement can be substantial. While one process waits for data to be returned by an I/O device, the CPU can process instructions of another process. When an interrupt is received indicating that the I/O operation is complete, the CPU can process the interrupt (e.g., retrieve the data) and then return to the original (i.e., suspended) process.

Interrupt Handlers

The interrupt handling mechanism is less a hardware feature than a means of interfacing to system software. Recall that the operating system service layer and kernel provide a set of routines (or service calls) for performing low-level processing operations. I/O operations comprise a large portion of these routines. Viewed in this light, an interrupt is a request by an I/O or storage device for service from the operating system. For example, an interrupt signal that indicates requested input is ready is actually a request to the operating system to retrieve the data and place it where the program that requested it can access it (e.g., in a register or in memory).

There are service routines—called **interrupt handlers**—to process each possible type of interrupt. Interrupt handlers that process requests from storage and I/O devices are implemented within device drivers in the operating system kernel. To process an interrupt, the CPU must load the first instruction of the interrupt handler for execution, but each interrupt uses a different interrupt handler and, thus, a different address for the first instruction.

The usual method for determining the proper set of instructions to process a given interrupt is to use a master interrupt handler and an **interrupt table**. The master interrupt handler (also called the **supervisor**) is a service routine that examines the value in the interrupt register and determines the proper service routine to process the interrupt. An interrupt table is maintained that contains each interrupt code and the starting address in memory of its interrupt handler. The value in the interrupt register is used as an index into this table, and a branch instruction to the corresponding address is executed.

Multiple Interrupts

The previously described interrupt handling mechanism seems adequate until one considers the possibility of multiple competing interrupts. What happens when an interrupt is received while the CPU is busy processing a previous interrupt? Which interrupt has priority? What is done with the process that doesn't have priority? The answers to these questions require the use of interrupt priorities and a stack.

Interrupts can be roughly classified into the following categories:

► I/O event.
► Error condition.
► Service request.

The interrupts discussed thus far fall primarily in the category of I/O events. These interrupts are used to notify the operating system that an I/O request has been processed and that data is ready for transfer. Error condition interrupts are used to indicate error conditions resulting from processing. They can be generated by software (e.g., attempting to open a nonexistent file) or by hardware (e.g., divide by zero).

In many operating systems, the interrupt processing mechanism is the means by which application programs make requests to the operating system service layer. In such systems, interrupt codes are assigned to each service call. An application program requests a service by placing the corresponding interrupt number in the interrupt register. The interrupt code is detected at the conclusion of the execution cycle, the requesting process is suspended, and the service program is executed.

An operating system must establish a hierarchy of interrupts in terms of their priority. For example, error conditions receive higher precedence than other interrupts, and critical hardware errors (e.g., power failure) receive the highest priority. The priorities determine whether an interrupt that arrives while another interrupt is being processed will be handled immediately or suspended until current processing finishes. For example, if a hardware error interrupt code was detected while an I/O interrupt was being processed, the I/O processing would be suspended and the hardware error would be processed immediately.

Stack Processing

Consider the following problem. While executing instructions in an application program, an interrupt is received from a pending I/O request. The interrupt is detected, and the appropriate interrupt handler is called. During the execution of the interrupt handler, several values in general purpose registers are overwritten by intermediate results generated by the interrupt handler. When the interrupt handler terminates, processing of the application program resumes, but an error occurs because a value in a register was altered during the execution of the interrupt handler.

How could this error have been prevented? How did the CPU know which instruction from the application program to load after the interrupt handler terminated? These problems arise from the need to "pick up where you left off" after a process or program is interrupted and suspended. It would be a waste of resources to restart the suspended process from the beginning. Therefore, a mechanism—called a **stack**—must be provided whereby processing can begin from the point of interruption.

A stack is an area of storage that is accessed in a last in first out (LIFO) basis. It is analogous to a stack of plates in a cafeteria or the bullets in the ammunition clip of an automatic pistol or rifle. As items are added to the stack, they are placed on the top; as items are removed, they are removed from the top and only from the top.

For the purposes of interrupt processing, the stack is a stack of register values. Whenever a process is interrupted, values in the CPU registers are added to the stack in an operation called a **PUSH**; the set of registers saved are referred to as the **machine state**. When an interrupt handler finishes, the set of register values on the top of the stack is removed and loaded back into the appropriate registers in an operation called a **POP**.

It is not always necessary to push the values of all registers onto the stack. At a minimum, the current value of the instruction pointer must be pushed because the instruction at that address will be needed to restart the interrupted process where it left off. If indirect addressing is being used, then the offset register also must be pushed. The general purpose registers also are pushed because they may contain intermediate results necessary for further processing. These registers might also contain values that must be output to a storage or I/O device.

Multiple machine states can be pushed onto the stack as high-precedence interrupts occur while processing lower-precedence interrupts. The stack can fill to capacity, in which case further attempts to push values onto the stack result in a **stack overflow** error. The size of the stack thus represents a limitation on the number of suspended processes.

The stack is physically implemented as a set of extra registers or as a special area of main memory. If the stack is implemented in memory, then a separate register within the CPU called the **stack pointer** must be provided. This register always contains the memory address of the value on the top of the stack. Its value is incremented or decremented each time the stack is pushed or popped.

Performance Effects

The sequence of events that occurs when processing an interrupt is summarized in Figure 7-5. This complex series of events is relatively expensive to implement in terms of CPU cycles consumed. The most expensive parts of this sequence are the PUSH and POP operations. Each requires the copying of a large number of register values from registers to main memory or vice versa. Wait states can be incurred if the stack is not implemented within a cache permanently maintained within the CPU.

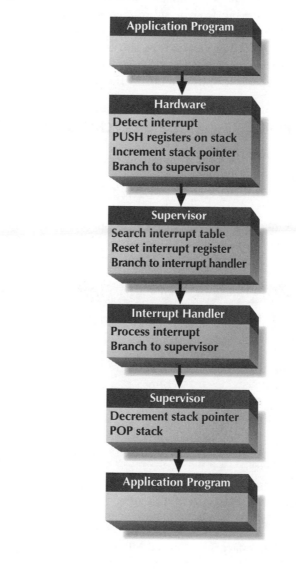

Figure 7-5

Sequence of events that occurs when processing an interrupt

Supervisor execution also consumes CPU cycles. The table lookup procedure is relatively efficient, but it still consumes several CPU cycles. Additional steps in the process consume further CPU cycles (discussed in Chapter 12). In total, the processing of each interrupt consumes at least 100 CPU cycles in addition to the interrupt handler's cycles.

It was common practice to minimize the use of interrupts in early computers with slow CPUs. Programmers considered that the overhead associated with interrupt processing was too high. In particular, programmers often avoided the use of subroutines and functions in their source code because the process of calling and returning from a subroutine or function is implemented through the PUSH/POP mechanism of the CPU. (Think of the calling process as the process that is being interrupted and the subroutine or function as the interrupt handler.) Use of subroutines and functions became commonplace as CPUs became faster and less expensive, thus minimizing the real cost of interrupt processing overhead.

I/O AND STORAGE DEVICE PERFORMANCE ISSUES

The speed of the CPU is many orders of magnitude greater than other devices in the system. Table 7-2 lists access times and data transfer rates for various types of secondary storage devices. Table 7-3 provides the data transfer rates of various I/O devices. Relatively long storage device access times lead to wait states when the CPU performs read operations or awaits acknowledgment of a completed write operation. Differences in data transfer rates among the system bus and attached devices create incompatibilities that must be resolved.

TABLE 7-2	Storage Device	Maximum Data Transfer Rate	Access Time
Performance differences between typical memory and storage devices	Dynamic RAM	66 MBps	60 ns.
	Hard Disk	5 MBps	8 ms.
	Floppy Disk	500 KBps	200 ms.
	CD-ROM	250 KBps	100 ms.
	DAT Tape	1 MBps	N/A

TABLE 7-3	Device	Speed
A comparison of typical speeds for a range of I/O devices	Video Display	10–50 MBps
	Ink Jet Printer	30–120 LPM
	Laser Printer	4–120 PPM
	Network Interface	10–100 MBps

Interrupt processing is the primary means of resolving speed differences among the CPU, storage devices, and I/O devices. From the CPU's standpoint, the timing of I/O events and responses to storage access commands is unpredictable. Because of differences in data transfer rate and timing, I/O and storage access operations must be performed asynchronously with respect to CPU cycles. Interrupt handling permits the current process in the CPU to be suspended and then resumed after the interrupt has been processed. When interrupts are used to signal I/O events, the only disruption to CPU operation is the actual time interval required to perform the action specified by the interrupt.

Other means can be employed to reduce the inefficiencies that arise from connecting a fast CPU and system bus to slower peripheral devices, including:

▶ The use of RAM to improve the efficiency of data transfer operations.

▶ The use of RAM to improve storage device access time.

▶ The use of compression to reduce the volume of data transferred.

Each of these topics is discussed in detail next.

Buffering

A **buffer** is a small storage area (usually dynamic RAM [DRAM]) used to hold data in transit from one device to another. Buffers are used whenever there is an incompatibility between sender and receiver in the unit of data transfer or the overall data transfer rate. Their use to resolve differences in data transfer unit size is required, whereas their use to resolve data transfer rate differences is not. Implementing them for this purpose generally results in substantial performance improvements.

As an example of incompatible units of data transfer, consider communication between a PC and a laser printer (see Figure 7-6 on the next page). A PC sends data to a printer through a parallel communication port that transmits 8 bits per message. Thus, the unit of data transfer is 1 byte. The print engine of a laser printer accepts and prints an entire page at once. The data that comprises a printed page can range from several hundred to several million bytes. Thus, the input unit of data transfer (from the PC) is a single byte, and the output unit of data transfer (from the laser printer) is a full page consisting of multiple bytes. A buffer is required to resolve this difference.

As single bytes are received from the parallel port, they are added to the buffer sequentially. When all of the bytes of a printed page have been received, they are removed from the buffer and transferred to the print engine as a single unit. The print engine requires a buffer equivalent in size to its unit of data output. If the buffer is not large enough to hold a full page of data, then an error called **buffer overflow** will occur. A buffer for an input or output device is implemented within the device, as it is in a laser printer.

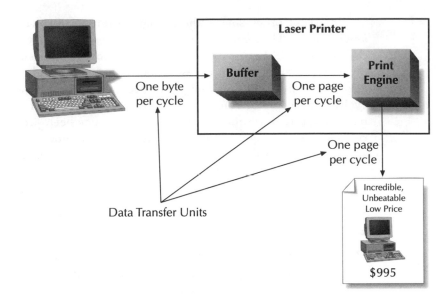

Figure 7-6

A buffer is used to resolve differences in data transfer unit size between a PC and a laser printer.

Laser Printer

Buffer

Print Engine

One byte per cycle

One page per cycle

One page per cycle

Data Transfer Units

Incredible, Unbeatable Low Price

$995

A buffer also can be used to improve system performance when two devices have different data transfer rates. Consider, for example, communication between a PC and an Internet service provider (ISP) by an internal modem. The modem transmits data to and from the ISP at a relatively slow rate (e.g., 56 Kbps). Data is sent to the modem from primary storage or the CPU at the data transfer rate of the system bus (e.g., 133 MBps). Thus, there is a data transfer rate difference of approximately 10^4.

Assume that the modem has a one-character buffer. Given that, data must be transferred from memory to that buffer one character at a time. Once the character is received, the modem sends a status signal (i.e., an interrupt) to indicate that no further data should be sent. This message prevents another character from arriving (i.e., causing buffer overflow) before the modem can transmit the current buffer content to the ISP. Once the buffer content has been sent, the modem transmits another status signal (an interrupt) to indicate that it is again ready to receive input.

How much data transfer capacity of the system bus is used to transmit 100,000 bytes through the modem? The answer is 300,000 bus cycles. Each byte transmitted results in two interrupts being sent across the bus—one to stop further transmission and another to restart it. Although the two interrupts are not data, they do consume bus data transfer capacity. Further, those interrupts must be processed by the CPU, thus incurring the processing overhead described earlier. If we assume 100 CPU cycles per interrupt, then a total of 20,000,000 CPU cycles are consumed. Thus, the use of a single-byte buffer degrades system performance by consuming two bus cycles and the CPU resources needed to process two interrupts for each byte sent to the modem.

Consider the same scenario with a 100-byte buffer implemented in the modem. A transfer from main memory to the modem would send as many bytes per bus cycle as could be carried on the data bus (e.g., 32 bits or 4 bytes). After 25 bus transfers (100 bytes ÷ 4 bytes per bus cycle), the modem would send an interrupt to stop transmission. After the buffer content was transmitted to the ISP, the modem would transmit another interrupt to restart data transfer from memory.

The buffer dramatically improves the performance of the entire computer system. Two interrupts are generated for each block of 100 bytes transferred, so, the total number of bus cycles consumed is:

$$100{,}000 \text{ data bytes} + \left(\frac{100{,}000 \text{ bytes}}{100 \text{ bytes per bus transfer}} \times 2 \text{ interrupts}\right) = 102{,}000 \text{ bus cycles}$$

This represents a 2% bus cycle overhead as compared to the previous 200%. The number of CPU cycles consumed falls to 200,000 (100,000 ÷ 100 × 2 × 100). These bus and CPU cycles are thus freed for other purposes (e.g., executing other programs).

The marginal improvement in system performance drops quickly as buffer size rises beyond the unit of data transfer. In the preceding scenario, an increase in buffer size from one to four would yield dramatic improvements. The data bus would be used more efficiently (i.e., to transfer 4 bytes at a time instead of one). CPU overhead for interrupt processing would drop to one-fourth of its previous value. Further increases in buffer size would not yield an equally dramatic improvement. Raising the buffer size to eight would improve bus utilization only by cutting the number of interrupts transmitted from 25,000 to 12,500. Similarly, CPU cycles consumed to process those interrupts would only be cut in half, as compared to a 4-byte buffer.

The conclusion is simple: A little bit of buffer space goes a long way toward improving computer system efficiency and performance. Further increases yield much less dramatic improvements. Given the relatively low cost of RAM, however, further increases may still be well worth their cost.

Cache

Like a buffer, a **cache** is a storage area (usually DRAM) used to improve system performance. However, a cache differs from a buffer in several important ways, including:

▶ Data content is not automatically removed as it is used.
▶ Cache is used for data transit in both directions.
▶ Cache is used for access to storage devices.
▶ Cache size is generally larger than buffer size.
▶ Intelligence (processing) is required to manage cache content.

These differences can lead to substantial performance improvements due to a reduction in transmission delays. Those performance improvements are gained at the expense of the cache itself and of the processing intelligence used to effectively manage its content.

The idea behind caching is simple. Accesses to data contained within a high-speed cache can occur much more quickly than accesses to slower storage devices or media. The speed difference is due entirely to the faster access speed of the RAM used to implement the cache. Successful use of caching requires a sufficiently large cache and the "intelligence" to use it effectively. The size of a buffer is typically equal to the size of a block of data transfer to or from a device. The size of a cache is typically large enough to hold tens or hundreds of data transfer blocks.

Write Caching

A cache is used in fundamentally different ways for data movement to and from a storage device. When used for writing to a storage device, a cache behaves similarly to a buffer (see Figure 7-7). The primary difference is in the confirmation signal returned to the sending device. Storage devices are designed to return a confirmation signal after the storage medium has been updated successfully. When a cache is used, the confirmation signal is sent as soon as data is written to the cache, before it is written to the storage device. This "early" generation of a confirmation has a significant performance advantage as well as a significant risk.

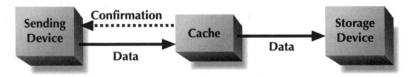

Figure 7-7

Use of a cache during write operations to a storage device. Confirmation of the write operation is sent as soon as data is written to the cache. Data is copied from the cache to the storage device after the confirmation is sent.

The performance advantage arises when a sending process must wait for the confirmation before it can proceed with other processing tasks. In that situation, returning a confirmation as soon as the cache is updated allows the sending process to continue with its next task immediately. If the confirmation isn't sent until the storage device is updated, then a delay occurs before the sending process can continue. If the storage device is relatively slow (e.g., a phase-change optical disk), then the delay can be substantial. The delay is particularly troublesome if the sending process is an interactive program (e.g., a word processor performing an automatic document save operation).

The performance advantage is not always realized. If the next operation the process performs is another write operation to the storage device, then performance improvement may not be realized. This is because the cache has much more limited capacity than the underlying storage device. If a write operation sends more data than the cache can hold, then the transmission of the excess must wait until cache content can be written to the storage device. As with a buffer, interrupts are used to coordinate the transmission activity in this case.

The risk associated with early write confirmation is that an error might occur when copying data from the cache to the storage device. In that case, the sending process thinks that something has happened that hasn't. This situation may be recoverable if the error occurs in the storage device. The larger danger is failure of the cache itself (i.e., because of a power failure) before data is written to the storage device. In that case, the data is permanently lost, and there is no mechanism to inform the sending process of that fact.

Data written to a cache during a write operation is not removed automatically from the cache as it is written to the underlying storage device. This can provide a performance advantage when data is reread shortly after it is written. A subsequent read of the same data item will be much faster unless the data item has been purged from the cache for some other reason (i.e., to make room for other data items).

Read Caching

Caching during a read operation is depicted in Figure 7-8 on the next page. A request for data from the storage device is first routed to the cache. If the data is already in the cache, then it is accessed from there. Performance is improved because access to the cache is much faster than access to the storage device. If the requested data is not in the cache, then it must be obtained from the storage device. Thus, a performance improvement is realized only if requested data is already waiting in the cache.

Figure 7-8

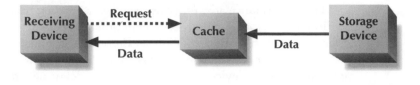

A **cache controller** is a processor that manages the content of a cache. It attempts to fill the cache with data that it expects the CPU to need. Thus, the cache controller "guesses" what the CPU will request next and reads that data from the device into the cache before it is actually requested. The methods by which so-called guesses are made range from the very simple to the very complex. Simple guessing methods often assume linear access to storage locations. Thus, a read request for location n causes the cache controller to prefetch location $n+1$ into the cache. More complex guessing methods require more complex processing circuitry. The expense of that circuitry is only justified if the guessing improves and if it results in improved performance.

When the requested data is already contained within the cache the access is called a **cache hit**. When the data needed is not in the cache, the access is called a **cache miss**. Cache misses require that a **cache swap** to/from the storage device is to be performed. The requested data is read from the storage device and placed in the cache. Other cache contents typically must be written to the storage device and purged from the cache to make room for the new content. The cache controller attempts to guess which data items in the cache are least likely to be needed in the near future and chooses those items to be purged. The swap process takes some time, but the delay incurred in a cache swap should be more than offset by the faster performance resulting from cache hits. This trade-off depends on the size of the cache and the time required to perform a cache swap.

Early implementations of caching implemented cache control (e.g., content checks, cache swaps, etc.) through system software. However, this resulted in a large amount of system overhead for every memory access. Modern implementations of cached memory rely on hardware-based cache control. That is, the cache and cache controller are implemented separately—within a device controller. Thus, the CPU and system software are unaware of the use of a cache or of the cache-management activities performed by the cache controller.

A surprisingly small amount of cache is required to realize considerable performance gains. Typical ratios of storage device capacity to cache size range from 10,000:1 to 1,000,000:1. Primary storage caches with a 10,000:1 size ratio typically achieve cache hits more than 90% of the time. The cache hit ratio also depends on the nature of access to the storage device. Large numbers of sequential read accesses tend to increase the hit ratio, whereas large numbers of random (scattered) read accesses tend to reduce the hit ratio.

Primary Storage Caches

Recall from Chapter 6 that current processor speeds generally meet or exceed the capability of dynamic RAM (DRAM) to keep them supplied with data and instructions. Although static RAM (SRAM) more closely matches the processor speed, it is too expensive to use for all of primary storage. The use of DRAM to implement primary storage results in a delay whenever the CPU loads data from memory. For example, a 200-MHz processor will incur 10 wait states each time it reads from 50-nanosecond DRAM.

One way to address the problem of wait states is to provide a limited amount of SRAM cache and use slower (less expensive) DRAM for the remainder of primary storage. The SRAM cache can be located on the same chip as the CPU or can be located outside of the CPU. Modern computers often use both approaches—a relatively small (e.g., 16-Kbyte) cache within the microprocessor and a larger (e.g., 512-Kbyte) cache outside. When both types are used, the on-chip cache is called a **level one (L1) cache** and the off-chip cache is called a **level two (L2) cache**. Some microprocessors (e.g., the Pentium II) implement both the microprocessor and the L2 cache on a single chip or small multiple-chip module, which reduces delays in accessing data from the L2 cache.

Cache control for primary storage can be sophisticated. The default sequential flow of instructions allows simple guesses based on linear storage access to work some of the time. But accesses to operands and branch instructions are not accounted for by this simple strategy. Modern CPUs use sophisticated methods to deal with these problems. Guessing the location of operands is often accomplished by fetching multiple instructions at a time. As the first instruction is being executed the cache controller examines the operands of the next few instructions. If they reference memory locations not currently held in the cache then those locations are prefetched into the cache.

Dealing with branch instructions is more complex. Fetching operands for the next instruction in linear sequence provides no benefit if the first instruction branches to some other part of the program. CPUs sometimes implement a technique called **branch prediction** to handle this situation. The stream of subsequent instructions is examined for conditional and unconditional branch instructions. The cache controller prefetches instructions from the program branch if the branch instruction is unconditional. If a conditional branch instruction is found, then the cache controller attempts to guess whether or not the branch condition will be true to determine whether or not to prefetch instructions from that portion of the program.

Secondary Storage

Disk caching is a commonly implemented strategy in modern computer systems. This is particularly true of network file servers and database servers. Sequential read-ahead strategies are effective for file servers because many files (e.g., executable program files and read-only documents) are read sequentially. File servers also can track file use and give higher priority for retention in the cache to those that are accessed frequently.

Applying file usage statistics to modify the cache control strategy requires that some portion of cache management be implemented within the operating system. The operating system also can provide information about file access modes (e.g., random, sequential, and write after read) that enable more intelligent cache management. For example, if a file is opened for read only access, it makes little sense to keep recently read records in the cache. If a file is opened for sequential reading, then a sequential, look-ahead strategy for subsequent records is likely to be advantageous.

Because the operating system can be a key to successful cache replacement and retention, there is a growing trend away from hardware-based cache implementation in disk controllers. Many computer-system designers feel that the cost of hardware-based caching systems is better spent on extra primary storage. The operating system—and, thus, the CPU—can then use this extra primary storage as a disk cache. This approach only makes sense if the computer system has "extra" CPU cycles to devote to cache management. This is generally the case with network file servers but may not be so with the more complicated processing requirements of database servers.

Data Compression

Data compression can be used to reduce the total number of bits used to encode a set of data items. Data compression requires a **compression algorithm** (program) and a processor to implement the algorithm. Depending on the type of data being compressed, a compression algorithm can reduce total data volume from 2 to 10 times or more. With a 4:1 **compression ratio**, a 100-Kbyte data set would be reduced to 25 Kbytes ($100 \div 4 = 25$).

Pentium II® Memory Cache

The Pentium II® microprocessor uses two levels of integrated primary storage caching. The L1 cache is divided into two 16-Kbyte segments: one for data and the other for instructions. A 512-Kbyte L2 cache is closely coupled to the microprocessor through a 64-bit dedicated cache bus. The L2 cache is implemented as a separate chip. The microprocessor and L2 cache chips are combined into a single integrated package and installed into a computer motherboard through a single connection point. The combined packaging allows the L2 cache bus to operate at one-half the clock rate of the CPU.

Figure 7-9 shows the relationship of the L1 and L2 caches to other CPU components, the system bus, and main memory. Both the L1 and L2 caches are divided into segments called cache lines. Each cache line is 32 bytes long. Reads and updates of cache contents always occur in units of a cache line. Thus, an update of a single byte in the L1 data cache causes the entire cache line that contains it to be flagged as modified, and the entire line is written to main memory during a subsequent cache swap or flush operation. The system bus is assumed to be a PCI bus with a 64-bit data path. Filling or flushing a cache line to/from main memory requires four system bus cycles.

Figure 7-9

Relationship of various architectural components of the Intel Pentium II® processor. Two levels of memory caching are used to improve processor performance.

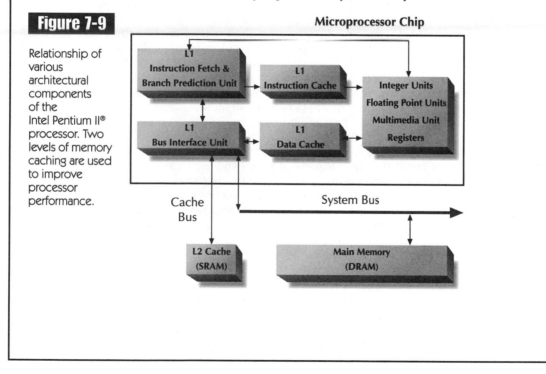

The bus interface unit serves several purposes—an interface to the system bus, a direct controller for the L1 data cache, and an indirect controller for the L1 instruction cache. Direct control over the L1 instruction cache is performed by the instruction fetch and branch prediction unit. This unit tracks the flow of instructions and maintains the contents of the instruction pointer, or program counter. By looking ahead of the current instruction using subsequent instructions contained in the L1 instruction cache, it predicts the path of future processing. It also interacts with the various CPU execution units to determine the likelihood of branch conditions being true or false. The addresses of predicted branch paths are passed back to the bus interface unit so that instructions and data can be loaded into the L1 caches before they are needed by the execution units.

Various caching modes can be assigned to 2-Kbyte, 4-Kbyte, or 4-Mbyte regions of main memory. Write-back caching enables full caching of both read and write operations. Main memory is updated only when necessary (e.g., when a modified cache line must be removed to make room for new content). Write-through caching causes all write operations to update the caches and main memory at the same time. This mode is useful for regions of memory that can be read directly (i.e., without knowledge of the CPU) by other devices on the system bus (e.g., a disk or video controller). All caching functions can be turned off for any memory region.

The use of separate L1 caches for instructions and data allows some streamlining of processor functions. For the instruction execution units, the L1 instruction cache is usually a read-only cache. The L1 data cache can be read or written as operands are loaded and stored. Cache controller functions can be optimized for each type of access; however, it is possible for a program to modify its own instructions. In this case, the code section being modified would be contained in both L1 caches at the same time. The processor implements specific circuitry to detect this condition and to ensure consistency between the two caches.

The Pentium II® was designed to participate in multiprocessor computer systems. Such systems introduce additional complexity into cache management because multiple processors can cache the same region(s) of memory at the same time. Inconsistency can result if two separate processors both modify their cached copy of the memory region and then, in separate write operations, flush the cache back to main memory. Program execution errors are possible if a program executing on one processor updates a cached memory location and another program, executing on another processor, subsequently reads that same memory location from an outdated cache copy.

Pentium II® processors use a technique called "memory snooping" to maintain consistency among their caches and main memory. Each processor monitors all other processor's accesses to main memory. Accesses to currently cached memory cause the corresponding cache line to be marked invalid, which forces the cache line to be reloaded from main memory the next time it is accessed. Write operations to memory locations that are currently cached and modified by another processor cause the two processors to exchange cache contents to ensure consistency.

Data compression is commonly used to reduce secondary storage requirements, as shown in Figure 7-10(a). A pre-storage compression algorithm reduces the volume of data sent to the storage device. Data received from the storage device is returned to its normal state by applying an inverse compression (decompression) algorithm. Data compression also can be used to expand the capacity of a communication channel, as shown in Figure 7-10(b). Data is compressed prior to entering the channel and decompressed as it leaves the channel. This technique is a standard component of high-speed (e.g., greater than 9,600 bps) analog modems.

Figure 7-10

The use of data compression with a secondary storage device (a) and a communication channel (b)

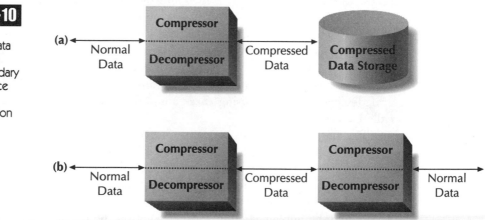

The use of data compression always represents a trade-off between processor resources and communication and/or storage resources. The implementation of the compression and decompression algorithm consumes processor cycles. Algorithms with high compression ratios frequently consume more processing resources than those with low compression ratios. The proper trade-off among these resources depends on their relative cost, availability, and degree of utilization. Implementing data compression with CPU-processing resources is not usually cost-effective; however, implementing data compression through special purpose processors is often advantageous. Such processors are now commonly available and are frequently used in disk controllers, video display controllers, and high-speed modems.

The bus is the communication pathway that connects the CPU with memory and other devices. The bus is implemented as a set of data lines, control lines, and status lines. The number and use of these lines and the procedures for controlling access to the bus are called the bus protocol. The CPU is assumed to control access to the bus. However, the implementation of a multiple master bus or direct memory access allows devices other than the CPU to control access to the bus temporarily, which improves performance.

The CPU communicates with storage and I/O devices through I/O ports on the system bus. To simplify bus and processor design, the commands and status signals used to access these devices are straightforward. Minimally, the commands "read" and "write" and the status signals "ready," "busy," and "result" are required to implement communication among the CPU and peripheral devices. An access using these simplified commands and status codes is called a logical access.

The physical processes needed to implement logical accesses are relatively complex and device-specific. To implement a logical access, it must first be translated into a specific set of instructions that are understood, and can be implemented, by the receiving device. The instructions and physical processes are referred to as physical access. The translation between physical and logical access may be implemented in software, within the device, or by a device controller. Implementation within software is highly inefficient because it uses excessive CPU resources to implement each access. Implementation within a device is efficient but entails redundant processing and hardware within multiple devices of the same type.

A storage or I/O device controller implements the bulk of logical to physical translation. In addition to performing this translation, device controllers serve as an interface to the bus's logic and communication protocol. They also can allow multiple devices to be attached to an I/O port. Access by multiple devices through a single device controller can be implemented in either dedicated or multiplex mode.

A mainframe channel is an advanced form of a device controller used in mainframe computers. It differs from a device controller in terms of the number of devices that can be controlled, the ability to communicate with different device types, and its maximum communication capacity. Mainframe channels can communicate with dozens of mass storage devices or hundreds of terminals simultaneously. They allow a mixture of attached device types and have communication capacity measured in hundreds of MBps.

An interrupt is a signal to the CPU that some condition requires its attention. Interrupts are identified by unique codes (numbers) and can be classified as I/O events, error conditions, or program service requests. Interrupts can be placed on the bus by an I/O device or can be placed directly in the interrupt register by a program or CPU component. The control unit checks for interrupts at the conclusion of each execution cycle. If an interrupt is detected, a system software program called an interrupt handler is executed to service the interrupt. The starting address of the interrupt handler for each interrupt is maintained in an interrupt table.

The detection of an interrupt causes suspension of the currently executing process. The stack is a mechanism provided to allow restarting of a suspended process at the point where it was interrupted. When an interrupt is detected, the current values of processor registers, which describe the current state of the process, are placed on the stack. When an interrupt handler finishes execution, the values on the stack are copied back into the processor registers, and the suspended process resumes.

Various measures can be taken to improve communication performance between storage and I/O devices and the CPU. Interrupt processing is a basic means of economizing on CPU resources dedicated to I/O and storage accesses. The use of interrupts allows physical accesses to occur asynchronously with CPU processing. Performance improvements also can be realized through buffered I/O, cached I/O, and data compression.

A buffer is a region of memory that holds a single unit of data transfer to or from a device. Use of a buffer allows rapid movement of entire blocks of data. Buffers are implemented with the computer system, under operating system control, and within devices or device controllers. A cache is a large buffer implemented within a device controller. When used for input, a cache can allow access to input more rapidly than the data transfer rate of the associated storage device if the cache controller can guess what data the CPU will request next and load that data into the cache before it is requested. Cache implementation is complex, requires extensive processing capability within the device controller, and may use additional processing by the operating system.

Data compression allows the data stored within a storage device or communicated through a channel to exceed device (or channel) capacity. Data entering the device or channel is reduced in size by a compression algorithm. Data leaving the device or channel is expanded to normal size (and representation) by a decompression algorithm. Data compression trades additional processing resources against reduced resources for data storage and/or communication.

Key Terms

address bus
branch prediction
buffer
buffer overflow
bus
bus arbitration unit
bus clock
bus clock rate
bus cycle
bus line
bus master
bus protocol
bus slave
bus transfer
cache
cache controller
cache hit
cache miss

cache swap
channel
compression algorithm
compression ratio
control bus
data bus
data compression
dedicated mode
device controller
direct memory access (DMA)
DMA controller
front-end processor
I/O channel
I/O port
I/O wait state
interrupt
interrupt code
interrupt handler

interrupt register
interrupt table
level one (L1) cache
level two (L2) cache
linear address space
logical access
machine state
multiple master bus
multiplex mode
peripheral processing unit
physical access
POP
PUSH
stack
stack overflow
stack pointer
supervisor
system bus

Vocabulary Exercises

1. The _____ is the communication channel that connects all computer-system components.

2. A(n) _____ is generally implemented on the same chip as the CPU.

3. The CPU is always capable of being a(n) _____, thus controlling access to the bus by all other devices in the computer system.

4. A(n) _____ is a small area of fast memory that can be used to minimize wait states.

5. A cache controller is a hardware device that initiates a(n) _____ when it detects a(n) _____.

6. The _____ is used to transmit command, timing, and status signals between devices in a computer system.

7. A(n) _____ is a small region of memory that can be used to improve the performance when moving data from main memory to an output device.

8. Under ideal conditions, the _____ rate of the system bus should be matched to the speed of the CPU.

9. The _____ is a register that always contains a pointer to the top of the _____.

10. Interrupt handlers are called by the _____ after it looks up the _____ in the _____.

11. The set of register values placed on the stack while processing an interrupt is also called the _____.

12. A(n) _____ is a software program that is executed in response to a specific interrupt.

13. During interrupt processing, register values of a suspended process are held on the _____.

14. A(n) _____ is a signal to the CPU or operating system that some device or program requires processing services.

15. A(n) _____ is a hardware device that intervenes when two potential bus masters want control of the bus at the same time.

16. The initiator of a bus transfer assumes the role of a(n) _____. The recipient assumes the role of a(n) _____.

17. The CPU incurs one or more _____ when it is idle pending the completion of an I/O operation by another device within the computer system.

18. The system bus can be decomposed logically into three sets of _____: the _____ bus, the _____ bus, and the _____ bus.

19. During a(n) _____, one or more register values are copied to the top of the stack. During a(n) _____, one or more values are copied from the top of the stack to registers.

20. The relative size of a data set before and after data compression is described in terms of the _____ of the _____.

21. A(n) _____ is a special purpose processor dedicated to managing the contents of a cache.

22. Storage and I/O devices are indirectly connected to a bus (or I/O) port through a(n) _____.

23. The _____ carries interrupts, read commands, and other control signals.

24. Primary storage receives a read command on the _____, the address to be accessed on the _____, and returns the accessed data on the _____.

25. The operating system normally views any storage device as a(n) _____, thus ignoring the device's physical storage organization.

26. Part of the function of a storage secondary device controller is to translate _____ into _____.

27. _____ allows the transfer of data directly between memory and secondary storage devices, without the assistance of the CPU.

28. A(n) _____ is a high-speed I/O port to which an I/O processor is dedicated.

29. An access to primary storage that is found within a cache is called a(n) _____.

30. The _____ defines the control, signals, and other communication parameters of a bus.

Review Questions

1. What is the system bus? What are its primary components?

2. What is a bus master? What is the advantage of allowing devices other than the CPU to be a bus master?

3. What characteristics of the CPU and of the system bus should be balanced to obtain maximum system performance?

4. What is an interrupt? How is an interrupt generated? How is it processed?

5. What is a stack? Why is it needed?

6. Describe the execution of the PUSH and POP operations.

7. What is the difference between a physical access and a logical access?

8. What functions does a device controller perform?

9. How are incompatibilities in communication parameters between the CPU and I/O and storage devices resolved?

10. What is a buffer? Why might one be used?

11. How may a cache be used to improve performance when reading data from a storage device? How may a cache be used to improve performance when writing data to a storage device?

1. Assume that a PC uses a 200-MHz processor and a system bus clocked at 50 MHz. An internal modem is attached to the system bus. Communication between the modem and telephone line occurs at continuous rate of 28,800 bits per second. No parity or other error-checking mechanisms are in use. The modem uses an 8250 UART chip containing a 1-byte buffer. After a character is received by the modem, it immediately sends an interrupt to the CPU.

    When the interrupt is received, the following actions are performed:

    a. The supervisor is called.

    b. The modem interrupt handler is called by the supervisor.

    c. The modem interrupt handler copies the character from the UART buffer to a memory buffer.

    d. The modem interrupt handler returns control to the supervisor.

    e. The supervisor returns control to the process that was originally interrupted (e.g., the program that is managing the file transfer).

    Sending the interrupt requires one bus cycle. A PUSH or POP operation consumes 30 CPU cycles. The supervisor consumes eight CPU cycles before starting to call the interrupt handler. The serial port interrupt handler consumes 10 CPU cycles before starting to return to the supervisor, including the time needed to transfer the character from the UART buffer to memory.

    Question 1:
    How long does is take to process one character (i.e., perform all of the steps outlined above)? State your answer in both CPU cycles and elapsed time.

    Question 2:
    Assume that the computer is executing a program that is downloading a large file using the modem. Assume that the file is held in memory until the transfer is completed. What percentage of the CPU capacity is being used to process input from the serial port UART during the transfer? Assume that the bus uses a simple request/response protocol for peripheral to memory transfers (i.e., no acknowledgment is sent by the peripheral, just the requested data). What percentage of available bus capacity is being used to process input from the serial port UART during the transfer?

    Question 3:
    Assume the same facts as in Question 2. Assume further that the 8250 UART has been replaced by a 16550A UART. The new UART is identical

to the old UART except that it contains a 16-byte buffer. Thus, an interrupt is generated for each group of 16 incoming data bytes. This increases the execution time of the interrupt handler from 14 to 50 CPU cycles. What percentage of the CPU capacity is being used to process input from the new serial port UART? Assume that 4 bytes at a time are transferred from the UART buffer to memory by the modem interrupt handler. What percentage of available bus capacity is being used to process input from the serial port UART during the transfer?

2. A video frame displayed on a computer screen consists of many pixels. Each pixel (or cell) represents one unit of video output. The resolution of a video display typically is specified in horizontal and vertical pixels (e.g., 800×600); and the number of pixels on the screen is simply the product of these numbers ($800 \times 600 = 480,000$ pixels). The data content of a pixel is one or more unsigned integers. For black and white (monochrome) display, each pixel is a single number (typically, between 0 and 255) that represents the intensity of the color white. Color pixel data typically is represented as either one or three unsigned integers. When three numbers are used, the numbers are usually between 0 and 255, and each number represents the intensity of a primary color (red, green, or blue). When a single number is used, typically it represents a predefined color selected from a table (or palette) of colors.

Motion video is displayed on a computer screen by rapidly copying frames to the video display controller. The video controller continuously converts frame data to an analog RGB signal that is output to the display device. Because video images (or frames) require many bytes of storage, generally they must be copied to the display controller directly from secondary storage. Each video frame is an entire picture, and its data content (as measured in bytes) depends on the resolution at which the image is displayed and the maximum number of simultaneous colors that can be contained within the sequence of frames. For example, a single frame at 800×600 resolution with 256 simultaneous colors contains $800 \times 600 \times 1 = 480,000$ bytes of data. Realistic motion video requires a minimum of 20 frames per second to be copied and displayed. The use of fewer frames per second results in perceived "jerky" motion because the frames are not being displayed quickly enough to fool the eye (and the brain) into thinking that they are one continuously changing image. Professional-quality motion display typically uses 24 or 30 frames per second to overcome this jerkiness.

Assume that the computer system being studied contains a bus mastering disk controller and a video controller that copies data to the video display at least as fast as it can be delivered over the bus. Assume that the system bus can transfer data at a sustained rate of 33 MBps, as are both of the controllers' bus interfaces. Assume that this system will be used to display motion video display capable of resolutions as low as 320 × 200 and as high as 800 × 600.

Assume that a single disk drive is attached to the disk controller and that it has a sustained data-transfer rate of 12 megabytes per second when reading sequentially stored data. Next, assume that the channel connecting the disk drive to the disk controller has a data transfer rate of 33 MBps. Finally, assume that the files containing the video frames are stored sequentially on the disk and that copying the content of these files from disk to the display controller is the only activity that the system will perform (i.e., no external interrupts, no multitasking, etc.)

Assume that the video display controller contains two megabytes of 70-nanosecond 8-bit buffer RAM (i.e., the video image arriving from the bus can be written to the buffer at a rate of 8 bits per 70 nanoseconds). Then, assume that the RAM buffer of the video display can be written to from the bus while it is simultaneously being read by the display device, sometimes called dual porting. Finally, assume that data can be received and displayed by the display device as fast as it can be read from the RAM buffer by the video controller.

Question 1:
What is the maximum number of frames per second (round down to a whole number) that can be displayed by this system in 256 simultaneous colors at a resolution of 320 × 200?

Question 2:
What is the maximum number of frames per second (round down to a whole number) that can be displayed by this system in 65,536 simultaneous colors at a resolution of 640 × 480?

Question 3:
What is the maximum number of frames per second (round down to a whole number) that can be displayed by this system in 16,777,216 simultaneous colors at a resolution of 800 × 600?

1. Investigate the implementation of modern PC buses such as the Peripheral Component Interconnect (PCI) bus. How wide are the data and address buses? What is the bus clock rate? What is the maximum data transfer rate? What is the set of commands and command responses? How is multiple master capability implemented? What is the bus arbitration strategy?

2. Fibre Channel (http://www.symbios.com/fclc/index.html) is a relatively new bus standard for high-speed connection to storage devices. It is currently used to implement high-speed storage arrays where storage devices are separated by large distances. It also can be used to allow multiple computer systems to share access to a single storage device or array. Investigate the implementation details of the Fibre Channel standard. What are its strengths and weaknesses as compared to SCSI-3? Which would be used in a typical LAN environment? Which would be used in a distributed transaction-processing environment?

Chapter

8

Data and Network
Communication Technology

A discussion of data and network communication could easily fill a book. This chapter concentrates on basic concepts and terminology. Discussions of data communication among internal computer system components were provided in Chapters 2 and 7, and an explanation of data representation using analog and digital signals was given in Chapter 4. This chapter builds on that foundation and extends the discussion to communication among computer systems, peripheral devices, and networks.

Successful and efficient communication is a complex endeavor that has challenged humans over time. Difficulties in communication include disparate methods of expression (e.g., words, pictures, and gestures), language (syntax and semantics), format (lectures and reports), and rules and conventions among cultures and specific communication contexts. The human ability to communicate successfully results from a long learning process covering all of these areas.

The complexities of computer communication are no less daunting in that parallel challenges with those that affect human communication are abundant. The breadth of methods for addressing these difficulties and the interdependence among them makes the topic of computer communication extremely difficult.

Communication Protocols

A **communication protocol** is a set of rules and conventions. Although this definition may sound simple, there is underlying complexity. As an example, consider the protocol used for communicating within a classroom. First and foremost, all participants must agree upon a common language. Once selected, the grammatical and semantic rules of the language become part of the classroom communication protocol. For spoken languages, the air within the room serves as the transmission medium. Large classrooms may require auxiliary means to overcome limitations of this transmission medium (e.g., sound insulation and voice-amplification equipment).

Certain command and response sequences are adopted to ensure the efficient flow of information. The instructor, generally assumed to control the communication process and the transmission medium, coordinates access to the shared transmission medium when others speak. Gestures such as raising a hand and pointing typically comprise the means of coordination. All participants adopt and observe the methods of requesting and granting access to the shared transmission medium.

Many conventions are adopted to ensure accurate transmission and reception of messages. The instructor often searches students' faces to determine whether the message is being correctly received and interpreted. Confused looks may lead to retransmission. Specific signals (e.g., a raised hand) are accepted as a means of requesting retransmission. The very act of entering the classroom signals the acceptance by students of conventions regarding the content and duration of discourse. Allowable messages are limited to those on a certain topic. Off-topic or overly long messages are ignored or cut off.

The complexity of computer communication protocols is even greater than that of human communication protocols. Figure 8-1 presents a hierarchical organization of the components of a computer communication protocol. Each node of this hierarchy has many possible technical implementations. Thus, a complete communication protocol is a complex combination of subsidiary protocols and the technologies by which they are implemented. It may be helpful to return to this hierarchy often while reading the remainder of this chapter so that topics can be placed in their proper context.

Figure 8-1

A hierarchical decomposition of a communication protocol

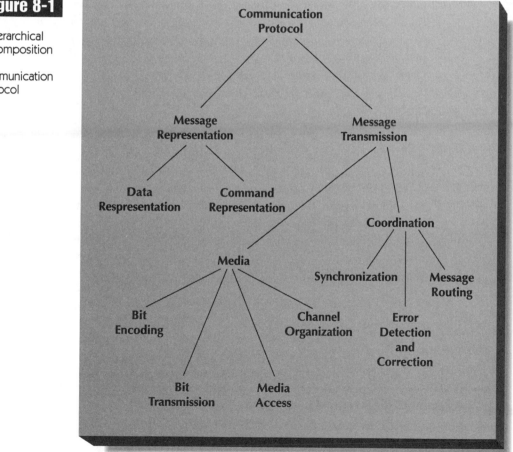

Messages

A **message** is a unit of data or information transmitted from a sender to one or more recipients. It can consist of a single byte or character (e.g., pressing the Escape key to terminate a program) or many bytes (e.g., a graphic image downloaded from an Internet server). Messages can be loosely categorized into two types—data and command. The content and format of data messages varies widely. Data messages can include any of the primitive CPU data types or any complex combination thereof. For purposes of computer communication, the content of a data message is passive; that is, it is simply moved from one place to another with no attempt to "understand" its content.

A command message contains instructions that control some aspect of the communication process. Examples of command messages include many of the ASCII device control characters, addressing and routing instructions, and error detection and correction information. The set of allowable command messages must be specified fully by the communication protocol in both syntax (allowable format) and semantics (meaning). Command messages also can be used to transmit information about data messages. Such information might include format, content, length, and other details needed by the receiver to successfully interpret the data being transmitted.

At the most primitive level, any message (data or command) is simply a sequence of bits. The communication protocol, which specifies a common method of encoding and transmitting bits among senders and recipients, may or may not specify the format of larger units of data. As discussed in Chapter 4, an encoding method for larger units of data (e.g., integers, characters, documents, and images) represents a trade-off among accuracy, size (number of bits stored or transmitted), processing efficiency, and standardized interpretation. The best design trade-off for a CPU or computer system may not be optimal for data transmission.

Generally, matters of processing efficiency are of less concern when evaluating representation methods for data transmission. Issues of standardized interpretation and size (and, therefore, speed of transmission) are of greater importance than for CPU design. Differences among data representation formats used within computer systems and used for data transmission create a need for conversion among formats. These conversions may be implemented within software as part of a communication interface program. They also can be implemented within the hardware devices that enable data transmission (e.g., a modem or network interface unit).

DATA ENCODING METHODS

Data encoding is the process of representing bit values in a form that can be propagated through a transmission medium. A primary division of data encoding methods can be made based on whether analog or digital signals are used.

Although digital signals are the norm for data encoding within a computer system, analog signals are often used for communicating between computers. The nature of many transmission media requires the use of analog signals. For media capable of digital transmission, analog signals may represent a more cost-effective trade-off among transmission capacity and reliability.

Analog Data Encoding

Optical, radio frequency, and some electrical carrier waves travel through space or guided media (i.e., cables) as an analog **sine wave**. Figure 8-2 illustrates a pure sine wave. The energy content of a sine wave varies continuously between positive and negative states. Three characteristics of a sine wave can be manipulated to represent data. The first is **amplitude**. Amplitude is a measure of wave power as the maximum distance between a wave peak and its zero value. Whether measured from zero to a positive peak or zero to a negative peak, amplitude is the same.

Figure 8-2

Various characteristics of a pure sine wave

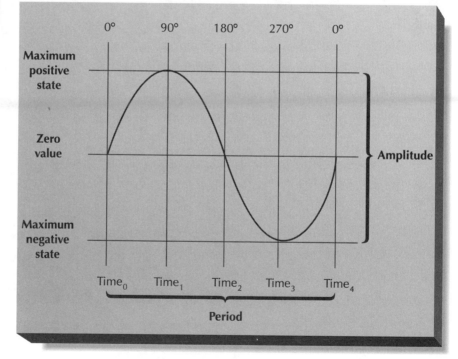

A complete **cycle** (also called a period) of a sine wave follows its full travel from zero to positive peak, back to zero, to its negative peak, and back to zero again. The **phase** of a sine wave is a specific time point within its cycle. Phase is measured in degrees, with 0° representing the beginning of the cycle and 360° representing the end. The point of positive peak is 90°, 180° is the zero point between a positive and negative peak, and 270° is the point of negative peak.

The **frequency** of a sine wave is the number of cycles that occur in one second. Frequency is measured in Hertz (Hz), where Hertz is interpreted as "cycles per second." Figure 8-3 illustrates two sine waves with 2-Hz and 4-Hz frequencies. Electrical and electromagnetic radiation covers a broad range of possible frequencies. Figure 8-4 shows the range of frequencies from 10^1 to 10^{19} Hz and the position of various types of radiation within that range. Figure 8-4 also shows the division of that range into various forms of broadcast communication.

Figure 8-3

Sine waves with 2-Hz and 4-Hz frequencies

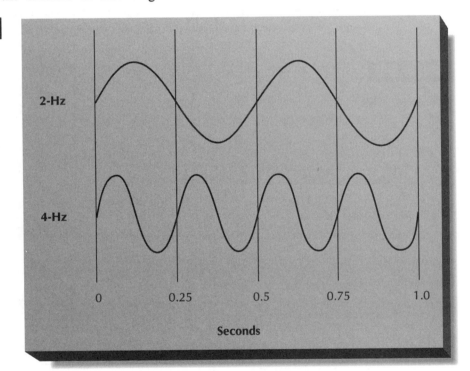

Data and Network Communication Technology

The spectrum of electromagnetic frequencies between 10^1 and 10^{19} Hz

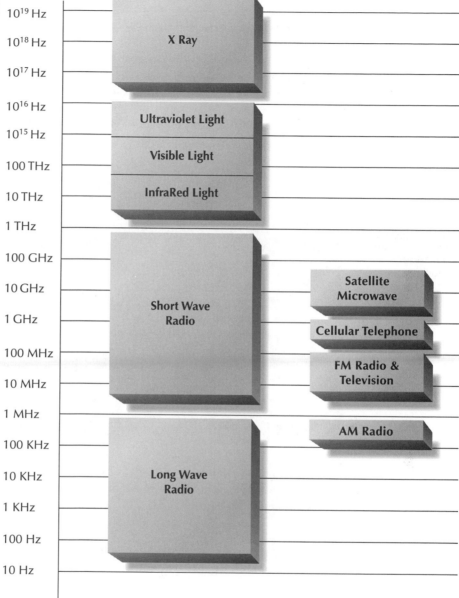

Data is encoded within a carrier wave by precisely manipulating the ampli-tude, frequency, and/or phase of a carrier wave. The process of manipulating one or more of these characteristics to encode data is called **modulation**. **Amplitude modulation** represents bit values as specific amplitude values. Amplitude modulation also can be called **amplitude shift keying (ASK)**. For example, data could be transmitted through a wire as an electrical signal with 1 volt indicating a 0-bit value and 10 volts indicating a 1-bit value (see Figure 8-5). Amplitude modulation holds frequency constant while varying amplitude to represent data. The amplitude must be maintained for at least one full wave cycle.

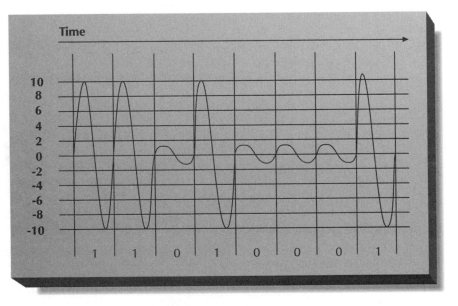

Figure 8-5

The bit string 11010001 encoded within a carrier wave using amplitude modulation

Frequency modulation encodes data by varying the frequency of the carrier wave while holding amplitude constant. Frequency modulation also can be called **frequency shift keying (FSK)**. For example, in a wire capable of carrying electrical frequencies between 1 and 5 Hz, 2 Hz could represent zero and 4 Hz could represent one (see Figure 8-6). A 1-bit value would be transmitted by generating a 4-Hz electrical current for at least one-fourth (0.25) of a second. Frequency modulation holds amplitude constant while varying frequency to represent data.

Figure 8-6

The bit string 11010001 encoded in a carrier wave using frequency modulation

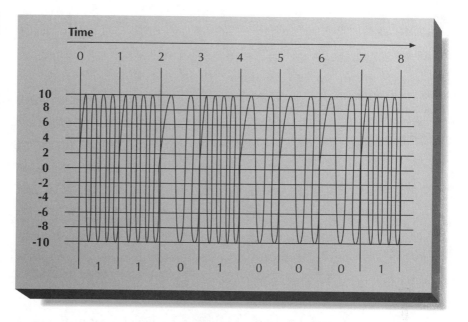

Phase is a fundamentally different wave characteristic from amplitude or frequency. The nature of a sine wave makes it impossible to hold phase to a constant value. However, phase can be used to represent data by making an instantaneous shift in the phase of a signal (or switching quickly between two signals of different phase). The sudden shift in signal phase can be detected and interpreted as data in a method of data encoding called **phase shift modulation**, or **phase shift keying (PSK)**.

Multilevel coding is a technique for encoding multiple bits within a single signal event. Groups of bits are treated as a single unit for purposes of signal encoding. For example, groups of 2 bits can be combined into a single signal event if four different levels of the modulated signal parameter are defined. Figure 8-7 on the next page illustrates a four-level amplitude modulation coding scheme. Bit pair values of 11, 10, 01, and 00 are represented by amplitude values of 8, 6, 4, and 2, respectively.

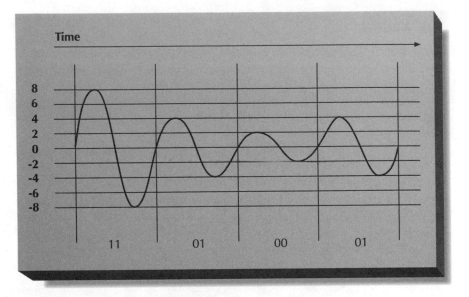

Figure 8-7

The bit string 11010001 encoded in a carrier wave using 2-bit multilevel frequency modulation

Multilevel coding schemes provide dramatic increases in data transfer rates at the expense of higher susceptibility to error. Error susceptibility increases due to the decreased "distance" between encoded data values. These smaller intervals between defined signal parameter values produced error rates that were too high in the early days of data communication. Modern equipment is built to much tighter tolerances and can reliably differentiate among closely spaced parameter values. Thus, multilevel coding schemes using as many as eight bits per signal event are commonly employed today.

Analog encoding schemes, commonly used in long distance communication, are more reliable for long distance communication by electrical signals than their digital counterparts. They are uniquely suited to local telephone communication lines because those lines were designed to carry electrical representations of analog voice signals. Analog signals are required for wireless transmission because radio waves are inherently analog.

Digital Data Encoding

Increasingly, the telecommunication industry is moving toward all-digital service. As discussed in Chapter 4, digital signals are inherently less subject to noise than are their analog counterparts. Generally, this resistance to noise-related problems comes at the expense of message-carrying capacity. However, modern methods of digital signal transmission considerably mitigate this disadvantage. In addition, digital signals can exploit digital switching technology (e.g., for routing messages over a network), which is inherently faster and more reliable than analog switching.

Electrical Digital Signals

Digital waveforms do not vary continuously between positive and negative states as do analog waveforms. Instead, they shift abruptly from one level to another, creating "squared off" edges in the wave graph—thus, the name **square waves** (see Figure 8-8). Electrical square waves can be generated with direct current. Data values are coded by modulating voltage. Several different modulations are commonly employed, including:

▶ Transistor-to-transistor logic (TTL).

▶ Zero crossing signals.

▶ Manchester encoding.

As with most other choices in data transmission, these methods include trade-offs between signal efficiency, cost, and susceptibility to transmission errors.

Digital signals within computer hardware can implement **transistor-to-transistor logic (TTL)**. As shown in Figure 8-8, TTL signals are discrete pulses of electrical direct current (DC). A DC signal has no negative component. Digital TTL signals vary between a high voltage state, such as +5 volts DC and a low voltage state, such as 0 volts DC. TTL signals are commonly used for communication over short distances (e.g., among transistors within a microprocessor and among devices attached to a system bus).

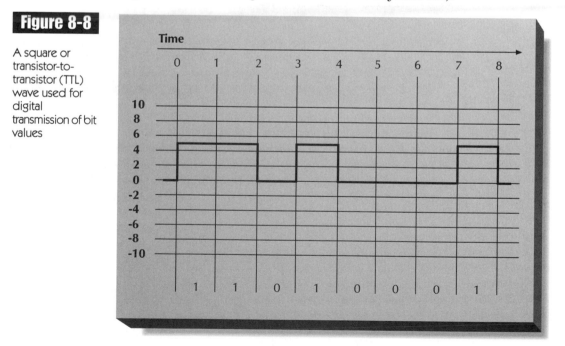

Figure 8-8

A square or transistor-to-transistor (TTL) wave used for digital transmission of bit values

Digital signals used in telecommunication typically have both positive and negative states. An example of such a **zero crossing signal** is shown in Figure 8-9. A zero crossing digital signal varies between a high positive voltage such as +5 volts and a low negative voltage such as -5 volts. In the zero crossing scheme shown, the negative state represents a bit value of 0, and a positive state represents 1. Zero crossing signals can be transmitted reliably over relatively longer distances than TTL pulses because the "distance" between signal states is relatively large.

Figure 8-9

A zero crossing digital signal: Positive voltages represent 1 bits, and negative voltages represent 0 bits

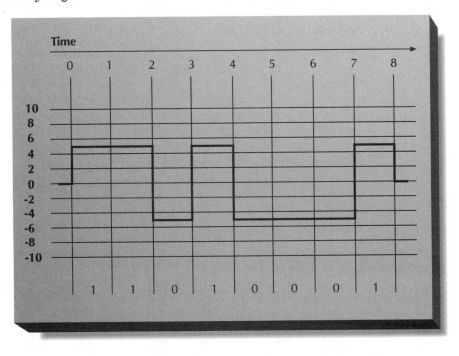

Another type of zero crossing digital signal is shown in Figure 8-10. The signaling method is called **Manchester coding**, which represents a bit value of 1 as an abrupt transition from positive to negative voltage. Zero values are represented with an opposite pattern. Manchester coding requires double the number of signal events as compared to TTL and ordinary zero crossing signals. Thus, data transfer capacity is reduced by one half.

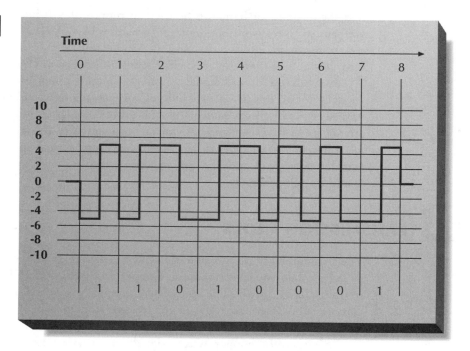

Figure 8-10

Manchester coding: One bits are represented by negative to positive transitions, and 0 bits are represented by positive to negative transitions.

Manchester coding is used commonly because it improves synchronization between sender and receiver. Consider sending a long sequence of 1 bits represented with ordinary zero crossing signals; the resulting message would be represented by a continuous signal of positive voltage. Without any transitions between negative and positive voltage, the internal clocks of the sender and receiver can drift out of synchronization with one another. With Manchester encoding, every bit contains a transition between positive and negative states, so the receiver can synchronize its clock with sender in the middle of each and every bit. There are no long sequences without transitions and, therefore, less opportunity for the receivers clock to drift out of synchronization.

Digital electrical signals are the preferred method of data transmission over relatively short distances (e.g., on a computer system bus), but they cannot be reliably transmitted over long distances (e.g., greater than a kilometer). Power loss, electromagnetic interference, and wire-generated noise all combine to "round off" the normally sharp edges of a square wave. Receivers cannot reliably interpret data values without sharp, abrupt changes in voltage level.

Optical Digital Signals

Digital waveforms can be represented optically in a manner similar to TTL. Short bursts of light generated from a laser diode (LD) can represent a 1 bit, and the absence of light can be used to represent a 0 bit. Optical signals are well-suited to transmitting digital information because they are: (1) immune to electromagnetic interference and (2) subject to signal distortion (orders of magnitude less than for electrical signals) generated by the optical cable.

As with all digital signals, optical digital signals sacrifice transmission capacity for accuracy and reliability. However, the extremely high frequency of optical signals (approximately 10^{14} Hz) gives them a many-orders-of-magnitude advantage in raw transmission capacity as compared to electrical signals. Only a fraction of this capacity is sacrificed when using digital signals. Thus, the data transfer capacity of optical digital signals remains far superior to that of electrical analog or digital signals.

TRANSMISSION MEDIA

The communication path used to send a message from sender to recipient is called the **transmission medium**. Copper wire and optical fiber are two types of transmission media. The atmosphere or space can also be used as a transmission medium (e.g., for radio or microwave transmissions). Messages must be encoded in signals that can be conducted or propagated through the transmission medium. In copper wires, signals are carried as streams of electrons, whereas in optical fiber, signals are pulses of light. Microwave and satellite transmissions are broadcast as **radio frequency (RF)** radiation through space.

A **communication channel** (see Figure 8-11) consists of a sending device, receiving device, and the transmission medium that connects them. In a less physical sense, it also consists of the communication protocol(s) used in the channel. Although Figure 8-11 implies a relatively simple channel design, most channels are of complex construction. They use segments of different transmission media with appropriate connections to convert signals and messages from one medium and protocol to another and generally use different communication protocols on each media segment.

Figure 8-11

Basic elements of a communication channel: a sender transmission medium, receiver, and message

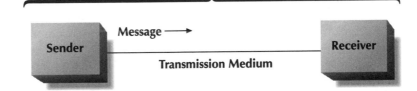

Consider a long-distance analog modem connection. A digital electrical signal sent by the system bus or computer serial port is converted to an analog signal by the modem and routed through twisted pair copper wiring to the nearest

telephone pole. From there, it travels by lower gauge (thicker) copper wiring or coaxial cable to the local telephone switching center. At that point, it is converted to a digital optical signal and routed by fiber-optic cable to a local connection to the long-distance grid. From there, it is routed across the long-distance grid (still as a digital signal) by fiber-optic lines. The conversion process is reversed at the other end of the long-distance grid. The channel from sender to recipient consists of the serial ports, the modems, and all the media segments, protocols, and conversion devices.

Several characteristics of transmission media affect their ability to successfully and efficiently transmit messages, including:

▶ Speed and capacity.

▶ Bandwidth.

▶ Noise, distortion, and susceptibility to external interference.

Transmission media are generally better when their speed, capacity, and bandwidth are high and noise, distortion, and external interference are low. However, such media tend to be costly. These factors also can be affected positively or negatively by the communication protocol used with the transmission medium. Different combinations of transmission medium and protocol are cost-effective for different types of communication links, which accounts for the different transmission segments described in the previous example.

Speed and Capacity

The fundamental speed limit for any transmission medium is the rate at which a carrier signal propagates through the medium. For wire-carrying electrical signals this limit is close to the speed of light. The speed of light also limits optical and radio frequency carrier waves, so the raw speed per distance unit varies little among all commonly used transmission media. What does vary is length of the media, the ways in which multiple media segments are interconnected, and the rate at which bits can be placed into and extracted from the media. It is these factors (primarily the last one, as discussed in detail next) that account for raw speed differences among different transmission media.

Speed and capacity are interdependent; that is, a faster communication channel has a higher transmission capacity per time unit than a slower channel. However, capacity is not solely a function of transmission speed. Rather, it is also affected by the relative efficiency with which the channel is used. Different communication protocols and different methods of dividing a channel to carry multiple signals alter the capacity implied by the raw speed of a channel.

The speed and data transmission capacity of a channel are generally stated jointly as a **data transfer rate**. Several different types of data transfer rates may be stated. A **raw data transfer rate** states the maximum number of bits or bytes per second that the channel can carry. Such a measure ignores any inefficiencies inherent in the communication protocol. In essence, raw data transfer rate is a measure of potential data transmission capacity.

Usable transmission capacity is generally stated as a net or **effective data transfer rate**. This measure covers not only the channel hardware but the communication protocol as well. It is typically less than the raw channel capacity. The reduction is a result of using part of the raw channel capacity to do things other than transmit raw data. These include transmission of command messages and retransmission of data not correctly received.

Bandwidth

The frequency of an analog carrier wave is a fundamental measure of data-carrying capacity, and this is true even when amplitude or phase shift modulation are used to encode data. Signal frequency limits analog data carrying capacity because a change in amplitude, frequency, or phase must be "held" for at least one sine wave cycle to be detected by a receiver. Thus, an individual data item (e.g., a bit) must occupy a time period at least as long as one sine wave period. If the unit of data transmission is a single bit, then the minimum frequency of a carrier wave is also the upper bound on transmission speed as measured in bits per second.

The difference between the maximum and minimum frequencies of a signal is called the signal **bandwidth**. The difference between the maximum and minimum frequencies that can be propagated through a transmission medium is called the medium bandwidth. Maximum frequency and bandwidth are important characteristics of any transmission medium because they determine fundamental limits on the raw data transfer rate of the medium.

As an example, consider the frequency range of an analog voice-grade telephone line (e.g., about 300 Hz to 3,400 Hz). These lines were designed originally to carry an electrical representation of the human voice. Sound frequencies are converted into directly corresponding electrical frequencies. Electrical signals below 300 Hz and above 3,400 Hz cannot be propagated across these lines.

The bandwidth of the channel is the difference between the highest and lowest frequencies that can be carried (3400 – 300 = 3,100 Hz). In contrast, the frequency range of human speech is approximately 200 Hz to 5,000 Hz (4,800 Hz bandwidth), and the frequency range of the human ear is approximately 20 Hz to 20,000 Hz (19,980 Hz bandwidth). Thus, the use of analog telephone lines provides only an approximation to human speech (i.e., some high and low frequencies are lost) and is incapable of carrying many signals (e.g., high-fidelity music) that can be received by humans.

Modulator-demodulator (modem) technology has developed over many years as a means of sending digital signals over analog voice-grade telephone wires. Digital signals from the computer bus are encoded in an analog carrier wave. Early modulation techniques used four separate frequencies to carry bit values (see Table 8-1). The lowest of these frequencies (1,070 Hz) created an effective upper limit on the speed of transmission. Assuming that a single bit is transmitted per signal cycle, the maximum transmission rate is 1,070 bps.

TABLE 8-1	Mode	Signal Frequency (Hz)	Binary Value
This table shows the pattern of frequency assignments for the transmission of binary data over 300-baud analog telephone lines.	Transmit	1,070	0
	Transmit	1,270	1
	Receive	2,025	0
	Receive	2,225	1

Subsequent modem standards have substantially exceeded this theoretical capacity, but they have had to employ ever-more-complex schemes to do so, including hardware-based data compression and the encoding of up to 32 bits into a single carrier wave cycle using an elaborate combination of amplitude and phase modulation. Each advance in speed has pushed the limits of modem technology to a new maximum and has increased the susceptibility of modem transmissions to noise and external interference. Extensive hardware-based error detection and correction is now a standard part of modem protocols to compensate for these problems.

Current modem transmission rates range as high as 56,000 bps. But the 3,100-Hz bandwidth of the transmission line and the use of analog sine waves is an effective barrier against substantially higher transmission rates. Interestingly, and somewhat counterintuitively, digital signaling methods using similar wire can achieve transmission rates of up to 100 Mbps. These rates are achieved primarily due to effective use of the higher (and more reliable) bandwidth of modern wiring.

Signal-to-Noise Ratio

Within a communication channel, **noise**[1] refers to any extraneous signals that can be interpreted incorrectly as data. Transmission media such as copper wires are affected by **electromagnetic interference (EMI)**. EMI can be produced by a variety of sources, including electric motors, radio equipment, and other communication lines. EMI can induce noise in nearby electrical circuits, including communication wires and equipment. In an area dense with wires and cables, this problem can be compounded because each transmission path can both radiate and respond to EMI.

Attenuation is a reduction in the strength (amplitude) of a signal as it passes through a transmission medium. It is caused by interactions of the signal energy with the material comprising the medium. Different types of signals and medium materials cause different amounts of attenuation to occur. Electrical resistance is a cause of attenuation. Optical signals transmitted through optical fiber and radio signals transmitted through space are also subject to attenuation. Attenuation is generally proportional to the length of the medium, so doubling the length of the medium doubles the amount of attenuation.

Another source of communication errors is **distortion**, or any characteristic of a communication channel that causes a data signal to be altered. The transmission medium itself is a primary source of distortion. Resonances within a transmission medium can amplify certain portions of a complex signal, and the medium may attenuate some parts of a signal more than others (e.g., depending on frequency). Echos and other unwanted artifacts can be introduced by the medium or by other components of the channel (e.g., an amplifier).

For a receiving device to interpret encoded data correctly, it must be able to distinguish the encoded signals from noise that can be present in the channel. Distinguishing valid signals from extraneous noise becomes increasingly difficult as the speed of transmission increases. The effective "speed limit" of any given channel is determined by the power of the message carrying signal in relation to the power of the noise in the channel; this relationship is called the **signal-to-noise ratio (S/N ratio)** of the channel. S/N ratio is measured at the (receiving) end of the channel, and thus, it also incorporates the effects of signal attenuation and distortion (see Figure 8-12).

---

[1] Noise can be heard in many common household devices. Turn on a radio and tune it to a frequency (channel) on which no local broadcasting station is transmitting. The hissing sound is produced by the radio receiving and amplifying background radio frequency noise. Noise and distortion also can be heard on any home stereo system. Set the amplifier of receiver input to a device that isn't turned on (e.g., a tape or CD player) and turn the volume up relatively high (be sure to turn it back down when you're finished!). The hissing sound is the result of amplifying noise in the signal transmission and amplification circuitry. Any other sound than hissing represents distortion introduced by the amplifier, speakers, or cabling among them.

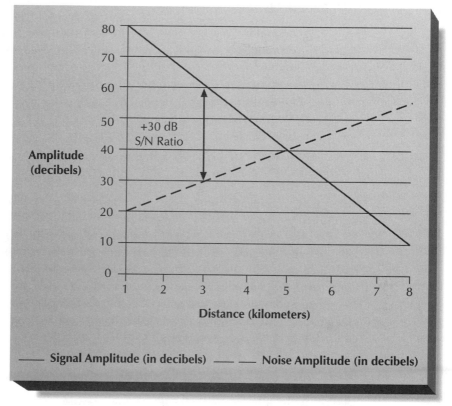

Figure 8-12

A plot of signal-to-noise ratio as a function of distance for a hypothetical channel. S/N ratio is positive for distances up to five kilometers (e.g., +30 decibels at three kilometers). Successful transmission is theoretically impossible beyond five kilometers because S/N ratio is negative.

As an example, consider the receipt (by a human listener) of spoken words during a speech. The difficulty (or ease) of understanding the speech is directly related to the speed at which it is delivered (spoken) and the volume (i.e., amplitude) of the speech in relation to background noise. Accurate reception of speech is impaired by sources of noise (e.g., other people talking, an air conditioner, or a construction project across the street). The speaker can compensate for noise by increasing the volume of the speech (i.e., increasing the signal-to-noise ratio). Accurate reception is also impaired if the speaker speaks too quickly. This is due to insufficient time for the listener to interpret one signal (e.g., a word or speech fragment) before the next is received.

In analog or digital signal transmission, a message is composed of individual bits. Each bit is a period of time during which a message representing a 0 or 1 is present on the channel. As transmission speed increases, the duration of each bit in the signal (i.e., the **bit time**) decreases. If signal-generating equipment could generate a full-amplitude signal instantaneously, then this would not be a problem. But no device (including the human voice) can go from a zero state to a full-power state instantaneously. Short bit times don't give the signal-generating device sufficient time to "ramp up" to full power before the end of the signal event. Thus, S/N ratio increases because the amplitude of each individual signal event is decreased. Eventually, a limit is reached at which the bit time is so short that it cannot be distinguished from noise—the point at which S/N ratio is small or zero.

If a higher speed is attempted, the rate of errors also increases. A transmission error represents a wasted opportunity to send a message, thus reducing the effective (or net) data transfer rate. A further difficulty is that noise usually is not constant. In electronic signals, noise usually occurs in short, intermittent bursts (e.g., due to a nearby lightning strike or the startup of an electric motor). As discussed in "Error Detection and Correction," the receiver can request retransmission of a message if errors are detected. If noise bursts are infrequent, retransmissions might not diminish the overall data transfer rate significantly and a relatively high raw transmission speed could be used.

An ideal communication channel would exhibit little or no noise. In copper wire or radio frequency channels, some noise usually is present, and the noise level increases with distance. To compensate for these effects, line-conditioning equipment (e.g., amplifiers and repeaters) can be used to boost the power of the signal and/or to suppress noise. Of course, this equipment adds to the cost of installing and maintaining the channel, and, not surprisingly, telecommunication vendors charge premium rates for high-quality service.

Electrical and Optical Cabling

Electrical signals are generally transmitted through copper wiring. Copper is used because it is relatively inexpensive, highly conductive, and a material from which wire is easily fabricated. Copper wiring can vary widely in construction details, including gauge (diameter), insulation (or lack thereof), purity and consistency, and number and configuration of conductors per cable. The two most common forms of copper wiring used for data communication are twisted pair and coaxial cables.

Twisted pair wire is a very common transmission medium for local area network (LAN) connections. It consists of two conductive wires that are twisted around one another. The wires may or may not be enclosed (shielded) by nonconductive material. The primary advantages to twisted pair are its low cost and ease of installation. Its primary disadvantages are a high susceptibility to noise and limited transmission capacity. These are due to low bandwidth (generally less than 1 MHz) and a relatively low amplitude (voltage) limit.

Various standardized forms of twisted pair cable and connectors are available. One of the most common is a cable called **10BaseT**. It consists of two pairs of conductors that form a complete channel in each direction. All four wires are bundled in a single thin cable and standardized modular (RJ-45) connectors are used at either end. It is quite similar to the modular phone cord used with most modern telephones. 10BaseT is commonly used for 10 Mbps LAN wiring. Equipment is currently available that can reliably transmit 100 Mbps over 10BaseT cables.

Coaxial cable consists of a single conductor surrounded by a metallic shield and wrapped in nonconductive material. Because of its construction, it generates very little external electromagnetic interference and is also relatively immune to external interference. Thus, several cables can be run together without interference. It also has a relatively high bandwidth (up to 50 MHz) and, thus, relatively high data transmission capacity. This has made it the transmission medium of choice where high bandwidth and low noise are essential (e.g., cable television).

The primary disadvantages of coaxial cable are its cost and ease of use. Because of its construction, special adapters and connectors are required to connect cables or attach cables to devices. It is comparatively costlier than twisted pair cable but not nearly as expensive as fiber-optic cable. It is somewhat less flexible than twisted pair and may be more difficult to install.

Fiber-optic cable consists of one or more strands of light-conducting filaments (e.g., fiberglass). Because light waves (rather than electricity) are used as the basic transmission medium, electromagnetic interference is not a problem, so fiber-optic transmissions are extremely resistant to noise. The transmission capacity of fiber-optic cable is substantially higher than either twisted pair or coaxial cables because of its much higher bandwidth (approximately 500 MHz).

The primary disadvantages of fiber-optic cable are its high cost and difficulty in installation. High-quality, thin fiberglass cables are expensive to produce. Plastic optical cable can be used for its lower cost but at some sacrifice in capacity. Connections between cables and devices require specialized, expensive hardware. Accidental breaks in fiber-optic cable are also more difficult to locate and repair.

No type of cable is capable of carrying signals more than a few dozen kilometers at acceptably high transmission rates. Long-distance communication requires the use of devices to increase signal strength and to remove unwanted noise and distortion. Such devices, installed at the junction between media segments, include amplifiers, repeaters, and line conditioners.

An **amplifier** is used to increase the strength (amplitude) of a signal. It is primarily used to extend the range of an electrical cable carrying analog signals. However, the effective length over which the signal travels cannot be increased indefinitely; rather, it is limited by two factors. The first is noise and interference introduced during transmission. An amplifier amplifies whatever signal is present at its input. If that signal includes noise along with the intended message, then the noise is amplified as well as the message. In addition, amplifiers are never perfect. Some distortion or noise is generally introduced in the amplification process. Thus, a signal that is amplified many times will contain noise from multiple transmission line segments as well as distortion introduced by each stage of amplification.

A **repeater** fills much the same role as an amplifier but operates on a different principle. It may be used with electrical or optical cable carrying analog or digital signals. Rather than amplifying whatever is sent to it, a repeater interprets the message it receives and retransmits (or repeats) it. Because of this mode of operation, noise is much less of a problem with repeaters than with amplifiers. As long as the noise introduced since the last transmitting or repeating device does not cause a misinterpretation of the message, the retransmitted message will be noise-free. Thus, the noise introduced in one transmission stage is not carried into the next. Repeaters are required every 4 or 5 kilometers for copper cable and every 40 or 50 kilometers for fiber-optic cable.

The term **line conditioner** is used to describe a number of different devices. In general, a line conditioner is used to suppress noise or to prevent damage to network components. For example, a line conditioner prevents strong power surges (e.g., as induced by a nearby lightning strike) from entering a telecommunications line.

Wireless Data Transmission

Wireless data transmission uses waves in the short wave radio frequency or **infrared light** bands transmitted through free space. The short wave radio spectrum covers frequency bands commonly used for FM radio, VHF and UHF broadcast television, cellular telephones, land-based microwave transmission, and satellite relay microwave transmission. Infrared transmission occupy higher-frequency bands that are generally usable over much shorter distances than short wave radio is.

The primary advantage of wireless transmission is its relatively high bandwidth and its avoidance of costly wired infrastructure. Another advantage in some situations is the inherently broadcast nature of transmission (many receivers receiving simultaneously). Its primary weaknesses are susceptibility to many forms of external interference and the cost of transmitting stations. In addition, most of the bands that can support high-speed communication are strictly regulated by national and international authorities and treaties. Use of these bands requires an expensive license.

The relatively high cost of transmission equipment and licenses makes short wave radio a rare method for a single user or organization; instead, companies are formed to purchase and maintain the required licenses and infrastructure. Users purchase capacity from these companies on a fee per use or bandwidth per time interval basis. Equipment costs have dropped dramatically over the last two decades. In addition, improved technology has resulted in much more efficient use of the available bandwidth. As a result, there has been a substantial decrease in cost to end users and a dramatic increase in the use of short wave radio as a telecommunication medium.

Wireless Data Communication

There are several short-haul data communication technologies in use today. Perhaps the simplest is analog modem transmission by cellular telephones—a method frequently used by salespersons and other "road warriors" to upload and download files, read and send e-mail, and interact with home office applications and databases from laptop computers. Internet links using point to point protocol (PPP) or similar protocols can be established by this method. Its primary drawbacks are the crowded cellular telephone network and the risk that data can be interrupted as users move from one location to another. Cellular telephone users share antennae that cover up to 50 square miles (i.e., a cell). As users move from one cell to another, they must establish a new connection with the antenna in the new cell, which temporarily disrupts the flow of data.

Cellular modem transmission rates are limited by the same telephone grid properties that limit ordinary modems; throughput is generally 56 Kbps or less. Many mobile users want network access equivalent to that of a hard-wired desktop computer from their cars, homes, airports, and client offices. This demand has driven a boom in metropolitan wide wireless networking technologies offering higher data transfer rates than analog cellular modems. Unfortunately, these methods are often incompatible and specific services tend to be available only in limited markets.

The Institute of Electrical and Electronics Engineers (IEEE) creates standards for many aspects of telecommunication and network system implementation. (These standards will be more fully discussed in Chapter 9.) A committee, formed in the early 1990s to develop a standard for metropolitan (and smaller) area wireless networks, has a standard under development called IEEE 802.11. The standard was published in 1997, and additional parts will be added.

Some of the goals of the IEEE 802.11 standard are as follows:

► Define a standard that meets the requirements of national and international regulatory bodies and pursue adoption by those bodies.
► Support communication by radio waves and infrared or visible light.
► Support stationary and mobile stations moving at vehicular speeds.
► Support data transmission rates of between 1 and 20 Mbps.

The current standard describes two forms of radio frequency (RF) transmission and one form of infrared transmission, all three of which share a compatible method of media access. One of the RF transmission methods, called frequency hopping spread spectrum (FHSS), is fully specified. The following discussion focuses on that part of the standard.

The committee looked to define at least one portion of the standard fairly quickly. It decided to focus on technologies that were already well-developed and RF bands that were available in most or all of the world. Attention quickly focused on the 2.5-GHz band because portions of it are not allocated to licensed transmitters in the United States, Europe, and Japan. A regulated but unlicensed band is required to meet the standard goals. Also, the properties of wave propagation in this band allow transmission though walls and other obstacles over sufficient distances.

A major problem in any RF data-transmission scheme is error detection and correction because RF transmissions are subject to many sorts of interference, and mobile transmitters/receivers encounter varying degrees of interference as they move. Many newer forms of analog data encoding (e.g., those used in 56-Kbps modems) are difficult to implement reliably over an RF band because of these problems. Thus, the committee settled on a data-encoding method that relied on older technology with less susceptibility to transmission error.

FHSS divides the available 2.4-GHz band into 79 1-MHz subchannels. Transmitters can "hop" among channels to find unused channels and/or those with reduced noise and interference using one of two modulation methods. The first encodes single bits using frequency modulation. Bit values correspond to frequencies 170 KHz above or below the channel midpoint frequency, with a raw data transfer rate of 1 Mbps. The second modulation technique uses multilevel coding of two bit groups. Signal frequencies are 75 and 225 KHz above and below the channel midpoint frequency, which achieves a raw data transfer rate of 2 Mbps.

Frequency hopping addresses a significant problem with RF transmission—multipath distortion, which occurs when an RF signal bounces off of stationary objects. A receiving antenna can receive multiple copies of the same signal at slightly different times (see Figure 8-13), which blurs or smears the signal content causing bit detection errors. The problem is especially severe indoors where there are numerous hard, flat surfaces to "bounce" RF signals.

Multipath distortion effects are highly frequency specific; that is, in a given location, multipath distortions for one frequency band may be severe whereas those in adjacent bands are absent. Frequency hopping allows a receiver that experiences multipath distortion to hop to another channel with less distortion. It also permits mobile receivers to change channels as they move into a position where multipath distortion occurs on their current frequency.

It remains to be seen whether this part of the IEEE 802.11 standard will enjoy wide acceptance. There are many political hurdles to overcome both within the committee membership and among international RF regulatory bodies. In addition, the 1- and 2-Mbps data transfer rates are disappointing and may be well below typical application requirements by the time actual network hardware is available. It is always possible that someone may develop a popular, technically superior proprietary method.

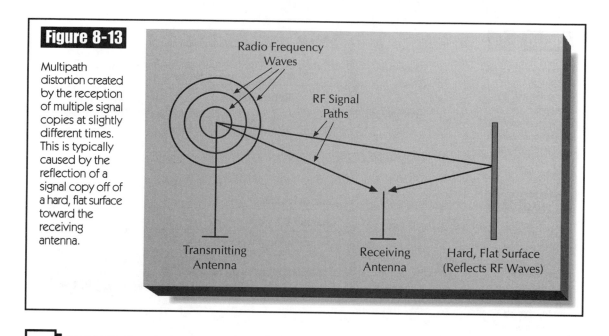

Figure 8-13

Multipath distortion created by the reception of multiple signal copies at slightly different times. This is typically caused by the reflection of a signal copy off of a hard, flat surface toward the receiving antenna.

Radio Frequency Waves

RF Signal Paths

Transmitting Antenna

Receiving Antenna

Hard, Flat Surface (Reflects RF Waves)

CHANNEL AND MEDIA ORGANIZATION

The configuration and organization of media components directly affects the cost and efficiency of data communication. Variables include the number of wires or bandwidth assigned to each channel, the assignment of those wires to carry specific signals or signal components, and the sharing (or lack thereof) of channel components among multiple senders and receivers.

Simplex, Half-Duplex, and Full-Duplex

The most basic form of channel organization for electrical transmission through wires is shown in Figure 8-14 on the next page. Note that a single transmission line requires two wires—a **signal wire**, which carries data, and a **return wire**, which completes an electrical circuit between the sending device and the receiving device. Optical transmission requires only a single optical fiber per channel because a complete circuit is not required. In some transmission modes, two or more communication paths may be necessary between sender and receiver. This type of channel organization is shown in Figure 8-15 on the next page. If multiple electrical transmission lines are used, they may share a single return wire to complete all electrical circuits.

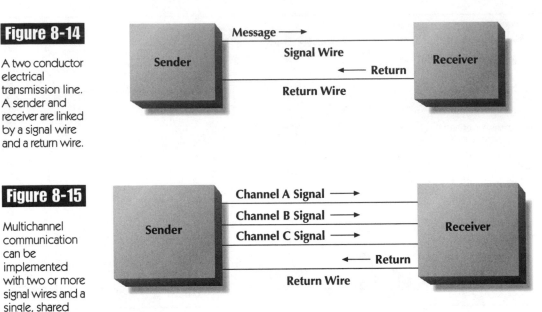

Figure 8-14

A two conductor electrical transmission line. A sender and receiver are linked by a signal wire and a return wire.

Figure 8-15

Multichannel communication can be implemented with two or more signal wires and a single, shared common wire.

A single transmission line can be used in either simplex or half-duplex mode. **Simplex mode** communication is an entirely one-way affair; that is, messages flow only in one direction (see Figure 8-16a). This is a useful mode of communication when data flows in only one direction and when the chance of transmission error is remote. However, if transmission error is expected or if the correction of error is important, then simplex mode is inadequate. If the receiver detects transmission errors there is no way to notify the sender or to request retransmission. A typical use of simplex mode is to send file updates or system status messages from a host processor to distributed data storage devices within a network. In such cases, the same message is transmitted to all devices on the network simultaneously, or in **broadcast mode**.

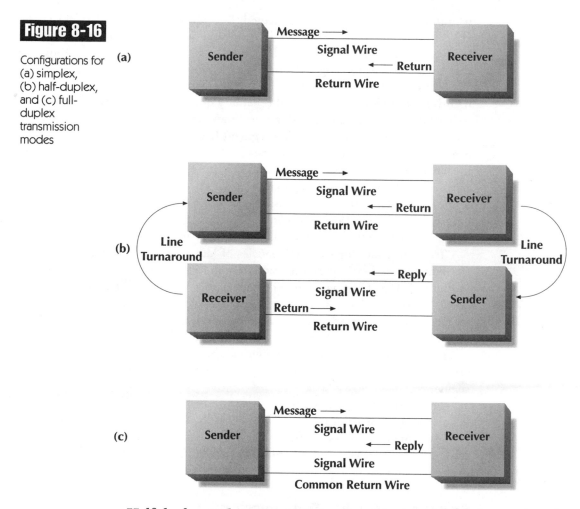

Figure 8-16

Configurations for (a) simplex, (b) half-duplex, and (c) full-duplex transmission modes

Half-duplex mode uses a single line shared between two nodes. Each node takes turns using the line to transmit and receive. There must be an agreement between the nodes as to which is allowed to transmit first. The transmitting node signals its desire to cease transmission by sending a special control message called a **line turnaround**. The receiver recognizes this message and subsequently assumes the role of transmitter. When its transmission is finished, the receiver sends a line turn around message and the original transmitter once again becomes the sender.

Half-duplex mode transmission allows the receiver to request retransmission of a message segment in which errors were detected (see Figure 8-16b). For example, in character-oriented ASCII communication, the receiver might perform a line turn around and transmit a status signal after each group of characters is received. A **negative acknowledge (NAK)** control character would be sent if errors were detected, and an **acknowledge (ACK)** control character would be transmitted if no errors were detected. In half-duplex mode, receipt of NAK at the sender causes it to retransmit the preceding message segment after turning the line around again.

The cost of the communication lines is essentially the same in simplex and half-duplex modes, but the added reliability of half-duplex mode is achieved at a sacrifice in overall data transfer rate. If an error is encountered, the receiver must wait until the entire message segment has been transferred before the line can be turned around and retransmission requested. If the error occurs near the beginning of a message segment, the transmission time used to send the remainder of the message segment is wasted. Also, because errors typically occur in bursts, many retransmissions of the same segment may be required before the entire message is received correctly. Thus, if a channel is persistently noisy, half-duplex transmission will be relatively inefficient and slow.

To compensate for the inefficiencies of half-duplex mode, reserve separate communication lines for transmission in both directions concurrently, as shown in Figure 8-16c. This two-channel organization permits **full-duplex**, or concurrently bidirectional, communication. In full-duplex mode, the receiver can communicate with the sender at any time by using the second channel, while the data transmission is underway in the first channel. If an error is sensed, the receiver can notify the sender, which can halt the data transmission immediately and retransmit. Thus, the sender avoids transmitting redundant message segments.

The speed of full-duplex transmissions may be relatively high, even if there is noise within the channel. Error bursts can be corrected promptly, with minimal disruption to message flow. Compared with single channel modes, the trade-off for the speed and reliability of full-duplex mode is the additional cost of the second channel.

Serial and Parallel Transmission

Parallel transmission requires a separate data line for each bit position (see Figure 8-17). The width, or number of lines, of a parallel link typically is 1 byte or one word as well as a common return line. Because of the number of channels required, parallel communication is comparatively expensive. Its advantage is the higher data transfer rate that results from combining the capacity of multiple transmission lines.

Data and Network Communication Technology

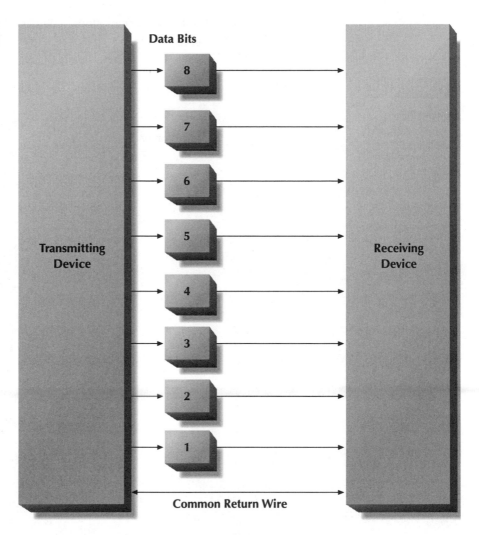

A configuration that supports parallel transmission of full data bytes

Data Bits

8

7

6

5

4

3

2

1

Transmitting Device

Receiving Device

Common Return Wire

The maximum distance over which data can be sent reliably by parallel transmission is limited by an effect called **skew**. Because of slight differences among parallel channels, data bits may arrive at the receiver at slightly different times. The timing difference, or skew, between bits increases with distance and transmission rate. At some point (usually less than 10 meters), skew is significant enough to cause errors in signal interpretation. Skew can be corrected by placing repeater equipment at intervals along the communication route, but this equipment adds to the expense of implementing parallel channels.

Serial transmission requires only a single transmission line. Data bits are sent sequentially through a single line and reassembled by the receiver into larger groups (see Figure 8-18 on the next page). Digital communication over any significant distance uses serial transmission, which eliminates skew and minimizes the cost of wiring. LAN, WAN, and dedicated telecommunication lines all use serial transmission.

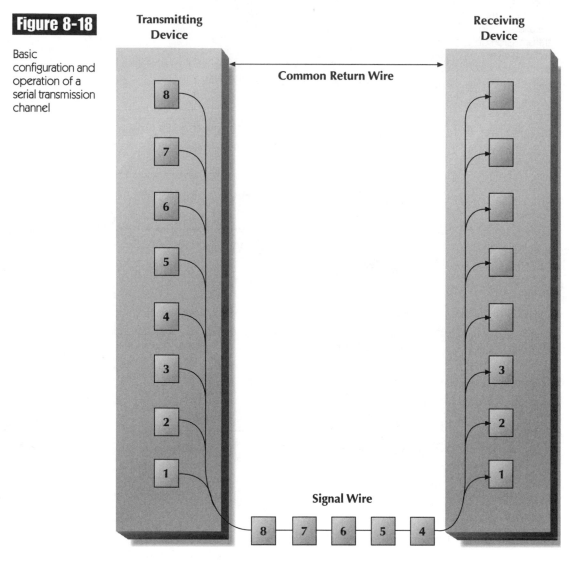

Figure 8-18

Basic configuration and operation of a serial transmission channel

Channel Sharing

Few users require large amounts of data transmission capacity continuously. Rather, transmission capacity is needed for short periods (or bursts) of time and idle in between these periods. Techniques for sharing communication channels allow the traffic of multiple users to be combined resulting in more efficient utilization of available data transfer capacity. As long as many users don't need large amounts of capacity at the same time, the combined load on the channel averages to an acceptable level.

Much of the voice telephone network is based on a sharing strategy called **circuit switching**. When a user establishes a connection to a remote site, the full capacity of a channel is devoted to that communication link. The user has a continuous supply of data transfer capacity, whether or not it is needed. The capacity is assigned exclusively to that user and is unavailable to other users until the link is terminated.

Packet switching, a generic technique for sharing channel capacity, is actually a specific example of **time division multiplexing (TDM)**. TDM refers to any technique by which data transfer capacity is split into time slices and allocated to multiple users and/or applications (circuit switching is also an example of TDM using this definition). Packet switching does this by dividing messages from all users/applications into relatively small pieces called packets (see Figure 8-19). Each packet contains a header that identifies the sender, recipient, sequence number, and other information about its contents. Then, packets are routed to their destination by the network as data transfer capacity becomes available. Packets can be held temporarily in a buffer while waiting for an available channel. If multiple channels are available, then multiple packets can be sent simultaneously across those channels.

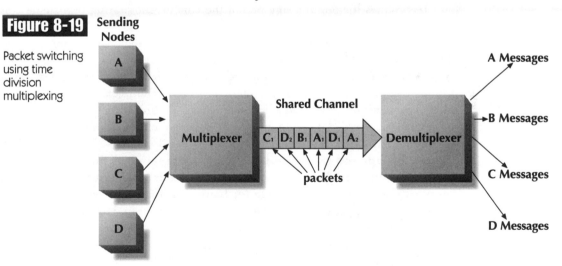

Figure 8-19

Packet switching using time division multiplexing

The primary advantage of packet switching is that the decision of how to most effectively use available data transfer capacity and channels is left to the telecommunication service provider. In most situations, the provider can make rapid (automatic) decisions that efficiently partition available capacity among users, which results in a substantially less expensive telecommunication service.

The primary disadvantages of packet switching are uncertain delays in message transmission and the complexity of using packets. Because a user does not have a dedicated channel, the message-sending time is unpredictable and rises and falls with the total demand for the available data transfer capacity. Delays can be substantial when many users simultaneously send messages and insufficient capacity is available immediately.

Packets themselves introduce substantial complexity into communication protocols. Messages must be broken into packet-sized chunks and appropriate headers must be created. The packets must be routed through the network to their intended destination, and the receiver must reassemble packets into the original message in their intended sequence even when they arrive in a different order. Errors in transmission must be localized to a specific packet and retransmission of that packet must be requested. The hardware and system software required to perform these tasks is complex and, often, expensive.

Nevertheless, packet switching is the predominant form of intercomputer communication. The added cost and complexity is more than offset by the reduction in the cost of providing data transfer capacity for most users and applications. Circuit switching is used only in situations where data transfer delay and available data transfer capacity must be within precise, predictable limits.

Techniques for multiplexing permit the concurrent sharing of a single communication line or channel by multiple users or transmissions. Multiplexing technology has grown from users' needs to use relatively expensive communication lines efficiently. Originally, these methods were developed to support high-volume long-distance telecommunications in order to use existing channels to near capacity. Also, managers of computer centers have used multiplexing to minimize the direct costs of leasing dedicated lines from telecommunication utilities.

Another method by which communication channels can be shared is **frequency division multiplexing (FDM)** (see Figure 8-20). Under FDM, a single **broadband** (i.e., high-bandwidth) channel is partitioned into multiple **baseband** (i.e., low-bandwidth) subchannels. Each subchannel represents a different frequency range (or **band**) within the overall bandwidth of the broadband channel. Signals are transmitted within each subchannel at a fixed frequency or frequency range. Digital information is generally encoded in an analog carrier wave through amplitude and/or phase modulation. Broadband LANs and cable television (CATV) are examples of systems that use FDM.

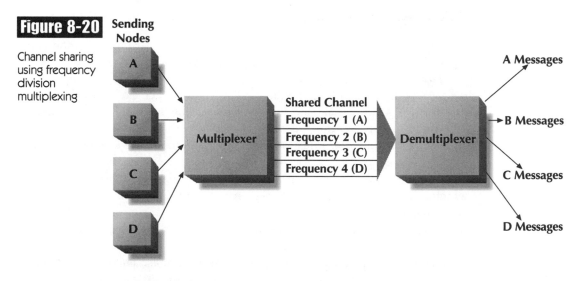

Figure 8-20

Channel sharing using frequency division multiplexing

The use of FDM may require that signals originally intended for one frequency band be moved (or remapped) to another frequency band. Cable television is an excellent example of this process. In most cable TV systems, channels 2 through 13 are carried across a coaxial cable at their original broadcast frequencies. Channels above that are remapped to frequencies lower than their assigned broadcast frequencies so that they occupy contiguous bands within the coaxial broadband channel.

Individual baseband channels within a single broadband channel aren't required to use identical encoding methods or transmission speeds. In addition, baseband channels can be shared using packet switching or other time-division multiplexing methods. Multiple baseband channels also can be combined to form a parallel transmission channel.

COORDINATING COMMUNICATION

Coordination schemes and protocols range from nonexistent to extremely complex. At one end of the scale are some broadcast protocols that require no coordination or feedback from receiver to sender; at the other end are complex schemes addressing sender/receiver synchronization, packet routing and sequencing, and extensive error detection and correction. The following sections describe some of the more common approaches to coordinating communication, but they do not provide an exhaustive treatment of the subject.

Clock Synchronization

An important part of any communication protocol is an agreed upon rate of transmission; from this, the duration of individual signal events is derived. Senders place encoded bits onto a transmission line by modulating specific signal parameters (e.g., phase, voltage, etc.). The receiver examines the signal at (or during) specific time intervals to extract the encoded bits. Sender and receiver must be coordinated by a common timing reference and must use equivalent transmission rates for any such communication scheme to succeed.

There are two primary problems to be addressed when synchronizing timing. The first involves keeping clocks in synchronization during transmission, which is important, especially for high-speed communication. The second problem concerns synchronizing the start of message transmission; that is, the receiver must know when a transmission has begun so that it can begin sampling the signal content to extract data.

Sending and receiving devices have internal clocks. These clocks generally use crystals that generate voltage pulses at a specific rate when a specific input voltage is applied. Unfortunately, such clocks are not perfect. The rate at which timing pulses are generated can vary with temperature and with minor variations in the power input. Individual clocks may run slightly fast or slow as a result of these effects. Further, the speed of clock pulses may vary up and/or down during a relatively short period.

Because two independent clocks are in use, there is no guarantee that their timing pulses are generated at exactly the same points. Most communication protocols provide for clock synchronization when a connection between sender and receiver is first established, but even if perfect synchronization is achieved at that time, the clocks may subsequently drift out of synchronization with one another. As clocks drift out of synchronization, bit interpretation errors (by the receiver) become more likely; see Figure 8-21 for an example of this effect. The receiving device clock pulses are being generated at a slower rate than the sender clock pulses. The result is a misalignment of bit time boundaries with respect to the signal. In Figure 8-21, the receiver receives fewer bits than were sent and incorrectly interprets several of them.

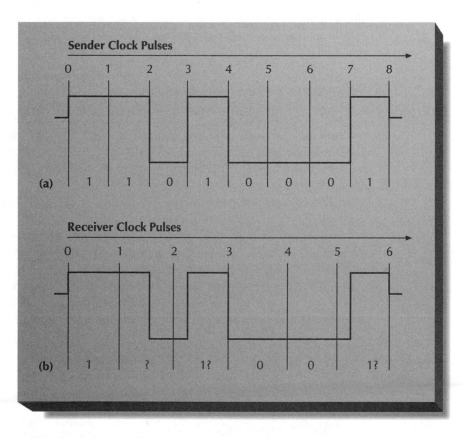

Figure 8-21

An example of communication errors resulting from lack of clock synchronization. The sender (a) encodes bits into the signal at a faster clock rate than the receiver (b) extracts them. This results in errors in interpreting specific bits and a difference between the number of bits sent and received.

Signal transitions between bit values can be used as a means for the receiver to synchronize its clock with sender, or at least recognize that the clock has drifted out of synchronization. A receiver expects signal transitions to occur at precisely the same instant as timing pulses are received from its clock. Small differences can be an implicit message to resynchronize the clock to the incoming signal. Clock drift is a particular problem when long sequences of the same bit value are sent and when the encoding method provides no signal transitions within a bit value. In this situation, there are no signal transitions to compare to the timing pulses, so the receiver has no means to determine whether synchronization has been lost. Certain types of data encoding methods (e.g., Manchester coding) combat this problem by providing a signal transition in the middle of every bit.

Sender and receiver must also agree on the boundaries of a message (i.e., when it begins and ends); to do so, they must agree on the length of a message. They may or may not use some sort of marker embedded in the signal to identify the beginning or end of a message. There are two common primary approaches to coordinating message boundaries—asynchronous transmission and synchronous transmission. These methods are sometimes referred to as **character framing** methods when character or byte streams are the basic message units.

Asynchronous Transmission

If serial communication is performed in **asynchronous** mode, the timing of transmission for each character within the data message can vary. From the receiver's standpoint, the pattern of transmission is intermittent, or apparently random. For example, data that are input through a keyboard typically would be sent asynchronously to an I/O port of the host CPU. As each key is pressed, a character (or byte) is sent. From the standpoint of the I/O port, the timing of keystrokes is unpredictable (i.e., the time interval between characters will vary).

For an asynchronous bit stream to be interpreted correctly, the same predetermined character framing scheme must be applied by both sending and receiving devices. The purposes of character framing are for the sender to indicate and for the receiver to detect character boundaries within a bit stream. As shown in Figure 8-22, character framing in asynchronous serial mode uses a **start bit** and a **stop bit** to indicate these boundaries. Accordingly, asynchronous mode also may be called start/stop communication.

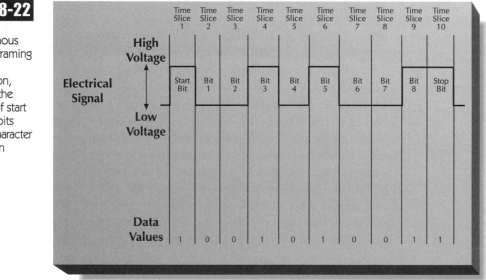

Figure 8-22

Asynchronous character framing for serial transmission, including the addition of start and stop bits used for character recognition

The number of data bits framed by the start/stop bits depends on the coding method used. If characters are coded in ASCII-7 with an error check bit, a frame contains 8 data bits. Clearly, the same coding method and number of data bits must be used by both the sender and the receiver. When sensing a start bit, the receiving device interprets the next bit received as the first of a character or byte. When a stop bit is received, the transmission of the character or byte is complete, and the communication line is returned to idle (or neutral) state, in anticipation of the next start bit.

In asynchronous mode, the internal clocks of sender and receiver operate independently. Start/stop bits are the means by which these independent clocks are synchronized. Asynchronous serial communication is used for low-speed, interactive data transfers between remote terminals and a central host processor, as well as among microcomputers.

Synchronous Transmission

If serial transmission is in **synchronous** mode, the clocks of sending and receiving devices are precisely coordinated. Messages are exchanged within predetermined time intervals. Framing of transmissions is by groups of bytes (or blocks) rather than by separate characters, as shown in Figure 8-23. The size of each block component is always the same so the receiver knows where one component ends and another begins.

Figure 8-23

A typical format for messages to be transmitted using synchronous character-framing techniques

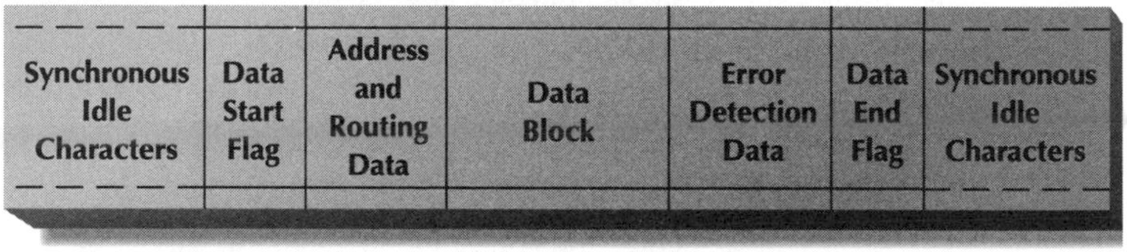

Synchronous Idle Characters	Data Start Flag	Address and Routing Data	Data Block	Error Detection Data	Data End Flag	Synchronous Idle Characters

The boundaries of each block are marked by time intervals between blocks (or gaps) during which clock synchronization is checked and readjusted, if necessary. Readjustment is performed by sending a continuous stream of **synchronous idle messages** during the time intervals between "real" messages. The synchronous idle message has a predetermined pattern of signal transitions designed to allow easy clock synchronization. Generally, synchronous mode is used to transfer large batches of data at high speed.

Parallel Transmission

Clock synchronization problems are not limited to serial transmission; they also exist with parallel transmission to a lesser degree. Parallel transmission protocols often assign a transmission line to carry a clock pulse simultaneously with the data. Thus, the receiver does not use an independent clock and explicitly monitors the senders clock. By definition, there cannot be a synchronization problem if only one clock is in use.

Determining message boundaries is not a problem if the number of parallel data channels equals the number of message components. For example, if each line carries one bit, eight data lines are used, and bytes are the unit of data transfer, then there is no need for any special character-framing techniques. When the unit of data transfer is larger than can be encoded simultaneously on the available data channels, then the communication protocol must provide some means of marking message boundaries.

Error Detection and Correction

A crucial component of any communication protocol is the means by which errors in data transmission, reception, or interpretation are detected and corrected. Under most protocols, the correction action is straightforward. On sensing an error, the receiver transmits a negative acknowledge (NAK) signal to the sender. On sensing a NAK from the receiver, the sender retransmits the preceding message unit.

All known error-detection correction methods are based on some form of redundant transmission. That is, a redundant message or message component is transmitted with or immediately after the original message. The receiver keeps a copy of the original and compares it to the redundant version. If the two match, then the original message is assumed to have been transmitted, received, and interpreted correctly. If the two don't match, then a transmission error is assumed to have occurred. The receiver then asks the sender to retransmit the message and error checking content.

Error detection and correction methods vary in:

► Probability that they will detect a real message error.

► Probability that they will identify a correct message as an error.

► Complexity of the error detection method.

► Efficiency of channel utilization.

Efficient channel utilization usually must be sacrificed to achieve greater success in identifying real errors, which results directly from increasing the level of redundancy used. As greater redundant content is sent, the proportion of channel capacity used to transmit "real" data goes down. At the same time, the chance of detecting an error increases.

The probabilities of detecting "real" and "imagined" errors can be mathematically or statistically computed. In fact, statistical terminology can describe these two probabilities. The probability of not detecting a real error is called **Type I error**, whereas the probability of identifying good data as an error is called **Type II error**. As in any statistical confidence interval, these two types of error are inversely related: that is, a decrease in Type I error (generally desired) is accompanied by an increase in Type II error.

Type II errors result in needless retransmission of data that was received correctly but assumed to be in error. Thus, an increase in Type II error decreases the efficiency with which the channel is used. A greater proportion of channel capacity is used to needlessly retransmit data. The decrease of Type I error thus has two negative effects on efficient channel use. Generally, it is achieved by increasing redundant transmission and at the expense of greater Type II error.

The desired degree of Type I error should be balanced against the probability that an error will occur. In some types of communication channels (e.g., a system bus), the probability of a transmission or reception error is remote. Thus, no error detection and correct scheme might be the best alternative. In other channels, the occurrence of an undetected error is not considered "important." For example, communication with character-based video display terminals often uses weak error detection methods (i.e., high Type I error). It assumes that a user can easily spot the result of an error and that application software provides a means for the user to request retransmission (e.g., a function key or command to redraw the screen content).

Commonly used methods of error detection include:

► Parity checking (vertical redundancy checking).

► Block checking (longitudinal redundancy checking).

► Cyclical redundancy checking.

Each of these is described in detail next.

Parity Checking

Error detection for character data can be implemented through **parity checking**. In a character oriented transmission, 1 bit (usually the eighth) in each character is designated the **parity bit** (or check bit). The value of this bit is set to correspond with a count of other bits within the character. Parity checking may be based on even or odd bit counts.

With **odd parity**, the sending device sets the value of the parity bit to 0 if the number of 1 valued data bits within the character is odd. Conversely, if the count is even, the parity bit is set to 1. Thus, the count of 1 bits in each byte received always should be an odd number. If the receiving device counts an even number of 1 bits in the byte, it is assumed that one or more bit values have been altered in transmission (i.e., an error has been detected). Under **even parity**, the sending device sets the parity bit to 0 if the count of 1 bits is even. If an odd number of 1 bits is found, the parity bit is set to 1. If no errors occur, the receiving device always should find an even number of 1 bits within each byte.

Parity checking provides some degree of assurance that errors will be detected, but this technique is not highly reliable. For example, it would be possible for the error to go undetected if several bits within a single character have been altered in multiples of two. That is, a pair of altered bits can act as **compensating errors** on one another. So, even though the received character contains errors, the value of the parity bit will correctly match the altered data bits.

Parity checking is a type of **vertical redundancy checking (VRC)**. VRC techniques are based on bit counts within individual characters. In asynchronous, character oriented transmissions, VRC may be the only practical means of error detection. Since characters are transmitted separately and at unpredictable times, it can be difficult to perform checks on groups of characters.

Block Checking

Parity checking on groups of characters, or blocks, can be performed under **longitudinal redundancy checking (LRC)**. To implement LRC, the sending device counts the number of 1 bits at each bit position within a block. After the block is sent, the sending device derives a **block check character (BCC)** from these counts and transmits it to the receiver.

As shown in Figure 8-24, each bit within the BCC is a parity bit that is set to correspond with the number of 1 bits at the same position within the preceding characters. Either odd or even parity may be used. The receiving device derives a separate BCC after receiving the block of characters. Then, it compares the BCC developed locally with the BCC that has come from the sender. If the values are not the same, an error is indicated. NAK is returned to the sender, and the entire block is retransmitted. Like VRC techniques, LRC is vulnerable to compensating errors. A higher degree of assurance can be provided by applying both VRC and LRC within the same protocol. However, even with this "double check" approach, some compensating errors might go undetected.

Figure 8-24

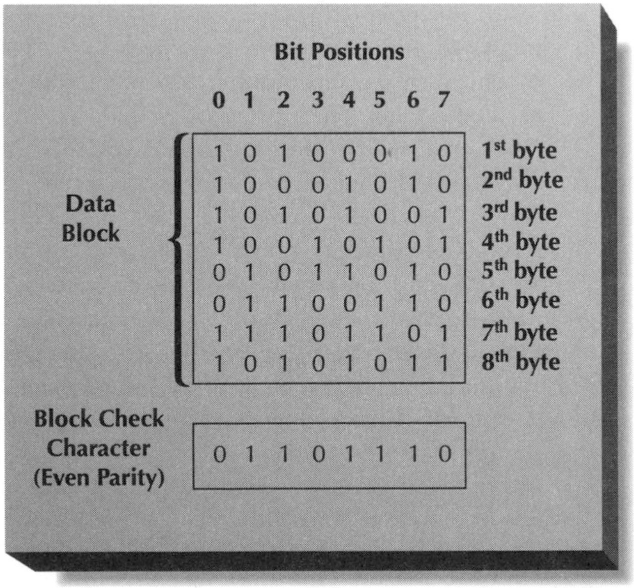

An example of longitudinal redundancy checking. An even parity bit is computed for each bit position of a block of 8 bytes. The set of parity bits forms a block check character that is appended to the block for error detection.

	Bit Positions								
	0	1	2	3	4	5	6	7	
Data Block	1	0	1	0	0	0	1	0	1st byte
	1	0	0	0	1	0	1	0	2nd byte
	1	0	1	0	1	0	0	1	3rd byte
	1	0	0	1	0	1	0	1	4th byte
	0	1	0	1	1	0	1	0	5th byte
	0	1	1	0	0	1	1	0	6th byte
	1	1	1	0	1	1	0	1	7th byte
	1	0	1	0	1	0	1	1	8th byte
Block Check Character (Even Parity)	0	1	1	0	1	1	1	0	

Cyclical Redundancy Checking

Another error detection technique is **cyclic redundancy checking (CRC)**, which applies a mathematical algorithm to a block of characters to generate a BCC. CRC uses a common check (bit) string known by both the sender and receiver and a separate block check bit string that is generated by the sender and appended to the end of each data block. The combined block (data block and appended block check string) may be interpreted as a single large integer. The check string appended to the data block is calculated by the receiver so as to make the large integer evenly divisible by the common check string. The receiver verifies the data block by dividing the large integer by the common check string.

In general, the accuracy of a particular method of CRC depends upon the length of a transmitted block, the length of the block check string appended to the transmitted block, and the value of the common check string used by the sender and receiver. Transmitted blocks generally consist of several hundred to a few thousand data bits. The length of the appended block check string is typically between 16 and 64 bits, depending on the length of the data block. Mathematical algorithms can be used to select a common check string (the divisor) that minimizes the chance of undetected transmission errors.

Like LRC, CRC is commonly used in synchronous character oriented protocols. With CRC and a sufficiently large check string, the assurance of error detection is substantially higher than with other redundancy checking methods. The trade-off for such extensive error detection is extra processing overhead. The calculation of the block check string is much more complex than the calculation of an LRC check string. To minimize processing delay, CRC is often implemented in dedicated electronic circuits (i.e., firmware) within sending and receiving devices.

Technology Focus

Asynchronous Transfer Mode

Asynchronous transfer mode (ATM) is a protocol for high-speed broadband data transmission. ATM was developed by a consortium of telecommunications companies as a standardized method for implementing broadband integrated services digital network (B-ISDN). ATM was designed to carry any type of data including voice and video. Design for voice and video data required the implementation of measures to guarantee minimum data transfer rates between senders and receivers. It also required minimal switching (packet relay) delay from sender to receiver.

ATM standards exist for twisted pair, coaxial, and fiber-optic cable. The twisted pair standard requires Category 5 (an industry standard) or better cable and provides 25 Mbps. Cable runs between switches are limited to 100 meters. ATM on coaxial cable provides 155 Mbps. ATM over single mode fiber-optic cable provides 622 Mbps.

ATM implements time division multiplexing using packets. ATM packets are referred to as cells. Each cell is 53 bytes in length; 48 bytes are allocated to data and 5 bytes to control information (i.e., a cell header). No error detection information (e.g., a CRC value) is added to the data portion of the cell, but there is error detection for the header. Thus, ATM performs no error checking of data content at the network level. However, software that uses the network may implement error detection and correction and include error detection information within the data messages. The decision to exclude low-level error checking was based on a desire for high-speed packet switching and the assumption that reliable physical connections would be used.

ATM is a connection-oriented packet transfer protocol. Connection-oriented means that a specific path between sender and receiver is established before transmission of any data (much like the connection established when dialing a telephone number). This connection is maintained for the duration of the data transfer session. ATM calls these connections virtual circuits (VCs). VCs are given certain characteristics at the time they are created. One of the more important is a guaranteed quality of service (including minimum data transfer rate). Thus, an ATM network allocates a portion of its data transfer capacity to each VC at the time it is created and guarantees the availability of that capacity for the duration of the VC. This capability is well-suited to audio and video data transmission.

The process of creating a VC is relatively complex. Senders and receivers are connected through intermediate devices called ATM switches. A VC creation request from a potential sender goes to the nearest switch; then that switch maintains tables of routes to other switches and network nodes. A complex series of requests passes from switch to switch until the intended receiver is located. Each ATM switch stores information about currently defined VCs and available capacity. As new VCs are created, this information is updated. A VC request is denied (or a lower quality of service is negotiated) unless there is sufficient uncommitted capacity within the switch and its attached transmission lines.

Once created, a VC is assigned a unique identifier. The identifier is added to the control message of each cell by the sender. Each switch stores VC routing information in an internal table and uses that information to send cells to their destination. Cells sent within a single VC are guaranteed to arrive in sequential order, which simplifies and speeds the task of reassembling them into larger data units (e.g., video frames).

ATM has been widely adopted by telecommunication vendors for long distance links. There is a good chance that your next long-distance telephone call will at least partly traverse an ATM network. Adoption for computer to computer links in both local area networks (LANs) and wide area networks (WANs) has been very slow, partly because of the relatively late development of standards for slower (and cheaper) transmission over copper cable and inexpensive fiber.

Cost has been a major barrier. Until recently, ATM switches cost $10,000 or more and the cost of connecting a desktop computer was as much as the computer itself. As more products have been brought to market and production quantities have increased, costs have dropped rapidly. ATM may or may not become the connection of choice for LANs, but it is clearly well-poised to grab a large share of the market for broadband and backbone connections. Its combination of guaranteed data transfer capacity, ability to simultaneously carry many types of data, and high overall data transfer rates is peerless among standardized and readily available telecommunication protocols.

Business Focus

Upgrading Network Capacity (Part I)

The Bradley Advertising Agency (BAA) specializes in the design and development of printed and video advertisements. Most of the company's work is performed under contract to other advertising agencies. Of the 30 people BAA employs, most are actively engaged in design and production work.

Printed ad copy is designed and produced almost entirely by computer. BAA uses Adobe and other software products on a mixture of approximately one dozen networked IBM-compatible and MacIntosh PCs. A variety of peripheral devices are used, including several scanners, various types of color printers, and a high-speed color copier. A dedicated file server holds most data and software files on its 40 Gbytes of disk storage. The file server also has an 8-Gbyte DAT tape drive and a 625-Mbyte magneto-optical drive.

Until recently, video production was done almost entirely using conventional methods (e.g., video cameras and video tape editing equipment). Over the last two years, BAA has been experimenting with an expensive ($125,000) video editing software package, and now BAA has two SUN workstations dedicated to video production and editing. BAA expects to add two more workstations this year and two or three more the following year.

The files used to store print ads are large—it is not unusual to use several dozen megabytes of storage for each project. BAA typically has several dozen print projects in process at a time. Video production files are much larger than print files—it is not unusual for several Gbytes of storage to be devoted to a single video project (e.g., a 30-second television ad).

File-sharing performance was already slow when the first SUN workstation was purchased. A consultant hired to analyze the problem determined that the file server was being used to only about 20% of its capacity. The bottleneck was discovered to be the network itself. The existing network is a 10-Mbps Ethernet LAN. The network wiring is unshielded twisted pair (UTP) that was installed in 1989.

Network bottlenecks have become more acute as the two SUNs have moved into daily use. Access to print ad files that took a few seconds a year or two ago can now take up to a minute during busy periods. Extra disk storage (4 Gbytes each) was purchased for the SUN workstations so that video files can be stored on the workstation temporarily, but downloading and uploading can take 10 minutes or more. This also has created data sharing problems when both workstations are used to work on the same project simultaneously.

BAA is studying alternatives to upgrade its network capacity and is unsure whether it should attempt to increase its network (relying on its existing cabling) or move to a more modern network with substantially higher performance. The two alternatives that seem most promising are an upgrade to 100-Mbps Ethernet and a partial or total conversion to ATM. BAA is uncertain whether its cabling will work with 100-Mbps Ethernet due to its age. For 155-Mbps ATM, coaxial cable would be required; 622-Mbps ATM would require fiber-optic wiring.

Questions
1. If BAA's current cable is tested and determined to be reliable with 100-Mbps Ethernet, should the company choose that option?
2. If the existing cable is tested and determined to be incompatible with 100-Mbps Ethernet, should the company consider installing coaxial cable?
3. Assume that the installation of fiber-optic cable will cost four times as much as coaxial cable ($60,000 vs. $15,000). Is fiber worth the extra money?

SUMMARY

A communication protocol is a set of rules and conventions for communication. It covers many aspects of communication, including message content and format, bit encoding, signal transmission, transmission medium, and channel organization. Also, it includes procedures for coordinating the flow of data including media access, clock synchronization, and error detection and correction. A message is a unit of data or information transmitted from a sender to one or more recipients. Messages contain data and/or control information.

Data bits can be encoded into analog or digital signals. Encoding within analog carrier waves is used often because of the requirements of the transmission medium and/or because it is more reliable over longer distances. Analog signals can be electrical, optical, or radio frequency. Bits are encoded in analog sine waves by modulating one or more wave parameters. Possible modulation techniques include frequency, amplitude, and phase modulation. Modulation methods can be combined and frequently are in high-speed transmission using multilevel coding.

Electrical digital bit encoding methods include TTL signals, zero crossing signals, and Manchester encoding. Each method encodes digital data within electrical square waves. Square waves are difficult to reliably transmit over long distances; thus, electrical digital signals aren't generally used for telecommunication or network links. Digital data can be directly encoded as pulses of light. Then, these optical signals can be transmitted reliably over long distances.

Important characteristics of transmission media include raw data transfer rate, bandwidth, and susceptibility to noise, distortion, external interference, and attenuation. Bandwidth is the difference between the highest and lowest frequencies that can be propagated through a transmission medium. Bandwidth is an important determinant of raw data transfer rate because it determines the maximum rate at which signal parameters can be modulated.

The effective data transfer rate may be much less than the raw data transfer rate because of noise, distortion, external interference, and signal attenuation. Noise can be generated internally or added through external interference. The power of internally generated noise is proportional to medium length and varies with media type. Signal attenuation is a loss of signal power and is also proportional to length. Distortion is an alteration of signal parameters because of normal passage through the medium. Signal-to-noise ratio is a measure of the difference between noise power and signal power. For reliable transmission to occur, signal power must be greater than noise power.

Electrical cables are of two primary types—twisted pair and coaxial. Both use copper but differ in construction and shielding. Twisted pair is relatively cheap but limited in bandwidth, signal-to-noise ratio, and (therefore) transmission speed because of weak shielding. Coaxial cable is more expensive but offers higher bandwidth, greater signal-to-noise ratio, and lower distortion. Optical cable provides very high bandwidth, little internally generated noise and distortion, and is immune to electromagnetic interference. It is much more costly than copper cables but provides much higher data transfer capacity.

Data may be transmitted without wires by radio waves and infrared light. Transmitters are costly, but receivers are relatively inexpensive. Radio frequency channels are in short supply because of their limited number and intense competition for licenses. Radio channels are leased from telecommunication companies on an as-needed basis. Infrared transmission has limited transmission distance.

Channel organization describes the number of lines dedicated to a channel and the assignment of specific signals to those channels. A simplex channel uses one optical fiber or copper wire pair to transmit data in one direction only. Wiring of a half-duplex channel is identical, but sending a specific control signal reverses the direction of transmission. Full-duplex channels use two fibers or wire pairs to allow simultaneous transmission in both directions.

Parallel transmission uses multiple lines to send several bits per signal event. Serial transmission uses a single line to send 1 bit at a time. Parallel transmission provides higher channel throughput but it is unreliable over distances greater than a few meters because of an effect called skew. Serial transmission is reliable over much longer distances. Serial channels are cheaper to implement due to the fewer number of wires or wireless channels used.

Channels are often shared among users and applications when no one user or application needs a continuous supply of data transfer capacity. Time division and frequency division multiplexing (TDM and FDM) are the primary means by which channels are shared. Circuit switching is a form of TDM where an entire channel is allocated to a single user for the duration of one data transfer operation. Packet switching allocates time slices on the channel by dividing many users' messages into smaller units called packets and intermixing them during transmission. FDM divides a broadband channel into several baseband channels and dedicates one or more of them to individual users.

Successful data transmission requires coordinated sender and receiver actions. Specific coordination tasks include clock synchronization, error detection, and error correction. Clock synchronization is required so that the receiver uses the same time periods and boundaries to decode a message from a signal as are used by the sender to encode the message into the signal. A single shared clock is the most reliable synchronization method, but it requires that clock pulses be sent continuously from sender to receiver.

Asynchronous transmission relies on specific start and stop signals to indicate the beginning and end of a message unit (usually a single character). These signals also provide clock-synchronization information to the receiver. Synchronous

Data and Network Communication Technology

transmission maintains a continuous flow of signals from sender to receiver to provide continual opportunities for clock synchronization. Self-clocking signals provide signal transitions within each bit to aid in synchronization.

Error detection is always based on some form of redundant transmission. The sender compares redundant copies of messages and assumes that they are correct if they match. The receiver requests retransmission if they don't match. Increasing the level of redundancy increases the chance of detecting errors at the expense of reducing channel throughput. Common error-detection schemes include parity checking, block checking, and cyclical redundancy checking.

Key Terms

10BaseT
acknowledge (ACK)
amplifier
amplitude
amplitude shift keying (ASK)
amplitude modulation
asynchronous
attenuation
band
bandwidth
baseband
bit time
block check character (BCC)
broadband
broadcast mode
character framing
circuit switching
coaxial cable
communication channel
communication protocol
compensating errors
cycle
cyclic redundancy checking (CRC)
data transfer rate
data encoding
distortion
effective data transfer rate
electromagnetic interference (EMI)

even parity
fiber-optic cable
frequency
frequency division multiplexing (FDM)
frequency shift keying (FSK)
frequency modulation (FM)
full-duplex mode
half-duplex mode
infrared light
line conditioner
line turnaround
longitudinal redundancy checking (LRC)
Manchester coding
message
modulation
modulator-demodulator (modem)
multilevel coding
negative acknowledge (NAK)
noise
odd parity
packet switching
parallel transmission
parity bit
parity checking
phase
phase shift keying (PSK)
phase shift modulation

radio frequency (RF)
raw data transfer rate
repeater
return wire
serial transmission
signal wire
signal-to-noise ratio (S/N ratio)
simplex mode
sine wave
skew
square wave
start bit
stop bit
synchronous
synchronous idle message
time division multiplexing (TDM)
transistor-to-transistor logic (TTL)
transmission medium
twisted pair wire
Type I error
Type II error
vertical redundancy checking (VRC)
zero crossing signal

Vocabulary Exercises

1. _____ transmission sends bits one at a time using a single transmission line.

2. During half-duplex transmission, sender and receiver exchange roles after a(n) _____ is transmitted.

3. _____ encodes data by varying the distance between wave peaks within an analog signal.

4. A(n) _____ converts a digital signal to an analog signal (or vice versa) to allow data transmission over analog phone lines.

5. In synchronous data transmission, a(n) _____ signal is transmitted during periods when no data is being transmitted.

6. _____ transmission uses all of a transmission medium's capacity to implement a single channel. _____ transmission divides the capacity of a transmission medium to implement multiple channels.

7. A pair of errors that combine to "fool" an error detection protocol is called _____.

8. _____ cabling is a twisted pair cable that normally carries transmissions at a rate of 10 Mbps.

9. Digital transmission with _____ signals uses square waves with no negative voltage states.

10. The _____ of an analog signal is measured in _____, or cycles per second.

11. _____ embeds a voltage transitition within every bit.

12. A local telephone grid uses _____ switching to route messages from sender to receiver. Most networks use _____ switching to route messages from sender to receiver.

13. _____ checking is a form of longitudinal redundancy checking in which bits of a single character or byte are used to derive a check digit.

14. With parity checking, sender and receiver must agree whether error detection will be based on _____ or _____.

15. The _____ of a channel describes the mathematical relationship between noise power and signal power.

16. _____ is any change in a signal characteristic caused by components of the communication channel.

Data and Network Communication Technology

17. Methods of error detection and correction represent specific trade-offs among effective data transfer rate, processing complexity, _____, and _____.

18. Wireless transmission uses the _____ or _____ frequency bands.

19. _____ cannot interfere with signals sent through _____.

20. A communication channel that uses electrical signals must have at least two wires—a(n) _____ and a(n) _____ to form a complete electrical circuit.

21. _____ signals are digital signals that contain both positive and negative voltage states.

22. The _____ of a transmission is a measure of its theoretical capacity. The _____ is a measure of its actual capacity when a specific communication protocol is employed.

23. Simultaneous transmission of multiple messages on a single channel can be accomplished by _____ or _____ multiplexing.

24. The _____ of a signal is a measure of its peak strength.

25. In asynchronous transmission, _____ is implemented through the use of start and stop bits.

26. The term _____ describes the encoding of data as variations in one or more physical parameters of a signal.

27. In _____ transmission, blocks or characters arrive at unpredictable times. In _____ transmission, the timing of character or block data transfers between sender and receiver is precisely coordinated.

28. The _____ of a communication channel is the difference between the highest and lowest frequencies that can be transmitted.

29. _____ transmission implements two-way transmission with two separate communication channels. _____ transmission implements two-way transmission with only one communication channel.

30. _____ encodes data by varying the magnitude of wave peaks within an analog signal.

31. _____ transmission uses multiple lines to send multiple bits simultaneously.

32. A(n) _____ extends the range of data transmission by retransmitting a signal.

33. _____ checking generates a(n) _____ consisting of a single check digit for each bit position within the bytes of a block.

34. _____ uses more than two signal parameters levels to encode multiple bits within a single signal event.

35. _____ cabling is generally the least expensive alternative for network wiring.

36. A(n) _____ increases the strength of a signal, including both the encoded message and any noise that may be present.

37. The length of a parallel communication channel is limited by _____, which can cause bits to arrive at slightly different times.

38. _____ waves are analog signals. _____ waves are digital signals.

39. A receiver informs a sender that data was received correctly by sending a(n) _____ message. It informs the sender of a transmission or reception error by sending a(n) _____ message.

40. _____ is used in some networks (e.g., cable television networks) as a transmission medium.

41. _____ is a term that describes loss of signal strength as a it travels through a transmission medium.

42. Messages to be transmitted by time division multiplexing are broken up into _____ prior to physical transmission.

Review Questions

1. What are the components of a communication channel?

2. What are the components of a communication protocol?

3. What are the comparative advantages and disadvantages of frequency, amplitude, and phase shift keying?

4. How is multilevel coding used to increase the effective data transfer rate of a communication channel?

5. What are the comparative advantages and disadvantages of digital transmission using TTL, zero crossing, and Manchester encoding?

6. Why is analog transmission of electrical signals generally preferred to digital transmission over long distances?

7. Describe the relationship among bandwidth, data transfer rate, and signal frequency.

8. How can noise and distortion be introduced into a transmission medium? How does a channel's signal-to-noise ratio affect the reliability of data transmission?

9. Compare and contrast twisted pair, coaxial, and fiber-optic cabling in terms of raw data transfer rate, cost, susceptibility to transmission error, and ability to transmit analog and/or digital signals.

10. What are the advantages of wireless transmission using RF waves as compared to infrared waves?

11. Describe simplex, half-duplex, and full-duplex transmission. Compare and contrast them in terms of cost and effective data transfer rate.

12. Why will the actual data transfer rate of a channel usually be less than the theoretical maximum implied by the technology used to implement the channel?

13. Compare and contrast serial and parallel transmission in terms of channel cost, data transfer rate, and suitability for long-distance data communication.

14. What are the differences between synchronous and asynchronous data transmission?

15. What is character framing? Why is it generally not an issue in parallel data transmission?

16. Describe the differences between even and odd parity checking.

17. What is a block check character? How is it derived and used?

18. Compare and contrast frequency and time division multiplexing. What physical characteristics of the communication channel are required by each type? Which provides greater data transmission capacity?

19. What is the difference between an amplifier and a repeater?

Problems and Exercises

1. Assume that a baseband channel can propagate signal frequencies up to 50 MHz. What is the raw data transmission capacity of the channel if data is encoded as binary TTL signals? What is the raw data transmission capacity of the channel if data is encoded using eight-level TTL signals? What is the raw data transmission capacity of the channel if data is encoded as binary Manchester signals?

2. Assume that data is transmitted on a dedicated channel in blocks of 48 bytes at a rate of 10 Mbps. Assume further that an 8 bit block check character is used for error detection and that an error is detected, on average, once every 1,000 transmitted blocks. What is the effective data transfer rate of the channel?

3. Assume that data is transmitted over twisted pair cable. The power of the signal when placed on the channel is 75 dB. Signal attenuation is 5 dB per hundred meters. Average noise power on cable is 0.1 dB per meter. What is the signal-to-noise ratio of a 50-meter channel? What is the signal-to-noise ratio of a 1-kilometer channel? What is the maximum channel length?

Research Problem

Several companies have recently begun to offer products that implement transmission speeds higher than 100 Mbps over twisted pair wiring. One of these, called Gigabit Ethernet, runs over wiring designated 1000BaseT. Investigate Gigabit Ethernet and its wiring requirements (www.3com.com). How does 1000BaseT cable differ from 10BaseT and 100BaseT cable? What other cabling options can be used with Gigabit Ethernet?

Data and Network Communication Technology

Chapter 9

Networks and Distributed Systems

Chapter Goals

▶ Describe the costs and benefits of distributing computer resources.

▶ Describe specific methods of resource sharing for I/O devices, mass storage (files), and processing (CPUs).

▶ Describe the structure and organization of network software and hardware.

▶ Describe common physical attributes of networks, including topology, access protocol, and communication hardware.

WHY DISTRIBUTE COMPUTER RESOURCES?

Consider the state of computer networking in the year 1970. A corporation with several regional offices operates a large mainframe computer at its home office. The majority of its application systems run in batch mode, but many of the newer ones use online data entry. To provide better service to the branch offices, a number of measures are taken, including:

▶ The provision of one remote job entry (RJE) computer at each branch office. Each RJE computer has a card reader, card punch, and line printer, and each is connected to the home office mainframe by a 9600-baud modem over a leased telephone line.

▶ The provision of staff (e.g., keypunch operator, computer operator, programmer/analyst, etc.) and supporting equipment (e.g., keypunches, card sorting and collating machines, etc.) at each branch office.

▶ Placement of three data-entry terminals in each branch office connected to the home office by 1200-baud modems over normal telephone lines and used for online data entry.

Although costs are high, turnaround time for many applications is vastly improved. Documents no longer need be transmitted by overnight mail. Data capture occurs locally and is transmitted to the mainframe by online terminals or through the RJE facilities. Output is sent directly to the RJE facilities. Local support personnel build and maintain a few custom applications at the branch offices using RJE facilities.

Now envision the same company 15 years later.

The home office operates four of the largest available mainframes, each of which has the maximum possible amount of disk storage, memory, and terminal connections. A minicomputer is in use at each branch office for both local processing and for RJE. These machines support many local applications, as well as data capture for applications at the home office. Some online terminals at the branches are connected directly to the local minicomputer, and some are connected by modem to one of the home office mainframes.

There are several problems with this configuration:

▶ Data communication capacity is near the saturation point. A great deal of data transmission occurs between the branch and home offices. Home office applications require access to local data stores. Large batch jobs (through the RJE facilities) copy files from the branch minicomputers to the home office.

▶ Maintenance of distributed applications is problematic. Changes to distributed programs must be made separately on each minicomputer. Installation of new or revised application system often takes weeks or months.

▶ Online terminal communication is insufficient to meet peak demand. Management wants to be able to shift terminals at the branch offices between computers as needed, but this is not feasible with the current configuration.

▶ Some branch offices are starting to use desktop computers. Mechanisms are needed to move data between these machines and the minicomputers and/or mainframes.

▶ The I/O and processing requirements of the largest application systems are nearing the capacity of the largest available mainframes. Hardware technology is advancing rapidly, and the company upgrades to the latest models quickly. Despite this, the corporation anticipates that its largest online applications will exceed the capacity of the largest available mainframes within a year or two.

These problems typified the late 1970s and early 1980s, and the response was always the same: Buy a bigger machine. When that didn't work, buy another big machine. Eventually, the problems overwhelmed even the largest computer installations. To address these dilemmas, improvements in data communication technology and the introduction of networked computing were required.

Distribution of Computing Resources

It is difficult to discuss the topic of distributed systems or networked computing without identifying the goals of networking and exactly what portions of a system are to be distributed. Unfortunately, the terms **distributed system** and **network** are too often used without context or specificity. Because there are many types of networks and many ways to distribute computing resources, these terms are too vague to be useful.

The goal of networking computer hardware and systems can be stated simply: to allow communication between computing devices. But this definition fails to address higher-level issues of networking, including:

▶ What is to be communicated?

▶ What additional capabilities does networking make possible?

▶ What are the benefits of these capabilities?

▶ What are the costs of these capabilities?

Networking provides a basic communication capability between computer systems. This communication may be an end in and of itself. For example, e-mail, news distribution, and bulletin boards are services based solely on communication capabilities. The communication capability also can be used as a basis for more complex services, which are described in detail next.

Benefits of Distribution

Most of the benefits of networking are derived from the additional capabilities made possible by intercomputer communication. These benefits arise from the ability to share resources among many networked computers. In particular, the following resources may be shared:

► Data.
► Programs.
► I/O devices.
► Mass storage devices.
► CPU services.

The ability to share programs and data across a network is normally implemented by a **distributed file system**. Access to files containing data or programs are provided by network connections. In general, distributed file systems provide for access to a common set of files, which may reduce redundant storage of data and programs. This, in turn, reduces the problems inherent in maintaining file consistency (e.g., redundant updates) and minimizes the total amount of mass storage space required within the network.

The ability to share common hardware devices allows a reduction in total hardware investment. Unnecessary redundancy in computer hardware can be avoided, which often saves a substantial amount of money. This is true of all types of computer hardware, but is often most pervasive with I/O devices. Often, it is difficult for individual users to justify the cost of expensive I/O devices such as high-capacity laser printers, color plotters, scanners, and high-speed communication devices. However, these devices become more affordable on a per-user basis when shared by many users through a network.

In addition to its use for distributed file systems, shared mass storage has other advantages. For example, many programs require large amounts of temporary storage during execution, and replicating this storage across multiple computer systems can be expensive. Providing a common pool of such temporary storage can substantially reduce total mass storage costs.

Access to shared CPUs can be a benefit in several ways. Recall from Chapter 2 that cost-effective computing solutions tend to use the smallest possible class of hardware. For example, providing 10 microcomputers is often a more cost-effective solution than providing a single minicomputer. Although such trade-offs are valid in an aggregate sense, they do not address issues such as peak demand and irregular need for access to specialized computing resources. At times, a user may need to execute an application that exceeds the capacity of his or her own microcomputer; at other times, a user may need access to a substantially more powerful machine (e.g., to execute a simulation program on a supercomputer).

Networking provides a basic capability to address these types of problems. A user needing access to a larger CPU might be able to access such a CPU over the network. Input data can be routed over the network to a process executing on another CPU and the processing results routed back to the user's computer. A user can access local machines that have excess capacity or a distant computer center that provides access to highly specialized and/or expensive hardware. In either case, the user can address most computing needs with local (i.e., relatively inexpensive) hardware and address peak and/or specialized needs through access to shared resources.

Costs of Distribution

Although the benefits of networking and distributed computing are numerous, there are many costs to be considered. First and foremost is the cost of the network itself. As with most other types of computer hardware, network cost rises nonlinearly in relation to its capacity. Thus, the cost of network for 500 computers may be substantially more than 10 times the cost of a network for 50 computers.

Networks consist of a variety of hardware and software elements, each of which must be purchased and maintained. Maintenance costs apply not only in the case of equipment failure, but also in response to changes in the composition of the network. Associated maintenance expenses tend to be high because networks are dynamic (i.e., resources are frequently added, changed, and deleted, etc.).

A less-tangible cost of networked computing lies in the loss of local control over hardware and software. One of the reasons for the tremendous growth in the use of smaller machines has been users' desire to exercise more control over their own computing, and yet networking often dilutes that control. Users require access to data, programs, and hardware devices that they themselves do not own or control. Access must be negotiated between resource owners, and changes in those resources require changes in user computing procedures.

Another cost of networked computing lies in the inefficiency inherent in intercomputer communications. Despite rapidly advancing data communication technology, communication between hardware devices within a single computer system is still much faster than communication across a network. As resources are further and further distributed, communication inefficiencies become a drag on the performance of each individual computer on the network. This may be addressed by adding additional network capacity, but the user then faces costs that rise nonlinearly, as described earlier.

Although the aforementioned costs are real and significant they are often overshadowed by the benefits of networking. In addition, the performance gap between intracomputer and intercomputer (i.e., network) communication continues to decline. The cost of network capacity is also falling rapidly. It is up to each network designer and administrator to determine the best trade-off for their particular computing needs.

NETWORK SERVICES

Whereas all network services rely on basic intercomputer communications, the type and complexity of these services varies substantially. Network services can be roughly classified into the following categories:

► Access to shared I/O devices.
► Terminal to host communications.
► User to user communications.
► File transfer.
► File sharing.
► Distributed processing.

These services are listed in order of increasing complexity. In this case, complexity refers to both the complexity of the software that provides the services and the hardware resources consumed by that software.

Extended Communication Services

Many common networking services are relatively simple extensions to the services available to the users of a large multiuser computer. Access to shared I/O devices is the best example of these. In a multiuser computer system, users compete for access to shared peripheral devices. For example, several users can generate output destined for a laser printer, but only one such user can access that printer at a time.

A multiuser operating system must provide software to control access to shared I/O devices. In the case of a printer, the operating system provides a mechanism for spooling users' printed output to temporary storage. Another mechanism continually monitors that storage for additions and sends them, one at a time, to the printer. The former mechanism is generally called a **spooler**, and the latter is called a **scheduler**. The organization of these components is depicted in Figure 9-1.

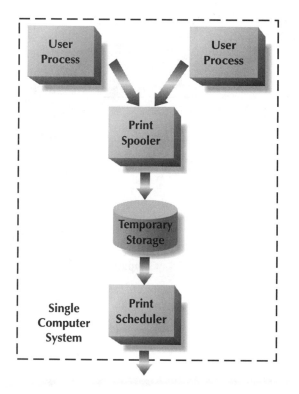

These components can also be referred to by a more general set of terms. The spooler and scheduler are designed to provide shared access services to a common resource (the printer). These types of processes are commonly called **servers** because they provide services to user (or other) processes on request. The user processes—referred to as **clients**—are users of the service provided by the server process(es).

In a large multiuser computer, both server and client processes are executed within the same machine. In a networked computer system, they may execute on different machines. For example, various user processes may be executing concurrently on individual users' microcomputers. The print spooler and scheduler may be executing on a print or network server. This organization, depicted in Figure 9-2 on the next page, is an example of a **client/server architecture**.

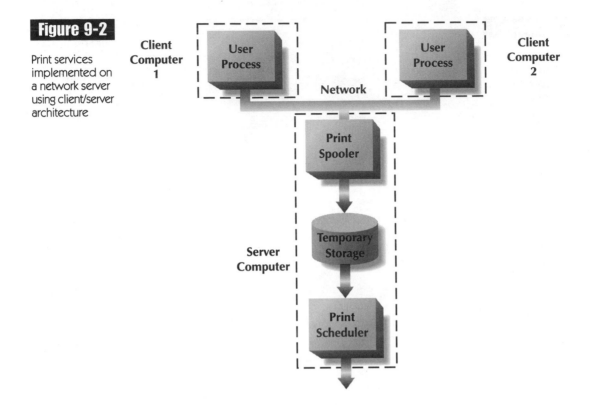

Figure 9-2

Print services implemented on a network server using client/server architecture

When executing on a single computer, client and server processes communicate through the interprocess communication facilities of the operating system. When executing on different computers, they communicate through the network's software and hardware facilities. Ideally, the use of a network is transparent to client processes. For example, a user process that generates output destined for a printer should be unaware that the printer is not directly connected to the local computer's system bus.

Network transparency requires modifications to the operating system of each computer on the network. In the example of remote printer access, the operating system must capture requests for printer access and reroute them to the appropriate node of the network. This requires a modification to the normal printer service routines provided by the operating system and the addition of network interface services. These software components are depicted in Figure 9-3.

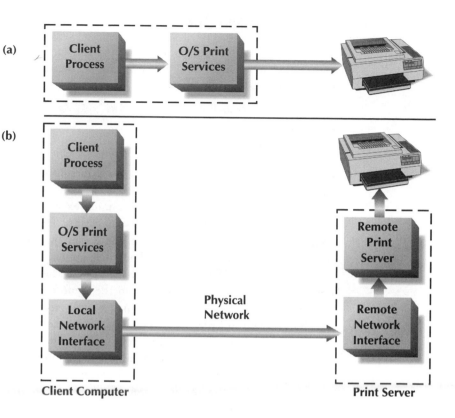

Figure 9-3

Comparison of the software components needed to implement print services in (a) a single machine architecture and (b) a client/server architecture

E-mail and bulletin board services present problems similar to those of I/O device access. Each of these services requires access to an area of storage where messages are stored. Users gain access to these storage areas by invoking a mail or message utility. When implemented within a single machine, these utilities combine aspects of both server and client processes. They are client processes to the extent that they provide user-oriented facilities (e.g., creation and editing of messages), and they are server processes to the extent they provide access to message-storage areas.

The distinction between server and client becomes much clearer when message passing takes place over a network. In a networked system, message storage may be provided on each machine (for that machine's users), or it may be provided on a single machine in the network. In the single-machine case, the machine that stores messages and provides access to them executes the server process(es). The machines that users use to send or receive messages execute the client processes.

As with remote printing, extensions to normal mail facilities are required to interact with the network. Outgoing messages must be re-directed from a local destination to the server, which requires mail processes to interact with network interface services in a manner similar to that depicted previously in Figure 9-3.

Distributed File Systems

Sharing of programs and data in a network requires the provision of a mechanism for sharing files. File sharing can be implemented in several different ways, including:

► Explicit file transfer.

► Transparent file transfer.

► Transparent file access.

The term distributed file system can be applied to any of these methods, though it most commonly describes transparent file access.

In **explicit file transfer**, the user must explicitly request the transfer of a file from a remote machine to a local machine. The user must know on which machine the file is stored and must use a specific software utility (e.g., a Web browser) to request the transfer. The file is copied from one machine to another over the network. If the user needs to update the shared copy of the file, the local version must be copied back to the remote machine after it is modified.

Transparent file transfer eliminates the need for the user to request a transfer or even to know that the file does not reside locally. When a user (or user process) requests access to a remote file, the transfer is initiated automatically by the operating system. The implementation of the transfer mechanism is similar to the implementation of the remote print server described earlier. File access requests to the service layer of the local machine operating system are re-directed to a remote file server.

In particular, a modification to the operating system service for opening and closing files is required. When an open operation is requested, the operating system must determine if the file exists locally. If not, a transfer request must be initiated to the remote machine that contains the file. This implies that the local file system contains information about files stored on remote machines or that it can request such information over the network.

When a file is closed by a user process, a transfer in the opposite direction must be initiated. Thus, a users' access to a remote file requires copying the file twice—once when the file is opened and again when it is closed. Transparent file transfer can be extremely inefficient, especially if the user needs only a portion of the file. For example, assume that a user needs to update a single record in a large file and that the file is organized for direct access (e.g., using an index). The mechanism just described requires that every record of the file be copied from the remote machine to the local machine and back again, yet the user needs access to only one record. Because of this inefficiency, transparent file transfer is seldom used in modern networks.

Transparent file access solves the efficiency problems associated with transparent file transfer. Instead of copying an entire file and then operating on it locally, only the needed records are transferred. In addition, modified records are transferred back to the remote location immediately. Thus, there is little or no local storage allocated to the file, and a minimal amount of network communication is required.

The advantages of transparent file access come at the expense of additional complexity in all file manipulation operations. The open and close functions for files must be modified, and every file access service call requires network access. Each read or write operation requires network access to the remote file. Thus, transparent file access reduces network data communication at the expense of increased complexity of O/S file service routines.

Distributed Processing

Distributed processing refers to access by a client process to the CPU resources of a server machine. It can be implemented in a number of ways and with various degrees of transparency. Some possible implementations include:

► Remote login.

► Remote process execution (explicit).

► Remote process execution (transparent).

► Load sharing.

Remote login facilities are provided in networks to allow users full access to the facilities of a remote machine. The user essentially interacts with the remote machine as if he or she were using a terminal attached directly to that machine. The network serves as the data communication facility, and a process on the local machine provides an interface to the user's I/O device.

The user must explicitly request a connection to the remote machine through a software utility specifically designed to support remote login. Once a connection has been established, all input from the user's input device (e.g., a terminal) is re-directed to the remote machine over the network. Any processes the user initiated are executed on the remote machine. Output from those processes is routed back through the network to the user's local machine and from there to his or her I/O device. Telnet is a widely used example of a remote login utility. The telnet program provides a software-based character terminal on the user's desktop machine, and all other processing occurs on the remote machine (i.e., a telnet host).

A comparison between this type of interaction and a direct connection is shown in Figure 9-4 on the next page. Notice that the local operating system command layer is not used when the remote login is in progress; instead, it is replaced by a local version of a remote login process. The function of this process is to translate terminal I/O between the normal local terminal interface and the network. Because of incompatibilities between terminal I/O devices and the I/O services available on various machines, some translation is necessary. The remote login process on the local machine and the remote terminal interface on the remote machine will negotiate an appropriate I/O protocol when a connection is first established. Each process will then perform any necessary translation between this protocol and local protocols during all I/O operations.

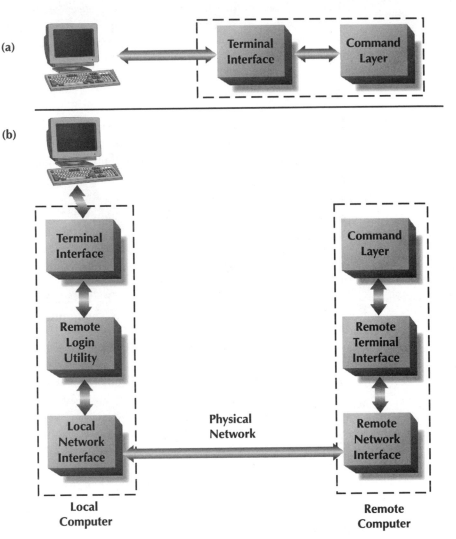

Figure 9-4

The software components needed to implement a terminal interface (a) on a single machine and (b) with a remote login utility under client/server architecture

A user can explicitly request the execution of a process on a remote machine. If such a facility is provided, the formalities of user login on the remote machine are bypassed. Thus, the interface to the command layer on the remote machine is also bypassed and a single process is loaded and executed.

As before, I/O protocols must be negotiated, and appropriate translation must take place, but the I/O connection exists only while the remote process is executing. The connection is terminated as soon as the remote process terminates and final output is received on the local machine.

Remote process execution may also take place transparently (i.e., without the user's knowledge). It is programmed directly into an application or system utility. Note that this is a different arrangement than the normal client/server relationship. In a client/server relationship, the server processor is continually active on the remote machine. It resides on the machine in an idle state pending the arrival of a service request on the network. In contrast, a remotely executed process is not loaded and executed until a request is received. The remote process is terminated once the request is satisfied.

Remote process execution facilities can be used to implement **load sharing** (or **load balancing**) across a network. Load sharing occurs when a busy processor asks another processor to execute a process for it. Thus, processing load is shared (or balanced) between the two machines. In the previous examples of remote process execution, the process to be executed resided on the remote machine, and only processing requests and I/O were transferred over the network. However, when remote process execution is used for load sharing, an executable image of the process is transferred between machines over the network.

In sophisticated load-sharing systems, it is possible to move a partially completed process between processors, which requires that the participating CPUs have compatible architectures (e.g., instruction sets, registers, etc.). It also demands operating system compatibility, to the extent that the executing process uses operating system services. For these reasons, load sharing is only implemented in tightly integrated networks of fully compatible computer systems.

NETWORK ARCHITECTURE

There are numerous issues to be resolved when implementing a computer network, including:

▶ Topology.
▶ Media access control (MAC) protocol.
▶ Communication channel hardware.
▶ Data transfer capacity.

There are several alternatives within each of these categories. Thus, it appears that a network designer faces a large number of possible combinations.

Fortunately, there has been some standards-setting activity in the area of network architecture. The **Institute of Electrical and Electronics Engineers (IEEE)** has taken a leading role in establishing such standards, which have been widely adopted by manufacturers of networking hardware and software. In essence, these standards provide a prepacked set of compatible network solutions addressing the design parameters above. (These standards will be discussed more fully in "Standard Network Architectures.")

Network Topology

Topology refers to the spatial organization of network devices, the physical routing of network cabling, and the flow of messages from one network node to another. Network topology can be referred to in logical or physical terms. Physical topology describes the physical placement of cables and device connections to those cables. Logical topology refers to the path that messages traverse as they travel from node to node. Logical topology may be different from the underlying physical topology as later examples will illustrate. In addition, physical networks can be composed of several segments and each segment may use a different topology.

There are three primary options for physical network topology:

► Star (or Hub).

► Ring.

► Bus.

The primary characteristics that differentiate these topologies are the length and routing of network cable, the nature of node connections, data transfer performance, and susceptibility of the network to failure.

The **star topology** uses a central node to which all other nodes are connected. The central node can be a computer system, a network server, or a specialized network device. The star topology requires all communication to pass through the central node. Because this can overload the processing and data communication capabilities of the central node, the star topology is seldom the primary network topology in large-scale networks. However, it is often used in local area networks (LANs) or in small (less than 100-node) network segments.

The **ring topology** connects each network node to two other nodes. The entire network must form a closed loop (see Figure 9-5). Messages are passed through the ring in one direction, and each node on the ring acts as a **repeater**. Messages received by a node are retransmitted to the next node. A message received by a node that is addressed to that node is not retransmitted. Thus, a message travels around the ring, retransmitted by intervening nodes, until it reaches its destination. Some networks use a dual ring with each ring carrying messages in a different direction.

Figure 9-5

A simple ring network. Each node is attached to two neighbor nodes. Messages flow from sender to recipient in one direction through the ring.

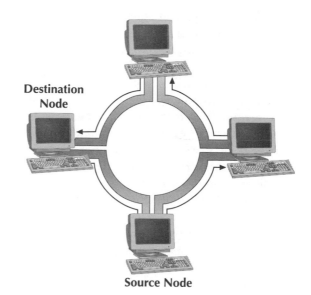

The **bus topology** requires that each node be connected to a common transmission line (see Figure 9-6). A node's connection to the bus is referred to as a **tap**. Each node detects messages passing on the bus through the tap. If the message is not intended for a node, it simply ignores it. A terminating resistor located at each end of the bus absorbs any signals it receives. Data movement is bidirectional. A node sends a message through its tap onto the bus. The message then travels in both directions until it reaches the terminating resistors at each end of the bus.

Figure 9-6

An example of a bus network topology. The arrows show the movement of a message from source to destination. Note that the message moves through the bus in both directions starting at the sender's bus connection.

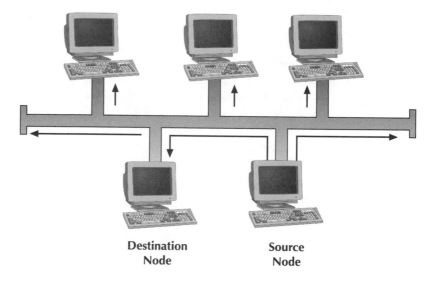

The advantages of the bus topology are: simplified wiring and network interface hardware. The bus location is not restricted by the location of nodes attached to the bus. The use of taps is an advantage when considering network stability. Taps are passive devices that simply pass part of any passing signal to the attached node. If a node on the bus fails, the flow of data across the bus continues uninterrupted. Thus, the network continues to operate, allowing uninterrupted communication among the other nodes.

The primary disadvantage of the bus topology arises from the use of taps and passive node interfaces. Each tap removes some signal energy from the bus, thus decreasing the signal strength beyond that tap. The more taps there are, the greater the loss of signal strength. At some point, signal strength degrades to the point where reliable communication is impossible. Thus, the number of nodes on the network and the total length of the bus and its taps must be limited. To counteract this, a bus network can be broken into segments and repeaters or amplifiers can be used to connect the segments. By using a sufficient number of properly placed repeaters, the effective length of a bus network is virtually unlimited.

The primary advantage of the ring topology is a relatively long maximum network length created by the use of repeaters at each node. Because each node interprets and retransmits every message, there is little or no degradation of signal strength as a message traverses the ring. The disadvantages of the ring topology are susceptibility to failure and difficulty in adding, deleting, or moving nodes. Because each node is connected to two others, the addition, deletion, or movement of a network node requires that two connections be changed. This fact, and the basic ring organization, make network wiring a difficult task. Also, because each node is an active repeater, the failure of any node breaks the ring, thus disabling the entire network.

Media Access Control

A **media access control (MAC) protocol** controls the timing of transmission by nodes on the network. In general, only one node can be actively transmitting at a time. If multiple nodes attempt to transmit simultaneously, their messages will mix, thereby producing noise or interference, which is referred to as a **collision**. Methods of dealing with collision fall into two broad categories: those that allow collisions but detect and recover from them and those that avoid collisions altogether. The most common media access protocols are:

► Carrier sense multiple access/collision detection (CSMA/CD).

► Token passing.

CSMA/CD (carrier sense multiple access/collision detection) is an access protocol commonly, although not exclusively, used in bus network topologies. The basic strategy used in this protocol is not to avoid collisions but, rather, to detect their occurrence. If a collision is detected, the transmitting nodes must retransmit their messages. The basic protocol is as follows:

► A node attempting to transmit first listens (carrier sense) until no traffic is detected.

► The node then transmits its message.

► The node listens during and immediately after its transmission. If abnormally high signal levels are heard (collision detection), then the node ceases transmission.

► If a collision was detected, the node waits for a random time interval and then attempts retransmission.

The **token passing** access protocol is commonly (but not exclusively) used in ring network topologies. It uses a control message called a **token**. The node that "possesses" the token is allowed to originate messages on the network. All other nodes can only receive or retransmit. A node gains possession of the token by receiving it from another node. It must relinquish the token (transmit it to another node) after a specified time interval. The order in which the token is passed between nodes can be different from the physical order of nodes on the network.

The advantage of CSMA/CD lies primarily in its simplicity. There are no control messages (tokens) to be passed between nodes and, thus, no inherent ordering of nodes in the network. Nodes can be added or deleted without any update to the specific media access control (MAC) method. The simplicity of the MAC method also simplifies associated hardware and software. Simpler hardware is generally cheaper and sometimes faster, whereas simpler software is smaller (memory and disk space), executes more quickly, and requires less tuning and configuration.

The disadvantage of CSMA/CD is its potentially inefficient use of available data transfer capacity. Because this method does not avoid collisions, a portion of the network transmission capacity is wasted. Each collision is, in essence, a missed opportunity to transmit a message. As network traffic increases, the number of collisions also increases. At sufficiently high levels of communication traffic, network performance starts to decrease because of excessive numbers of collisions and message retransmissions. The effect of this phenomenon on effective communication capacity is depicted in Figure 9-7 on the next page. Networks that use CSMA/CD typically achieve an effective data transfer rate no greater than one-half their potential data transfer rate.

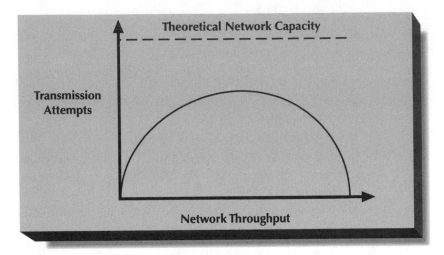

Figure 9-7

The relationship between network throughput and attempted packet transmissions using the CSMA/CD media access protocol

Token passing avoids the potential inefficiencies of CSMA/CD. Because collisions are prevented, no transmission capacity is wasted due to collisions and retransmissions. A small portion of network capacity is needed to transmit the token between nodes, but it is generally insignificant compared to total data transmission capacity. Thus, token ring networks can achieve an effective data transfer rate nearly equal to their potential data transfer rate.

Other advantages of token passing include the ability to tune network performance and a superior ability to support certain types of network applications. Network performance can be tuned by varying the time interval after which a node must relinquish the token. This can be used to increase the network capacity allocated to nodes that need to transmit large amounts of data (e.g., file servers). Each node has a limited time during which it can possess the token. Thus, the maximum time interval that a node must wait between message transmission opportunities is simply the sum of the maximum times that all other nodes may hold the token. This predictability is essential in many applications such as video conferencing and network telephones.

The disadvantages of token passing arise from its complexity. For example, it is more susceptible to failure and requires special procedures when the network is first started. Some device must be responsible for creating and transmitting the token when the network is initiated. There must be a strategy for generating a new token if a node "dies" while in possession of the original token. Furthermore, each node must know the next node in the token passing sequence. The failure of a node requires the previous node to bypass the failed node. Startup, failure recovery, and token routing require additional control strategies to be implemented within each node and/or the intervention of a dedicated network control device.

Communication Channel Hardware

The hardware required to implement network connections consists of several types of devices: those to conduct messages; those to connect transmission lines network nodes; and those to connect one network to another. The exact hardware required to implement a network depends on the type of network (e.g., local or wide area), the network topology, number and type of attached computer equipment, method of data transmission, and other factors. Each class of hardware is discussed in detail next.

Transmission methods and media for data communications, including twisted pair coaxial copper cable, fiber-optic cable, and wireless transmission by many different frequency bands, were discussed in Chapter 8. Fiber-optic cable provides performance superior in every way to other forms of transmission. But other methods persist because of the relatively higher cost of fiber-optic cable and the significant installed base of other transmission media. Twisted pair cable is the medium of choice for small networks and fiber-optic cable is the preferred medium for large networks and long-distance transmission.

Computer networking almost always implies the use of packet switching (as compared to circuit switching), which adds a layer of complexity to computer networks in the form of hardware (and software) to connect network segments and to route packets from one segment to the next. This equipment is often more costly and more difficult to maintain than the transmission media it controls.

Several types of devices are used to connect network segments and to route packets, including:

► Network hubs.
► Bridges.
► Routers.
► Switches.

These devices can be confusing because of some overlap in functionality and the existence of devices that perform more than one function.

An understanding of the physical and logical architecture of a typical network is helpful in sorting out the roles of various network hardware devices. Consider, for example, a typical large college campus network, which might encompass dozens of buildings and hundreds, or thousands, of network nodes. Creating a single network to which every node was directly attached would be impractical because the network would need a very high data transfer rate to accommodate all of the nodes. The network interface devices needed to attach the nodes directly to

---

the high-speed network would be very expensive. In addition, a failure of the network would disable network functions for the entire campus.

Larger networks are organized in a manner similar to that depicted in Figure 9-8. Nodes within a building or on one floor of a building are connected with a local area network (LAN). The architecture of these LANs can be tailored to the users' needs. They have low data transfer capacity and are relatively inexpensive unless users need more network power. Their use is particularly advantageous when large amounts of network messages are sent within a single floor or building, which commonly occurs when user workstations and departmental servers are located in close proximity.

Floor and building LANs are part of a larger zone network. Individual nodes aren't directly attached to the zone network; rather, they are connected indirectly through network hardware that is part of their own LAN. The zone network must have data transfer capacity sufficient to carry the traffic of all of the attached LANs. The higher-capacity zone network requires more expensive network interface devices, but because only a few network interface devices (e.g., the LAN hubs) are directly connected, the expense is minimal.

Zone networks are, in turn, connected to a larger campus backbone network, which must have sufficient data transfer capacity to carry campuswide traffic. However, traffic on this network is often minimized by filtering out messages between nodes in the same zone. That is, the campus backbone network only carries messages between nodes in different campus zones and between on-campus and off-campus nodes. The campus backbone network is, in turn, part of a larger network (typically, the rest of the Internet).

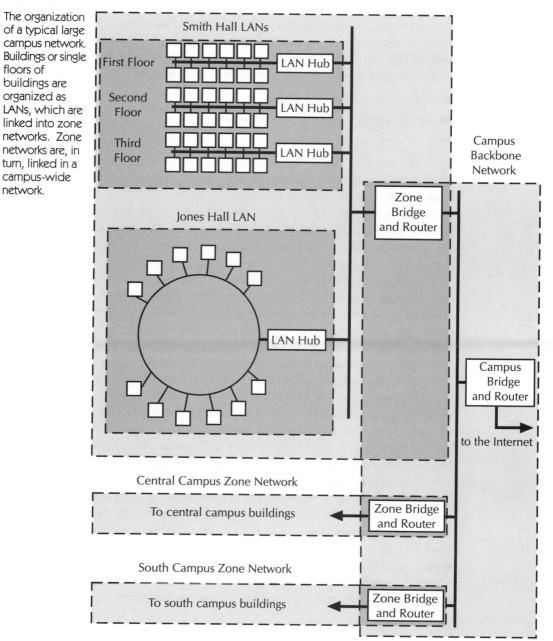

Figure 9-8

The organization of a typical large campus network. Buildings or single floors of buildings are organized as LANs, which are linked into zone networks. Zone networks are, in turn, linked in a campus-wide network.

North Campus Zone Network

Smith Hall LANs

First Floor — LAN Hub

Second Floor — LAN Hub

Third Floor — LAN Hub

Jones Hall LAN

LAN Hub

Campus Backbone Network

Zone Bridge and Router

Campus Bridge and Router

to the Internet

Central Campus Zone Network

To central campus buildings — Zone Bridge and Router

South Campus Zone Network

To south campus buildings — Zone Bridge and Router

Network Interface Units

Hardware devices are required to connect individual computers and related equipment (e.g., network printers) to a network transmission cable. The generic terms for such a device are **network interface unit (NIU)** and **network interface card (NIC)**. An NIU for a single computer is usually a printed circuit board (or card) attached directly to the system bus. It performs the following functions:

- Detection of incoming transmissions.
- Generation of outgoing transmissions.
- Low-level media access.

An NIU requires one or more device drivers to enable communication with other devices (e.g., the CPU and primary storage). Device drivers are also used to implement higher levels of software.

Both incoming and outgoing network traffic require translation or data conversion functions between the bus protocol and the physical network protocol. These translation functions can include character framing, serial to parallel conversion, analog to digital conversion, and other translation functions.

Incoming traffic in a bus network is detected by scanning a field within each packet header for a physical address. Each network interface unit is assigned a unique address during manufacture or installation. Packets that match the physical address are placed on the computer system bus, whereas other packets are simply ignored.

The physical connection between a computer system's network interface unit and a bus network is implemented with a **transceiver**. Its function is to detect any data passing on the network bus and to re-direct a portion of that signal to the network interface unit. In some types of bus networks (e.g., those using coaxial cable), the transceiver is part of the NIU; in such cases, physical connection to the network bus is implemented with a **T connector**. Bus networks that use twisted pair cable may place the transceiver within the network hub; in such cases, the only packets received by the NIU are those addressed to it.

In ring networks, all incoming traffic must be actively processed. If the packet is destined for the local computer, then the NIU removes it from the ring (it isn't repeated), stores it in an internal buffer, and sends an interrupt to the CPU. In some ring network protocols, another packet is created and transmitted to acknowledge receipt of the original packet. If an incoming packet contains a node address other than that of the local computer, then it is simply echoed (repeated) onto the outgoing network connection.

Network Hubs

A **network hub** serves as the first connection point between a network node and the network. Network hubs vary widely in their capability, with the simplest hubs serving only to concentrate network wiring connections at a single point. Hubs are sometimes called **wiring concentrators**, particularly when that is their only, or primary, function. Hubs also can incorporate routing, bridging, or switching functions.

LANs are typically implemented on a single floor of a building. Although ring and bus networks are the predominant logical topologies for LANs, they are usually physically wired in a star configuration. Network hubs are the key component that enables this dual topology. A network hub is typically located in a central part of a building or floor (e.g., a **wiring closet**), and cables from all of the other rooms are routed to the wiring closet. Standard modular connectors, similar to those used on modular telephone cable, are used at each end of the cable to connect workstations and other network devices to the hub.

The logical topology of the network is implemented within the hub. For a bus network, each hub connector is attached to a transceiver. The network bus is implemented within the hub itself (see Figure 9-9 on the next page). For a ring network, each hub connector implements two connections: one for messages transmitted to the network node and another for messages transmitted by the network node. The hub uses internal switches to route the outgoing connection of one node to the incoming connection of the next node (see Figure 9-10 on the next page). Thus, a complete ring is implemented by appropriate connections made within the hub.

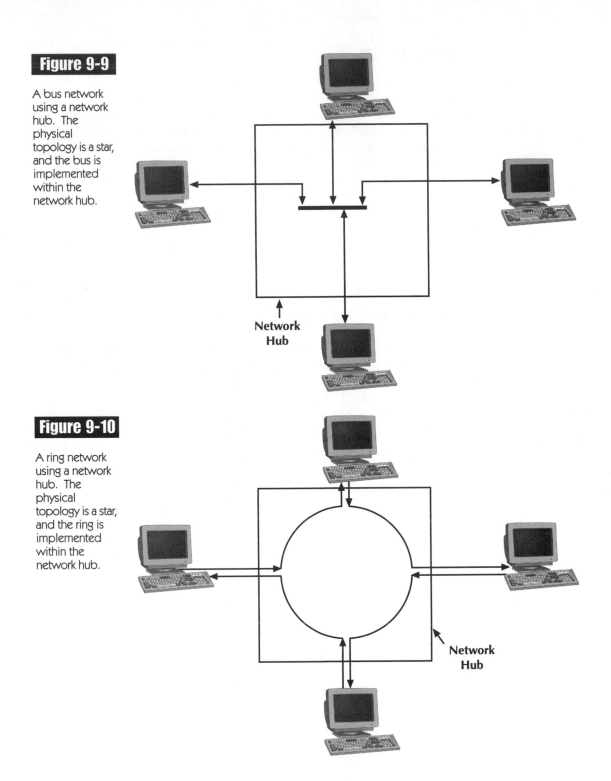

Figure 9-9

A bus network using a network hub. The physical topology is a star, and the bus is implemented within the network hub.

Network Hub

Figure 9-10

A ring network using a network hub. The physical topology is a star, and the ring is implemented within the network hub.

Network Hub

A ring network hub usually performs some important network-management functions such as creating a token when the network is restarted. It also monitors the network for "dead nodes" and reroutes traffic around them to keep the ring intact. Also, it generates a new token if the old one becomes "lost." Ring network hubs cost more than bus network hubs because they perform these extra functions. Such hubs eliminate most of the wiring disadvantages inherent in the ring topology.

Bridges

A **bridge** is a device used to connect one network segment to another. The exact function and operation of a bridge depends on the types of networks being connected. They forward traffic between a LAN and a wide area network (WAN) or another LAN. They also are useful devices for extending the length of an existing LAN. Most LANs have a limit on the number of attached nodes and the total length of the transmission line. The use of a bridge between two LANs effectively makes them one large "extended length" LAN.

A bridge constantly scans packets on each network for physical addresses that correspond to nodes on the other network. Any such messages are echoed onto the other network. Bridges can connect network segments of the same type (i.e., topology and protocol) or of different types. Bridges that connect dissimilar networks are especially common in connections between small local (or zone) networks and larger (or backbone) networks. In this situation, a bridge must be capable of translating frame and packet formats from one protocol to another. Bridges often can be customized to this purpose by the addition of optional hardware or specific software.

Routers

A **router** performs the same function as a bridge but does so in a more intelligent manner. A router constantly scans the network to monitor the pattern of message traffic and the addition, modification, or deletion of network nodes. This information is stored in internal tables and used to make decisions about packet forwarding from one network or segment to another. As changes are made to the network, a router records these changes and modifies its behavior accordingly. It can use its knowledge of the network to improve message transfer efficiency.

A router is an active device that requires information exchange with network nodes and other routers. Nodes must notify a router of their existence, their network address, and the existence (and addresses) of other nodes to which they are directly connected. Routers use this information to build an internal "map" of the network. Also, routers communicate with one another. For example, they periodically exchange information in their internal tables so that they may have knowledge of networks beyond those to which they are directly connected. They use this information to forward messages from local nodes to distant recipients.

Communication among network nodes, local routers, and remote routers is fairly complex. A specific language and protocol are required to enable and control this information exchange. This language is called a **routing protocol**. Routers can speak more than one of these "languages." As with bridges, they also must implement multiple network protocols when they connect different types of networks.

Bridges and routers are generally implemented as special purpose network devices. They are often implemented within a network hub. However, they also can be implemented as a general purpose computer executing appropriate software. Any computer system with multiple NIUs (connected to different networks or segments) can be a router

Routing software, which is provided as part of a network operating system on a server computer, may or may not be enabled by the server administrator. Although routing is not a computationally complex task, it does require extensive I/O capabilities. Every network packet must be examined, and many packets must be forwarded. In a busy network, this amount of I/O might occupy most or all of the bus capacity of a general purpose computer, which might leave insufficient capacity to carry on server functions such as file transfer and printer access.

Switches

A **switch** performs functions similar to that of a router. However, its primary distinction lies in the number of networks (or segments) to which the devices are connected. Bridges and routers are connected to only two networks, whereas a switch can be connected to as many as a dozen networks at once. The task of forwarding packets from one network to another is clearly much more complex in this case. Thus, a switch requires much greater processing power and more primary storage as well as electronic components (switches) that can rapidly route signals among any pair of its network connections. Because such devices are complex and expensive, switches are used only in high-speed networks where their speed is essential and their cost is effectively spread across a large number of network users.

Data Transfer Capacity

The data transfer capacity of a network depends on far more than the transmission cable's raw or effective data transfer rate. It also relies on the capabilities and speed of the various attached hardware devices and the protocols used to communicate among them, which may interact in complex ways to produce unexpected results.

One reason that the campus networking architecture (see Figure 9-8) is so commonly employed is its cost-efficiency. High data transfer rates imply high cost. The layered structure of a campus network implements the majority of network connections using low-cost network equipment. Higher-cost (and -capacity) devices are used only for the connections between networks, and the cost of interconnecting networks is spread out among all the users of those networks.

The data transfer capacity of an interconnecting network (e.g., a zone network or campus backbone) does not have to equal or exceed the sum of the capacities of all the attached networks. Generally, it can be much less because of the intermittent nature of network traffic from individual nodes and the filtering of messages among nodes within a LAN or zone. Message volume to/from a large number of network nodes averages out to a level well below the theoretical maximum. Thus, for example, it is common to implement the zone network for a few dozen 10-Mbps LANs with a 100-Mbps network. Similarly, a 500-Mbps backbone network might provide sufficient capacity to interconnect a dozen zone networks.

OSI NETWORK LAYERS

Most modern networks are implemented in layers. One such layering method has been issued as a standard by the International Standards Organization (ISO). The ISO network layer model, referred to as the **Open Systems Integration (OSI) model** (see Figure 9-11 on the next page), was conceived to standardize networks. It is organized into layers in a manner similar to the software layers described in Chapter 2.

Each layer uses the services of the layer immediately below it for all network accesses. Thus, layer n uses the services of layer n-1 and only those services. The services and implementation of layers n-2 through layer 1 are unknown to layer n. In addition, the implementation of layer n-1 is unknown. Only the services and the mechanism(s) by which services are requested are known. The function of each layer is discussed in detail in the following sections.

Whereas most currently used networks do not adhere strictly to the OSI standard, most are at least loosely based on its concepts. Thus, the OSI model is useful as a general model of networks and as a framework for comparing networks.

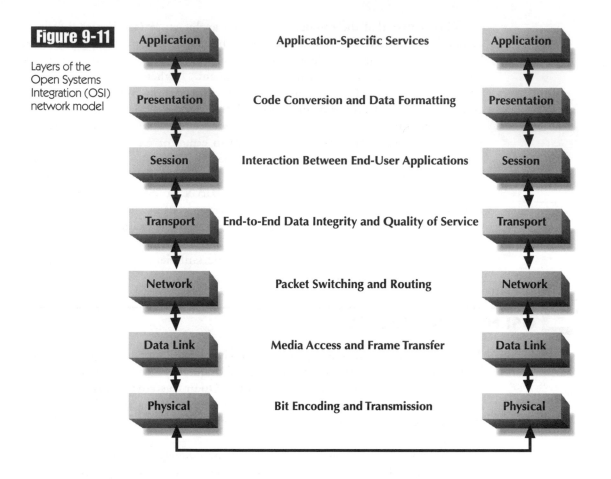

Figure 9-11

Layers of the Open Systems Integration (OSI) network model

Application	Application-Specific Services	Application
Presentation	Code Conversion and Data Formatting	Presentation
Session	Interaction Between End-User Applications	Session
Transport	End-to-End Data Integrity and Quality of Service	Transport
Network	Packet Switching and Routing	Network
Data Link	Media Access and Frame Transfer	Data Link
Physical	Bit Encoding and Transmission	Physical

Application Layer

The **application layer** refers to any software that generates high-level requests for network services. The application layer consists of network utilities used directly by end users (e.g., remote login and mail programs) as well as network services embedded in the operating system service layer (e.g., access to remote files). The OSI application layer accepts requests for service and forwards them to the appropriate network node for service. Each command accepted from a user or program is translated into an equivalent set of **network messages**, which are transmitted to the application layer of a server process over the network. The result of processing the message is sent back over the network.

The application layer of the client communicates with the application layer of the server and vice versa, which is true for each layer of the OSI model—not just the application layer. Thus, equivalent layers on the client and server are, in fact, cooperating processes or hardware devices.

Presentation Layer

The **presentation layer** is responsible for all communication with the user's I/O device. Some translation is required to display output generated by a server on a user's I/O device. Translation is required because of incompatibility between local I/O devices and remote servers. Typically, a network protocol defines a generic display device. All communication destined for display on a user's terminal is encoded for this generic device. The presentation layer is responsible for translating user input into generic device input and for translating generic device output into output specific to the user's display device.

The presentation layer is not required for some network services. For example, reading or writing to a remote file requires no direct interaction with the user's local I/O device. Similarly, a request by a user's application program for a remote process execution may not require user-oriented I/O. Network services that do require presentation layer services include remote login and interactive user application programs.

Session Layer

The **session layer** is responsible for establishing a connection with a server. This connection is initiated by sending the server a message requesting a connection. As part of this process, the client and server session layers will establish one or more parameters of the communication protocol, which can include character coding, asynchronous or synchronous message passing, and other communication parameters. Once a connection is established and the protocol negotiated, the session layer simply passes messages received from layers above to layers below and vice versa.

Transport Layer

The **transport layer** is responsible for translating messages according to the transport protocol of the network. In most networks, the major portion of this translation involves the division of messages into packets. A **packet** is a unit of data transmission on a network, whose format can vary widely, depending on the transport protocol in use. However, the logical content of packets is usually similar, including:

▶ The message (or portion thereof).
▶ Identification of the sender and receiver.

- Error detection data.
- Packet sequencing information.

Identification of the sender and receiver is included in coded form. Senders and receivers are identified in terms of their physical network location and a process identifier, and each is identified by a symbolic or numeric field within the packet. This information can be used by intermediate nodes of the network to route messages to their destination and to route error notification back to the sender. Process identifiers allow low-level network operating system processes to route packets to the transport layer of the appropriate process.

To protect against errors in network transmission, packets usually have encoded error checking data. This may take numerous forms, including parity checking, cyclical redundancy checks, and others (refer to Chapter 8). Error checking information is usually encoded in a header or trailer field of the packet by the sending transport layer. The receiving transport layer recomputes the error checking data and compares it against the version stored in the packet to determine if a transmission error has occurred. If an error is detected, the receiver will transmit a message to the sender requesting retransmission of the packet.

Because messages can be decomposed into many packets, some method must be provided to correctly reassemble them at their destination. A number of schemes can be used for this, of which the following is typical:

- Packets are sequentially numbered, and the number is stored in the packet header field.
- A coded field in the last packet identifies it as the last packet.

This information is sufficient to allow the sender to determine if all packets in a message have been received and to reassemble them in the proper order.

In many networks, there is no guarantee that packets will arrive in the same order that they were sent, so sequencing information is necessary. Packets are simply decoded in sequential order, according to their identification numbers. Specially coded first and last packets indicate the start and end of a message. By examining the sequence numbers of packets received, the receiver can determine the identity of missing packets and request their retransmission.

Network Layer

The **network layer** is responsible for routing individual packets to their proper destination. Remote locations are known to users or applications by a symbolic name. The network layer must convert this name into a physical address before a packet can be sent. In some network topologies, this is a trivial exercise. The sender simply looks up the address of the receiver in a table or file, places that address in the packet, and transfers the packet to the data link layer. This scenario typifies routing within small networks.

In large networks, routing can be much more complicated. Because of the large number of nodes in such networks, it is impractical for every machine to maintain a local data store containing the name, address, and routing path of every machine connected to the network. There are several strategies to resolving this, including the following:

► Certain machines on the network can be designated as **name servers**. Every network node knows the physical address of one or more name servers and can request an address lookup from any of them.

► Certain machines on the network may be designated as message forwarding machines for a range of symbolic names. Packets can be sent to these machines which, in turn, send them to their final destination or to another forwarding machine.

In either case, knowledge of symbolic names and addresses is limited to a relatively small number of machines on the network.

The routing layer also can be responsible for forwarding messages to other machines. In the second example previously cited, the network layer of the intermediate machine would be responsible for forwarding the message to its final destination. In certain types of networks, all machines must be capable of message forwarding. Such a network usually implies the use of name servers for address lookup functions.

Data Link Layer

The **data link layer** is responsible for transmitting a packet to the network interface device of the computer (i.e., the physical layer). The data link layer implements the MAC protocol of the network. Thus, it handles issues such as media access, collision detection, collision recovery, and token passing.

The physical layer may or may not support the packet format used by the network layer. If a different data format is required, it is called a **frame**. Thus, the data link layer is responsible for translating packets into frames (at the sending end) and translating frames into packets (at the receiving end). Matters of network interface device control (e.g., interrupt handling and direct memory access) also are implemented within this layer.

Physical Layer

The **physical layer** is the layer at which communication between devices occurs. The network interface device is responsible for interacting with the data link layer and for placing packets (or pieces thereof) on the network. This can involve a number of physical transmission functions, including blocking, character or block framing, digital to analog conversion, bus interface, and media access protocols.

TCP/IP

The **transmission control protocol (TCP)** and **Internet protocol (IP)** were originally developed for use in the U.S. Department of Defense (DOD) Advanced Research Projects Agency network (ARPANET) in the late 1960s and early 1970s. ARPANET was a test bed for DOD networking technology and was actively used by researchers working on DOD projects. Many of the researchers worked at universities and the networking techniques were soon applied throughout these institutions. The use of **TCP/IP** grew and, through a series of organizational changes, the once-loose network of universities became what we know today as the Internet.

Most of the facilities normally associated with the Internet are delivered by TCP/IP. This includes file transfer through the file transfer program (ftp), remote login through the telnet protocol, e-mail distribution through the simple mail transfer protocol (SMTP), and access to the World Wide Web (WWW) through various Web browser programs (e.g., Microsoft Internet Explorer and Netscape Navigator and Communicator). It is fair to say that TCP/IP is the language of the Internet, or the glue that binds networks together and makes the Internet and WWW possible.

The correspondence between the layers of the TCP/IP protocols and the OSI model is poor. IP is roughly equivalent to the OSI transport and data link layers. TCP is roughly equivalent to the OSI session layer. TCP/IP predates the OSI model by a number of years, so the lack of correspondence is understandable. Much of what was learned in the development and use of TCP/IP was incorporated into the OSI model, but there are numerous differences as well, which make it extremely difficult to integrate TCP/IP with a network strictly based on the OSI standard.

IP provides a data transfer service to higher network layers by routing and forwarding packets to their ultimate destination. IP is independent of the underlying physical network layer and effectively hides it from the layers above. An IP implementation exists for every known type of network. TCP passes network packets called **datagrams** to IP for transmission.

The IP layer is responsible for translating datagrams into a format suitable for transport by the physical network. Frequently, datagrams are larger than the physical frames and packets used in the physical layer. In this case, the IP layer is responsible for dividing the datagram into appropriate-sized frames and transmitting them individually. It attaches header information to each piece describing its sequence within the datagram. The IP layer at the receiving end is responsible for verifying the receipt of datagram components and reassembling them in the proper order to be passed up to TCP.

IP assumes that a datagram will traverse multiple networks by nodes called **gateways**. It determines the transmission route by a number of related protocols, including the **Internet Control Message Protocol (ICMP)** and the **Routing Information Protocol (RIP)**. A gateway is simply any node that connects two or more networks or network segments. A gateway might be physically implemented as a workstation, server, hub, bridge, router, or switch. Figure 9-12 shows the routing of datagrams thorough multiple gateways. Notice that only the IP layer is needed on the gateways.

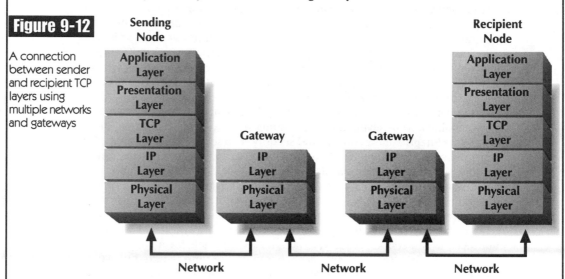

Figure 9-12

A connection between sender and recipient TCP layers using multiple networks and gateways

All IP nodes are identified by a unique 32-bit address of the form nnn.nnn.nnn.nnn, where nnn is a decimal number between 0 and 255. The internal tables of an IP layer contain correspondences between these addresses and physical network (e.g, Ethernet) addresses. Each IP layer also knows the physical network addresses of one or more gateways. An IP address that doesn't appear within the internal tables causes the datagram to be sent to a default gateway.

An IP layer also can know of a gateway corresponding to a partial IP address. For example, all nodes within the University of New Mexico have the 129.24 as the first 16 bits of their IP address. An IP layer may not know the network address for a specific node (e.g., 129.24.200.69), but its internal tables may contain the address of a gateway to any node with 129.24 as the first 16 bits. IP nodes periodically exchange routing information to keep their tables current. New nodes announce their presence to the nearest gateway, which propagates the message throughout the Internet.

TCP implements a number of network services that IP does not provide, including error checking of datagrams received and the management of connections and sessions between Internet nodes. IP is an example of a **connectionless protocol**; that is, it does not attempt to verify the existence of a datagram recipient or ask its permission before sending it data. IP simply sends the datagram to the appropriate gateway and assumes that it will successfully find its way to the intended recipient.

Because it is connectionless, IP cannot implement some forms of error control. Datagrams are simply "launched" toward a recipient IP layer without telling it that data is coming, so there is no way for the recipient to know if a datagram is "lost" in transit. Thus, the recipient IP layer has no mechanism to trigger a negative acknowledgment or request for retransmission. TCP provides the required framework by explicitly establishing a connection with an intended recipient before transmitting any "real" datagrams. Thus, TCP is called a **connection-oriented protocol**.

A number of functions can be performed by TCP, depending on the needs of the sending and receiving nodes, including verification of receipt, verification of data integrity, flow control, and security. The sender and recipient TCP layers maintain information about one another, including routes, errors encountered, time delays in transfer, and the status of ongoing data transfers, which allows TCP to implement transmission-management functions more efficiently.

TCP uses a positive acknowledgment protocol to ensure data delivery. The recipient TCP layers send ACK signals to the sender TCP layer to acknowledge data receipt. The sender TCP layer waits for ACK signals for a specific time interval. The time interval is initially set based on the time required to establish a connection. It can be extended or shortened during the life of the connection to adapt to changing network conditions. If an ACK is not received, it assumes that the message has been lost and retransmits it. If a number of messages in a row are lost, the connection "times out" and the sender assumes that the connection itself has been lost (e.g., because the receiver is down).

TCP connections are established through mechanisms called **ports** and **sockets**. A port is simply a TCP connection with a unique integer number, whereas a socket is the combination of an IP address and a port number. Thus, socket numbers are unique within the Internet. Many sockets are "prewired" to specific higher-level applications. The default port numbers for these applications are an Internet standard. For example, port 21 is the default TCP connection for an ftp service and port 80 is the default TCP connection for an http (Web browser) service. The socket 129.24.200.69:80 is the TCP connection point for the Web server used to access the Web site for this book. It is the TCP equivalent of the WWW Uniform Resource Locator (URL) http://averia.unm.edu.

STANDARD NETWORK ARCHITECTURES

The IEEE has drafted a number of standards concerning the hardware, transmission method, and protocols of network communications and control, which are collectively referred to as the **IEEE 802 standards**. The current and developmental standards are listed in Table 9-1.

TABLE 9-1	IEEE Standard	Functional Description
Various IEEE network standards	802.1	media access control (MAC)
	802.2	logical link control (LLC)
	802.3	CSMA/CD
	802.4	token bus
	802.5	token ring
	802.6	metropolitan area networks (MANs)
	802.7	broadband LANs
	802.8	fiber-optic LANs
	802.9	integrated data and voice networks
	802.10	LAN/MAN security
	802.11	wireless networks
	802.12	fast Ethernet (demand priority access method)
	802.13	fast Ethernet (CSMA/CD)

The 801.1 and 802.2 standards correspond roughly to the data link layer of the OSI model. The methods defined in these standards are incorporated into most of the other standards. The 802.1 standard addresses a number of issues, including media access and bridging. It also describes an architectural framework into which the other standards fit. The 802.2 standard addresses issues such as routing, error control, and flow control.

The most commonly used standards for LANs are 802.3 and 802.5. The 802.3 standard describes the physical implementation of bus networks using the CSMA/CD media access protocol. The 802.5 standard describes the physical implementation of ring networks using the token passing media access strategy. Each of these standards has a number of subsidiary standards

---

corresponding to the use of different cabling, bit-encoding schemes, and other physical implementation parameters (i.e., different implementations of the OSI physical layer).

A number of different commercial products are based on the IEEE standards. For example, Ethernet is based on one of the subsidiary standards of 802.3, and IBM's token ring network is based on a subsidiary standard of 802.5. Many other LAN products also conform to one of the definitions within 802.3 or 802.5. Switched multimegabit data service (SMDS) is based on standard 802.6.

The IEEE standards-development process is significantly influenced by companies and organizations in the telecommunications and networking industry, in that standards are developed by committee members largely drawn from those organizations. These committees usually work closely with other standards and regulatory agencies such as the American National Standard Institute (ANSI), the ISO, and the Federal Communications Commission (FCC) (and its international counterparts). The standards-development process can be political, although technological considerations should predominate.

The committees set criteria for evaluation and then invite proposals for adoption as standards. Proposals are usually supplied by vendors based on products and technologies already under development. Thus, the lag time between standard publication and the appearance of related commercial products is less than might otherwise be assumed. Occasionally, the process does backfire because adopted standards sometimes represent significant compromises among very different proposals. Companies may decide not to adjust their developmental products to meet the compromise standard, thus generating an "orphan" (e.g., token bus). A company also might choose to release its products despite their lack of adherence to any published standard.

Ultimately, the marketplace decides which technologies and products will succeed. The standards-setting process simply provides a (usually) accepted target for implementation and—in the case of a widely implemented standard—a degree of compatibility among competing products.

Ethernet

Ethernet, a local area network (LAN) technology closely related to the IEEE 802.3 standard, was developed by Xerox in the early 1970s. Digital Equipment Corporation (DEC) was the driving force in developing commercial products in the late 1970s. DEC, Intel, and Xerox jointly published a specification for Ethernet networks in 1980 in a version sometimes referred to as the DIX standard. Ethernet has been very successful in the marketplace and is the basis for implementation of the vast majority of LANs in use today because it is inexpensive and well adapted to networks and applications with low data-transfer requirements.

Ethernet uses a bus topology and has a number of different cabling options, including 10BaseT twisted pair, "thin" Ethernet coaxial, "thick" Ethernet coaxial, and fiber-optic cable. Fiber-optic cable is seldom used because Ethernet transmission speeds only require it for very long distances. Thick Ethernet cable has been largely abandoned in favor of thin Ethernet and 10BaseT because of lower cost and easier installation. Bus length is limited to 500 meters for coaxial cable, and transceivers must be located at least one meter apart. Thus, there is a limit of 500 transceivers on a single bus, which can be extended by the use of multiple cable segments and repeaters.

Ethernet transmits data at a rate of 10 Mbps. CSMA/CD is used as the media access protocol. Because of the inefficiency of this protocol, it is rare for an Ethernet network to achieve 10-Mbps throughput. Heavy network traffic causes a large number of collisions, which result in retransmissions that waste available data transmission capacity. Throughput approaching 10 Mbps is generally achieved only in LANs with a single file server and with client nodes that primarily download data from the server.

Bits are encoded using Manchester encoding, and only a single packet can be transmitted at a time. Thus, the transmission line is generally used as a single baseband channel. The format of an Ethernet packet is shown in Figure 9-13 on the next page.

▶ The first field contains a bit pattern that is used to synchronize the clock of the recipient.
▶ The second field contains an 8-bit pattern that marks the end of the synchronization data.
▶ The next two fields contain the 48-bit Ethernet addresses of the recipient and sender.
▶ The next field contains a 16-bit integer containing the number of data bits in the following field.

► The next field contains the packet data. This field is variable in length between 576 and 12,208 bits.

► The last field contains a 32-bit CRC computed from all the preceding fields.

Figure 9-13

Format and content of an Ethernet packet

Synch (56 bits)	Start Flag (8 bits)	Recipient Address (48 bits)	Sender Address (48 bits)	Length (16 bits)	Data (12,208 bits maximum)	CRC (32 bits)

The bandwidth of Ethernet cable is capable of broadband transmission using the latest equipment. Some Ethernet hubs (switching hubs) can divide a single cable into multiple channels, thus implementing several independent Ethernet buses. Network addresses are set within the hardware of each NIU during manufacture. Each address is a unique 48-bit sequence. Ethernet routers use these addresses to route frames to and from network nodes.

The original Ethernet standard is starting to show its age because many LANs now require data transfer rates above 10 Mbps. Two different groups of companies have formed to develop 100-Mbps Ethernet standards. One of these groups is primarily responsible for the development of the IEEE 802.12 proposed standard called **100BaseVG** or **100Base-AnyLAN**. The other group is heavily involved in the 802.13 standard called 100BaseX.

100BaseVG does not use CSMA/CD as the media access protocol. Rather, it requires the use of an intelligent network hub that routes signals among local nodes directly from one node to another. Other nodes on the network never see messages not addressed to them. The standard also implements a strategy called **demand priority**, which allows priority to be given to certain packets and nodes. It is especially useful when transmitting packets containing audio or video data.

Upgrading Network Capacity (Part II)[1]

The Bradley Advertising Agency (BAA) has decided to implement a fiber-optic based ATM network, based on the following reasons:

▶ 100-Mbps Ethernet will minimally meet current demands for video file sharing. Demand will likely exceed capacity when additional video-editing workstations are added.

▶ ATM will provide the throughput guarantees needed for time-critical applications.

▶ Fiber-optic cable will allow current network capacity of up to 625 Mbps. Future upgrades to this capacity are expected as fiber-optic technology matures.

▶ BAA foresees the need to deliver video to customers by direct connections. ATM will provide the needed internal network capacity and a standard method of interfacing to telecommunication providers.

Although the decision has been made to adopt fiber-optic ATM, the exact method of deployment has not. It has been decided that BAAs entire building will be wired with fiber-optic cable (i.e., one or more connections to each room from a central wiring closet). The video-editing workstations and file server will be directly attached to an ATM switch in the wiring closet. The estimated cost for the ATM hardware (exclusive of cabling) is $15,000.

BAA is uncertain of how to integrate the microcomputers into the new network and is considering three options:

1. Leave the existing Ethernet network as is. The existing Ethernet hub would continue to function as the network backbone, and the Ethernet NIC would remain in the file server. The file server would be configured to respond to service requests through both networks (Ethernet and ATM).

2. Upgrade the Ethernet network to 100 Mbps (assume that the existing cable is adequate for this purpose). The Ethernet hub would be replaced with a hub that implements both 10-Mbps and 100-Mbps connections (approximately $1,000). The file server NIC would be replaced with 100-Mbps NIC (approximately $400). The microcomputer NICs would be upgraded on an as-needed basis (approximately $150 each).

[1] Part I of this Business Focus was presented in Chapter 8, page 327.

3. Replace all existing Ethernet connections with ATM connections. This will require a more powerful ATM switch ($4,000 more) to increase the number of switch ports to 24. ATM NICs for the microcomputers will cost approximately $750 each.

Questions:

1. What are the advantages and disadvantages of each option? Which is the best option?

2. The first two options route Ethernet traffic directly to the server. It is also possible to bridge the Ethernet traffic to the ATM switch and route it from there to the file server. What are the advantages and disadvantages of this approach? Would you recommend it for BAA if they choose the first or second options?

SUMMARY

The goal of distributed computing systems is to provide low-cost and/or high-quality services to users. Such systems allow the distribution of various computer resources among multiple computer systems. Resources that can be distributed include access to mass storage (e.g., data, programs, and temporary storage), I/O devices, and CPU services. A computer network provides access to distributed resources. The network is a collection of hardware and software components that allows communication between computer systems.

Sharing hardware resources among computer systems often allows a reduction in total hardware cost, which may result from sharing access to expensive I/O devices, consolidating duplicate storage of data and files, or from distributing peak demands for CPU capacity among multiple computers. Disadvantages of distributed systems include the cost and complexity of network hardware and system software. Networks require substantial investments in communication capacity, which must be maintained and updated in a dynamic environment. Users also face a loss of control over shared resources.

Network (distributed) services can be classified loosely as I/O device access, extended communication, file sharing, and distributed processing services. I/O device access and extended communication services are direct extensions of services provided in multiuser operating systems. Both of these services require modifications to normal operating service routines to allow users and resources to be located on different computer systems. Service requests by users or application programs (i.e., clients) must be re-directed to the remote machines that provide those services (i.e., servers). The results of processing service requests must be routed back to client processes in an approach to resource sharing called a client/server architecture.

Distributed file systems allow the sharing of files among computer systems. Their implementation requires modifications to operating system services for file manipulation. Accesses to remote files must be re-directed to appropriate server processes, and I/O to and from those files must occur over the network. Distributed file systems can be implemented by explicit file transfer, transparent file transfer, and transparent file access. Explicit file transfer requires a user to know the location of a file and to explicitly request that the file be copied to the local machine for manipulation. Transparent file transfer also copies a file to the local machine but does so without the user's direct knowledge. Transparent file access performs all read and write operations over the network. Only those needed portions of the file are transferred between client and server. Although this method of file sharing is the most common, it requires the most complex system software support.

Distributed processing, which allows sharing of CPU resources among computers, can be implemented in several ways, including remote login, remote process execution, and load sharing. Remote login allows the user to execute an interactive dialog with a remote computer. The local computer redirects all I/O to and from the remote computer. Remote process execution allows a user or application to request the execution of a single process on a remote machine. Input, output, and interprocess communication for that process are routed across the network. Load sharing allows any executing process to be moved from one CPU to another. Generally, it occurs without the knowledge of a user or application program. Load sharing requires that each machine have compatible hardware architecture and operating system services.

As with other system software, network services are implemented in layers. A standard layering scheme, defined by the ISO, is referred to as the Open Systems Integration (OSI) model. The layers of the OSI model are—from most logical to most physical—the application, presentation, session, transport, network, data link, and physical layers. Each layer uses the services of the layer below, and each is unaware of the implementation and the existence of any lower layers. Each layer on a client machine has an equivalent layer on the server machine. Coordination between equivalent client and server layers is achieved through message passing over the network.

The application layer is the layer at which requests for network services are generated. Applications can include specific network utilities or embedded operating system services. The presentation layer, responsible for interactive I/O between client and server, resolves incompatibilities between high-level I/O facilities such as graphic display and full-screen terminal functions. The session layer, responsible for initiating connections between client and server, also establishes initial communication parameters.

The transport layer is responsible for the delivery of messages between client and server as well as for the division of messages into packets and for the reassembly of packets into messages. The network layer, responsible for packet routing within the network, chooses a communication path and, in intermediate network nodes, is responsible for packet forwarding. The data link layer transmits individual bits of a packet and checks for errors and requests retransmissions, if necessary. The physical layer refers to the hardware implementation of the communication medium and the devices that directly interact with that medium. Issues such as timing, character framing, and digital/analog conversions are addressed in this layer.

Various options exist for implementing the physical layer of the network. Important implementation choices include network topology, access protocol, channel hardware, and communication capacity. There are many possible combinations of these parameters. To address this complexity, a number of standards have been developed by the IEEE. Any of these specify an integrated set of choices in the preceding categories. These standards are collectively referred to as the IEEE 802 standards.

Network topology refers to the physical connections between network nodes. Nodes are connected in either a bus, star, or ring. A star network connects every node to a central point, which is a network hub or wiring concentrator. Star configurations are the easiest to wire but are prone to complete failure if the central connection point fails. A bus network uses a common communication channel with a single connection (tap) for each node. A node sends messages by placing them on the bus, and then they propagate in both directions simultaneously. Messages are received by detecting a packet with an appropriate address. Bus networks require relatively simple wiring and interface devices. However, they are limited in length and number of nodes because of noise and signal degradation.

A ring network connects each node to two others. The nodes form a ring, and packets travel in one direction around the ring from sender to receiver. Each node receives every packet on the network and retransmits (repeats) the packet if its own address does not appear in the packet. Ring networks are not as susceptible to noise and signal degradation because of the use of active repeaters. However, they are more difficult to wire and maintain, and the ring may be broken by the failure of any network node.

When two packets are placed on the network simultaneously, the result is noise and is referred to as a collision. A media access control (MAC) protocol ensures that collisions do not occur or that they are appropriately remedied. The two most common access protocols are carrier sense multiple access/ collision detection (CSMA/CD) and token passing. CSMA/CD does not prevent collisions from occurring, but it does detect them. Rules for detection and retransmission are used to ensure accurate communication. Token passing regulates network access by passing a special control packet (the token) among nodes. Only the node that possesses the token is allowed to place packets on the network. CSMA is less efficient than token passing because of network capacity wasted during collisions. Token passing is more efficient but also substantially more complex.

Network channel hardware includes the medium for signal propagation, hardware to interface with the transmission medium, and devices used to connect one network to another. A network interface unit is the device that computer systems use to interact with the transmission medium. Network nodes within a LAN typically are connected to a network hub, which acts as a central connection point and also can implement internetwork communication.

A bridge is a device that connects two networks or network segments, copies messages from one network to another. Bridges also can implement protocol conversion if the connected networks are dissimilar. Routers are intelligent bridges that actively monitor the network and continually update their internal database of network information. This database allows them to more effectively manage network traffic. Switches are high-speed routers connected to more than two networks.

Key Terms

100Base-AnyLAN
100BaseVG
application layer
bridge
bus topology
(CSMA/CD) carrier sense
   multiple access/collision
   detection
client
client/server architecture
collision
connection-oriented protocol
connectionless protocol
data link layer
datagram
demand priority
distributed file system
distributed processing
distributed system
Ethernet
explicit file transfer
frame
gateway
IEEE 802 standards
Institute of Electrical and
   Electronics Engineers (IEEE)

Internet Control Message
   Protocol (ICMP)
Internet protocol (IP)
load balancing
load sharing
media access control (MAC)
   protocol
name server
network
network hub
network interface card (NIC)
network interface unit (NIU)
network layer
network message
Open Systems Integration
   (OSI) model
packet
physical layer
port
presentation layer
remote login
remote process execution
repeater
ring topology
router
routing protocol

Routing Information Protocol
   (RIP)
scheduler
server
session layer
socket
spooler
star topology
switch
T connector
tap
TCP/IP
token
token passing
transceiver
transmission control protocol
   (TCP)
transparent file access
transparent file transfer
transport layer
wiring closet
wiring concentrators

1. Standards for many types and aspects of network communication are defined by the _____.

2. The _____ layer of the OSI model establishes connections between clients and servers.

3. A network that connects two separate networks is called a(n) _____.

4. A(n) _____ is a special purpose packet, which represents the right to originate network messages.

5. The _____ layer determines the routing and addressing of packets.

6. A TCP/IP _____ is the combination of an Internet address and a service _____ number.

7. The _____ layer refers to programs that generate requests for network services.

8. A network using a physical _____ topology connects all nodes to a central point (e.g., a hub or wiring concentrator). The central connection point is located in a(n) _____.

9. A physical connection between two different networks is implemented using a(n) _____, a(n) _____, or a(n) _____.

10. The generic term for a node's connection to a bus network is a(n) _____, which may be implemented physically with a(n) _____ or a(n) _____.

11. In a network that supports _____, a node with an overburdened CPU may ask other nodes to perform processing tasks for it.

12. Packet loss can't always be detected by a receiver if a(n) _____ protocol is in use.

13. A(n) _____ provides token initialization and recovery from node failures in a(n) _____ network.

14. Using a(n) _____ program, a user connects to a remote system over a network as if he or she were using a directly attached terminal.

15. TCP uses a(n) _____ as the basic unit of data transfer.

16. _____ transmits at 10 Mbps over twisted pair cabling.

17. In a ring network, each node contains a(n) _____ which retransmits incoming messages that are addressed to other nodes.

18. A(n) _____ is a network node that supplies the physical network address associated with a node's symbolic name.

19. Messages to be transmitted on a network are normally formatted into _____ by the data link layer prior to physical transmission.

20. Under the _____, an application program that requests a resource and the systems software that provides that resource may be located on different computer systems.

21. The _____ defines a generic set of software and hardware layers for networks and distributed systems.

22. In a(n) _____, message are passed from one node to the next until they reach their destination.

23. When two messages are transmitted at the same time, a(n) _____ is said to have occurred.

24. Routing information in a TCP/IP network is exchanged according to the _____.

25. Under the _____ media access strategy, collisions are allowed to occur, but they are detected and corrected.

26. A(n) _____ protocol defines the methods by which a network node can access a transmission medium.

27. A microcomputer or workstation hardware interface to a network transmission medium is generally called a(n) _____.

28. The _____ layer is responsible for negotiating parameters for interactive I/O between a server and a client.

29. The _____ standards defined many parameters of network design and implementations.

Review Questions

1. What is a distributed system? What specific resources may be distributed?

2. Why do distributed computer systems often have lower total hardware costs than centralized systems of equal capacity?

3. Why is communication between devices in the same computer system usually less efficient than communication between devices across a network?

4. Define the term client/server architecture.

5. What is network transparency? Why is it desirable? What are its implications for the implementation of operating system services?

6. By what methods can a distributed file system be implemented? What are the comparative advantages and disadvantages of each method?

7. What options exist for implementing distributed processing?

8. What is load sharing? What does it require of the processors and operating systems among which load is shared?

9. Describe the function of each layer of the OSI network model.

10. Describe the movement of messages on a bus network. Describe the movement of messages on a ring network.

11. What are the comparative advantages and disadvantages of ring and bus networks?

12. Describe the operation of the CSMA/CD media access protocol.

13. Describe the operation of the token passing media access protocol. Why do networks that use token passing generally realize higher effective data transfer rates than networks that use CSMA/CD?

14. What is the function of a bridge? How does it differ from a router? How does it differ from a switch?

15. What is a network hub? What functions does it normally implement?

16. What functions are normally implemented within the network interface hardware of a single computer system?

Research Problem

The IP address format has become a severe restriction on Internet growth. A 32-bit address can theoretically represent 4 billion (2^{32}) addresses. However, the number of IP addresses is substantially more limited due to their segmentation and the reservation of certain values for specific purposes. Investigate the details of the 32-bit format (www.internic.net, keyword IPv6). What are class A, B, and C networks? Does their use limit the total number of addresses? What changes are planned to increase the addressing capacity? Will they be compatible with the original 32-bit format?

Chapter 10

Input/Output Technology

MANUAL INPUT DEVICES

Manual, or "hand operated" input devices include keyboards, mice, and other pointing devices. Keyboards have been used for computer input almost since computers were invented, but pointing devices have come of age only recently. Speech recognition is expected to supplant keyboards as the primary means of human input to computer systems, although the timing of this is uncertain. Even when speech-recognition hardware and software are reliable and affordable, keyboards (for persons with speech impairments) will be necessary. Pointing devices also will be required for certain types of processing such as image manipulation.

Keyboard Input

The predominant mechanism for human input to computer systems is a character keyboard. The exact mechanisms by which character data from a keyboard is captured for processing and storage have changed substantially over time. Early computer systems accepted input by punched cards or tape. A stand-alone device such as a keypunch machine was used to generate punched input based on manual keyboard input.

Keypunch devices captured manual keystrokes and converted them into mechanical motion, which was, in turn, used to punch holes in a cardboard card or paper tape. A punched card was divided into 80 columns and 12 rows, and each character typed at the keyboard was converted to a sequence of punched holes in one column of the card. Another device, called a card reader, was attached to the computer system bus. Punched cards were passed over a light source within the card reader. Then, light shining (or not shining) through each row of a single column was detected and interpreted as an input character. Once interpreted, the character was converted to electrical signals that the CPU could recognize.

Modern keyboard devices translate keystrokes directly into electrical signals, eliminating the need for intermediate devices such as punched card or paper-tape readers. In keyboards manufactured during the 1960s and early 1970s, keystrokes generated an electrical input to a simple electrical circuit dedicated to the key pressed. This circuit generated an output bit stream that was routed to the device's output circuitry. These outputs represented printable characters in ASCII or EBCDIC or a specially coded control signal.

Current keyboards use an integrated microprocessor (also called a **keyboard controller**) to generate bit-stream outputs. Pressing a key sends a coded signal to the controller. The controller generates a bit-stream output according to an internal program or lookup table. Also, the controller recognizes multiple key inputs (e.g., Shift A) as valid input signals and generates a single bit-stream output.

Input/Output Technology

A modern keyboard contains a large number of special purpose keys, including function keys (e.g., F1), display control keys (e.g.,↑, Page Down, and Scroll Lock), program control keys (e.g., Print Screen and Escape), and modifier keys (e.g., Shift, Caps Lock, Control, and Alt). The exact set of keys can vary widely from one keyboard to another, although those provided with an IBM-compatible PC keyboard have become a defacto standard.

There are many combinations of two or more keys (e.g., Control+Alt+Shift+F8) that can be recognized as valid input by software. The large number of keys and key combinations makes coding keyboard outputs using ASCII or EBCDIC impractical because there are simply not enough unused codes in those coding tables to accommodate all of the possible combinations.

A keyboard controller generates an output called a **scan code**. A scan code is a 1- or 2-byte data element that represents a specific keyboard event. For most keys, the event involves simply pressing a key. For some keys, pressing and releasing the key can represent two different events (and, thus, generate two different scan codes). There are few standards for the coding of keyboard events into scan codes. Once again, the widespread use of IBM-compatible PCs has created some defacto standards (e.g., those based on keyboard controllers used in the original IBM PC, the IBM AT, and the IBM PS/2).

Transmission between a keyboard controller and a computer system is implemented in serial mode. The physical connector for the transmission cable and the signal levels generally follow one of the IBM microcomputer hardware standards. Asynchronous character framing is typically used. Synchronous character framing is used only for directly connected displays (e.g., the system console) with some mainframe computer systems.

Pointing Devices

Devices that can be used to capture graphic and image data include the mouse, track ball, joystick, and digitizer tablet. Essentially, all of these devices perform the same function: that is, they translate the spatial position of a pointer, stylus, or other selection device into numeric values within a system of two-dimensional coordinates. These devices can be used to enter drawings into the system or to control the position of a **cursor,** or physical pointer, on a display device. The cursor control function typically is used to indicate selections from visual menu displays that are presented to the system user.

A **mouse** is a pointing device meant to be moved on a flat surface (e.g., a table, desk, or rubber pad). The position of the mouse on a surface is an analog to the position of a cursor or pointer on a video display. As the mouse is moved left or right, the pointer on the screen goes to the left or right. As the mouse is moved toward or away from the user, the pointer goes to the bottom or top of the display.

Mouse position is translated into electrical signals by a roller ball and two wheels mounted on pins. The roller turns as the mouse is moved on a surface. Pins attached to the wheels are in contact with the ball and turn as the ball turns. This action turns the wheels attached to the pins. Wheel movement can be translated into electrical signals by a number of means, including magnetic deflection and disturbances in the path of a light emitting diode (LED) toward a photodetector. The electrical signals are sent to the CPU by a serial transmission line, and a device driver interprets them as horizontal and/or vertical motion.

A mouse has one or more buttons mounted on the top. With these, the user indicates selection of a display item being pointed to (i.e., a menu item or an icon on a virtual desktop). The buttons—usually one, two, or three—are implemented with simple electromechanical switches. User commands can be indicated by a single button press, a double button press, or a rapid double-press (i.e., a double-click) of a single button. The device driver interprets each action as a distinctly different command.

A track ball is similar to a mouse laid upside-down. The roller ball is much larger than in a mouse, and the selection buttons are mounted on the same plane as the roller. The roller is moved by fingertips or the palm of the hand. These devices are considered easier to use for persons lacking fine motor control (e.g., young children and persons with certain physical disabilities).

Digitizers use a pen (or stylus) and a digitizing tablet, which is sensitive to the placement of the stylus at any point on its surface. Recognition of stylus position can be implemented by pressure sensitive contacts beneath the surface or by disturbance of a magnetic field in the tablet. Two uses of digitizers are common. The first is to trace the outline of an object or drawing. The drawing is placed over the tablet, and the stylus is manually moved around objects or lines on the drawing.

The other use of a digitizer is as an alternative form of menu input, which is common in drawing or computer-aided design programs when the user wishes to continuously view a drawing on the video display device while entering commands from the digitizer. A menu or command template is placed over the digitizing tablet, and the user selects commands by pointing the stylus to one of the command positions. In either case, the digitizer communicates stylus position (or position changes) within the tablet coordinate system to a computer. Software programs interpret the meaning of those positions or changes.

BASIC CONCEPTS OF PRINT AND DISPLAY

Communication by the printed page is several thousand years old, whereas communication by video display devices is less than a century old. Both share many common features, including character-representation methods, measurement systems, and methods of generating color. These issues are covered before discussions of specific output devices because they are common to many different I/O technologies and devices.

Matrix-Oriented Image Composition

A display surface can be of any size and varied composition. The most commonly used display surfaces are paper and video, or television screen. Display surfaces have an inherent background color—usually white for paper and black for video display. Display surfaces can be divided into rows and columns similar to a large table or matrix. Each cell in that table represents one component of an image. A generic term for one of these cells is a **pixel**, a shortened form of "picture element." For printed output, a pixel can be empty or can contain one or more inks or dyes, whereas for video display, a pixel can be empty or can contain various intensities of light.

The number of pixels into which a display surface is divided varies widely, depending on the size (height and width) of the surface and the size of individual pixels. Pixel size for a modern video display is between 0.2 and 0.3 millimeters. Pixel size for a video display is sometimes called **dot pitch**. This range of sizes exists because of limitations in the capabilities of modern display technology. It is difficult and very expensive to manufacture video display devices with smaller pixel sizes.

Technology also limits pixel size for paper output. Minimum pixel size corresponds to the smallest drop of ink that can be accurately placed on the page. Years ago, printers adopted ½ inch as the standard for the smallest pixel—called a **point**. The terminology and its use as a measuring system endures, even though modern printing techniques can apply ink to much smaller areas.

Images, printed characters, and other display elements are typically described by their height and width in pixels or points. An image might be said to be 300×200, which means that image fits within a rectangular area consisting of 300 rows and 200 columns of pixels or points (pixels are usually implied). If the measurement unit is pixels, then the actual displayed size of this image will vary depending on the pixel size used. The image will be 90 millimeters wide and 60 millimeters high if the dot pitch of the display is 0.3 millimeters, assuming that there are no gaps or spaces between the pixels.

The **resolution** of a display is a measure of the number of pixels displayed per linear measurement unit. The unit of measure is stated as **dots per inch (dpi)**, where a dot is assumed to be equivalent to a pixel. For example, the resolution of a laser printer might be said to be 300×300 dpi, which describes a device with 300 pixels per linear inch both horizontally and vertically. Higher resolutions correspond to smaller pixel sizes. To a human observer, the quality of a printed or displayed image is directly related to the pixel size. In general, smaller pixel sizes correspond to higher-quality print and display images. Finer details can be incorporated into an image to make it more pleasing to the eye.

Fonts

Written Western languages are based on systems of symbols called **characters**. Individual characters can be represented as a matrix of pixels, as shown in Figure 10-1. Humans do not require an exact match to a specific pixel map to recognize a specific printed character. For example, the symbols **E**, E, $\mathcal{E}$, E, and **E** are all interpreted easily by a human as the letter "E," even though their pixel composition varies.

Figure 10-1

Representation of the letters "p" and "A" within an 8 X 16 pixel matrix

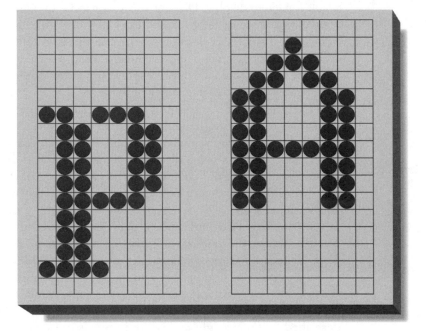

A collection of characters of similar style and appearance is called a **font**. Figure 10-2 shows a sample of characters from several different fonts. The characters of a font can be displayed in a number of different sizes; for example, characters can be large or small or an almost-infinite variety of sizes in between. The most common unit for measuring font size is points. Figure 10-3 shows a single font printed in several different point sizes. Notice that the measurement refers to the height of the characters but not their width, although width is scaled to match height. Because the size of characters within a font varies, the font point size is a measure of the distance between the top of the highest and bottom of the lowest characters in the font.

Figure 10-2

A sample of
characters
printed in various
12-point fonts

AaBbCc XxYyZz 123 !@#   News Gothic
AaBbCc XxYyZz 123 !@#   Braggadocio
AaBbCc XxYyZz 123 !@#   Century Schoolbook
AaBbCc XxYyZz 123 !@#   Garamond Condensed
AaBbCc XxYyZz 123 !@#   Tahoma

Figure 10-3

A sample font
(Times Roman)
printed in various
point sizes

8 point Times Roman
10 point Times Roman
12 point Times Roman
16 point Times Roman
24 point Times Roman

Color

Color physically corresponds to the light frequency or mixture of frequencies that reach the eye. For video displays, these colors are generated directly by the display device. For printed display, these colors represent the frequencies of light reflected from the page. Any color in the visible spectrum can be represented as a mixture of varying intensities of the primary colors red, green, and blue. For example, mixing equal intensities of red and blue produces the color magenta (or purple). The absence of all three is black (i.e., no light generated or reflected). Full intensity of all three results in white. The primary colors are sometimes referred to by their first letters—**RGB**. Thus, a video display that generates color using mixtures of the primary colors is sometimes called an RGB display.

The printing industry commonly generates color by using the inverse of the primary colors. In this context, the primary colors are called the primary **additive colors**, and their inverse colors are called the primary **subtractive colors**. The primary subtractive colors are cyan (absence of red), magenta (absence of green), and yellow (absence of blue), and they are often referred to by the acronym **CMY**. Black can be generated by combining all three, but a separate black dye (identified by the letter K) is generally used. Although colored dyes are expensive to produce, pure black dye is relatively inexpensive. The four-dye scheme is called **CMYK** color.

Pixel Content

Because computers are digital devices, the content of a pixel is described numerically. A stored set of numeric pixel descriptions is called a **bitmap**. The number of bits required to represent an image depends on whether the display is monochrome (black and white), grey scale, or color. Monochrome display uses a single bit to indicate whether each pixel is "on" or "off." For print display, "on" generally means black. For video display, "on" means whatever foreground color is generated by the device (usually white).

Grey-scale display can generate shades of grey between pure black and pure white. The number of shades that can be generated depends on the number of bits used to represent a pixel. Assume, for example, that 4 bits are used. Interpreting 4 bits as an unsigned integer provides numeric values between 0 and 15. Zero can be interpreted as no generation of light (i.e., the black background is displayed). Fifteen can be interpreted as maximal generation of light (i.e., bright white). Values between 0 and 15 can represent successively higher levels of generated light. Thus, 3 could represent a very dark grey, and 12 could represent a very bright grey.

The number of shades of grey that can be displayed is increased if the number of bits used to represent a pixel is increased. If 8 bits are used per pixel, then 254 shades of grey are defined in addition to pure black (0) and pure white (255). The number of distinct colors or shades that can be displayed is sometimes referred to as **chromatic depth** or **chromatic resolution**.

For color display, three separate numbers are required to represent a pixel. One represents the intensity (or amount) of each of the primary additive or subtractive colors. Chromatic depth depends on the total number of bits used to represent the pixel. For example, if 8 bits are used to represent each of the three colors, then the chromatic depth is $2^{(3 \times 8)}$, or approximately 16 million. This is sometimes called **24-bit color** because of the use of 24 bits to represent each pixel. Each group of 8 bits, represents the intensity (0 = none, 255 = maximum) of one primary color. With RGB color, for example, the numbers 255:0:0 represent bright red and 255:255:0 represent bright magenta.

Color display using less than 24 bits is problematic. Computers deal with data in byte-size increments so multiples (or even fractions) of 8 bits are most efficient (e.g., 4, 8, or 16 bits). Because these numbers aren't evenly divisible by three, there is no equal division of the available bits among the primary colors.

The normal approach to this problem is to define a color **palette**, which is simply a table of colors. The number of bits used to represent each pixel determines the table size. If 8 bits are used, then the table has 2^8, or 256, entries. Each table entry can contain any color (e.g., 200:176:40), but the total number of colors that can be represented is limited by the size of the table. Table 10-1 lists the 4-bit color palette used for the Computer Graphics Adapter (CGA) video driver of the original IBM PC/XT.

TABLE 10-1	Table Entry Number	RGB Values	Color Name
The 4-bit (16-color) palette of the IBM PC/XT Color Graphics Adapter	0	0:0:0	black
	1	0:0:191	blue
	2	0:191:0	green
	3	0:191:191	cyan
	4	191:0:0	red
	5	191:0:191	magenta
	6	127:63:0	brown
	7	223:223:223	white
	8	191:191:191	grey
	9	127:127:255	light blue
	10	127:255:127	light green
	11	127:255:255	light cyan
	12	255:127:127	light red
	13	255:127:255	light magenta
	14	255:255:127	yellow
	15	255:255:255	bright white

Color also can be produced by a process called **dithering**, which mixes small dots of different colors in close proximity to one another. If the dots are small enough, the human eye interprets them as being a uniform color representing a mixture of the dot colors. Thus, a tight pattern of 50% red and 50% blue dots is interpreted as magenta. A smaller percentage of red, and higher percentage of blue, would be interpreted as violet. A tight pattern of black and white dots is interpreted as grey. (Look at a grey area of one of the figures in this book with a magnifying glass.)

Dithering is commonly used when a device cannot apply fine differences in color to an area. For example, if an ink-jet printer could only generate four different amounts of a single dye in one pixel area, then only 64 (4^3) different colors could be generated. With dithering, the same printer would be able to simulate several shades in between those 64 colors, which would effectively extend the range of colors that could be printed to as many as a few thousand.

Image Storage Requirements

The number of bits used to represent a pixel and the height and width in pixels determine the amount of storage required to hold an image. For example, assume the use of a video display with a 800 × 600 pixel matrix. There are 480,000 pixels in the matrix. If each pixel is represented by a single bit (monochrome), then a full-screen image requires 60,000 bytes of storage (480,000 pixels × 1 bit/pixel ÷ 8 bits/byte). The use of 24-bit color for the same device would require approximately 1.5 MB of storage.

The storage requirements apply for disk or other mass storage devices used to store an image file as well as to the size of buffers used within I/O devices. These storage requirements can be reduced through the use of image compression techniques, including **Joint Photographic Experts Group (JPEG)** for single images and **Motion Pictures Experts Group (MPEG)** for moving images.

Image Description Languages

There are two significant drawbacks to the use of bitmaps for representing high-quality images. The first is their size. A full page (8.5 × 11 inch) 24-bit color image with 300-dpi resolution requires approximately 25 MB of storage. A good compression algorithm might achieve compression ratios of 10:1, or 20:1, thereby reducing the storage requirements to a few MB. Files of this size take a great deal of time to move over a slow communication link such as the typical parallel port of a PC.

The second drawback to bitmaps is their device dependence. There is no standard storage format for raw bitmaps, and output devices vary widely among their storage formats. Thus, image files stored by a program must be converted into a different format before they are printed. This process can be computationally complex and memory-intensive. A large number of conversion utilities (or output device drivers) may be required.

An approach that addresses both problems is the use of device-independent **image description languages**, which provide compact means of describing primitive image components. They reduce required storage space because a symbolic description of an image component is often much smaller than a bitmap of that same component. The simplest languages use vector lists and are limited to describing image components that consist of (or can be built from) straight-line segments. More advanced languages incorporate the ability to describe complex images such as shapes and fonts.

In mathematical terms, a **vector** is a quantity that has both direction and magnitude. In graphics, a vector is a line segment that has a specific angle and length with respect to a baseline and point of origin, as shown in Figure 10-4(a). Line drawings can be described as lists of concatenated, or linked, vectors. Such an image resembles a "connect the dots" drawing, as shown in Figure 10-4(b).

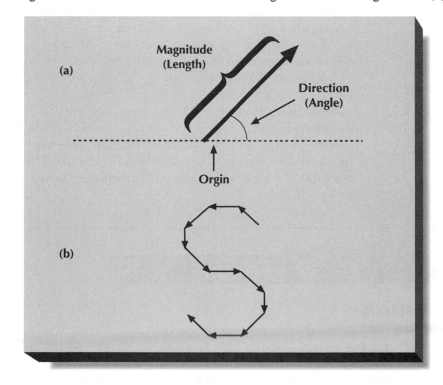

Figure 10-4

Two types of vector list data representations for graphic displays: (a) the elements of a single vector and (b) how a more complex image can be built from concatenated vectors

A group of high-level commands for generating graphic output is called a **display list**. A display list describes an image as a collection of **display objects** (or shapes) and **display attributes** such as color and spatial position. An output device that accepts display lists as input (e.g., a Postscript printer) must contain an embedded processor and software for processing the display list to compose a bitmap within its internal buffer.

The advantage of graphic output based on vectors and display lists lies in the compactness of the message sent to the display device. Consider a graphic display composed of a square, a circle, and a triangle. The output of such an image to a bitmapped display must include one or more numbers representing every pixel on the display. A monochrome display with 640×480 resolution requires 307,200 bits (38,400 bytes) to be sent to the display device. Gray scale and color output will require substantially more bits, depending on the number of possible gray scales or colors.

A vector list describing the same three images is substantially more compact. The square requires sending vector information for four vectors, and the triangle requires three. The information for each vector must include origin, length, thickness, and color (or gray scale), which can be coded in a few bytes per vector. The circle, however, presents a problem because it is not composed of straight lines. Images containing curved lines, and irregular patterns must still be communicated as a bitmap to be represented accurately.

A display list representation of the same three images is even more compact. If the object's circle, square, and triangle are defined in the display language, then the description of those objects can be sent directly. Such descriptions include position or origin, size, line thickness, line color, and fill pattern (if any). A display list representation is the most economical alternative in terms of communication bandwidth consumed; however, this economy comes at the expense of substantial complexity in the display device itself to process the image.

Technology Focus

Postscript

Postscript is a display language designed primarily for generating printed documents, although it also can be used to generate video display outputs. Postscript, which can be considered as both a display language and a programming language, is in essence a program (procedure) for producing graphic images. These images can be composed of various graphic objects, including characters, lines, and shapes. Graphic objects can be manipulated in a variety of ways, including rotation, skewing, filling, and coloring. The language is sufficiently robust to represent any imaginable graphic image.

A Postscript display list consists of a set of commands composed of normal ASCII characters. Thus, transfer of a display list between devices (e.g., a computer and a printer) can be performed by the same means as normal character-based communication. Commands can include numeric or textual data, primitive graphic operations, and definitions of procedures. Data is held within a stack. Primitive operations and/or procedures remove data from the stack and use that data as control parameters.

For example, consider the following short program:

```
newpath
400 400 36 0 360
arc
stroke
showpage
```

The first line declares that a new path (straight line, curve, or complex line) is being defined. The second line contains a list of numeric data items that are pushed onto the stack. These are control parameters used by the primitive graphic operation arc (the third line) and then removed from the stack as they are used. The left two parameters specify the origin (center) of the arc as row and column coordinates; the center parameter specifies the radius of the arc in points; and the remaining parameters specify the starting and ending points of the arc in degrees. In this case, the starting and ending points represent a circle.

Primitive graphic operations are specified in an imaginary drawing space. Individual locations within this grid are specified as row and column coordinates in points. The first three commands specify a tracing within this space. The fourth command states that this tracing should be stroked, or drawn. Because no width is specified, the circle is drawn using the current default line width. The final command instructs the output processor to physically display the contents of the page as defined thus far. Thus, a Postscript printer receiving the previous program as input would generate a single page containing one circle.

To improve efficiency and reduce the amount of code in a program, Postscript provides for the definition of procedures. The following program defines a procedure named "circle" composed of the primitive commands and parameters used to draw a circle:

```
/circle %stack = x-center, y-center, radius
{ newpath
0 360
arc
stroke } def
```

The first line defines the name of the procedure and includes a comment (i.e., the text following the '%' character) that explains the parameters used by the procedure. The definition of the procedure appears between the curly brackets and may be arbitrarily long. Notice that the procedure only defines the final two parameters needed to draw a circle. The first three must be placed on the stack before the procedure is called. Figure 10-5 shows a program that draws three concentric circles using this procedure.

Figure 10-5

A Postscript program that draws concentric circles using the previously defined procedure "circle"

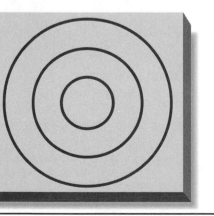

```
400 400 36 circle
400 400 72 circle
400 400 108 circle
showpage
```

Figure 10-6 shows a more complex program for drawing squares that includes the definition of a procedure called square. It draws a square as a sequence of four connected line segments. Because some parameters to the procedure are used more than once, the procedure must duplicate and reorder some of them with the dup (duplicate), exch (exchange), and sub (subtract) operations. These operations manipulate the top one or two elements of the stack, pushing the result (if any) onto the stack. Notice also that the procedure controls the width of lines with the setlinewidth command.

Figure 10-6

A Postscript program that defines a procedure for drawing squares and uses it to draw the squares shown

```
/square % stack = x-origin, y-origin,
        % line-thickness, size
{newpath
moveto
setlinewidth
dup 0 exch rlineto         % up
dup 0 rlineto              % right
0 exch sub 0 exch rlineto  % down
closepath                  % left
stroke}def
216 3 180 324 square
144 2 216 360 square
72 1 252 396 square
showpage
```

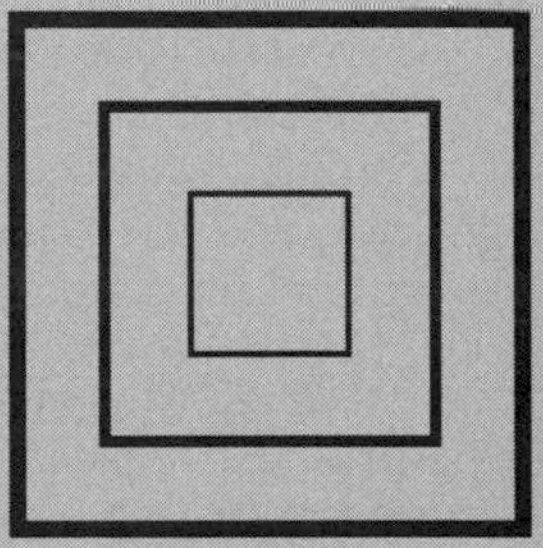

Figure 10-7 shows a sample Postscript program used to generate a pop-up menu, as might be used in a window-based command interface. A few computers have used Postscript for video display (e.g., the NeXT), although it is not commonly done because of the time required to process a Postscript program. A video interface would probably encode many of the command sequences in the sample program as predefined procedures to improve video performance.

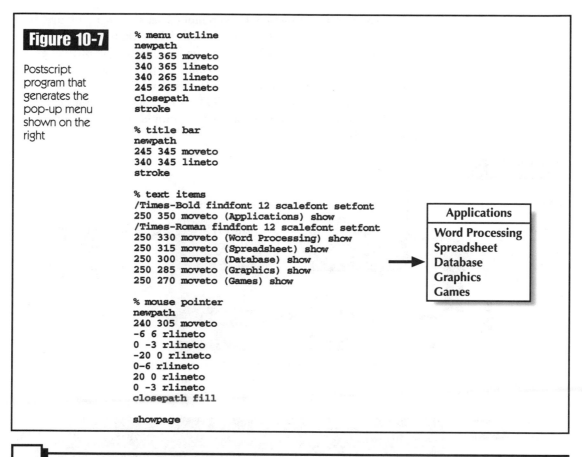

Figure 10-7

Postscript program that generates the pop-up menu shown on the right

```
% menu outline
newpath
245 365 moveto
340 365 lineto
340 265 lineto
245 265 lineto
closepath
stroke

% title bar
newpath
245 345 moveto
340 345 lineto
stroke

% text items
/Times-Bold findfont 12 scalefont setfont
250 350 moveto (Applications) show
/Times-Roman findfont 12 scalefont setfont
250 330 moveto (Word Processing) show
250 315 moveto (Spreadsheet) show
250 300 moveto (Database) show
250 285 moveto (Graphics) show
250 270 moveto (Games) show

% mouse pointer
newpath
240 305 moveto
-6 6 rlineto
0 -3 rlineto
-20 0 rlineto
0-6 rlineto
20 0 rlineto
0 -3 rlineto
closepath fill

showpage
```

Applications
Word Processing
Spreadsheet
Database
Graphics
Games

VIDEO DISPLAY

Video display devices have changed greatly since their first widespread use in the mid- to late 1960s. Early devices, which were capable only of displaying characters in monochrome, were little more than modified television sets with functions similar to that of a teletype. Modern video displays have evolved into much more sophisticated devices. For example, they are larger, have higher resolution, and can display colors and graphic images in addition to characters.

Character-Oriented Video Display Terminals

The first video display devices used widely for computer I/O consisted of an integrated keyboard and television screen. Such a device is normally called a **video display terminal (VDT)**, or simply a **terminal**. These devices were the most common form of video display during the 1970s and much of the 1980s. Their use has shrunk considerably since PCs came into common use. In the 1990s, they are seen primarily in single purpose systems such as retail checkout counters and factory floor environments.

A VDT consists of five functional components, as shown in Figure 10-8. The physical keyboard is attached to a keyboard driver that generates output signals according to scan codes received from the keyboard. The display generator generates the image viewed by a user. Display generators are available in a variety of types and capabilities. The output format for a VDT is illustrated in Figure 10-9. A typical character-oriented VDT can display 80 columns and 24 rows, although larger display dimensions are commonplace. Characters are displayed using a matrix of pixels.

Figure 10-8

Functional components of a VDT

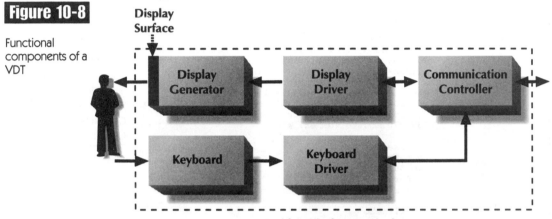

Figure 10-9

Coordinate system used for character positioning on a VDT

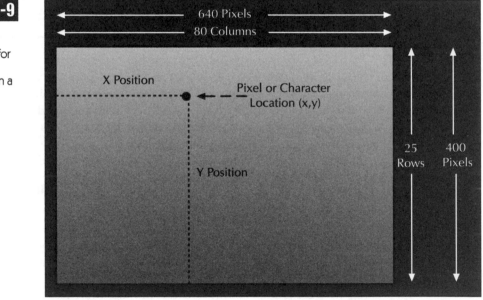

Communication between the terminal and an external device is performed through a communication controller, which sends and receives coded character input to/from a computer system. It implements any required aspects of the terminal to host communication protocol such as speed conversion, serial to parallel conversion, buffering, and flow control. Signals received by the communication controller are sent to a display driver, which consists of electronic circuitry that translates coded character input into commands to the display generator. The exact nature of this translation depends on the type and capabilities of the display generator and the display functions supported by the display driver. Transmission between a VDT and a computer system is usually serial.

In early video displays, relatively simple electronic circuitry translated coded character input (usually ASCII) into the control signals of a corresponding pixel matrix. These pixel control signals, in turn, were input to a display generator that lit individual pixels of the display surface. In modern video display devices, the translation between coded character input and pixel patterns is controlled by a microprocessor, which allows a great deal of flexibility in character display. Microprocessor-based video displays are able to alter fonts, sizes, intensity, and other display features.

Graphic Video Display Devices

A few VDTs historically were endowed with substantial graphic display capabilities. Manufacturers of these devices—often used for displaying for various types of plan diagrams (e.g., construction blueprints, wireframe models, and wiring diagrams) and to visually represent (i.e., graph) mathematical models— included Tektronix and Digital Equipment Corporation. Their production and use faded quickly with the advent of PCs that used graphic display devices. After the mid-1980s, it was generally cheaper to manufacture a general purpose microcomputer with a high-quality video display than to manufacture a graphics terminal. Higher-production volume and the low cost of microcomputer components account for the cost differential.

Modern video displays, decoupled from the keyboard and called **monitors**, operate as independent devices under control of a video controller attached to the system bus of a workstation. Input is no longer in the form of coded characters transmitted by digital signals. Instead, analog signals similar to those generated by video cassette recorders are generated by the video controller. These signals are continuously transmitted at high frequency so that the content of the display can be updated quickly. The communication of video display data is depicted in Figure 10-10 on the next page.

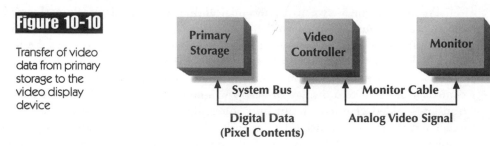

Figure 10-10

Transfer of video data from primary storage to the video display device

Primary Storage — System Bus — Video Controller — Monitor Cable — Monitor

Digital Data (Pixel Contents)   Analog Video Signal

Monitors are generally implemented using **cathode ray tubes (CRTs)**. CRTs are based on the same highly refined and relatively inexpensive display technology used in conventional television sets. A CRT is an enclosed vacuum tube. An electron gun in the rear of the tube generates a stream of electrons that are focused in a narrow beam toward the front surface of the tube, and the interior of the display surface is coated with a phosphor. When struck by a sufficient number of electrons, the phosphor reaches an excited state and emits light. Pixel illumination is controlled by pulsing (i.e., turning on and off) the electron beam. The color of the display is determined by the chemical composition of the phosphor coating. The intensity of individual pixels, or the entire display, can be varied by varying the intensity of the electron beam.

The electron beam is directed toward individual pixels on the interior display surface by magnetic deflection. The magnetic deflection is shifted constantly so that the beam moves continuously between pixel locations in a left to right, top to bottom motion. Color images are generated using three different electron beams directed toward a grid of three different phosphor coatings (i.e., red, green, and blue). Different levels of intensity for each color combine to produce a continuous, or nearly so, range of color.

The phosphor atoms remain in an excited state for only a fraction of a second. To continually display light, a pixel area must be bombarded by the electron gun many times per second. The number of times per second that the entire surface is scanned by the electron gun is called the **refresh rate** (measured in Hertz [Hz]) of the monitor. Typical refresh rates range from 50 to 100 Hz. Images displayed with higher refresh rates are regarded as being of higher quality and causing minimal eye strain.

CRT technology is relatively old and has some disadvantages. Physical size and weight increase as the size of the display area increases. The glass vacuum tube is relatively heavy, and larger screen sizes require larger (and deeper) vacuum tubes. A typical CRT monitor occupies 0.5 cubic meters or more—a substantial fraction of the available space on a typical desk. Another drawback to CRT monitors is their power consumption. CRTs consume up to several hundred watts of power and dissipate much of that as heat. Power requirements and heat generation increase with screen size. The power consumption of a typical workstation monitor is at least half that of the entire computer system.

Many alternative video display technologies are under development or in current production. As a group, they are called **flat panel displays** because of their characteristic shape. None has yet proven superior to CRT technology at equal or less cost, but extensive research and development activities promise that the cost/performance ratio of CRTs will be surpassed within the next decade or sooner.

A **liquid crystal display (LCD)** panel is a matrix of encapsulated liquid crystals sandwiched between two polarizing panels. Exterior light passes through the top polarizing layer and into the liquid crystal layer. If no power is applied to the liquid crystal, the polarized light reflects back toward the front panel, appearing as a white or light gray spot. If power is applied to the liquid crystal, light passes through the crystal and is absorbed in the rear polarizing panel. Thus, the spot appears dark (e.g., usually blue or black).

Modern LCDs use a technique called backlighting for improved contrast and brightness. In this approach, a flourescent light source is placed behind the display panel. Liquid crystals can allow or block the passage of this light through the front of the display, depending on their electrical state. Color output is generated by using three LCD matrices with a different color filter for each.

Switching individual pixels on and off requires transistors and wiring. Active matrix LCDs use one or more transistors for every pixel, whereas passive matrix LCDs share transistors among rows and columns of pixels. Additional transistors increase the speed at which pixel elements can be changed. They also allow a continuous charge to be applied to each pixel thus generating more (brighter) light, and increase the complexity and cost of the display panel.

Current LCD displays have less contrast than do CRT displays. The use of color filters reduces the total amount of light that passes through the front of the panel. They also have a much more limited viewing angle than CRTs do. Best contrast and brightness is achieved only within a 30- to 60-degree angle. Outside of that angle, contrast drops off sharply and colors shift. In contrast, CRT viewing angles usually approach 180° with little loss in contrast or brightness.

The advantages of an LCD are substantially reduced size, weight, and power consumption as compared to a CRT. Power consumption is less than 100 watts. Weight is relatively small because of small size and the use of only two flat-glass panels. The cost of an LCD panel is considerably higher than a CRT of similar size and capability. Manufacturing methods have evolved slowly, and production yields are often poor.

Gas plasma displays are constructed using a matrix of electrodes, which are sandwiched between two transparent plates, and bubbles of neon gas trapped in the spaces between the electrodes. Applying sufficient voltage to an electrode junction excites the gas into a plasma state, causing it to emit a reddish-orange light. Through proper voltage control, the intensity of the light may be regulated, producing up to 16 different discernable light intensities. Gas plasma displays are seldom used any more because of their inability to generate more than one color.

Electroluminescent (EL) displays are similar in construction to gas plasma displays. However, instead of a volatile neon gas, these displays use a solid state phosphor to generate light. Color is generated by using three different matrices with different-colored phosphors. There are no liquids or gasses and very little glass so the devices are stable, rugged, and lightweight. The primary drawbacks of EL displays are color quality and cost. Color quality suffers because of the difficulty in developing "true blue" phosphors that operate at low voltages and have long life.

The manufacturing process is similar to that for microchips. The matrices of pixel activation transistors and related electrodes are built up on a layer of clear silicon. Phosphors are deposited directly on the matrices, and another layer of clear silicon is overlaid. Problems with material purity and low production yield, similar to other semiconducting fabrication processes, are exacerbated by the relatively large size of the product.

PRINTERS AND PLOTTERS

Despite repeated promises of a paperless office, printers remain an important part of most computer systems. Early printers were little more than electric typewriters with a communication port substituted for a keyboard. Printed output was generated by moving paper past a ball or wheel on which all the characters of a font were embedded as raised metal images. The ball or wheel was rotated to place the desired character in a "ready to print" position, and the character was forced against the paper at high speed. A cloth ribbon soaked in ink was positioned between the print head and paper to generate a character image.

As with video displays, character oriented output has given way to graphic-oriented devices. Dot matrix printers represent the last generation of printers that generate images through mechanical contact with ink and paper. The clicking and clacking of metal gears, mechanical print wheels, and print head pins has been replaced by quieter paper movement mechanisms and the silent flow of ink to paper from liquid and powdered ink cartridges.

The functional components of a character printer are shown in Figure 10-11. Notice the similarity to the components of a VDT shown in Figure 10-8. The communication controller and the print driver perform essentially the same functions as their counterparts in a VDT. The exact nature of the translation process performed by the printer driver depends on the type and capabilities of the print generator. Commonly used types of print generators include dot matrix, ink-jet, and laser print generators.

Input/Output Technology

Functional components of a typical character printer

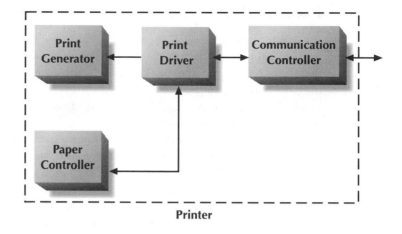

Printer

Dot Matrix Printers

Control of a **dot matrix printer** is similar to that of a VDT. Character codes received by the print generator are transmitted as a series of pin control commands to the print head. The print head consists of one or more vertical lines of pins that can be forced outward from the head a fraction of an inch. An inked ribbon is moved in front of the pins and the paper is on the opposite side of the ribbon. The print head is physically moved from left to right on each print line. Control signals force pins to be pushed out from the print head, impressing a dot of ink from the ribbon onto the paper. A dot matrix printer thus generates individual characters one or two vertical rows at a time, from left to right.

Paper control is provided by one or more motors controlling rubber rollers and/or sprockets. Paper is normally advanced past the print head one line at a time. The printer prints characters on the page in a left to right, top to bottom order. Advanced control capabilities can allow paper movement in increments of less than a single line. Reverse movement of paper and/or print head also can be implemented.

Dot matrix printers have all but disappeared from the marketplace because of poor output quality, noisy operation, and competition from other printer technologies. The few models still available are customized for high-speed printing of preprinted forms. Print quality is sacrificed for overall speed. Fast dot matrix printers can achieve print speeds of several hundred characters per second. Sprockets and fan folded paper with matching holes along the edges allow rapid paper movement through the print mechanism. The combination of fast print heads, rapid paper movement, and preprinted documents with few characters added by the printer allows up to a dozen pages per minute to be printed, which is faster than equivalently priced alternative technologies.

Ink-Jet Printers

An **ink-jet printer** uses a paper movement mechanism similar to that of a dot matrix printer. Individual sheets of paper are used instead of fan fold paper. Another similarity is the use of a single print head drawn horizontally across the width of the paper. The primary difference is the operation of the print head: There is no inked ribbon and no direct physical contact between elements of the print head and the paper.

The print head of an ink-jet printer consists of an ink cartridge, a set of ink chambers, and a set of ink nozzles. A row of nozzles corresponds to a short vertical row of pixels in the printed image. Each nozzle has a small ink chamber immediately behind it. At the back of the chamber is a flexible piezoelectric membrane. Small wires are attached to the membrane. When an electrical charge is applied to the membrane, it flexes toward the nozzle, which creates pressure within the ink chamber that forces liquid ink out through the nozzle. When the electrical charge is removed, the membrane flexes back to its original shape. The now-enlarged chamber has a partial vacuum that draws fresh ink in from the cartridge.

The print head nozzles are placed in very close proximity to the paper surface. Paper is fed in from an input hopper and positioned so the print nozzles are along its top edge. As the print head is drawn across the paper width by a pulley, the print generator rapidly modulates the flow of current to the ink-chamber membranes, which generates a pattern of pixels across the page. The paper is then advanced and the process repeated until the entire sheet has been drawn past the moving print head. The ink dries very quickly so there is little or no opportunity for it to smear.

Ink-jet printers use very small chambers and nozzles capable of producing ink dots $\frac{1}{300}$ inch in diameter (i.e., resolution is up to 300 pixels per inch). Color output can be generated by using three (CMY) or four (CMYK) inks and an equal number of nozzle/chamber rows. As each row passes a horizontal point on the page, different color ink can be applied. The inks mix to form pixels of various colors. Variable charge can be applied to the chamber membranes to place variable amounts of ink on the page. This allows CMY colors to be mixed in different proportions, thus producing a wide range of possible pixel colors.

Ink-jet printer technology has advanced rapidly. Typical output speeds for inexpensive models are two pages per minute for black and white and one-half page per minute for full color. Black and white print quality is good to excellent, and color print quality is often better than color laser printers. Ink-jet printers are the only type of printer capable of producing high-quality color output within the budget of a typical home computer or small office user. Clogged print heads were a problem with early models, which was solved by integrating the print head and the ink cartridge. Both are disposed of as a unit when the cartridge is empty.

Laser Printers

A **laser printer** operates quite differently from other types of printers. No print head is used, nor is any inked ribbon required. The print driver of a laser printer does not transmit individual characters or lines of an image to the print generator. Instead, it stores an internal image of an entire printed page in an internal buffer. Once filled, the buffer is sent to the print driver for page generation.

The print generator operates using electrical charge and the attraction of ink to that electrical charge, as illustrated in Figure 10-12. A rotating drum is first lightly charged over the width of its surface. The print driver then reads rows of pixel values from the buffer and modulates a tightly focused laser over the width of a metal drum. The drum is then advanced and the process repeated with the next line of pixels. The laser removes charge wherever it shines on the drum, and thus the drum contains an image of the page with charged areas representing black pixels and uncharged areas representing white pixels. After charging, the drum then passes a station where fine particles of toner (i.e., a dry powder ink) are attracted to the charged areas.

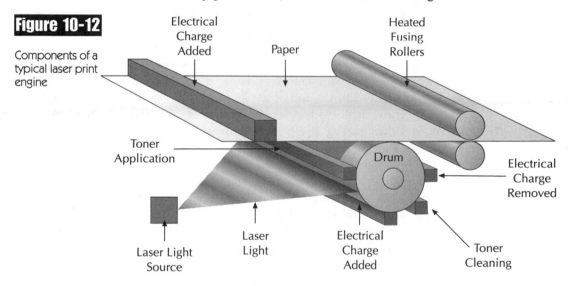

Figure 10-12

Components of a typical laser print engine

In synchronized motion with the drum, paper is fed through a series of rollers and given a high electric charge. As the paper passes over the drum surface, toner on the drum is attracted to the paper due to its higher charge. The paper and attached toner are then fed through heated rollers that fuse the toner to the paper surface. A light source then removes all charge from the drum, and excess toner is removed from the drum by a fine blade and/or vacuum.

Color output can be achieved by using three separate print generators (i.e., laser, drum, and laser modulators), one for each of the primary colors, which requires precise alignment of the paper as it passes over the three drums. Also, the complexity (and expense) of the printer is increased substantially. Color also can be generated by three passes over the same print generator with a different color ink applied each time, which substantially reduces maximum printer speed as well as the complexity and cost of the printer.

Laser printers are nearly ubiquitous in modern offices, and lowered prices have made them affordable for most home office users. Print speeds range from a few pages per minute to several dozen pages per minute. The devices are well-developed and reliable.

Plotters

A **plotter** is an output device that generates line drawing on sheets or rolls of paper. Paper is mounted within a paper control mechanism that can precisely move the paper up or down. One or more pens are mounted on a motor controlled station that may be moved left or right over the paper. Each pen may be raised or lowered to contact, or not contact, the paper.

Images are drawn on paper by moving the pen and/or the paper. Drawing a line from left to right requires the paper to be positioned so that the line position is directly below the pen. The pen is lowered onto the paper at the beginning of the line and moved to the end of the line and then raised. A vertical line can be drawn by correctly positioning the paper and pen, lowering the pen, moving the paper, and then raising the pen. Nonperpendicular lines, curves, and shapes require the simultaneous movement of both paper and pen.

The operation of a plotter is a sophisticated example of digital to analog conversion. The initial representation of an image within a computer (and its communication to the plotter) is usually in digital form, which is translated into analog motor control signals that physically draw the image on paper. Plotters differ primarily by number of pens (colors), size of paper, and the presence (or absence) of local digital-to-analog processing.

If the original image to be printed is a bitmap, it must be converted into a series of control signals for the various plotter motors. A sophisticated plotter accepts an image as a bitmap, vector list, or display list and performs this conversion process internally. Other plotters require that input be in the form of digitally encoded direct motor control signals. For such a plotter, all translation from bitmap or other representation formats is performed by software within the computer system.

CHARACTER-BASED DISPLAY CONTROL

Character-based output devices that accept ASCII input provide a set of specialized display functions that somewhat depend on the type of device. For printers, specialized control functions include the ability to change output fonts, font sizes, character intensity (e.g., black or various shades of grey), and character color. Video displays usually incorporate these capabilities as well as formatted output, cursor positioning, and inverse or blinking character display.

As described in Chapter 4, the ASCII character coding scheme provides a set of nonprintable **control characters**, including carriage return (new line), form feed, bell, and a few dozen other codes. These codes can be used by I/O devices to activate (or deactivate) various device-control functions. However, the set of undefined characters is not large enough to control all of the specialized display functions available in modern output devices.

To address this problem, most manufacturers implement specialized device control with sequences, or strings, of control characters. In most devices, a control character string always begins with an escape character (i.e., ASCII 27). Thus, such control strings are sometimes called **escape sequences**. There is considerable variation between devices and manufacturers over the exact format, content, and meaning of escape sequences. Table 10-2 shows a small sample of the escape codes used with a Hewlett-Packard LaserJet printer.

TABLE 10-2	Function	Control String (Esc=ASCII 27_{10})	Parameters
A subset of the escape codes used by the Hewlett-Packard LaserJet printer	Reset	Esc E	none
	Portrait mode	Esc & 1 0 O	none
	Landscape mode	Esc & 1 1 O	none
	Line spacing (per inch)	Esc & 1 # D	# = lines per inch (decimal)
	Select Courier font	Esc (s 3 T	none
	Select CG Times font	Esc (s 4 1 0 1 T	none
	Set font height	Esc (s # V	# = height in points (decimal)
	Set font style	Esc (s # S	# = style 0, normal 1, italic 4, condensed 5, condensed italic

An ANSI standard exists for a limited set of VDT escape sequences, including escape sequences to control foreground color, background color, inverse video, underlined text, and blinking text. VDTs from most manufacturers adhere to these standards. Most implement a large number of control functions in addition to the ANSI standard escape sequences.

OPTICAL INPUT DEVICES

Devices that convert patterns of light into character or image data are categorized as **optical sensors**. Such devices include digital still and video cameras, optical scanners, mark sensors, bar-code readers, and optical character recognition (OCR) devices. The operating principles behind all of these devices are fairly similar. A light source (usually a laser) is shone onto a printed surface, and light is reflected from the surface into a photodetector. High light intensities induce relatively large current outflows from the photodetector, whereas low intensities produce little or no current.

Optical scanning devices can be differentiated by the following criteria:

► Input format requirements.
► Normal and maximum spatial resolution.
► Normal and maximum chromatic resolution.
► Amount of required local processing.

High-resolution devices may be capable of spatial resolutions as high as 1,200 dots per inch, whereas simpler devices may require relatively large marks (e.g., no smaller than five millimeters) for reliable recognition. Certain devices require marks to appear at specific locations on a page or within specific distances of one another. Devices that attempt any form of image recognition or classification (e.g., optical character recognition devices) must embed processing capabilities within the device or must employ software on the host computer.

Mark and Bar-Code Scanners

Mark sensors and **bar-code scanners** are relatively simple optical recognition devices. A mark sensor scans for light or dark marks at specific locations on a page. These devices (or the input to them) are familiar to students who take standardized multiple-choice tests. Input is from a page preprinted with circles or boxes, and these locations can be marked by filling them with a dark pencil. The mark sensor uses preprinted bars on the edge of the page to establish reference points (e.g., the row of boxes corresponding to the possible answers to question 5). It then searches for dark marks in specific locations with respect to those reference points. Marks must be of a minimum size and intensity to be recognized correctly.

Bar-code scanners operate similarly to mark sensors, but input does not have pre-established reference marks. A bar code is a series of vertical bars of varying thickness and spacing. The order and thickness of bars is a standardized representation of numeric data. Bar-code readers use one or more **scanning lasers** to detect the vertical bars. A scanning laser sweeps a narrow laser beam back and forth across the bar code. The laser requires precise thickness and distances as well as high-contrast or accurate decoding. The use of multiple scanners at oblique angles allows for variation in the bar-code position and orientation.

Bar-code readers are used to track large numbers of inventory items. Common uses include grocery store inventory and checkout, tracking of packages during shipment, warehouse inventory and picking, and zip code routing of postal mail. The U.S. Postal Service uses a modified form of bar coding with evenly spaced bars of equal thickness but varying height. These appear along the lower edge of the envelope and are used to encode five- and nine-digit zip codes.

Optical scanners are used to generate bitmap representations of printed images. A laser is scanned across the width of the page, and variations in reflected light are converted to numeric pixel values. Multiple passes by different colored lasers can be used to detect color. Manual scanners (sometimes called hand scanners) require a user to move the scanning surface over a printed page at a steady rate of speed. Automatic scanners use motorized rollers to move paper past a scanning surface or to move a scanning device across the paper. Multiple scans can be used to compensate for scanning errors.

Optical scanners have dropped dramatically in price during the 1990s. Full-page automatic scanners are now commonly available for under $1,000. These scanners can handle high resolution (e.g., 300 DPI or greater) and color recognition. The need to accurately position the image to be scanned and the time required for the scanning laser to move across the width of a page limit the scanning speed. High-resolution scans take more than one minute per page with inexpensive scanners and no more than a few pages per minute for more expensive models.

Optical character recognition (OCR) devices combine optical scanning technology with intelligent interpretation of bitmap content. Once an image has been scanned, the bitmap representation is searched for patterns corresponding to printed characters. In some devices, input is restricted to characters in prepositioned blocks. In these devices, as with mark sensors, preprinted marks provide a reference point for locating individual characters, which greatly simplifies the character-recognition process.

More sophisticated OCR devices place no restriction on the position and orientation of symbols. Such a device faces the dual problem of both finding and correctly classifying printed symbols. This problem is greatly simplified when text is printed, in a single font and style, not mixed with nontextual images, and all oriented in the same direction on the page. As each of these factors is relaxed, accurate recognition becomes more and more difficult. Many inputs are simply beyond the capabilities of current hardware and software, which is not surprising, given estimates that approximately one-quarter of human brain power is devoted to processing visual input.

Artificial intelligence techniques including fuzzy logic and neural networks are applied routinely in OCR software, but error rates of 10 to 20% are still common with mixed font text and errors with handwritten text are higher. The field has experienced rapid improvement in accuracy and flexibility of input, but progress needs to be made before error rates are acceptable for many types of applications.

AUDIO I/O DEVICES

The use of computers to generate and/or recognize sound is a relatively recent phenomenon. Sound is an inherently analog signal and must be converted to digital form for computer processing or storage. The process of converting analog sound waves to digital representation is called **sampling**. The content of the audio energy spectrum is analyzed many times per second. For sound reproduction that sounds natural to humans, frequencies between 20 Hz and 20 KHz must be sampled at least 40,000 times per second. A rate of 48,000 samples per second is a common standard (e.g., it is used for CD recordings). Each individual sample is referred to as a sample point.

Sound varies by frequency (i.e, pitch) and intensity (i.e., loudness). By various mathematical transformations, a complex sound (e.g., a human voice) consisting of many pitches and intensities can be converted to a single numeric representation. For natural-sounding reproduction, 16 bits must be used to represent each sample point. Thus, one second of sound sampled 48,000 times per second requires 96,000 bytes for digital representation (48,000 × 16 bits × .0125 bytes/bit). A full minute of sound requires over 5 MB.

Computer generation of sampled sound is relatively simple. First, the digital information is input to a digital to analog converter. Then the output of the digital to analog converter is input to a conventional analog amplifier and then to a speaker. Sampling is also relatively simple. However, the processing and communication power necessary to support both of these operations has only recently become affordable enough to allow widespread application. Compact audio disk players and digital audio tape recorders are two consumer-oriented examples of this technology.

Within a computer system, sound generation and recognition are potentially useful for the following purposes:

► General purpose sound output (e.g., warnings, status indicators, and music).

► General purpose sound input (e.g., digital recording).

► Voice command input.

► Speech recognition.

► Speech generation.

Limited general purpose sound output has existed in computer systems for many years. In early teletypes, a bell was rung physically to signal events that required operator intervention (e.g., adding paper). Microcomputers, workstations, and VDTs typically include the ability to sound a single audible frequency at a specified volume for a specified duration of time. This type of sound output is called **monophonic** output because only one frequency (i.e., note) can be output at a time.

As computers have been adapted for multimedia presentation, greater sound generation power has been added. Many computers now employ **polyphonic** (i.e., multifrequency) sound-generation hardware with built-in amplification and high-quality speakers. This capability is required for multimedia presentations and to use the computer as a general purpose communication device.

Most current research in computer sound processing is focused on voice recognition and generation. Devices resulting from this research include voice-command-recognition devices and speech generators. For all processing of human speech, problems of word recognition, or generation, must be addressed.

Speech Recognition

Human speech consists of a series of individual sounds called **phonemes**, roughly corresponding to the sounds of each letter of the alphabet. A spoken word is a series of interconnected phonemes. Continuous speech is also a series of phonemes interspersed with varying amounts of silence. Recognizing individually voiced (i.e., spoken) phonemes is not a difficult computational problem. Sound can be captured in analog form by a microphone, converted to digital form by digital sampling, and the resulting digital pattern compared to a library of patterns corresponding to known phonemes, as shown in Figure 10-13 on the next page.

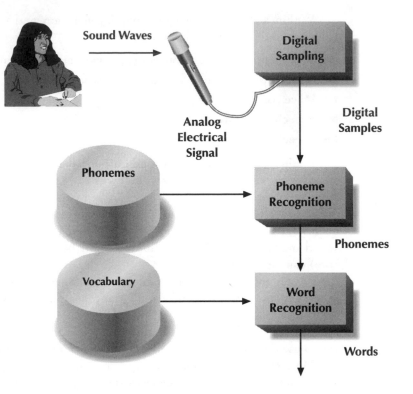

Figure 10-13

Process of speech recognition. Sound waves are converted to an analog electrical signal by a microphone. That signal is sampled to produce a digital equivalent, and the digital signal is converted to phonemes and words.

Although the process is conceptually simple, a number of factors complicate it, including:

▶ The need for realtime processing.

▶ Variation between speakers.

▶ Phoneme transitions and combinations.

Speech recognition is only useful for computer I/O if it can be performed in realtime, which implies that the capture, conversion, and comparison functions must occur fairly quickly. Variation between speakers in the generation of phonemes makes comparison an inexact process. Comparison to a library of known phonemes cannot be based on an exact match; rather, the closest approximation must be determined and a decision between multiple possible interpretations must be made. This processing difficulty requires complex software and a relatively powerful CPU to execute it.

General purpose speech-recognition systems can be used for command and control or for the input of large amounts of textual material. Current technology has produced inexpensive single speaker systems with large vocabularies (tens of thousands of words) and relatively low (below 5%) error rates.

The computational complexity of sound and speech processing has led to development of processors designed specifically for such processing—**digital signal processors (DSPs)**. These processors, highly optimized for processing continuous streams of digitized audio or video, are not designed for general purpose processing tasks. They are commonly embedded within audio and video hardware such as PC sound cards and dedicated audio/video workstations.

CPU manufacturers have recognized the increasing importance of audio/video processing capabilities. Some of the latest generation of general purpose microprocessors include additional instructions specifically designed for audio processing. In essence, these manufacturers have incorporated DSPs within a general purpose CPU. The Intel Pentium MMX processor family and many of its direct competitors exemplify this trend.

The most difficult problem in speech recognition arises from the continuous nature of speech. Individual letters sound relatively similar when repetitively voiced by the same speaker. However, when combined with other letters in different words, their voicing varies considerably. In addition, a computer must determine where one phoneme ends and another begins, as well as where one word ends and another begins. Once again, the complexity of such processing requires complex software and powerful CPUs.

Most currently available speech-recognition systems (hardware and associated software) are not capable of unrestricted speech recognition among multiple speakers in realtime. They must be "trained" to recognize the sounds of individual speakers; these systems are sometimes referred to as **speaker dependent**. They also are restricted to relatively limited vocabularies (e.g., a few thousand words). The most limited of these are command-recognition systems. They are designed to recognize up to a few hundred spoken words from a single speaker. Such devices are useful when a single user will use the device and when manual (e.g., keyboard) input is impractical. These systems have been applied in airplane cockpits, manufacturing control, and input systems for the physically handicapped.

Speech Generation

One type of device that generates spoken messages is an **audio response unit**, which is used to deliver limited amounts of information over conventional telephone instruments (e.g., automated telephone bank tellers). Thus, any telephone becomes a potential output device for the system. Audio response is implemented by storage and playback of words or word sequences. Single words or entire messages are stored digitally. Output of the message involves accessing the corresponding storage location and sending it to a device that converts the digitized voice into analog signals that can produce sounds in electronic speakers or earphones.

A more general approach to speech generation is called **speech synthesis**. In speech synthesis, individual vocal sounds, or phonemes, are stored within the system. Character outputs are sent to a processor within the output unit, which assembles corresponding groups of phonemes to generate synthetic speech. The quality of speech output from such units varies considerably. Problems are encountered when combining groups of phonemes to form words. Transitional sounds must be generated between the phonemes and composite phonemes must be taken into account. However, the difficulties of speech generation are far less formidable than those of speech recognition. Speech-generation devices are available commonly with prices starting as low as a few hundred dollars.

General purpose audio hardware also can be used for speech generation. Digital representations of phoneme waveforms must be combined mathematically to produce a continuous stream of digitized speech. This digital data stream is then sent to the digital-to-analog converter of the audio hardware. Once converted to an analog waveform, it is amplified and routed to one or more speakers.

General Purpose Audio Hardware

General purpose audio hardware is typically packaged as an expansion card that attaches to a workstation's system bus. Common names for such hardware include **sound card**, **sound board**, and **multimedia controller**. The latter term is used often when a CD controller and drive are part of the purchased package.

At minimum, sound cards include an **analog to digital converter (ADC)**, a **digital-to-analog converter (DAC)**, a low-power amplifier, and connectors (jacks) for a microphone and a speaker or headphones (see Figure 10-14). More elaborate cards can provide:

► Two pairs of converters for stereo input and output.

► General purpose MIDI synthesizer.

► MIDI input and output jacks.

► More powerful amplifier to drive larger speakers and generate more volume.

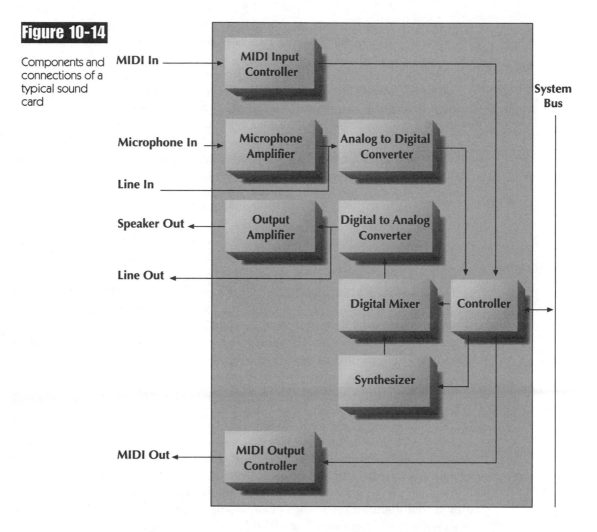

Figure 10-14

Components and connections of a typical sound card

MIDI In → MIDI Input Controller

Microphone In → Microphone Amplifier

Line In

Speaker Out ← Output Amplifier

Line Out ←

Analog to Digital Converter

Digital to Analog Converter

Digital Mixer ← Controller

Synthesizer

MIDI Out ← MIDI Output Controller

System Bus

Other differences in sound card capability include degree of polyphony (i.e., number of simultaneous sounds that can be generated), supported sampling rates, and the power and speed of embedded DSPs.

As stated earlier, sampling rates of 40,000 Hz or greater are required to capture and reproduce waveforms that sound accurate to the human ear. However, lesser sampling rates can be used when full range sound isn't required (e.g., speech generation) or when processing power or disk space is at a premium. Cutting the sampling rate in half reduces the required processing power and required storage space in half, which reduces cost at the expense of full-frequency range (i.e., 20 to 20,000-Hz) sound quality.

Musical Information Digital Interchange (MIDI) is a standard for the storage and transport of control information for synthesizers. MIDI has been used since the late 1970s to allow computers and music keyboard equipment to control remotely connected music synthesizers. MIDI defines a serial communication channel and a standard set of commands that can be sent through that channel. Commands are 1- to 3-byte sequences and include turning individual notes on and off, pitch modulation, various instrument controls (e.g., the sustain pedal of a piano), and the selection of specific instruments for note generation. MIDI synthesizers recognize this incoming stream of data and treat it as commands to be processed. Up to 16 channels of MIDI data can be sent over the same serial transmission line, thus allowing 16 separate instruments or instrument groups to be played at once.

Most sound cards include some MIDI capability. Minimally, they provide a simple synthesizer capable of simulating the sound of a dozen or so instruments. Files containing MIDI control data are transferred to the sound card over the system bus and the synthesizer processes the commands and generates digital waveforms for each instrument. These digital waveforms are combined (mixed) and sent through the digital to analog converter, amplifier, and speakers.

More complex MIDI capabilities include the provision of more synthesized instruments, better quality of those sounds, and the ability to accept or transmit MIDI data through external MIDI connections. The quality of synthesized sound varies widely from one synthesizer and/or sound card to another. The poorest-quality sounds are generated by inexpensive DSPs executing simple instrument simulation programs and generating low sample rate output. The best quality is typically obtained with sample based synthesis. This method uses stored digitized recordings of actual instruments. These are accessed and replayed when a corresponding MIDI "note on" message is received.

The primary advantage of MIDI is its compact storage format. A command to sound a single note on a piano for one second requires 6 bytes — 3 bytes each for the "note on" and "note off" control messages. A 16-bit digital recording of that same sound sampled at 48,000 Hz requires 96,000 bytes of storage. The disadvantage of MIDI is a lack of control over the exact nature of the generated sound. A MIDI command simply selects instruments and notes but doesn't control how instrument sounds are generated. The sound quality of an instrument can vary widely from one synthesizer to another depending on sample rate, synthesis technique, and hardware capability. The sound for a given instrument may not even exist on the synthesizer receiving the MIDI commands.

SUMMARY

Manual (hand-operated) input devices include keyboards, mice, and other pointing devices. The predominant mechanism for human input to computer systems is a character keyboard. Modern keyboard devices translate keystrokes directly into electrical signals. Pressing a key sends a coded signal to a microprocessor that generates a bit-stream output according to an internal program or lookup table.

Devices used to capture graphic and image data include the mouse, track ball, joystick, and digitizer tablet. A mouse is a pointing device that is meant to be moved on a flat surface. The position of the mouse on a surface is an analog to the position of a cursor or pointer on a video display. A track ball is similar to a mouse laid upside down. They are considered easier to use for persons lacking fine motor control. Digitizers use a pen (or stylus) and a digitizing tablet. The tablet is sensitive to the placement of the stylus at any point on its surface.

Display surfaces can be divided into rows and columns similar to a large table or matrix. Each cell (or pixel) in that table represents one simple component (dot) of an image. Images, printed characters, and other display elements are described by their height and width in pixels or points. Pixel size varies from device to device. A point is $\frac{1}{72}$ inches. The resolution of a display is a measure of the number of pixels displayed per linear measurement unit. The unit of measure is stated as dots per inch (dpi) where a dot is assumed to be equivalent to a pixel.

Written Western languages are based on systems of symbols called characters. A collection of characters of similar style and appearance is called a font. The most common unit for measuring font size is points.

Any color in the visible spectrum can be represented as a mixture of varying intensities of the primary additive colors red, green, and blue (i.e., RGB). The printing industry commonly generates color by using the primary subtractive colors (i.e., CMY): cyan (absence of red), magenta (absence of green), and yellow (absence of blue). Any color can be produced as a combination of either the primary additive or subtractive colors.

A stored set of numeric pixel descriptions is called a bitmap. Monochrome display uses a single bit to indicate whether each pixel is "on" or "off." Multiple bits can be used to represent varying levels of intensity (i.e., a gray scale). Three separate numbers (sets of bits) are required to represent a pixel in full color. One represents the intensity, or amount, of each of the primary additive or subtractive colors.

Significant drawbacks to the use of bitmaps include their large size and device dependence. Compression can be used to reduce size. An image-description language is another method of reducing image file size. Image-description languages may describe images using vectors, display objects, or both. Required storage space is reduced because a symbolic description of an image component is often much smaller than a bitmap of that same component.

VDTs consist of an integrated keyboard and television screen. A typical character-oriented VDT can display at least 80 columns and 24 rows. Characters are displayed using a matrix of pixels. Modern video displays are separate from keyboards and are called monitors. They operate under control of a video controller attached to the system bus of a workstation. Analog signals similar to those generated by video cassette recorders are generated by the video controller.

Monitors are implemented using cathode ray tubes (CRTs), which are enclosed vacuum tubes. An electron gun in the rear of the tube generates a stream of electrons that are focused in a narrow beam toward the front surface of the tube. The interior of the display surface is coated with a phosphor that generates light when struck by a sufficient number of electrons. CRT technology is relatively old and has some disadvantages including size, weight, and power consumption.

Flat panel display technologies include liquid crystal displays (LCDs), gas plasma displays, and electroluminescent displays (ELDs). LCDs have less contrast than and reduced viewing angle as compared to CRT displays. Their advantages include reduced size, weight, and power consumption. The cost of an LCD panel is considerably higher than a CRT of similar size and capability. Gas plasma displays use bubbles of neon gas trapped within an array of electrodes. ELDs use an array of electrodes with phosphors deposited directly on the array. ELDs overcome the weaknesses of LCDs but at considerably higher cost.

Commonly used paper output devices include dot matrix printers, ink-jet printers, laser printers, and pen plotters. A dot matrix printer uses a print mechanism with vertical lines of pins that can be forced outward from the head. Control signals force pins to be pushed out from the print head, impressing a dot of ink from a ribbon onto paper. Dot matrix printers are inexpensive but are also slow, noisy, and produce relatively poor-quality output. The print head of an ink-jet printer consists of an ink cartridge, a set of ink chambers, and a set of ink nozzles. Electrical charge sent to the ink chambers forces ink to be ejected onto a page. Ink-jet printers produce excellent-quality output but do so relatively slowly.

A laser printer operates using electrical charge and the attraction of ink to that electrical charge. A rotating drum is first lightly charged and a laser then shines a pattern over the drum to selectively remove charge. Dry ink is attracted to the remaining charged areas. Charged paper is passed over the drum to collect the ink and the paper and attached toner are then fed through heated rollers that fuse the toner to the paper surface. Laser printers are relatively fast and produce excellent quality output.

Character display on video displays and printers can be modified through the use of control codes called escape sequences. For printers, common escape sequences include those that change output fonts, font sizes, character intensity, and character color. Video displays also can format output, position the cursor, and generate inverse or blinking character display. A limited set of video escape sequences are standardized by ASCII. All other control codes are device-dependent.

Optical input devices include optical scanners, mark sensors, bar-code readers, and optical character recognition (OCR) devices. All use a laser reflected off of a printed surface into a photodetector. A mark sensor scans for light or dark marks at specific locations on a page. A bar code is a series of vertical bars of varying thickness and spacing. The order and thickness of bars is a standardized representation of numeric data. Bar-code readers use one or more scanning lasers to detect the vertical bars.

Optical scanners generate bitmap representations of printed images. A laser is scanned across the width of the page and variations in reflected light are converted to numeric pixel values. OCR devices combine optical scanning technology with intelligent interpretation of bit-map content. Once an image has been scanned, the bitmap representation is searched for patterns corresponding to printed characters. Artificial intelligence techniques are routinely applied in OCR software, but error rates of 10% to 20% are still common.

General purpose speech-recognition systems can be used for command and control or for the input of large amounts of textual material. Human speech consists of a series of individual sounds called phonemes. Continuous speech is a series of phonemes interspersed with varying amounts of silence. Recognizing individually voiced phonemes is not a difficult computational problem. Sound can be captured in analog form by a microphone, converted to digital form by digital sampling, and the resulting digital pattern compared to a library of patterns corresponding to known phonemes. Recognition of continuous speech is much more difficult.

Current technology has produced inexpensive single speaker systems with large vocabularies (tens of thousands of words) and relatively low error rates. Error rates below 5% are common. The computational complexity of sound and speech processing has led to development of processors designed specifically for such processing. Such a processor is called a digital signal processor (DSP). These processors are highly optimized for processing continuous streams of digitized audio or video. They are commonly embedded within audio and video hardware such as PC sound cards.

General purpose audio hardware includes an analog-to-digital converter (ADC), a digital-to-analog converter (DAC), a low-power amplifier, and connectors (jacks) for a microphone and a speaker or headphones. MIDI inputs, MIDI outputs, and a music synthesizer also can be provided. Computer generation of sampled sound is relatively simple. Digitized sound wave information is input to a digital to analog converter and the output is input to a conventional analog amplifier and then to a speaker. General purpose audio hardware also can be used for speech generation. Digital representations of phoneme waveforms are mathematically combined to produce a continuous stream of digitized speech.

Key Terms

24-bit color
additive colors
analog-to-digital converter (ADC)
audio response unit
bar-code scanner
bitmap
cathode ray tube (CRT)
character
chromatic depth
chromatic resolution
CMY
CMYK
control character
cursor
digital signal processor (DSP)
digital-to-analog converter (DAC)
digitizer
display attribute
display list
display object
dithering
dot matrix printer
dot pitch
dots per inch (dpi)
electroluminescent (EL) display
escape sequence
flat panel display
font
gas plasma display
image description languages
ink-jet printer
Joint Photographic Experts Group (JPEG)
keyboard controller
laser printer

liquid crystal display (LCD)
mark sensor
monitor
monophonic
Motion Pictures Experts Group (MPEG)
mouse
multimedia controller
Musical Information Digital Interchange (MIDI)
optical character recognition (OCR)
optical sensor
optical scanner
palette
phoneme
pixel
plotter
point
polyphonic
Postscript
refresh rate
resolution
RGB
sampling
scan code
scanning laser
sound board
sound card
speaker dependent
speech synthesis
speech recognition
subtractive colors
terminal
vector
video display terminal (VDT)

1. Color display can be achieved by using separate (but closely spaced) elements colored _____, _____, and _____.

2. The display device used in most computer monitors (and televisions) is a(n) _____.

3. A(n) _____ is used to control the location of the _____ (or input pointer) on a video display.

4. A color display system that uses 8 bytes to represent each RGB color intensity is called _____.

5. The printing industry uses inks based on the _____ colors, which are _____, _____, and _____. A(n) _____ ink also can be used as a fourth color.

6. A(n) _____ generates output by moving one or more colored pens in contact with a moving roll of paper.

7. _____ is the most commonly used display language.

8. The print head of a(n) _____ forces liquid ink onto a page through a small nozzle.

9. The _____ format is a common method of compressing graphic images.

10. The _____ format is a commonly used method of compressing motion video.

11. A(n) _____ synthesizer can generate multiple notes simultaneously. A(n) _____ synthesizer can generate only one note at a time.

12. The _____ of a display device describes the number of colors that can be displayed simultaneously.

13. Graphic images are stored as a(n) _____, where each pixel is represented by one or more numbers.

14. General purpose sound cards use a(n) _____ to generate the signal sent to speakers, headphones, or an internal or external amplifier.

15. A(n) _____ is a primitive component of voiced human speech.

16. A(n) _____ display is similar in construction to _____ and _____ displays except that it employs solid state elements for light generation.

17. A(n) _____ recognizes input in the form of alternating black and white vertical lines of varying width.

18. A(n) _____ converts analog sound waves to a digital representation.

19. A(n) _____ list differs from a(n) _____ list in that it may describe graphical objects other than lines.

20. An individual display element of a video display surface is called a(n) _____.

21. Many speech recognition programs are _____ (i.e., they are trained to recognize a single person's voice).

22. The output resolution of a graphic display device (e.g., a laser printer) is measured in terms of _____.

23. An electronic keyboard generates a(n) _____, which may be translated into ASCII character output by a keyboard controller.

24. _____ and bar-code scanners use a(n) _____ to detect marks on a page.

25. The size of an individual pixel displayed on a monitor is called the monitor's _____.

26. The _____ of a(n) _____ describes the number of times per second that the screen contents are redrawn.

27. A(n) _____ is ½ inch.

28. A group of symbols of a similar style is called a(n) _____.

29. Text pages can be converted to bitmaps by a(n) _____. These bitmaps are then input to a(n) _____ program that generates an ASCII or Unicode representation of the text.

30. _____ creates a pattern of primary colored pixels that fools the eye into thinking a composite color is being displayed.

31. ASCII _____ are used to control some aspects of communication with output devices.

32. _____ is a standardized method of encoding notes and instruments for communication with synthesizers and sound cards.

33. _____ are used for specialized control of printers and video display terminals (e.g., font changes and cursor positioning).

1. Describe the process by which software recognizes keystrokes.

2. What is a font? What is point size? How are output devices instructed to use specific fonts and point sizes?

3. What are the additive colors? What are the subtractive colors? What types of I/O devices use each type?

4. What is a bitmap? How does the chromatic resolution of a bitmap affect the size of a bitmap image?

5. What is a display list? What advantages does it have as compared to a bitmap?

6. What is JPEG encoding? What is MPEG encoding?

7. What is/are the difference(s) between a video display and a monitor?

8. Describe the various technologies used to implement flat-panel displays. What are their relative advantages and disadvantages?

9. How does the operation of a laser printer differ from a dot matrix printer?

10. Describe the various types of optical input devices. For what type of input is each device intended?

11. Describe the process of automated speech recognition. What are the sources of interpretation error inherent in this process?

12. Describe the components and functions of a typical sound card? How is sound input captured? How is speech output generated? How is musical output generated?

1. Many organizations have begun to store documents as graphic images rather than encoded (e.g., ASCII) textual and numeric data. Examples of applications that use such storage methods include medical records, engineering data, and patents. Identify one or more commercial software products that support this form of data storage. Examine the differences between such products and more traditional data-storage approaches. Concentrate on issues of efficient use of storage space, internal data representation, and method(s) of data search and retrieval. What unique capabilities are provided by image-based storage and retrieval? What additional costs and hardware capabilities are required? (e.g., http://www.belmont.com, http://www.psrw.com, http://www.virage.com)

2. Many computer scientists are researching the problem of textual data understanding. The visual recognition of character data and its conversion to ASCII- (or other-coded) representation is largely solved, but basic understanding and classification of the textual data remains poorly developed. Examples of applications that could be built with such capabilities include automatic indexing, abstracting, and summarization of news articles or other large textual inputs. Several products that address these needs are emerging from research labs into the commercial domain. Identify one or more of these products. What limitations must be placed on the text input to ensure reasonable success in automated recognition? What methods are used to perform classification and summarization? What is the relative performance (speed and error rate) of such systems compared to humans performing the same tasks? For what traditional information processing applications might this technology be used? (http://ciir.cs.umass.edu/)

Chapter

11

Applications Development

Chapter Goals

▶ Describe common methods of application system development and software support for those methods.

▶ Describe software to support the development and validation of system models.

▶ Describe software to support the development and translation of application programs.

▶ Describe programming languages and the process of program translation.

PROCESS OF APPLICATIONS DEVELOPMENT

The process of designing and constructing application software consists of translating the user's information processing needs into application software capable of fulfilling those needs. User needs are stated at a high level of abstraction in a natural language, whereas software programs are statements of formal logic using an artificial language with a low abstraction level. Thus, software development is both a translation from a natural to artificial language and a translation from an abstract concept to a finely detailed implementation of that concept.

Translation is an inherently complex process that requires an investment of considerable resources. Software has surpassed hardware to become the most costly component of an information system. Information system owners expect the resources they invest in software development to be repaid.

Unfortunately, the complexity of software development creates many possibilities for error. In software development, the result of translation errors is application software that fails to meet user needs. The economic costs of poor software can extend well beyond the cost of software development. Reduced productivity, dissatisfied customers, and poor managerial decisions are just a few of the indirect costs associated with poorly designed and implemented software.

Software developers attempt to minimize software development errors by using proven methodologies and tools. A systems or software development methodology is a detailed procedure for constructing software. It consists of general and detailed processes, rules for selecting among alternative processes, and rules for evaluating the quality of each process' output. Tools are devices used by software developers to implement development processes. Software development tools are programs that implement (or assist users to implement) one detailed process of a development methodology.

There are many different methodologies for software development and many variations of each methodology in current practice, including:

▶ Number of system models used.

▶ Characteristics of each model.

▶ Tools used to build and test each model.

A common feature of all software development methodologies is the **system model**, which is an abstract representation of information processing requirements. As such, it is an intermediate product in the translation process from user processing requirements to application software (see Figure 11-1).

Figure 11-1

Process of translating user processing requirements into application software

Application Development Translation Model

Structured Systems Development

Structured Systems Development

The **structured systems development** methodology models user processing requirements by constructing two separate system models—the **analysis model** (or **logical model**) and the **implementation model** (or **physical model**). The analysis model is an abstract representation of flows in from external sources, through internal processing logic and data storage, and out to users of the processed data or information. An implementation model is a blueprint for software construction that describes one specific software structure that implements the more abstract components of the analysis model.

The primary component of an analysis model, called a **data flow diagram (DFD)**, is a graphical depiction of the flow of data through an information system. A data flow diagram also can be used for the implementation model. However, other model formats are more commonly used for the implementation model, including systems flowcharts, program structure charts, and program flowcharts.

A DFD for a simple payroll system is shown in Figure 11-2. Model components include processes, data flows, files (or data stores), and external entities. Processes are represented by circles or rectangles with rounded corners. External entities—sources or destinations for data that are outside of the system—are depicted as squares, usually at the edges of the diagram. Files are represented by shallow rectangles that are open on the left or right side. Data flows are represented by named arrows that show the movement of data between processes, files, and external entities.

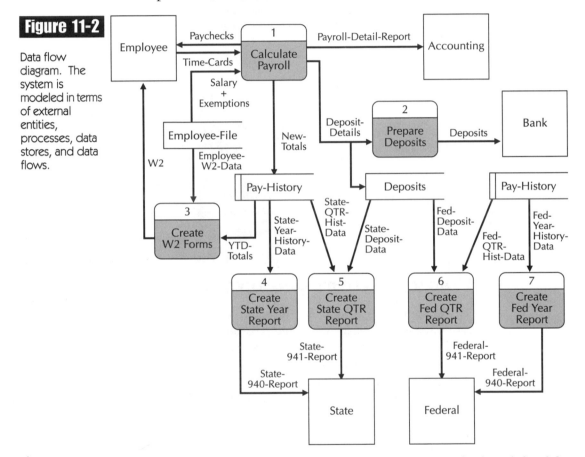

Figure 11-2

Data flow diagram. The system is modeled in terms of external entities, processes, data stores, and data flows.

Individual components of the data flow diagram are further defined by graphical or textual descriptions. DFD component descriptions provide enough detail to allow users to read and validate their content, but they are not sufficiently detailed for programs or file construction. Implementation issues, such as data (or file) formats and specific processing algorithms are ignored. The analysis model focuses on the content of data, its movement, and the rules by which data inputs are transformed into data outputs. Only after the user has validated these system features will a detailed implementation model be prepared.

Processes on a DFD are defined further by written **process specifications** that describe rules used to transform data inputs into data outputs. Data flows and data stores are defined in terms of their data content (i.e., data elements) and their internal structure. Information such as output format, file organization, and method of implementation for processes is purposely omitted. Definitions of all DFD components are stored in a **data dictionary**. Sample data dictionary entries are shown in Figure 11-3.

Figure 11-3

Sample data dictionary entries for a data flow diagram

```
Employee-File=
    1{  Social-Security-Number + Last-Name + First-Name +
        Middle-Initial + Street-Address + City + State +
        Zip-Code + Salary + Salary-Type + Filing-Status +
        Number-of-Exemptions + Additional-State-WH +
        Additional-Deferal-WH  }n

Time-Card =
    Social-Security-Number +
    1{ Date + Start-Time + End-Time + Hours-Worked }n +
    Total-Hours-Worked + Supervisor-Initials
```

The analysis model is a basis both for describing user processing requirements and for evaluating alternative methods of implementing a system to meet those requirements. Implementation options are developed and examined after the analysis model has been created and validated. Various alternatives for automated or manual processing, form of input, form of output, file organization, and grouping of processes into application programs are considered. These high-level design decisions are the starting point for establishing a detailed system design (i.e., physical model) and a plan for implementation.

The implementation model describes the application requirements in terms of specific manual and automated procedures. This model can be represented as a data flow diagram or as a system flowchart (see Figure 11-4). Individual components of this model are further defined by textual descriptions, including program specifications (see Figure 11-5), file specifications (see Figure 11-6), and descriptions of the user interface (e.g., the format and content of screens and printed reports). The implementation model also can be represented with data flow diagrams and their associated textual descriptions. In this case, processes can be used to represent specific programs, or manual procedures. Process specifications are used to describe program logic, or algorithms, and data dictionary definitions describe physical as well as logical content.

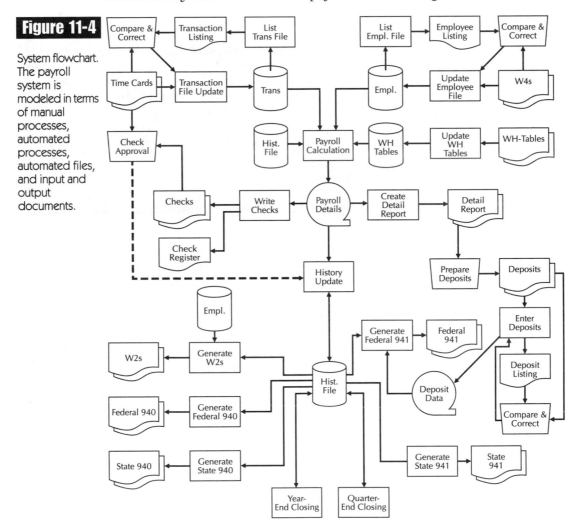

Figure 11-4

System flowchart. The payroll system is modeled in terms of manual processes, automated processes, automated files, and input and output documents.

Figure 11-5

Program
specification for
the Update-
History program in
the payroll system

```
Program Name:              Update History File
Inputs:                    Payroll Details File
Outputs:                   History File
Description:
    For each record in Payroll Details File do
        read History File record matching
            Social-Security-Number in
            Payroll Details File
        YTD-Gross-Pay = YTD-Gross-Pay + Gross-Pay
        YTD-Federal-WH = YTD-Federal-WH + Federal-WH
        YTD-State-WH = YTD-State-WH + State-WH
        YTD-SS-WH = YTD-SS-WH + SS-WH
        update History File record
```

Figure 11-6

File specification
for the employee
file in the payroll
systems

```
File Name:              Employee File
Access Method:          Indexed on Social-Security-Number
Content:                Fixed Length Employee Records

Record Definition:

    Social-Security-Number   String      999-99-9999
    Last-Name                String      X (30)
    First-Name               String      X (30)
    Middle-Initial           String      X (1)
    Street Address           String      X (60)
    City                     String      X (60)
    State                    String      X (2)
    Zip-Code                 String      99999
    Salary                   Real        99999.99
    Salary-Code              String      X
    Filing-Status            String      X (2)
    Number-Of-Exemptions     Integer     99
    Additional-State-WH      Real        9999.99
    Additional-Federal-WH    Real        9999.99
```

The entire process of structured system development is shown in Figure 11-7 on the next page, as a series of translation steps. The user's description of processing requirements is first translated into the analysis model in a process called **system analysis**. The analysis model is then translated into an implementation model in a process called **system design**. The implementation model is then translated into one or more programs. Finally, these programs are translated into executable machine instructions by program translators, and the entire translation process is thus completed.

Figure 11-7

Translation steps used in the structured development life cycle

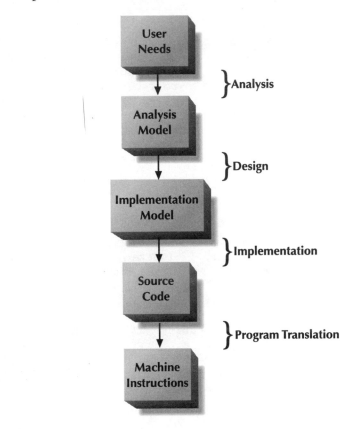

Rapid Prototyping

Structured development makes extensive use of abstract paper models. Abstract models are used to describe and validate all of the details of the system before any programs are written. The **rapid prototyping** methodology takes an opposite approach. Rather than concentrating on developing and validating an abstract model, the method focuses on developing, validating, and debugging a prototype system. Thus, the user can view and use a working model.

The steps of rapid prototyping are shown in Figure 11-8. The first step is to develop an initial set of requirements. This can be accomplished through traditional analysis tools (e.g., data flow diagrams) or by less formal interaction with the user. Often, these initial requirements are for only a subset of the full system. Once initial requirements are identified, a prototype is created as quickly as possible, which necessitates the use of powerful application development software that allows application programs to be developed rapidly (e.g., in a matter of hours or days) and modified easily. Once the prototype is created, user(s) validate it.

Figure 11-8

Process of prototype development and evaluation

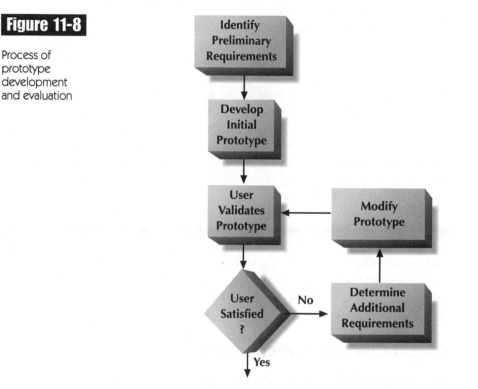

During validation, the user(s) may identify deficiencies in the prototype, possibly including missing functions, poor interface design, or any number of other problems, which are the basis for revisions to the prototype. Once revised, the prototype is again given to the user(s) for validation. These steps can be repeated many times until all requirements have been incorporated into the prototype and all problems resolved.

For relatively small systems, rapid prototyping may be the development methodology exclusively used (i.e., a "standalone" or "full life cycle" methodology). For larger systems, rapid prototyping is usually combined with various steps of other methodologies in two ways:

► Use of a prototype to define user requirements.

► Prototyping as an alternate or supplemental methodology for detailed design.

When used to define user requirements, rapid prototyping is, in essence, an alternative form of system analysis. It is especially useful in this role when users are not entirely sure of their needs or desires. In such situations, rapid prototyping provides a mechanism for adding concrete reality to abstract needs and desires. The prototype is thus used both to define and refine user requirements.

Prototyping also can be used as a design and implementation methodology. In structured development, most of the low-level details of an implementation program are specified before any software is written. Examples of these details include structure of menus, screen layouts, order of data entry and validation, and so forth. With rapid prototyping, design and development are combined through the iterative development, validation, and refinement of a prototype. The use of rapid prototyping in this manner is very common for the interactive portions of an information system.

Often, the prototype cannot be used as the actual implementation of the system, perhaps because of software or hardware incompatibilities with existing applications, inability of the prototyping tool to implement systems that can deal with anticipated processing volume, or other factors. In this case, the prototype can be considered a tangible model of the users' processing requirements. These requirements can be converted to paper, if desired, and design and implementation may then proceed by structured development methods.

Object-Oriented Development

Object-oriented (OO) system development is an extension of the **object-oriented programming (OOP)** paradigm. Structured systems development and many related methods use process and data items (flows or stores) as the basic units of modeling. Object-oriented system development uses objects as the basic unit of modeling. An object is a combination of data describing some physical or conceptual entity (e.g., a customer or a bank account) and all of the processing actions, or methods, that can be applied to that data. Structured development views data and methods as separate and distinct model components. Object-oriented development views data and methods as an integrated and inseparable whole.

The process of developing an OO system model is the process of identifying objects of interest within the system and the relationships among them. Several relationships are possible, including **message passing** relationships (e.g., a customer passing a withdrawal message to a bank account) and **superclass-subclass** relationships among objects (e.g., a checking account is a subclass of the superclass called bank accounts).

The methods by which objects and relationships among them are described is still evolving. OO design is relatively new and, as such, no one method of modeling and utilizing the model(s) has gained widespread acceptance. However, there are several conceptual overlaps between commonly used OO development methods and structured development methods. The most important similarity is the use of distinct life cycle phases for systems analysis, design, and implementation. Although the models that are the input and output of these phases differ, the basic goals and purposes of each phase are similar.

Object models are inherently designed to be implemented by OO programing languages and other construction tools. This implementation bias increases with each life cycle phase. Thus, an OO analysis model could potentially be implemented with a non-OO programming language, but OO design models are highly customized to implementation with OO development tools (e.g., OO programming languages and OO database management systems [DBMSs]).

Methodology Comparisons

Early methods of system development (i.e., those that preceded structured methods) failed to distinguish between analysis models and implementation models. The lack of a clear distinction between the analysis and implementation models made it difficult for users and system designers to discuss alternative methods of meeting users' processing requirements. A thorough examination of implementation alternatives and their associated costs and benefits was not possible because no method was provided to state system requirements independently of specific implementation methods.

Structured and OO development methods provide a clear transition point between analysis and design. Although the user and analyst may have preconceptions about implementation alternatives, the analysis process and the model used to state requirements do not convey (or require) any decisions regarding implementation. At the conclusion of the analysis phase, a clear model exists that states what the system must do with no information or bias as to how the system will be implemented. This distinction between "what" and "how" is the essential difference between a logical model and a physical model.

In structured and OO system development, analysis is followed by a thorough examination of important (high-level) design decisions made with all user requirements fully stated and understood. The costs, benefits, and trade-offs associated with each design decision can thus be made based on a complete understanding of all user needs, which is especially important with large systems due to the large number of trade-offs and dependencies between their various components.

A primary criticism of structured system development arises from its extensive use of abstract models. User needs must be assessed initially and documented as an abstract model. Development and validation of DFDs and related component descriptions consumes a substantial amount of time (between one-third and one-half of the total project schedule).

The quality of user feedback is suspect because of the abstract nature of analysis models. Consider, for example, purchasing an automobile based only on pictures, a list of features, and detailed design and performance specifications. This is similar to what the user is asked to do when structured or classical system development is used. A "test drive" is a far better method for eliciting user response and determining his or her level of satisfaction with the product. Thus, rapid prototyping is often preferred as an analysis method or full life cycle process because it provides numerous opportunities for the user to "test drive" the system.

OO development is subject to many of the same criticisms concerning the use of abstract paper models. However, OO design models directly represent implementation objects if OO development tools are used. Thus, the transition from design to implementation is easier and the use of rapid prototyping is more common with OO development than with structured development.

The reliance of older methodologies on paper models is based on a common underlying assumption—that application software is costly to develop and difficult to change once implemented. In the analysis phase, this assumption implies that all user requirements should be stated completely and precisely, which requires a substantial level of detail as well as a great deal of user involvement. The extra effort expended on analysis is supposed to ensure that the system that is designed and implemented is exactly what the user wants and needs.

In the design phase, costly development and maintenance implies that a substantial amount of time should be spent working out all possible technical and implementation problems ahead of time. The internal aspects of application programs should be specified in great detail so that the design can be evaluated in terms of ease of initial programming and future changes. With respect to both analysis and design, resources are added to planning processes so as to reduce resources consumed during implementation.

Until relatively recently, the assumption of costly software development and change was well-founded. The task of writing, testing, and debugging application programs was time-consuming and costly, even when the best of available tools were used. If development is costly, then it makes good economic sense to work out the details of the system to the greatest extent possible before programming begins, which results in savings during implementation that more than offset the cost of developing and validating paper models.

Although application software is still costly to develop and change, a number of factors have altered the trade-off between these and other costs. The most important of these factors is the availability of improved tools for programming and other application development tasks. Newer tools for application program development, such as fourth generation programming languages (4GLs), database management systems (DBMSs), and report generators have drastically reduced the time needed for programming, testing, and debugging. Costs of program development have been reduced as a result. Thus, the extra resources used for extensive planning in the structured methodologies no longer provide the same magnitude of savings during implementation, which is not to say that analysis and design are no longer needed or that programming is inexpensive. Rather, the economic balance between these activities and programming has shifted.

System Software Support for Application Development

The last three decades have seen intensive development and proliferation of automated tools to support the development of application software. The compilers, interpreters, and operating system service layers of the 1950s and 1960s have been supplemented by a wide range of tools, including:

► Intelligent program editors.
► Interactive program debuggers.
► Source code generators.
► Interactive report and screen designers.
► Greatly expanded operating system service layers.
► Database management systems (DBMSs).
► Computer assisted software engineering (CASE) tools.
► Reverse engineering tools.
► Integrated program and application development workbenches.

The role that these tools play in the application development process is summarized in Figure 11-9 on the next page. Most of these tools are described in detail later in the chapter.

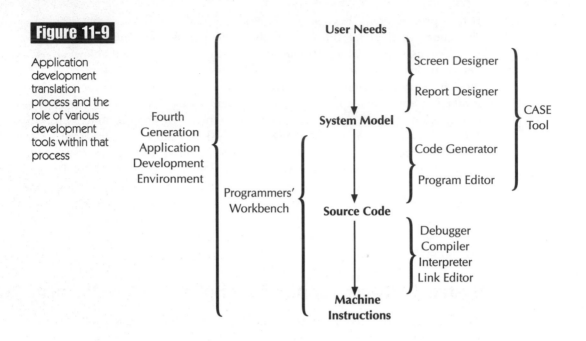

Figure 11-9

Application development translation process and the role of various development tools within that process

The declining cost of computer hardware and the increasing cost of application development labor have spurred the rapid development and widespread adoption of automated system and software development tools. Adoption has accelerated as users have demanded increasingly more powerful and sophisticated applications. The substitution of automated tools for manual methods represents a classic economic trade-off between labor (e.g., analysts, designers, and programmers) and capital resources (e.g., sophisticated application development tools and the computer hardware required to run them).

In the early days of computers, hardware was so expensive that its use could only be justified for high-value applications. The use of the computer to directly support the development of applications was extremely limited. The analysis and design tools of the typical application developer were paper, pencil, and a flowcharting template, and source code was developed by hand. Compilation (or assembly) and testing was performed in the wee hours of the morning or during other times when the computer was not busy executing application programs.

It made more economic sense to use labor-intensive processes for application development because labor was relatively cheap compared to computer hardware. Machine time was too costly to allow programmers to "waste" it executing and debugging developmental software. As hardware costs decreased, this economic balance shifted, which led to the introduction of systems software designed to support the development of application software.

Assemblers, high-level programming language compilers, and operating system service layers are the earliest examples of this shift. They are all methods by which computing hardware may be used to directly support the development (as opposed to the execution) of application programs. As the economic balance continued to shift, the number, power, and complexity of these tools continued to grow.

The proliferation of application development tools today is a continuation of this trend. New programming languages, CASE tools, and programmers' workbenches all examplify relatively recent tools for application development. Each of them directly supports application development and consumes large amounts of hardware resources while reducing the consumption of labor resources. If the decline in hardware cost relative to labor cost continues, the proliferation and extension of these tools also will persist.

COMPILATION

A **compiler** translates a program written in a programming language (e.g., C or FORTRAN). The original program (the compiler input) is called **source code**. The result of translating (or compiling) source code is called **object code**. Object code consists of CPU instructions and other components, which will be described in "Support Libraries." Source code is read from a file by the compiler and object code is written to a newly created file. By convention, the source code file is named to indicate the programming language (e.g., program.f where ".f" indicates a FORTRAN program). Object code filenames generally have a ".o" or ".obj" appended to the original root filename (e.g., program.o or program.obj).

Source code statements are read into the compiler one at a time. Different actions can be taken by the compiler, depending on the type of statement encountered. Three classes or types of source code statements are recognized:

▶ Data declarations.
▶ Data operations.
▶ Control structures.

Data Declarations

As **data declarations** are encountered in source code, memory locations are set aside to store the declared data. The amount of memory allocated depends on the type of the data (e.g., integer, real, character array, etc). The compiler builds an internal table, called a **symbol table**, to keep track of the data names, types, and assigned memory addresses.

For example, consider the following portion of a COBOL data division:

```
77      A      PIC 9 (6)
77      B      PIC 9 (6)
77      C      PIC 9 (6)
```

This example declares the existence of three working storage variables (named A, B, and C), which are each integers of six significant (decimal) digits. Upon encountering these declarations, the compiler would create an entry for each of these variables in the symbol table. Information stored in the table includes the variable name, the fact that the name refers to a data element (as opposed to a program subroutine), the type of the data element, and its memory location. The compiler would assign memory locations based on the type of the data and the number of bytes needed to store the variable. Sample entries are shown in Table 11-1.

TABLE 11-1	Name	Type	Length	Address
Sample table entries for data declarations	A	Integer	4	1000
	B	Integer	4	1004
	C	Integer	4	1008

Data Operations

As **data operations** are encountered in source code, they are translated into the sequence of machine instructions necessary to implement those operations. These instruction sequences include primitive data manipulation instructions as well as any necessary data movement instructions. When data must be moved (e.g., from memory to registers), the compiler looks up the name of the data item in the symbol table and uses the corresponding address to build appropriate instruction operands and/or data movement instructions.

For example, consider the following COBOL source code statement:

ADD A TO B GIVING C

Assume that data items A, B, and C are the same as those in the earlier COBOL data division example and that they have been assigned memory addresses 1000, 1004, and 1008, respectively. Also assume that the CPU instruction set requires operands of computation instructions to be stored in registers (e.g., as in a RISC processor). The compiler would translate the preceding statement into a sequence of machine instructions similar to the following:

```
MOV 1000 R1      ; move A to register 1

MOV 1004 R2      ; move B to register 2

IADD R1 R2 R3    ; add the contents of registers 1 and 2

                 ; store the result in register 3

MOV R3 1008      ; copy register 3 to C
```

More complex operations (e.g., implementing a complex formula) might require the generation of a series of data manipulation instructions and the movement of data between them by general purpose CPU registers. The compiler is responsible both for generation instructions to implement the formula and for keeping track of intermediate results stored in registers.

Control Structures

Control structures are programming language statements that control the execution of other language statements. Typical examples of these include unconditional branches (e.g., a GOTO statement or a subroutine call), conditional branches (e.g., an IF-THEN-ELSE statement), and loops (e.g., WHILE-DO and REPEAT-UNTIL). The common thread among all control structures is the transfer of control from one code segment to another.

To implement transfer of control, the compiler must keep track of where code segments are located within memory. Certain features of programming languages may make this task easier. For example, some programming languages (e.g., BASIC) require that every program line be numbered, whereas other programming languages (e.g., FORTRAN and C) allow the use of optional labels. Within the source code of such languages, transfer of control is represented using specific statements and line numbers or labels. For example, the BASIC statement:

GOTO 1500

directs the unconditional transfer of control to source code line number 1500.

Compilers for languages that use statement labels record the statement labels within the symbol table. As a source code line is read, an entry is made in the symbol table containing both the label and the memory address of the first machine instruction that implements the source code line. For example, when the BASIC source code line:

1500 LET Customer_Balance = Customer_Balance - 2.50

is read by the compiler, it will record the label 1500 and a memory address, such as 045F in the symbol table. When a GOTO 1500 is later read by the compiler, it will look up the label 1500 in the symbol table. Then it will generate an unconditional branch (or jump) instruction with 045F as its operand to implement the transfer of control.

A similar symbol table recording process occurs for named code segments. Most programming languages allow or require the naming of individual code segments such as **functions, subroutines**, and **procedures**[1]. These names are processed similarly to variable names when they are encountered; that is, they are added to the symbol table. The address field of the table stores the address of the first executable instruction of the function.

For example, consider the following C function:

```
float farenheit_to_celsius(float F) {

    /* convert F to a celsius temperature */

    float C;

    C = (F - 32) / 1.8;
    return(C);

}
```

When the compiler reads the first line of this code, it will record an entry in the symbol table for the symbol "farenheit_to_celsius." The address field will store the memory address of the first line of executable code within the module (the first machine instruction that compiler generates to implement the formula $C=([F-32])/1.8$). Other information is also stored, such as the function type (float) and the number and type of input parameters.

When the compiler reads a later source code line that refers to the function:

c_temperature = farenheit_to_celsius(f_temperature);

it retrieves the corresponding symbol table entry. The compiler first verifies that the types of the input variable (f_temperature) and result assignment variable (c_temperature) match the types of the function and function arguments. Then it uses the address of the function to generate code to transfer control to and from the function.

---

[1] The terms **function, subroutine**, and **procedure** all imply a code module that receives (and possibly modifies) parameters and to which control is transferred by a call/return mechanism. For the remainder of the chapter, the term **function** is used exclusively although everything said about functions also applies to procedures and subroutines.

Implementing transfer of control for a function call requires implementing mechanisms to:

► Pass input parameters to the function.

► Transfer control to the function.

► Pass output parameters back to the calling module.

► Transfer control back to the calling module at the statement immediately following the function call statement.

Implementation is complicated by the possibility of calls to the function from many different points within a program. Each of these points corresponds to a different return address and a different set of parameter sources and destinations.

The complexities of implementing function calls and returns are similar to those that the operating system faces when processing interrupts. The calling process must be suspended, the function must be executed, and the calling process must then be restored to its original state so that it can resume execution. Interrupt processing is implemented by means of PUSH and POP instructions. PUSH and POP instructions are also the means of implementing transfer of control to and from functions. A call operation is implemented as a PUSH instruction followed by a branch to the function, and the return is implemented as a POP instruction. The stack serves as a temporary holding area for the calling module so that it can be restored to its original state after the function completes execution.

Table 11-2 shows a set of symbol table entries for the function farenheit_to_celsius, its internal parameters, and the variables within the function call:

c_temperature = farenheit_to_celsius(f_temperature);

TABLE 11-2	Name	Context	Type	Address
Symbol table entries for the function, function parameters, and calling program variables	farenheit_to_celsius	global	float, executable	3000
	F	farenheit_to_celsius	float	2000
	C	farenheit_to_celsius	float	2004
	c_temperature	global	float	1000
	f_temperature	global	float	1004

The compiler reads data names used in the function call and return assignment and looks up each name in the symbol table to extract the corresponding address. It also looks up the name of the function parameter and extracts its addresses and lengths. Then the compiler creates a MOVB (copy a specified number of bytes from one memory address to another) instruction using this address and places it immediately before the PUSH instruction:

```
MOVB 1004 2000 4    ; copy farenheit temperature

PUSH                ; save calling process registers

JMP 3000            ; transfer control to function
```

The return from the function call is handled similarly:

```
MOVB 2004 1000 4    ; copy function value to caller

POP                 ; restore caller register values
```

The compiler generates code to copy the function result to a variable in the calling program. Control is then returned to the caller by executing a POP instruction.

Control structures such as loops and if-then-else statements require the compiler to generate code to evaluate a simple or compound condition. Consider the following COBOL example:

IF A IS EQUAL TO 5 THEN ADD A TO B GIVING C

The IF statement is a control structure that governs whether or not the subsequent computation is performed. It can be implemented as a conditional branch as follows:

```
0100    MOV 100C R2     ; move the constant '5' to register 2

0104    MOV 1000 R1     ; move A to register 1

0108    XOR R1 R2 R3    ; compare the contents of registers 1 and 2

010C    CJMP R3 011C    ; branch around IF clause if true

0110    MOV 1004 R2     ; move B to register 2

0114    IADD R1 R2 R3   ; add the contents of registers 1 and 2 and

                        ; store the result in register 3

0118    MOV R3 1008     ; move the contents of register 3 to C

011C                    ; next instruction past the IF statement
```

In this example, the memory location of each CPU instruction is listed in the left column. The constant 5 (assumed to be stored at 100C) and the value of A are moved from memory to registers, and an XOR instruction in executed to compare them. The result of the comparison is stored in general purpose register R3, which is tested by the CJMP (conditional jump, or branch) instruction and a branch occurs if R3 contains a nonzero, or True, value.

SUPPORT LIBRARIES

From the examples in the previous section, it might seem that object code consists entirely of executable machine instructions and data (i.e., **executable code**), but this is not the case. Certain types of language statements, including data declarations, control structures, and many computational functions, are converted directly into executable code. Other language statements, such as file manipulation and I/O operations are translated differently. The usual result of translating such statements is a call to an external subroutine. These external subroutines can be found in one of two places. The first place is a library of machine language subroutines supplied with the compiler (i.e., the **compiler library**). The second place is the operating system service layer. A request to execute either type of subroutine is referred to as a **library call**.

The use of libraries and library calls is not a necessary feature of compiler design and operation. It is possible to construct a compiler that directly translates all programming language statements into executable code. The primary advantage of using libraries is to simplify the compilation process—this makes compilers easier to write and faster to execute—and to provide a measure of portability and flexibility.

As an example, consider the use of complex mathematical functions in a source language program (e.g., trigonometric functions such as sine and cosine, statistical functions such as factorial and standard deviation, and fractional exponents such as $10^{-2.34}$). Few CPU architectures provide single instructions to implement such functions. Thus, application programs that use these functions must implement them as complicated sequences of simpler computation instructions.

It is common to include support for complex mathematical functions within a separate compiler library and to translate program statements that use these functions into calls to these library subroutines. This approach increases compiler simplicity and flexibility for the following reasons:

► The complex translation of these functions into simpler computational operations is not implemented within the compiler. The size and complexity of the compiler is thus reduced, and the speed of compilation increases.

► Alternative libraries can be provided for the same functions. When implementing complex mathematical functions, there is often a trade-off between computational accuracy and speed. Two separate libraries can be provided— one that uses slow but highly accurate algorithms and another that uses fast but less accurate algorithms. The programmer can choose which library best suits the requirements of the application.

▶ Alternative libraries can be provided for different CPU architectures. For example, one library can be provided that only uses simple computation instructions and another can be used to take advantage of more complex instructions (e.g., the limited set of trigonometric functions implemented as Pentium Pro instructions). The programmer would select the library that corresponds to the CPU architecture of the computer system on which the application program will execute. This approach allows compiler writers to write a single compiler that allows for some variation in CPU architecture.

When a program statement is encountered that requires the use of a library subroutine, the compiler inserts a call that includes the name of the library routine as well as the addresses of any data that must be passed between the program and the routine in the object code. For example, consider the COBOL statement:

ADD A TO SINE(B) GIVING C

Assume that SINE() is a trigonometric function. A call, such as the following, would be inserted into the object code:

call sine(1004,R3)

The name sine refers to a library routine, and the numbers between the parentheses are the memory addresses of the operand and result. Unlike calls to internally defined subroutines, the compiler will not attempt to translate this call into executable instructions.

Linking

Note that the preceding library call is not a CPU instruction. Rather, it is a reference to a set of executable code that has been previously compiled and stored in a library for later use. Therefore, it must be replaced by the corresponding library code before the application can be executed. This replacement is performed by a program called a **linker**, or **link editor**. The process of replacement is called **linking**, **link editing**, or **binding**. Figure 11-10 shows how a compiler and link editor are used to produce executable code.

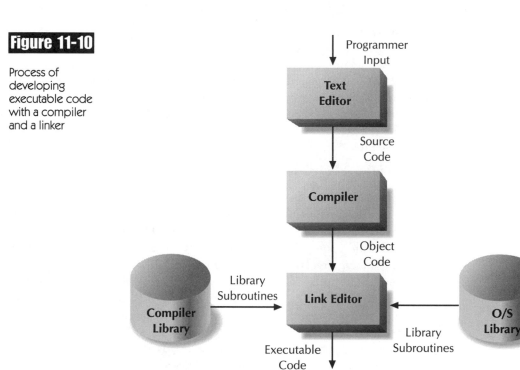

Figure 11-10

Process of developing executable code with a compiler and a linker

Programmer Input

Text Editor

Source Code

Compiler

Object Code

Compiler Library

Library Subroutines

Link Editor

Library Subroutines

O/S Library

Executable Code

The function of the linker is to combine separate sets of object code into a single executable program. Each object code module contains executable code and can contain calls to other object code modules. The call to the library routine sine() shown above is an example of such a call. In the terminology of linking, it is known as an **unresolved reference** and is a request for a subroutine or for data by name where that name is unknown within the symbol table of the associated object code. The process of locating the reference and connecting it to another object code module is called **resolving the reference**.

The linker searches object code modules for unresolved references. When an unresolved reference is found, the linker searches an index of one or more libraries for the names of those references. If it finds those names in a library, it extracts the corresponding executable code and/or data from the library and inserts (links) it to the object code. The routine can be inserted by physically placing it among the existing machine instructions (often referred to as an **in-line function call**). The routine also can be inserted by appending the library routine executable code to the object code, adding PUSH, branch, and POP instructions for the subroutine call and return, and adding data movement instructions to implement parameter passing and function return values.

If all of the unresolved references are resolved by the linker, then the resulting program consists entirely of executable code and data. This program then can be loaded into memory and executed. If some references are left unresolved, then the linker generates appropriate error messages and doesn't produce a file of executable code. Instead, it produces a new object code module containing the original object code modules integrated with the library modules that could be found.

Dynamic and Static Linking

The linking process described thus far is often called **static linking**, or **early binding**. The term **static** is used because library and other subroutines cannot be changed once they are inserted into object code. The term **early** denotes the relative point within the program development cycle at which binding occurs.

Dynamic linking (or late binding) is linking that is performed during program loading or execution. Calls to operating system service routines are usually implemented with dynamic linking. The compiler or link editor inserts code to call a predetermined interrupt handler and passes pointers to the service routine name and parameters in general purpose registers. The interrupt handler locates the service routine by name, loads it into memory (if it isn't already there), and executes it. When the service routine terminates, control is passed back to the calling program.

Notice that the use of dynamic linking implies that all operating system service routines are available at the time of execution. One way to ensure their availability is to load them all into memory when the operating system is started. Unfortunately, this consumes a great deal of primary storage, which is often wasted holding subroutines that are seldom or never executed. A more common approach is to hold service layer subroutines on disk until they are needed. When a service layer function is called, the operating system checks to see if it is already in memory. If the subroutine isn't in memory, then the operating system loads it from disk before transferring control. The operating system typically keeps "high demand" subroutines in memory if enough is available. "Low demand" subroutines are overwritten at first opportunity.

Dynamic linking provides two key advantages as compared to static linking. The first is a reduction in the size of application program executable files. Static linking incorporates copies of service layer subroutines directly into executable programs. Depending on the nature and number of subroutines included, this can add up to several million bytes to each program file. Much of this additional storage is redundant because commonly used subroutines (e.g., those to open and close files or windows) are used by many application programs. With static linking, these routines are redundantly stored hundreds or thousands of times (e.g., once per application program). Dynamic linking avoids this inefficient use of memory.

The second advantage of dynamic linking is flexibility. Service layer subroutines are often updated in new operating system versions. If application programs use dynamic linking, then updating the operating system effectively updates those application programs with the new service layer subroutines. The new subroutines are loaded and executed the next time an application program is executed after an upgrade. Static linking locks the application program to a particular version of the service layer subroutines. These routines can be updated only by providing a new link editor library and relinking the application object code module.

The primary advantage to static linking is speed of execution. When a dynamically linked program calls a service layer subroutine, the operating system incurs overhead to find the requested subroutine and transfer control to it. If the subroutine is not in memory, there is also a delay while the executable code is loaded. Neither the search overhead nor the loading delay are incurred if the application program is statically linked, which makes those applications execute more quickly than their dynamically linked counterparts.

Static linking also improves the reliability and predictability of executable programs. A program is developed and tested with a specific set (or version) of operating system service routines. Updates to those service routines are designed to be backward-compatible with older versions, but there is always the risk that new routines dynamically linked into an old program will result in unexpected, or "buggy," behavior. Static linking ensures that this cannot happen. However, the reverse situation also can occur. That is, a new library version may fix bugs that were present in an older version. Static linking prevents these fixes from being automatically incorporated into existing programs.

INTERPRETERS

Compilation is performed on a source code file at once. Link editing and program execution cannot occur until the entire source code file has been compiled. In contrast, **interpretation** interleaves source code translation and execution. An **interpreter** reads a single statement of the source code, translates into machine instructions, and immediately executes those instructions before the next program statement is read. If a source code statement corresponds to a library subroutine, then the corresponding executable code is dynamically linked and immediately executed. In essence, one program (the interpreter) translates and executes another (the source code) one statement at a time.

The primary advantage of interpretation over compilation is the flexibility—a direct result of dynamic linking—to incorporate new or updated code into an application program. Dynamic linking can be used for operating system services and for other program code. Program behavior can be updated easily by installing new versions of dynamically linked code. It also is possible for programs to modify themselves directly. This flexibility is required for certain types of applications (e.g., expert systems and some distributed transaction processing systems).

Resource Utilization

The primary disadvantage of interpretation as compared to compilation/linking is increased consumption of memory and CPU resources during program execution (see Table 11-3). With compilation and link editing, each program used to develop executable code resides in memory for only a limited time. When a program is loaded and executed, all of the software components that were used to create it (i.e., the text editor, compiler, and link editor) are no longer in memory. Thus, the memory requirements of a compiled and linked application program consist only of the memory required to store its corresponding executable code.

TABLE 11-3	Resource	Interpretation	Compilation
Summarized memory and CPU resources consumed during the execution of an application program	Memory Contents (during execution):		
	Interpreter or Compiler	Yes	No
	Source Code	Partial	No
	Executable Code	Yes	Yes
	CPU Instructions (during execution):		
	Translation Operations	Yes	No
	Library Linking	Yes	No
	Application Program	Yes	Yes

An interpreter is a relatively large and complex program that must occupy memory during the entire execution of a source code program. Essentially, there are two programs in memory at run-time—the interpreter itself and the application program it is translating and executing. Thus, the memory requirements of an interpreted program are substantially greater than those of a compiled program.

For compiled and linked programs, the CPU resources needed for translation and library code insertion are consumed before the program is executed (i.e., at compile-time). At run-time, a compiled and linked program only consumes CPU resources to execute its own instructions. An interpreted program consumes CPU resources for translation, linking, and execution at run-time. Thus, the elapsed time (i.e., wall clock time) required to execute an interpreted program is greater than the time required to execute a compiled and linked program. This is a critical performance difference for CPU-intensive applications (e.g., numerical modeling) and for online transaction processing.

Increased CPU resource consumption also results from executing an interpreted program multiple times. Source code is translated and linked every time the program is executed. For example, consider an application program that is run every day (e.g., a daily batch reporting program). If the program is used for a year without changes, then interpreted execution will translate the entire program 365 times. Compilation and link editing are performed only once, thus saving the computer resources necessary to perform the other 364 translations.

SYMBOLIC DEBUGGING

Once a program has been compiled and linked, the relationship between source and executable code is difficult to establish. Variable names are replaced by memory addresses and symbolic source code instructions are replaced by CPU instructions. Run-time errors (e.g., divide by zero and memory protection fault) generate interrupts that transfer control to an operating system error reporting subroutine, but this subroutine can only report errors in terms of the memory addresses and data values contained within registers at the time the error occurred (e.g., the contents of the instruction pointer and general purpose registers). Determining the source code lines and variables that correspond to these memory addresses is difficult.

A compiler can be instructed to incorporate its symbol table into object code. A link editor can be instructed to produce a **memory map** (or **link map**). A memory map lists the memory location of every object (code segment or data item) after all object code modules have been linked, and a programmer can use a memory map to trace error messages containing memory addresses to corresponding program statements and variables. However, tracing those addresses to particular source code statements or data items requires a detailed memory map and a thorough understanding of machine code and of the compiling and linking processes.

A **symbolic debugger** is an automated tool for testing and debugging compiled and linked programs. It provides a number of features, including:

▶ The ability to trace calls to specific source code statements or subroutines.

▶ The ability to trace changes to variable contents.

▶ The ability to detect run-time errors and report them to the programmer in terms of specific source code statements.

A symbolic debugger uses the symbol table, memory map, and source code files to trace memory addresses to specific source code statements and variables. It also requires debugging code to be generated by the compiler and/or linked in by the editor. This code implements checkpoints within the program that allow execution to be halted and variable contents to be displayed. An executable program containing symbol table entries and debugging checkpoints is sometimes called a **debugging version**. In contrast, the **production version** (or **distribution version**) of a program omits the symbol table and debugging checkpoints to reduce a program's size and/or increase its execution speed.

Interpreted programs have an inherent advantage over compiled programs in detecting and reporting run-time errors because the contents of the symbol table and program source code are always available to the interpreter at run-time. If an error occurs, it can be reported in terms of the most recent source code line translated. Memory addresses also can be converted to source code names by looking them up in the symbol table. Most interpreters directly incorporate symbolic debugging capabilities (i.e., they don't need a standalone symbolic debugging program).

Technology Focus

Java

Java is an object-oriented programming language and program execution environment. It was developed by SUN Microsystems during the early and mid 1990s. It first debuted in Netscape's Navigator Web browser (version 3.0) and, as a result, has been widely described as a "Web" or "network" programming environment. Although it satisfies that description, Java is much more. It is a programming environment that supports execution of its programs on virtually any combination of hardware platform and operating system. This characteristic makes it unique among modern programming environments and accounts for much of its rapid emergence as a programming language of choice.

The Java programming language shares many similarities with C++, including both syntax and capability. The syntactic similarity to C++ is purposeful and allows programmers familiar with C++ to be quickly trained in Java. Similarities in capability arise from the common implementation of object-oriented concepts in both languages. Java incorporates support for object-oriented features, including encapsulation, inheritance, and polymorphism (multiple inheritance is not supported). Java is designed to maximize reliability of applications and reusability of existing code.

A unique feature of Java is the standardized target language for Java interpreters and compilers. Compilers and interpreters for other programming languages translate source code into executable CPU instructions and library calls to the service layer of a specific operating system. A Java compiler or interpreter translates a Java program into executable instructions and service layers calls for a hypothetical computer system and operating system. That hypothetical execution environment is called the **Java virtual machine (JVM)**. Instructions and library calls to the JVM are called Java byte codes.

The JVM can be simulated with the hardware and operating system of a conventional machine (see Figure 11-11). CPU instructions to the JVM are translated into CPU instructions for the actual CPU (e.g., an Intel Pentium or SUN SPARC). Calls to the JVM service layer are translated into equivalent calls to the installed operating system (e.g., Microsoft Windows or UNIX). The JVM, actually a software program that performs these translation (emulation) functions, must be installed prior to executing Java programs.

Figure 11-11

Interactions among a Java application program, the Java virtual machine (JVM), the native operating system, and the CPU

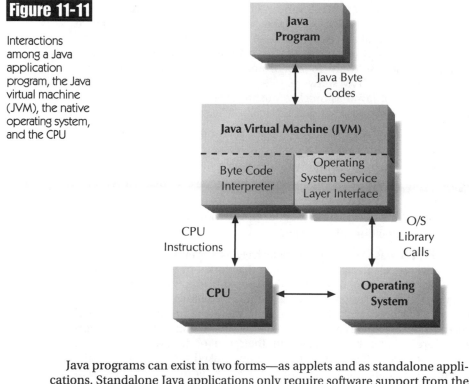

Java programs can exist in two forms—as applets and as standalone applications. Standalone Java applications only require software support from the JVM. All security considerations for standalone applications are handled within the JVM. A Java **applet** is designed to run within another program. Typically, they run within a Web browser program such as Netscape Navigator or Microsoft Internet Explorer. Applets run within a protected area of the Web browser called the **sandbox**, which implements extensive security controls to prevent "rogue" applets from accessing unauthorized resources or damaging the hardware, operating system, or file system.

SUN Microsystems has developed hardware and operating system components that directly implement the JVM. The first major step was taken in early 1997 with the development of a Java microprocessor. This CPU accepts Java byte codes as its native instruction set and is highly optimized to their execution. The initial target market for this chip consists of embedded applications such as device control for televisions, VCRs, and other appliances. Versions of the chip suitable for use in general purpose computers have been developed, but it remains to be seen whether they will see wide use. Existing CPUs (e.g., the Pentium, Alpha, and PowerPC) are well able to handle the demands of executing interpreted Java byte codes.

Native support for Java OS calls has been added to the SUN Solaris operating system (a highly customized version of UNIX). Other operating system vendors are expected to follow suit. Incorporating the JVM into the native operating system will likely reduce some of the inefficiency inherent in the current O/S translation processes. It also will minimize the need for Java-enabled application software to install and use private (and sometimes incompatible) copies of JVM software.

The popularity of Java has quickly mushroomed, thanks to the intersection of a number of factors, including:

- ► SUN's strategy of providing Java compilers and virtual machines at little or no cost.
- ► Incorporation of JVMs into Web browsers.
- ► Ability of Java to substantially expand the scope of applications that can be built using a Web browser as the primary I/O device.
- ► Ability of Java programs to execute on any combination of computer hardware and operating system.

Although Java was not designed exclusively as a Web programming language, it quickly and effectively penetrated the market by using the popularity of the Web as a stepping stone. This exposed it to a wide audience that then came to know and appreciate Java's other capabilities.

Java has a few drawbacks. The first is reduced execution speed because of the use of interpreted byte codes and operating system translation. Emulation processes are inherently inefficient in their use of hardware resources. "Native applications" (i.e., those compiled and linked for a particular CPU and operating system) typically execute 10 times faster than interpreted Java byte code programs. Some JVMs employ a just-in-time compiler to reduce the speed difference to approximately 3:1, but Java programs will always be slower than "native" programs when executed on anything other than a "real" Java machine. This is an important consideration for programs that make maximal demands of available hardware resources (e.g., computer aided design [CAD] applications, simulation applications, and database servers).

The security model of Java is also troublesome. Early versions of Java made several restrictions on the behavior of programs to enhance security, including a complete lack of access to the native file system and the inability of a program to read or write arbitrary memory locations. These restrictions enhanced security at the expense of program functionality and led to early perceptions of Java as a "toy" programming language that was only capable of implementing "cute" animated images in Web pages. Current Java standards have relaxed these restrictions and, as a result, have endowed the language with the capability to implement "industrial strength" applications, but this has created additional security problems for which complete solutions have yet to be developed.

PROGRAMMING LANGUAGES

The earliest shift of computer hardware resources to a direct support role in application development came with the development of programming languages. Because any programming language other than machine language must be translated, the use of a programming language requires the use of an automated program translator. Any hardware resources consumed by a program translator represent capital resources shifted from executing existing applications to developing new applications.

The history of programming language development has been driven largely by a desire to make application programs easier and easier to develop (i.e., less and less labor-intensive). To achieve this goal, it is necessary to substitute the use of software and computer hardware for labor in the translation from implementation model to executable code. The exact distribution of effort, and thus resources, between the programmer and the program translator depends on characteristics of the programming language and the support tools used for program development.

Instruction Explosion

A programming language is translated by an interpreter or compiler into many machine instructions. Depending on the type of statement and the programming language, a source code statement may be translated into one to thousands of machine instructions. This one-to-many relationship between programming language statements and the machine instructions needed to implement them is called **instruction explosion**.

Different programming languages have different degrees of instruction explosion. An older programming language such as C or FORTRAN has a relatively low degree of instruction explosion. This degree of explosion can be described as a ratio of source language statements to machine instructions. For example, if 1,000 machine instructions result from translating a 20-statement FORTRAN program, then the degree of instruction explosion can be stated as the ratio 50:1.

However, such a ratio is only an average. Within the same programming language, different types of statements can have vastly different degrees of instruction explosion. In general, statements that describe mathematical computation have relatively low instruction explosion, whereas statements that describe I/O operations have relatively high instruction explosion.

The degree of instruction explosion in a programming language has a direct effect on the distribution of translation effort between the human programmer and the compiler or interpreter used to translate the program. The total amount of effort required to perform the translation is constant. For a given application processing requirement and a particular machine instruction set, a fixed amount of translation effort is required. However, programming languages vary in how they distribute this effort between a human programmer and compiler or interpreter.

Figure 11-12 shows two programs that perform similar functions—one in C and the other in SQL. Notice that the number of program statements is much larger in the C program, which indicates lower instruction explosion. This implies a relatively high degree of effort on the part of a human programmer and a relatively small degree of effort for the program compiler or interpreter.

Figure 11-12

Equivalent programs in SQL (a programming language with relatively high instruction explosion) and C (a programming language with relatively low instruction explosion)

```
          Structured Query Language Example
open database banking;
select
    customer.acct_num, customer.name,balance+sum(transaction.
    amount)
from         customer,transaction
where        customer.acct_num=transaction.acct_num
group by   customer.acct_num,customer.name;


        C Example
balance_report () {

FILE  *cust_file, *trans_file;
int   status,acct_num,a_num,balance,amount;
char name [256];

cust_file=fopen ("customer","r";
trans_file=fopen ("transaction","r");
status=scanf (cust_file, "%d%s%d\n", &acct_num,name, &balance);
while (status! =EOF{
status=scanf (trans_file, "%d%d\n," &a_num, &amount);
while (status != EOF){
    if (acct_num == a_num) {
      balance+=amount;
    }
    status=scanf (trans_file,"%d%d\n", &a_num,&amount);
}
printf("%d  %s  %d",acct_num,name,balance) ;
trans_file=freopen ("transaction","r") ;
   status=scanf (cust_file,"%d%s%d\n", &acct_num,name,&balance);
}
close (trans_file) ;
close (cust_file) ;
exit (0) ;
} /* end balance_report */
```

The situation is reversed in the SQL program. There, the program has only a few statements and is far easier for a human to write. However, the translation effort of a compiler or interpreter would be substantially greater for the SQL program as compared to the C program. For example, the C program states an explicit procedure for searching the files and matching records based on account number. The SQL program merely states the record matching criteria (i.e., customer.acct_num equal to transaction.acct_num). The compiler or interpreter must determine the correct procedure to implement record selection and generate appropriate machine instructions and/or library calls to do so.

The division of translation effort and its relationship of instruction explosion is represented in Figure 11-13. If a programming language has a low degree of instruction explosion, then considerable human effort is required to program in that language because it takes a relatively large number of source language statements to state processing instructions. It also implies a lower level of detail in those statements and more opportunity for programmer error. The human programmer assumes more responsibility for the translation effort and the work required by the program translator is correspondingly reduced.

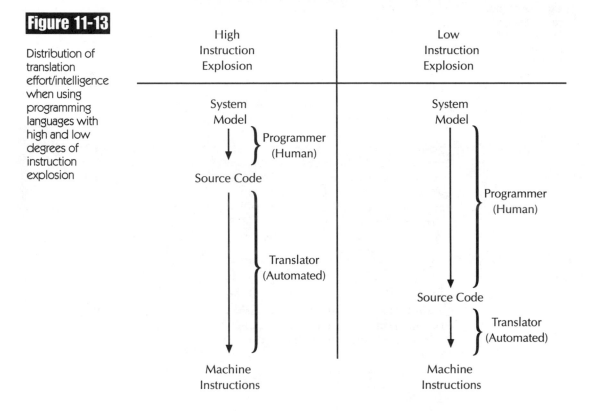

Figure 11-13

Distribution of translation effort/intelligence when using programming languages with high and low degrees of instruction explosion

Programming languages that exhibit significant degrees of instruction explosion are generically referred to as **high-level programming languages**. High-level programming languages are used to develop both application and system software. Rather than requiring the programmer to specify primitive machine actions, these languages provide the programmer with a set of language constructs (or statements) that each implement a sequence of primitive machine actions. As a result, a programmer needs to write substantially fewer program instructions than would be necessary if the same function were programmed using assembly or machine language. High-level languages also insulate the programmer from much of the physical detail of the hardware (e.g., the specific CPU instruction set and the methods used by the operating system to interact with storage and I/O devices).

ANSI Standard Programming Languages

Many high-level programming languages are standardized by the American National Standards Institute (ANSI) and other international standards agencies. Examples of ANSI languages include FORTRAN, COBOL, C, and C++. Standard behavior is defined by the behavior of the language compiler or interpreter and the behavior of the machine language programs produced. Developers of compilers and interpreters submit their programs to ANSI for certification.

Test programs are part of the standard. These test programs are input to a compiler or interpreter. Some test programs have syntax or other errors that must be detected by the compiler or interpreter. Correct test programs must be recognized as such. The object code of correct test programs is linked, and the resulting machine code is executed. The behavior of the program is compared to the behavior defined by the standard to determine whether the compiler or interpreter complies with the standard.

The use of ANSI-standard programming languages provides guaranteed portability of source code among operating systems and application programs. An application program can be moved to a new environment (operating system and/or CPU) by recompiling and relinking it with an ANSI-certified compiler for the new environment. The primary impediments to such portability are lack of a compiler in the new environment and/or the use of nonstandard language features within the program. Lack of a compiler isn't usually a problem because operating system and hardware vendors have strong market inducements to provide them. Few people will purchase a computer system or operating system that doesn't provide a compiler for the commonly used ANSI-standard programming languages.

The ANSI standards specify a set of test programs and required run-time behavior of those programs. However, they do not prohibit the implementation of additional language constructs and capabilities. For example, the C programming language contains no primitive data type for strings, but a compiler writer is free to define a string data type and implement processing operations on that data type. Neither the data type nor its associated processing operations exist in any of the ANSI test programs because they aren't part of the defined standard. Thus, the compiler can be certified as ANSI standard even though it recognizes language constructs that aren't part of the standard.

Many compiler vendors develop such extensions, which are often useful tools for overcoming limitations in the language and making application program development easier. Thus, the extensions are often incorporated into source code programs by application programmers. The use of such extensions removes the guarantee of portability inherent in the use of an ANSI standard language. Each compiler defines its own extensions and the processing behavior of programs that use them. There is no guarantee that those extensions will result in equivalent behavior—or that they will even exist—in another compiler.

Third Generation Programming Languages

Many high-level programming languages—for example, COBOL, FORTRAN, BASIC, C, and Pascal—are called **third generation programming languages (3GLs)**. These languages were developed between the late 1950s and the early 1970s to overcome the limitations of assembly and machine language programming. They also were the first programming languages to exhibit any significant degree of instruction explosion. The term **third generation** arises from considering CPU instructions sets (machine language) and assembly languages to be the first and second generations of programming languages.

The limitations of computer hardware capability during the late 1950s, 1960s, and early 1970s are reflected in the basic capabilities (or lack thereof) in these languages. In many cases, the rapid advance of hardware capability and software development tools has left serious gaps between language capabilities and the expected capabilities of a modern application program. COBOL, for example, was developed in the late 1950s. At that time, secondary storage devices were primitive and expensive. The computing power and technology needed to support interactive user interfaces was virtually nonexistent. Most application software systems consisted of batch programs that used cards or tapes for input and produced punched cards, tapes, or printed output.

The developers of COBOL were forward thinking in many respects, especially with regard to file manipulation. They foresaw the increased use of large disk files in information processing applications and defined many capabilities in the language to manipulate these files. For example, the ability to define and manipulate indexed file structures was part of the original language specification. They did not, however, endow the language with capabilities for database processing, graphical user interfaces (GUIs), or distributed computing. Although later adaptations of the language did address some of these issues, they were an incomplete solution.

In general, 3GLs suffer from serious design limitations with respect to modern hardware capabilities and modern requirements in information-processing applications. Two areas are of particular note—mass storage management (or manipulation) and interactive I/O capabilities. The traditional 3GLs have failed to evolve sufficiently in these areas to meet the needs of modern information processing applications.

Until the 1970s, the majority of applications were designed as single programs or as a relatively small set of integrated programs. These systems operated on a common set of files, and the responsibility for those files was localized to the department or organizational unit that owned the application. As the number of automated applications grew, the amount of overlap in data use also grew. It became common for files owned by one organizational unit (e.g., marketing) to contain data also stored within files owned by another organizational unit (e.g., manufacturing). Ensuring consistency of multiple stores of data and providing access to them across organizational and application boundaries was a difficult problem.

Database management systems (DBMSs) were developed partly to address issues of overlapping data requirements among application programs and software systems. The primary purpose of a database management system is to provide a common (integrated) repository for an organization's data. A DBMS allows data to be shared by many different application programs using a common interface and permits data to be more effectively managed and controlled.

To provide a basis for implementing application software, a DBMS must provide a set of tools that can be used to develop application programs. When DBMSs were introduced, none of the available programming languages had the capability to manage and interact with a database; instead, their capabilities were directed toward individual files.

Although database manipulation functions could have been implemented by extending the capabilities of programming languages, they were implemented as separate, or add-on, software components. These were usually composed of general purpose database definition facilities, data manipulation facilities, and interactive query facilities. The data manipulation facilities were implemented as libraries of subroutines that could be linked into application programs written in 3GLs. The other two components were implemented as standalone programs or systems. The entire set of facilities was developed and marketed as a package. Early examples of DBMSs include DB2 and IDMS, and modern examples include Oracle, Sybase, Microsoft Access, and Microsoft SQL Server.

There are advantages and disadvantages to extending programming language capabilities with add-on software components. One of the primary advantages is that the add-on software can be used with many different programming languages. With DBMSs, a single set of data manipulation library routines can be used with the compilers for many different languages. Variations among programming languages in data structure definitions must be accounted for, and a mechanism must be provided for moving data between program data structures and the add-on software package.

The primary disadvantage of using add-on software is a lack of standardization. DBMS enhancements to a 3GL are like any other language feature that isn't part of the ANSI standard. They are not guaranteed to exist or to produce equivalent behavior on other ANSI-certified compilers and interpreters. Thus, programmers who use add-on DBMS capabilities sacrifice the portability of the software they develop.

Programming language limitations similar to those described for database manipulation also exist for user interfaces and interaction with I/O devices. Older 3GLs (e.g., Fortran) were designed to accept batch input as streams or blocks of individual characters from devices such as tape drives and card readers. Similarly, these languages were designed to use character-based output to printers and other devices.

Newer 3GLs (e.g., BASIC and C) were designed with some facilities for interactive I/O using video display terminals (VDTs). However, these languages still tend to process both input and output as streams or blocks of characters. Thus, they interact with a keyboard in much the same manner as a tape drive and interact with a video display unit in much the same fashion as a printer. These similarities are also exploited by older 3GLs.

This approach to interactive I/O ignores some of the unique capabilities of modern I/O devices. Full-screen I/O, for example, is poorly handled if at all. The concept of cursor motion (e.g., using a TAB key to move from one input field to the next) is difficult to implement with character-based I/O. Window-based display and graphical user interfaces (GUIs) cannot be implemented in most 3GLs. There are no language constructs to describe user interface features such as window frames, pop-up menus, tool bars, fonts, and mouse movement and button clicking.

The approach to addressing the I/O limitations of older programming languages has been much the same as the approach to implementing advanced data manipulation. Rather than extend the I/O capabilities of these programming languages—or develop entirely new languages—add-on software components and language extensions have been designed to add these capabilities to existing languages. Examples of these include ADDS-ON-LINE, CICS, and modern object libraries such as the Microsoft Foundation Classes. These packages provide a library of I/O subroutines that can be linked into object code. Most include a standalone user interface definition program. As with standalone DBMSs, portability of application programs is sacrificed to gain increased functionality.

As add-on tools for database and I/O functions grew in popularity the complexity of application programming increased substantially. Some of this complexity was unavoidable because of the additional demands placed on the application programs. Some of it, however, was a direct result of the number of different tools used to develop application programs. Increased complexity also resulted from the interfaces between 3GL code, standalone tools, subroutine libraries, and operating system service layer subroutines. Portability became a distant memory because of the number of nonstandardized tools and software components that had to be available in each new operating environment.

FOURTH GENERATION APPLICATION DEVELOPMENT

The practice of software development limped through the 1970s and into the 1980s because of the shortcomings of programming languages and software development tools. Several trends and technology advances came together in the 1980s and 1990s to finally offer some solutions to the problems, including:

▶ The creation or enhancement of programming languages with capabilities for sophisticated data manipulation and user interface capabilities.

▶ The rapid proliferation of personal computers (PCs) and workstations.

▶ The development and widespread adoption of integrated sets of software development tools designed to be used on PCs and workstations.

▶ The development and widespread adoption of the object-oriented programming paradigm.

▶ The widespread adoption of "life cycle" system development models such as structured systems development.

▶ The development and widespread use of automated tools based on life cycle system development models.

These trends and advances were by no means coordinated, but in fact, they were often in conflict (e.g., structured system development and object-oriented development). But they did coalesce into a relatively well-defined approach to software development that leveraged human programming effort with improved languages, improved tools, improved computer hardware, and a model driven methodology to help organize all of the pieces and processes of software development.

There is no generally accepted term to describe the current state of software development. For the remainder of this chapter, it is referred to as a **fourth generation application development (4GAD)**. This term is not widely recognized and alternate definitions and terms are widespread, but it will suffice for the purposes of this text.

The key components of a 4GAD environment are as follows:

▶ A programming language capable of manipulating databases, implementing GUIs, interacting with modern peripheral hardware, and promoting the reuse of existing source code.

▶ An integrated set of automated tools for program development and testing (e.g., editors, code generators, compilers, and debuggers).

▶ A guiding methodology that defines a process of system development and one or more methods for modeling the various components of a complete information system.

▶ A set of automated tools based on the modeling and development methodology.

There is an abundance of software packages and automated tools that fit within one or more of these categories. Some of the more common ones are described in the remainder of this chapter.

Software packages that provide the first two components are called **programmers' workbenches**; examples include Microsoft Visual BASIC, Borland C++, and Oracle Power Objects. Software packages that provide the last two components are commonly called **computer assisted software engineering (CASE)** tools; examples include Oracle Designer/Developer and Application Development Workbench (ADW).

The term **CASE tool** is a bit of a misnomer because nearly any software covering any of the four categories fits within the literal definition of CASE, which has led to widespread confusion and widely varying usage of the term. Some vendors provide a complete 4GAD environment by integrating a programmers' workbench with a CASE tool (e.g., a combination of Oracle Designer/Developer and Power Objects). These are also called CASE tools.

Fourth Generation Programming Languages

The term **fourth generation programming language (4GL)** describes any programming language that addresses most or all of the shortcomings of 3GLs. Thus, the following features typify a 4GL:

▶ High instruction explosion.

▶ Native capability to implement interactive GUIs.

▶ Native capability to interact with recent I/O devices (e.g., laser printers and scanners).

▶ Native capability to interact with one or more databases.

▶ Native ability to interact with resources through a network.

4GLs vary widely in the number and degree to which they possess these capabilities. More troublesome is the lack of ANSI standardization for 4GLs. Most 4GLs are proprietary languages developed by operating system vendors (e.g., Visual BASIC) or DBMS vendors (e.g., Oracle Forms). A few are based on ANSI standard 3GLs with a large number of nonstandard add-on components and language extensions.

With one or two exceptions (e.g., Visual BASIC), 4GLs have largely faded from the software-development scene in recent years and have been replaced by a new class of languages based on a newer programming paradigm. However, these newer languages usually provide all of the features listed above and are, thus, sometimes identified as 4GLs.

Object-Oriented Programming Languages

Traditional 3GLs and 4GLs utilize a similar underlying programming paradigm. Under this paradigm, data and the procedures by which data is manipulated are two distinct things. Data is viewed as a passive entity. It is an object (or collection of objects) to be passed from place to place and processor to processor. Data can be acted upon in these places and/or by the processors before it is moved to the next place or processor. Data is static except when a program manipulates it to change its content or form.

Programs, considered the dynamic element of a system, exist to manipulate data and move it from source to destination. Data is processed by passing it among code modules, each of which performs one well-defined manipulation function. A data flow diagram is a direct representation of the program/data dichotomy. Data flows from place to place (represented as named arrows) and is transformed by processes (represented as circles or bubbles). Data definitions are static statements of format and content. Process specifications describe dynamic actions by which input data is transformed into output data.

Software researchers in the late 1970s and early 1980s began to question the efficacy of traditional views of data and programs. In particular, they questioned the expediency of redundant definitions of data form and content within every processing module that might manipulate the data. These redundant definitions were a major problem when maintaining software systems because a single change to a data definition might necessitate changes to dozens or hundreds of software modules.

Problems were also increasingly encountered with access to data by networks. Transporting data from site to site was often found to be less than equivalent to transporting information from site to site. Without detailed knowledge of data format and content, it was often impossible to extract meaningful information from the data. The problem was that knowledge of format and content was not stored as part of the data. It existed only in implicit form within the programs that manipulated the data, yet these programs were not transportable because of dependencies on hardware, operating system, and other specifics of local processing environments.

A new programming paradigm called object-oriented programming (OOP) was developed to address these and other problems. The fundamental difference between OOP and the traditional paradigm is that OOP views data and programs as two parts of an integrated whole, which is called an **object**. Objects consist of data and **methods** (programs or procedures that manipulate the data). Objects are said to encapsulate data because external programs cannot manipulate object data directly; rather, they must use the object's methods to do so.

Unlike data, objects are not passed from program to program. Rather, they reside in a specific location and wait for **messages**—requests for a specific method or methods to be executed—to be sent to them. The results of executing a method are sent to the requestor as a response. Object data may be sent as part of a response, but data sent in this fashion is only a copy. Manipulation of original object data can only be performed by object methods. Figure 11-14 shows the relationship among data, methods, messages, and responses.

Figure 11-14

Objects encapsulate data within methods. Objects receive messages requesting method execution and send the results of executing those methods as responses.

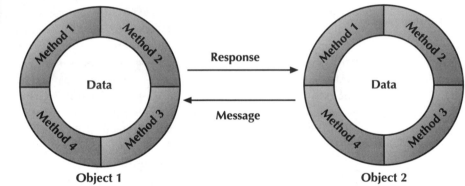

The first widely used OOP language was SmallTalk. It is an excellent programming language but has been perceived by many to be difficult to learn due to its radical departure from more traditional programming languages. To provide a more familiar programming environment, an extension to the C programming language was defined that incorporated most OOP concepts. The extended language is called C++ and exists as an ANSI standard. Other languages have been extended with OOP concepts (e.g., Visual BASIC) and a large number of proprietary OOP languages have been developed (e.g., Oracle Power Objects).

Applications Development

Object-oriented programming and design is uniquely suited to the development of realtime programs such as operating systems and interactive user interfaces. Current versions of most operating systems and window environments (e.g., Windows NT, UNIX, and X-Windows) were constructed with OOP languages. OOP languages and good object-oriented software design promote reusability and portability of source code. General purpose objects can be defined and access to their methods provided by the message handling mechanism. A feature called inheritance allows new objects to be defined that incorporate all of the data and methods of existing objects. Large libraries of predefined objects are available in a form called a **class library**.

Class libraries are starting to fill the role once played by compiler and operating system subroutine libraries. Application programmers build user interfaces, databases, and software that use operating system service layer routines by using objects and classes from these libraries. Thus, the program development process primarily consists of modifying existing source code and using precompiled objects rather than developing source code from scratch. The result has been a substantial increase in programmer productivity and a reduction in the time required to develop new software.

Programmers' Workbenches

A programmers' workbench or toolbox is an integrated set of automated support tools to speed the process of human-directed program development and testing. It consists of the following components:

▶ "Smart" program editor.

▶ Compiler and/or interpreter for a 4GL or OOP language.

▶ Link editor and a large library of classes or subroutines.

▶ Interactive facility for prototyping and designing user interfaces.

▶ Symbolic debugger.

▶ Integrated window-oriented graphical user interface (GUI).

The key aspect of a programmers' workbench is less the specific tools provided than the level of integration among them. The general idea is to leverage human effort so highly that complex software can be developed and tested in minutes or hours. Such rapid development enables the use of a prototyping life cycle and allows many program versions to be built and tested. Leveraging programmer effort to such a high degree is achieved by automating tedious and time-consuming development tasks, streamlining most other tasks, and maximizing the reuse of existing source code.

Workbench editors assist the programmer in writing syntactically correct code. They often perform rudimentary syntax checking as the code is being typed and visually flag errors, which allows programmers to immediately identify errors without having to compile the code. Program templates and skeletons are provided also. These can be stored in a library or generated from the output of other workbench components (e.g., data declarations generated from a stored user interface definition).

Program compilers have sophisticated error detection and reporting features. They are integrated with program editors to allow rapid discovery and correction of errors. A window containing a compiler error listing allows individual error messages to be highlighted. As errors are highlighted, the program editor positions the source code display at the appropriate source code location. Some compilers provide suggested corrections along with the error message.

Workbenches are typically endowed with large libraries of predefined subroutines or classes. These include routines for user interface construction, database manipulation, interaction with printers, and interaction with network resources. These routines can be incorporated as source code by the program editor or linked into object code by the link editor. Large libraries provide rich functionality to application programs and allow programmers to reuse source and executable code contained therein.

Testing and debugging facilities include the ability to trace calls to program components and changes to program data. These can be generated in report form or displayed as they occur during program execution. Programs can be started and stopped at programmer specified locations to allow testing of specific code segments. Debugging and tracing also can be performed at the level of machine code (e.g., CPU instruction and register tracing) and operating system service calls. Test executions occur in a simulated environment so that program crashes do not disable the workbench or the entire workstation.

Hardware and software requirements for a programmers' workbench are high. Complex tools and simulated execution environments require a great deal of CPU power and memory. The tools themselves are large, as are the applications being developed. Thus, much disk space is required. Large video displays are needed to display all of the windows used in the development process. Workbenches designed to run on microcomputers typically require two to three times the CPU power, memory, and disk storage of a "typical" business workstation.

Microsoft Developer Studio

Microsoft Developer Studio is an integrated set of tools to support program development. Its goal is to improve programmer productivity by a number of means, including:

► Providing all needed tools within a single integrated environment (the "studio").

► Ensuring maximal compatibility and cooperation among the tool set.

► Automating many of the tedious tasks of program and component development.

► Supporting a systematic and efficient approach to program development.

Developer Studio executes on PCs running various versions of the Windows operating system. It supports a number of programming languages, including C, C++, Visual Basic, FORTRAN, and Java.

The components of an application are represented within a Developer Studio workspace. The internal content of the workspace is a database of components and options used by Developer Studio for tasks such as editing, compilation, and code generation. The programmer interacts with a user-oriented view of the workspace in a window (see the upper-left subwindow in Figures 11-15 and 11-16 on the next page). This view organizes all of the program components so that the programmer can see or access them at a glance.

Developer Studio can consist of dozens to hundreds of individual software components (depending on configuration options), including compilers, link editors, program text editors, symbolic debuggers, and interactive programs for designing user interface elements such as menus, tool bars, icons, and forms. Developer Studio largely hides the complexities of interacting with this large and diverse set of tools. It knows which tools are used to create and modify each type of programmer-defined object.

A programmer selects the desired object from the workspace and indicates a desired action (e.g., double-clicking a source code file icon is an implicit request to modify the code contained therein). Developer Studio selects the tool appropriate to the selected object and starts execution of the tool with appropriate arguments and options. As is typical with window-based interfaces, a set of overlapping menus, tool bars, and icons provides access to various features of the tools.

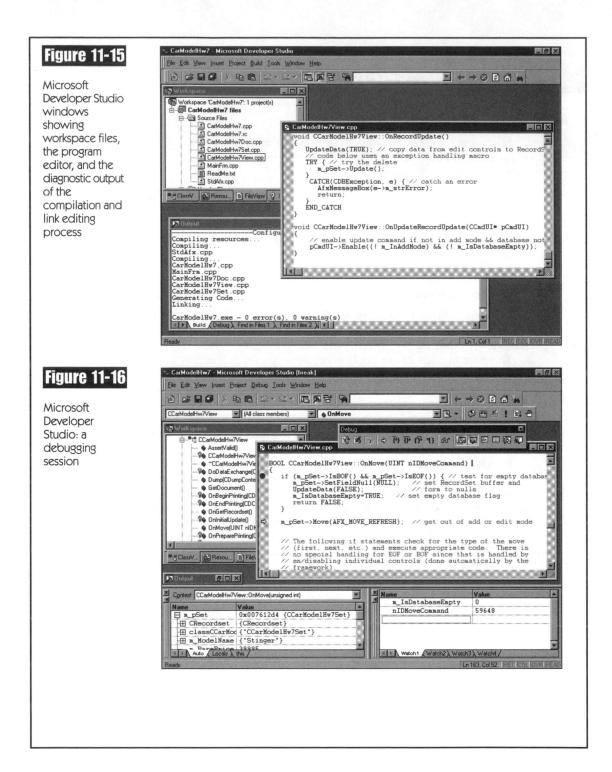

Figure 11-15

Microsoft Developer Studio windows showing workspace files, the program editor, and the diagnostic output of the compilation and link editing process

Figure 11-16

Microsoft Developer Studio: a debugging session

The functions of separate tools are integrated as much as possible. For example, a command to build a complete program executes all of the required resource compilers, the program compiler, and the link editor in the appropriate order. File dependencies are tracked (e.g., the inclusion of one source code file within another) so that changed files cause all dependent files to be rebuilt. Developer Studio manages the movement of data and files from one tool to the next and displays progress and results within an output window (see Figure 11-15).

Developer Studio provides extensive facilities for examining and debugging programs during execution (see Figure 11-16). Programs can be stopped during execution at programmer-specified source code lines. Further execution can be directed on a line-by-line or function-by-function basis. The contents of program variables can be examined while the program is paused (see lower subwindows of Figure 11-16). Changes to variables can be traced during program execution.

Tool and programming language documentation is an important component of Developer Studio. Over 500 MB of documentation is provided on the compact disc and can be loaded on a hard drive for faster access. A sophisticated browser is a standard Developer Studio tool. Extensive search and indexing capabilities are provided. The documentation itself contains many cross-references and hyperlinks to glossary terms, examples, and tutorials. The browser and documentation are designed for easy access during program development of debugging.

CASE Tools

Computer assisted software engineering (CASE) tools are designed primarily to aid in the development and efficient use of system models. It is common for CASE tools to be broken into modules that address specific life cycle phases. CASE tools—loosely classified into two types—**front-end CASE tools** and **back-end CASE tools**—are differentiated by the stages of system development for which the provide support. Front-end CASE tools provide support for early development phases such as the survey, analysis, and design phases of the structured systems life cycle. Back-end CASE tools provide support for later development phases such as implementation, evaluation, and maintenance.

Front-end modules also can be differentiated by the system development methodology (or methodologies) that they support. Examples include structured systems development, rapid prototyping, object-oriented development, and various combinations of those methodologies. Back-end modules can be differentiated by the implementation tools (e.g., DBMSs and programming languages) and operating environment (e.g., operating systems and hardware platforms) they support. Modules can be combined as needed for a particular development project. Thus, for example, front-end tools that support structured analysis and rapid prototyping might be combined with back-end modules that support COBOL, DB2, and MVS in the IBM mainframe environment. Another development project might combine those same front-end modules with back-end modules that support C and Oracle in a distributed environment of Pentium-based workstations and servers running Windows NT.

Front-end CASE tools—which originated in the late 1960s and early 1970s—were designed to address problems and complexities associated with developing and documenting implementation models for large systems of application programs. Examples of these tools included drawing tools for developing system and program flowcharts and data dictionary facilities (the latter were often components of DBMSs). Such tools allowed models and documentation to be developed and analyzed quickly. In addition, because they made development easier, they also made changes easier to make. Thus, alternative implementations could be developed and refined on paper.

As originally introduced, front-end CASE tools were little more than automated drawing and text editing programs. Some of their features (e.g., standard flowcharting symbols, templates for file definitions, etc.) were geared specifically toward developing system models. In practice, they often were used to document systems after they were built rather than to support the evaluation of system models before the system was built.

The improvement in productivity for analysts and designers that automated assistance for drawing and text editing provides is very limited. Further productivity increases require a tool that "understands"' what is being modeled and that can assist the analyst or designer to refine a model and evaluate its quality. Certain quality checks are universally applicable to the evaluation of system models. Examples of these include checks for completeness (e.g., does a textual specification exist for each program on a system flowchart?) and checks of consistency (e.g., is it possible to generate the specified program outputs based on the specified program inputs?).

Both completeness and consistency are relatively simple to verify, but the processes are extremely tedious if performed manually. The need to perform such checks became even greater in the 1970s and 1980s as structured system development became widely practiced, but the drawing and text editing tools of the time were simply not capable of performing these checks. Although these verification processes are simple, they require some knowledge of the objects that are being modeled. That knowledge is absent from a general purpose drawing or text editing tool.

Modern front-end CASE tools incorporate the textual and graphic editing features of older tools as well as the specialized knowledge required to manipulate the objects being modeled. Thus, a CASE tool "knows" that a rectangle on a system flowchart is a program and that it should have a corresponding program specification. It "knows" that an arrow flowing from a file symbol to a rectangle implies that the program reads data from that file. This knowledge of what is being modeled makes it possible to perform error and quality checking as well as sophisticated query processing and report generation.

For example, a CASE tool can compare the definitions of a program's outputs to the definitions of its inputs and listing those data items in the outputs that are not read into the program. It is also capable of answering queries such as "list all programs that read or write to file XYZ." Facilities such as these can be used to find errors in the model, to generate initial data declarations for a program, or for numerous other purposes. All of these factors lead to improved system models.

Front-end CASE tools make it possible to consider more model alternatives and to produce better models. The rapid query and display facilities allow models to be viewed and modified with ease. Their error-checking facilities allow hardware resources to be substituted for labor in many "straightforward but tedious" model validation procedures. All of these factors lead to a reduction in labor input to the analysis and design phases of application development. In other words, the productivity of analysts and designers is increased.

Although front-end CASE tools can increase the productivity of analysts and designers, they do little to address the productivity of programmers. Although high-quality models can provide a better basis for writing application programs, they don't directly support that activity. The translation from implementation model to a set of files and application programs still requires a substantial labor input.

Back-end CASE tools are designed to bridge the gap between analysis/design models and the source code that implements those modules. Various types of support can be provided, including generation of:

▶ Files or database descriptions.

▶ Data declarations for various programming languages.

▶ Source code skeletons.

▶ Complete source code subroutines or programs.

▶ "First cut" user interfaces.

Back-end CASE support ranges from providing a programming "starting point" to complete generation of all required source code. "Starting points" include database descriptions, screen and other user interface definitions, program data declarations, and program skeletons.

A **code generator** is a tool that generates application programs and/or file (or database) definitions from all or part of an implementation model. Although code generators have existed for decades, the technology did not see widespread use until CASE tools were developed and refined. The model descriptions generated by front-end CASE tools are the input for back-end code generation. The specification of model components must be complete, specific, and in a standardized format. Thus, for example, the automated generation of a report-writing program requires complete specification of the inputs, outputs, and the rules that describe the transformation between them. This implies a substantial amount of detail in the implementation model regarding file content, format and content of outputs. It also requires rigorous specification of program logic (i.e., data transformation rules). The consistency and completeness checking procedures of the front-end CASE tool can be used to validate the input prior to code generation. Problems can be brought to the attention of the designer for correction to ensure that valid programs are generated.

Code for many different programming languages can be generated by a code generator. Although it is possible to directly produce executable machine code (i.e., a code generator combined with an interpreter or compiler), it is more common to generate code in a traditional third (3GL) or fourth (4GL) generation programming language. This makes it much easier for a human programmer to validate and modify the generated programs, if necessary. The generated code is then translated with a conventional interpreter or compiler.

A recent addition to the suite of CASE tool modules is the **reverse engineering** module. As its name implies, a reverse engineering module reverses the normal direction of software engineering. Rather than developing source code from a system model, a system model is developed based on existing source code. Programs, database descriptions, and other source code descriptions are scanned, and an equivalent system model is constructed. Once constructed, the model can then be input to a back-end CASE module to re-develop the system with different tools and/or for a different operating environment in a process that is sometimes called **software reengineering**.

Reverse engineering tools—highly specific to a particular operating environment—are complex and expensive. However, their use saves thousands of labor hours that would be required to re-develop an existing system from the starting point of a system development methodology. They are particularly useful when a large number of legacy applications need to be moved to a new operating environment incompatible with the old one. Common examples include the reimplementation of character-oriented mainframe applications (e.g., COBOL/CICS/DB2 applications running on an IBM 360-compatible computer) in client-server environments with graphical user interfaces (GUIs). Organizations with large numbers of such applications can often justify the cost of reengineering tools based on hardware and system software cost savings alone.

Building the Next Generation of Application Software

Southwestern Gifts, Incorporated (SGI) is a small gift retailer that produces a catalog of gifts with a southwestern flavor (e.g., art and jewelry produced by Native Americans and imported from Mexico and other Central American countries). SGI distributes two catalogs per year (in May and October). The bulk of orders are received by telephone, although some are received by mail. SGI employs approximately 50 people including up to 15 warehouse staff and 30 order-processing clerks (highest number during the Christmas shopping season), 4 accountants and bookkeepers (to handle payroll and accounts payable), 3 buyers, 1 catalog designer, 2 managers (one for finance/accounting and one for operations), 3 IS staff, and the owner-operator.

SGI currently has a minicomputer running the UNIX operating system, a PC-based LAN server running Windows NT, 6 microcomputers, and 12 VT220 video display terminals (VDTs). The VT220 terminals are hardwired to the minicomputer and are used for order entry. The LAN server is a file and print server for the PCs and a print server for the minicomputer. Two PCs are used by the accountants, one is shared by the managers, one is shared by the buyers, one is used by the catalog designer, and one is shared by the IS staff. A 10-Mbps Ethernet network connects the PCs, network server, and minicomputer.

Software executing on the PCs includes an accounting package (Quicken), various productivity tools (e.g., Microsoft Office), and graphical design software for catalog production. SGI purchased an order entry and inventory control package eight years ago from an out-of-state vendor. The package stores data in ordinary UNIX files, and uses a proprietary development system called DERQS (Data Entry, Retrieval, and Query System) to implement and execute the various application programs.

DERQS provides forms-based data entry, a forms-based data query facility, a data (file) definition tool, and a simple report definition capabilities. Screen forms are defined using an interactive layout tool. This tool stores a compiled definition of the screen layout and content as a screen definition file (SDF). File descriptions also are defined using an interactive tool and descriptions of file format and content are stored in a data description file (DDF).

DERQS supports a simple interpreted script language. Scripts in this language can define and process full-screen menus, display SDF forms and manage data entry, update data files, and generate simple reports. Most of the existing application programs are written as DERQS scripts. Some application programs are written in C. DERQS includes a C compiler library with functions to interact with SDF and DDF files, which provides access to DERQS screens and file I/O routines and allows the development of programs that are beyond the capabilities of the DERQS script language.

Two of the IS staff are responsible for the DERQS software. One specializes in maintenance and training of the data entry functions and the development of script-based applications, and the other specializes in the query language and the development of new applications in C. The third IS staff member manages the LAN and PCs.

SGI would like to implement improvements to their information system infrastructure, including:

- ► Web site to provide customer information (a subset of the printed catalog) and order entry.
- ► Direct interface to financial information in the DERQs data files (to be uploaded to Quicken and Excel).
- ► Automated system to support the buyers and to interact directly with accounts payable.

Also, SGI would like to modernize their older systems with current GUIs, which they expect to simplify training of their seasonal employees.

SGI has just learned that the company that developed and supported DERQS has filed for bankruptcy. Because no company has shown an interest in purchasing the rights to DERQS, it appears that no further upgrades or technical support will be available.

Questions
1. Should SGI attempt the development of any new software using DERQS? If not, what tools should they acquire for new system development?
2. Should SGI reimplement their existing DERQS-based applications using more modern development tools?
3. What risks and hidden costs might SGI face if they acquire new tools to reimplement old software and/or develop new software?

SUMMARY

Information systems (i.e., related groups of application programs) are developed using a system development methodology, or life cycle. Currently used methodologies include structured system development, object-oriented development, and rapid prototyping. A common principle to all system development methods is the use of a system model, which is an abstract or protoype representation of information processing requirements.

Structured system development uses two separate models—the analysis, or logical, model and the physical, or implementation, model. The analysis model consists of a data flow diagram and a written specification for each of its components. An analysis model is designed to convey requirements for processing, data flow, and data storage. Information about possible methods of implementing a system to meet those requirements is purposely omitted from the model.

The implementation model consists of data flow diagrams and/or system flowcharts. Each model component is further defined by a written or graphic description. An implementation model describes the movement of data between system components, including application programs, files, and manual processes. The implementation model represents a specific plan for building a system to satisfy user requirements. The development of application programs does not begin until the implementation model has been completely specified.

Object-oriented development also uses an analysis model and an implementation model. The components of the models are objects and relationships among them. Objects are groups of related data items and the procedures by which those data items are accessed and modified. Relationships among objects include message passing and superclass-subclass. Object-oriented system models are specifically designed for implementation with object-oriented programming languages and system development tools.

Rapid prototyping is a methodology for the rapid development and evaluation of application programs. Initial requirements—used to quickly create a prototype system—are determined either using structured system analysis or by less formal means. The prototype is given to the user for evaluation. Changes or additions requested by the user are incorporated into the prototype. This process proceeds interactively until all user requirements are incorporated into the prototype.

Rapid prototyping avoids the use of abstract paper models. The methodology assumes that the best model of a system is a working prototype that can be directly manipulated by a user. Successful use of rapid prototyping requires application development tools that can quickly generate and refine working prototypes. Structured and object-oriented development methods devote substantial effort and resources to developing abstract models before any programs are developed. The underlying assumption is that program code is expensive to develop and change and that detailed planning is required to build correct code the first time.

All high-level language programs must be translated into machine instructions before execution. There are two basic approaches to this translation—compilation and interpretation. A compiler is a program that translates source code into object code. Object code consists of machine instructions and calls to previously defined machine language programs, or library routines. Library routines are provided either by the compiler or the operating system.

A link editor is a program that converts object code into executable code. First it searches object code for calls to library routines and then searches one or more libraries for the corresponding machine language programs. The original object code and the library routines are combined to create executable code. Compilation and link editing translate an entire source code file at once. The resulting executable code can be used many times without further translation.

An interpreter performs program translation, linking, and execution one statement at a time. A source code statement is read from a file, translated, linked (if necessary), and immediately executed. Interpretation provides superior error reporting and correction capabilities because errors can be reported in terms of source code statements and variables. It also provides flexibility to incorporate new library routines without recompiling or relinking existing code. However, interpretation consumes substantially more memory and CPU time during program execution than does the execution of compiled and linked programs.

Library routines may be connected to object code by static or dynamic linking. Static linking combines library code with object code to produce an integrated executable program. Executable program files are relatively large but execute quickly and efficiently. Dynamic linking delays the connection of object code to library routines until the program is executed. Library routines are found and linked to object code as they are needed during execution of the program. This reduces program size and allows new versions of library routines to be used with older object code; however, it reduces execution speed due to the need to search for and link library routines while a program is executed.

Modern programming languages are classified as either third generation (3GL) or fourth generation (4GL) languages. 3GLs are characterized by hardware capabilities and limitations at the time of their development (i.e., late 1950s to early 1970s). In general, they are poorly equipped to handle some modern application processing tasks, including database management and complex interactive I/O. Many 3GLs are standardized by ANSI, which guarantees application program portability from one hardware platform and operating system to another.

A number of add-on software systems have been developed to augment the capabilities of traditional 3GLs, including database management systems (DBMSs) and I/O management systems. These software packages extend the capability of 3GLs to handle the complexity and demands of modern applications. However, they do so at the expense of complexity in application development and portability of application programs. The extensions are not ANSI standard and, thus, their use removes the guarantee of portability inherent in an ANSI-standard 3GL.

All high-level programming languages exhibit a characteristic called instruction explosion. Each line of source code is translated into a much larger number of machine instructions. Programming languages with high instruction explosion allow a programmer to write programs more quickly because they contain fewer source code instructions. 4GLs generally exhibit higher degrees of instruction explosion than 3GLs do.

4GLs incorporate database management, complex interactive I/O, and other advanced functions. They are more powerful than 3GLs because of these added capabilities and a higher level of instruction explosion. These features make them especially well-suited to rapid prototyping. There are no "full featured" ANSI standard 4GLs, and thus, there is no guarantee of portability for application programs written in a 4GL.

Object-oriented programming languages are designed to promote code reuse. Data and related procedures (called methods) are packaged together in a unit called an object. Objects perform work in response to messages passed to them by other objects. Objects can inherit data and methods from one another. Libraries of predefined objects (called class libraries) are supplied with object-oriented programming tools. Application programs are developed by adding to or modifying the objects in the class library.

A programmers' workbench is an integrated set of programming tools designed to dramatically increase programmer productivity. Typical components include an intelligent program editor, a compiler and link editor, a large library of reusable software components, and a symbolic debugger. Interactive tools are also provided to design database and user interfaces.

CASE tools are programs designed to support many phases of a system development life cycle. They provide tools to interactively define system models (i.e., front-end CASE tools). These system models are, in turn, input to code generators (i.e., back-end CASE tools) that generate source, or executable, code.

Key Terms

analysis model
applet
back-end CASE tool
binding
CASE tool
class library
code generator
compiler
compiler library
computer assisted software engineering (CASE)
control structures
data declarations
data dictionary
data flow diagram (DFD)
data operations
database management systems (DBMSs)
debugging version
distribution version
early
early binding
executable code
fourth generation application development (4GAD)
fourth generation programming language (4GL)
front-end CASE tool
function
high-level programming language
implementation model
in-line function call
instruction explosion
interpretation
interpreter
Java
Java virtual machine (JVM)
library call
link editing
link editor

link map
linker
linking
logical model
memory map
message passing
messages
methods
object
object code
object-oriented programming (OOP)
object-oriented (OO) system development
physical model
procedure
process specifications
production version
programmers' workbenches
rapid prototyping
resolving the reference
reverse engineering
sandbox
software reengineering
source code
static
static linking
structured systems development
subroutine
superclass-subclass
symbol table
symbolic debugger
system analysis
system design
system model
third generation
third generation programming languages (3GLs)
unresolved reference

1. A compiler allocates storage space and makes an entry in the symbol table when a(n) _____ is encountered in source code.

2. A(n) _____ is produced as output of the analysis phase in structured system development.

3. A link editor searches object code for _____.

4. The structured SDLC produces a(n) _____ as output of the design phase.

5. A fourth generation programming language (4GL) has a greater degree of _____ than a third generation programming language (3GL).

6. _____ contains executable code and _____ calls.

7. A(n) _____ produces a(n) _____ to show the location of object code modules and library routines within executable code.

8. The process of _____ produces a system model from existing source code.

9. A(n) _____ is a model or system component that combines a set of related data and the methods that manipulate that data.

10. The names of data items and program functions are added to the _____ by the compiler when they are encountered in source code.

11. A(n) _____ translates an entire program before linking and execution, whereas a(n) _____ interleaves translation and execution.

12. A(n) _____ describes the content of data flows and files within a(n) _____.

13. The result of compilation is _____, whereas the result of link editing is _____.

14. A Java _____ runs within the _____ of a Web browser.

15. _____ in a source code program are translated into machine instructions to evaluate conditions and/or transfer control from one program module to another.

16. Relationships among objects that are modeled in an object-oriented system model include _____ and _____.

17. A(n) _____ uses the content of the _____ to establish correspondences between data items and program fragments in executable and source code.

18. _____ do not provide a programmer with the ability to manipulate databases and graphical user interfaces (GUIs).

19. A(n) _____ tool aids in the creation of system models. A(n) _____ tool generates program _____ from systems models.

20. _____ linking is generally performed by a link editor. _____ linking is performed by an interpreter.

21. Unlike _____, _____ proceeds under the assumption that a working model of the system is easy to develop and modify.

22. Java programs are compiled into object code for a hypothetical hardware and system software environment called the _____.

23. A compiler library is similar in function to an interpreter's _____.

Review Questions

1. What types of system model(s) are used in the structured, rapid prototyping, and object-oriented approaches to application system development?

2. What characteristics of application development software are needed to support rapid prototyping?

3. How can rapid prototyping be incorporated into the structured and/or object-oriented system development methodologies?

4. What are the basic components of an object-oriented system model?

5. What assumptions justify the extensive time and resources devoted to analysis and design in the structured system development methodology?

6. What does a compiler do when it encounters data declarations in a source code file? Data (manipulation) operations? Control structures?

7. What are the differences between source code, object code, and executable code?

8. Compare and contrast the execution of compiled programs to interpreted programs in terms of CPU and memory utilization.

9. What is a linker? What is a compiler library? What are the advantages of using them?

10. What types of programming language statements are likely to be translated into machine instructions by a compiler? What types are likely to be translated into library calls?

11. How does the process of interpretation differ from the process of compilation?

12. Compare and contrast the error detection and correction facilities of interpreters and compilers.

13. Compare and contrast static and dynamic linking in terms of efficient use of hardware resources during program execution.

14. What is instruction explosion? What are the advantages and disadvantages of using a programming language with a high degree of instruction explosion?

15. With respect to the requirements of modern applications, what are the shortcomings of third generation programming languages (3GLs) in the areas of data manipulation and user interface?

16. What happens to application program portability when a programmer uses non-ANSI programming language extensions? Why do so many programmers use these extensions?

17. What are the primary differences between object-oriented programming languages and more traditional programming languages?

18. What components are normally part of a programmers' workbench? In what ways does a workbench improve programmer productivity?

19. What is a CASE tool? What is the relationship between a CASE tool and a system-development methodology?

20. What is the difference between a front-end CASE tool and a back-end CASE tool?

21. What is a code generator? What is required to efficiently use a CASE tool and code generator to support application development?

Problems and Exercises

1. Develop a set of machine instructions to implement the following source
    code fragment:

    ```
    a = 0;
    i = 0;
    while (i < 10) do
      a = a+i;
      i = i+1;
    endwhile
    ```

2. Some RISC processors do not provide PUSH and POP instructions. The
    rationale is that these instructions are complex instructions that can be
    built from simpler components. Write a short machine language program
    to implement the PUSH and POP operations using only data movement
    and branching instructions.

Research Problems

1. Investigate a modern CASE tool such as those offered by Sterling
    Software (www.cool.sterling.com) and Oracle (www.oracle.com). On
    what system development methodology or methodologies are the tools
    based? What types of system models can be built with the tool? How is
    an analysis model translated into an implementation model? What pro-
    gramming languages, operating systems, and database management
    systems (DBMSs) are supported by the back-end CASE tool? How can
    the tool(s) be used to support rapid prototyping?

2. Investigate a modern programmers' workbench such as Microsoft Visual
    BASIC (www.microsoft.com) or Borland Delphi (www.borland.com).
    Are application programs interpreted, compiled, or both? What program
    editing tools are provided? What tools are available to support runtime
    debugging? What DBMSs can be accessed by application programs?

Chapter 12

Operating Systems

Chapter Goals

- ► Describe the components and functions of an operating system.

- ► Describe the resource allocation functions of an operating system.

- ► Describe the mechanisms by which an operating system manages programs and processes.

- ► Describe the mechanisms by which an operating system manages the CPU.

- ► Describe the mechanisms by which an operating system manages memory.

OVERVIEW

The operating system—the most important system software component of an information system—plays a dual role as service provider and hardware manager. In its service role, it has a critical impact on the capabilities that application programs can provide to users. In its hardware-management role, the operating system directly controls the allocation of hardware resources to user tasks as well as the efficiency of hardware utilization. Thus, the selection and configuration of an operating system largely determine the services available to end users and the efficiency with which those services are delivered.

The primary purpose and functions of an operating system can be summarized as follows:

► Provide an interface for users and application programs to low-level hardware functions.
► Efficiently allocate hardware resources to users and their application programs.
► Provide facilities for loading and executing application programs.
► Provide facilities for managing secondary storage.
► Provide controls over access to hardware devices, programs, and data.

Operating system functions can be fulfilled in many different ways. Differences in the set of functions that are implemented, the method and efficiency of implementation, and the hardware being controlled account for the wide variety of operating systems available.

Some basic criteria by which operating systems may be distinguished include:

► Single tasking versus multitasking.
► Single user versus multiuser.
► Batch versus timesharing.

Single tasking operating systems—designed to execute only one process at a time—are restricted to microcomputers and certain technical applications. **Multitasking operating systems** are designed to execute multiple programs concurrently.

Multiuser operating systems are designed to be used by more than one user at a time. By necessity, they are multitasking because there must be at least one active process per active user. Multiuser operating systems employ fairly elaborate mechanisms to account for resource use by individual users (e.g., logins, logouts, and related accounting procedures). They also provide mechanisms to implement ownership of and access controls over both programs and data.

Batch operating systems are designed to execute programs that do not require active user intervention (i.e., **batch processes**). Such processes use noninteractive I/O devices (e.g., document scanners) or secondary storage for input and return results to those same devices. They are unusual today because of the widespread use of online processing. Such systems are still in use, however, especially in large batch transaction-processing environments (e.g., nightly processing of checks in a bank, preparation of credit card statements, etc.) or as components of a distributed network of computers.

Operating System Functions

A functional decomposition of an operating system is shown in Figure12-1. These functions can be loosely divided between those directed toward hardware resources (e.g., CPU control, memory control, mass storage control, and I/O control) and those aimed at users, their processes, and their files. Notice that control of mass storage hardware and of user files is combined within the file-management subsystem.

Figure 12-1

Functional view of an operating system

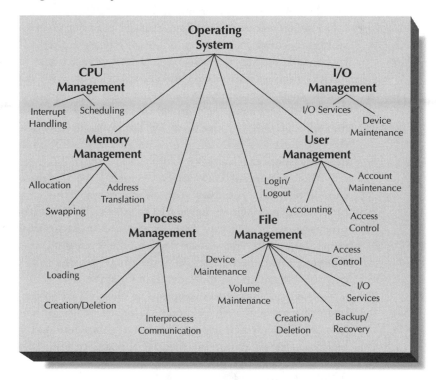

Management of the CPU requires allocating its time among various modules of system and application software. The operating system must decide which process will receive control of the CPU and for how long it will retain that control. The operating system must balance the CPU processing needs of all active processes and ensure efficient use of the CPU. (CPU management is discussed in detail in "CPU Allocation.")

Processes (also called **tasks**) are the basic unit of software to which resources are allocated. The operating system handles many aspects of process management, including creation, destruction, allocation of resources, and inter-process communication and synchronization. Process management is discussed in detail in "Process Management."

Management of memory (i.e., primary storage) resources is a critical task because its quantity is limited. Processes require main memory to hold their instructions and current data. The operating system must dynamically allocate a fixed pool of memory to multiple processes. In cases in which the total demand for memory exceeds physical memory, the operating system may employ techniques such as virtual memory management. Such techniques require the operating system to coordinate the loading and swapping of memory pages to/from secondary storage. Memory management is discussed in detail in "Memory Management."

Management of secondary storage is easier than management of primary storage because its capacity is not nearly as limited. The operating system is responsible for allocating secondary storage space and for allocating the time of devices that provide access to that space (e.g., hard drive controllers and the system bus). Secondary storage-allocation capabilities are implemented within the file-management portion of the operating system. Files can be created, moved, deleted, or modified by applications through the service layer or directly by users through the command layer. File management is covered in detail in Chapter 13.

Management of I/O devices, which have widely different functions and capabilities, is a relatively complex process. Thus, management and control procedures vary widely among I/O devices. The operating system must provide processes with access to I/O devices and must prevent those processes from interfering with one another. Also, the operating system must manage the movement of data among I/O and storage devices (i.e., primarily storage). I/O control is discussed in more detail in Chapter 14.

Management of user accounts is closely tied to the implementation of the operating system security model. User identification and authorization are the foundation on which access controls are implemented. User accounts also can be a basis for accumulating records of resource utilization and charging for that utilization. User accounts and operating system security are discussed in more detail in Chapter 14.

Operating System Layers

The operating system can be described as a layered set of software. (See Figure 12-2.) The **kernel** is the operating system layer that provides the most basic functions, including resource allocation and direct control over most hardware resources. Most of the interaction between hardware and software is localized to this layer. The kernel is also responsible for implementing the operating system security model.

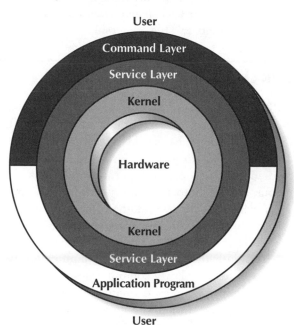

Figure 12-2

Model of an operating system as software layers

The **service layer** consists of a set of programs that are used by application programs or by the command layer. These programs provide the following types of services:

► Low-level access to I/O devices (e.g., the movement of data from an application program to a printer or video display).

► Low-level access to mass storage devices (e.g., the movement of data from a hard drive to an application program).

► File manipulation (e.g., open and closing files, reading and writing records, etc.).

► Basic process and memory control functions (e.g., sending signals between application programs, allocating additional memory to a program during execution, etc.).

Additional services also can be provided such as window management, access to communication networks, and basic database services. The entire service layer can be considered as a set of subroutines that are available to application programs on demand.

The **command layer** provides a direct interface to the service layer for end users, which allows many service functions to be requested by the user without the need to use an application program. This layer also provides the user with a means of managing hardware and files independently of application programs and gives some basic management and access controls through facilities such as user accounts and file ownership.

RESOURCE ALLOCATION

The management of processes, or programs, and the allocation of hardware resources to those programs is primarily the responsibility of the kernel. Processes must be provided with memory to hold data and instructions, access to the CPU to execute instructions, and use of I/O and secondary storage devices as needed. The operating system is responsible for managing these various resources to meet the needs of each process, preventing processes from interfering with one another, and guaranteeing efficient use of hardware and other information-processing resources.

The complexity of operating system resource-allocation procedures has risen with the complexity of computer hardware and the demands placed on that hardware. Early computer systems were relatively simple devices with few hardware resources to manage. Limited processing and storage capabilities made it unnecessary and impractical to implement complex resource-allocation schemes. Application programs "took control" of all hardware resources when they executed. Because only one application program was active at a time, there was no need for operating system software to mediate among competing demands for hardware resources.

Over time, computers became more powerful and the hardware resources of a typical computer system became more numerous and diverse. Extra processing power and storage capacity made multitasking and multiuser operating systems feasible. However, operating systems that implemented these features also needed to implement elaborate mechanisms to allocate resources to multiple processes and users.

Single Process Resource Allocation

Allocation of resources within a single tasking operating system is relatively simple. This simplicity arises from the coexistence of only two active programs within the computer—the operating system itself and a single application program. Resource allocations to the application program are made on demand if such resources exist within the computer system.

In single tasking operating systems, there is rarely any contention between the operating system and the application program for resources. The operating system reserves whatever resources it requires when it first boots up. All resources not allocated to the operating system are available for allocation to an application. CPU cycles are the only resources that need explicit allocation procedures.

When an application program is loaded and begins execution, control of all hardware resources except those reserved by the operating system is given to that program. The program retains this control until it terminates, an error occurs, or it needs service from the operating system. Errors and service calls are processed through interrupts. When an interrupt is detected, control of the CPU is passed back to the operating system. If an error has occurred, the operating system may attempt to correct it so that the program can continue. If this is not possible or if the operating system is not programmed to correct errors, then the application program is terminated and the operating system retains control of the hardware.

Most service calls can be considered an indirect request by an application program for system resources. For example, file-manipulation service calls represent implicit requests for access to secondary storage hardware. I/O service calls are implicit requests for access to I/O hardware. Because there are no other processes competing for access to these resources, the operating system simply grants access when asked. No elaborate mechanisms are needed to check for competing demands for those same resources.

The MS-DOS operating system is an example of a single tasking operating system that operates as just described. When MS-DOS is first loaded, it reserves enough memory for its kernel, service layer, and command layer. The command layer then awaits an explicit request by the user to execute an application program. When that request is received, the application program is loaded into memory, which implicitly allocates that memory. Control of the CPU is then turned over to the application program. The application program retains control unless an error occurs or a service request is made. Errors and service requests are signaled by an interrupt, the processing of which causes control of the CPU to revert to the operating system. CPU control passes back and forth between the operating system and the application program as services are requested and provided.

The lack of complex resource-allocation procedures allows a single tasking operating system to be small and efficient. MS-DOS requires less than 100 KB of memory and consumes relatively few CPU cycles during its execution. Most hardware resources are available to execute application programs.

Multiple Process Resource Allocation

Although single tasking operating systems are still in use in some situations (e.g., dedicated realtime processors) multitasking operating systems are the norm today. The flexibility of multitasking capabilities facilitates construction of application and system software. Modern application and system software programs are large and complex. Multitasking provides the flexibility to break these large programs up into smaller, independent modules that are easier to develop and can be suspended and resumed as needed.

For example, a typical word-processing program often needs to send output to a printer. A multitasking operating system allows the printer interface portion of the word processor to be implemented as a separate module, which can be loaded and executed when printing is required. Once the printing module is started, the remainder of the word-processing program can continue with other processing (e.g., formatting another document). There is no need for the document-formatting routines to wait for the printer control routines to complete their task. The result is a more efficient use of end-user time because the user can edit one document while another is printing.

System software also benefits from multitasking capabilities. Network interface units, for example, have a dedicated operating-system process that detects and responds to incoming data. Multitasking allows this process to be loaded into memory when the operating system is booted and placed in a suspended state until data arrives. Other I/O devices such as printers and terminals are managed similarly. Division of the operating system into multiple tasks allows operating-system components to be independently managed, resulting in more efficient use of hardware resources.

Resource allocation within a multitasking operating system is substantially more complex than for a single tasking operating system. The operating system must allocate resources to multiple processes and must ensure that processes do not interfere with one another. CPU time, storage (primary and secondary), and access to I/O devices all must be managed so that each process receives the resources it needs.

Resource-Allocation Goals

Traditionally, the efficient utilization of computer hardware has focused on the management of two key resources—storage space and CPU time. That is, storage space and processor instruction cycle time must be used as productively as possible. In the early years of computers, memory devices and processor hardware were so costly that considerations of machine efficiency were paramount. Elaborate schemes, spanning several generations of system software, were developed for sharing a single computer system among many users and applications.

Since then, hardware costs have plummeted while labor costs have continued to rise. Processors, storage devices, and I/O units have dramatically increased their capacity and capability—changes that have affected operating system implementation. For example, the prevalence of multitasking operating systems on desktop computers is a clear recognition of the trade-off between machine and labor resources. Multitasking capabilities improve labor productivity by allowing multiple applications and/or multiple parts of a single application to execute concurrently. Extensive service layer support for graphical user interfaces (GUIs)—and their considerable consumption of machine resources—is another example of an operating system-based substitution of machine for labor resources.

Modern operating systems are similar to early operating systems in their resource-allocation goals. That is, efficient use of machine resources (CPU and I/O cycles, in particular) remains the primary goal. Related goals include maximizing the amount of work performed per time unit, ensuring reliable program execution, protecting resource security, and minimizing the resources consumed by the operating system itself.

Resource-Allocation Processes

Consider a resource-allocation task such as operating the audio visual support service for a school. Most schools cannot afford to place expensive audio visual equipment (e.g., motion-picture projectors and laptop computers) in every classroom. Such schools operate audio visual equipment centers that maintain a pool of equipment and allocate it to instructors and students on an as-needed basis. Requests for equipment are received by the service, and equipment, if available, is delivered to the classroom when needed and retrieved when it is no longer needed.

Managing the inventory of equipment and responding to user requests is a complex task that requires detailed knowledge of equipment availability. A system that performs this task must maintain detailed records of equipment capabilities and status, schedules for its use over some period of time, and schedules for moving equipment from one location to another. Detailed procedures must be implemented to ensure the correctness and currency of these records and to make scheduling decisions. Policies and procedures are needed to prioritize conflicting demands for service.

The resource-allocation functions of an operating system are similar to this audio visual service scenario. The operating system must keep detailed records of every available resource and must know which resources are used to satisfy which requests. It must schedule resource availability to meet present and anticipated demand. Records of resource availability—or lack thereof—must be updated continually to reflect the commitment and release of resource by processes and users.

Like application software, the operating system is simply a program. Thus, resource-allocation goals must be defined operationally as a set of procedures, or algorithms, implemented in software. The record keeping required to implement these algorithms must be implemented as a set of data structures. Thus, the resource-allocation functions of an operating system are similar to any other program—a set of algorithms and data structures organized to accomplish a particular purpose.

Computer scientists have long studied the efficiency and efficacy of various data structures and algorithms. A large body of knowledge exists concerning which algorithms and data structures are most efficient for particular types of processing tasks. This body of knowledge finds extensive application in the design and implementation of operating system resource-allocation procedures.

Operating-system designers take great pains to implement resource-allocation procedures that consume as few machine resources as possible. The reason is simple—every resource consumed by the operating system resource-allocation function is unavailable to application programs. Think of the resource-allocation function as similar to middle-management functions in a business. Every business has a limited set of resources. Every resource devoted to a management function is unavailable for other business functions (e.g., producing and selling products).

Businesses strive to define management procedures that do what they need to do (i.e., allocate resources where needed and guarantee correct functioning of the business) while consuming the fewest resources possible. Similarly, an operating system seeks to allocate resources where needed and to ensure reliable program execution while consuming as few resources as possible. The resources consumed by the resource-allocation procedures are sometimes referred to as **system overhead**. A primary design goal for most operating systems is to minimize system overhead while achieving acceptable levels of throughput and reliability.

Notice that these goals are in direct conflict. "Better" allocation decisions (i.e., those that provide higher levels of resource availability to application programs) typically require more elaborate and complex allocation procedures. Ensuring higher levels of reliability also requires more extensive oversight procedures, yet increases in the complexity of allocation and oversight procedures imply greater consumption of hardware resources when those procedures are implemented in software. Thus, operating systems designers must determine the right balance among these competing objectives.

Real and Virtual Resources

The physical capabilities (i.e., devices and associated system software) of a computer system are called **real resources**. As allocated by the operating system, the resources apparent to a process or user are called **virtual resources**.

Modern operating systems are designed to make the set of virtual resources available to active processes or users equal to or greater than the available set of real resources.

Although it may seem counterintuitive to have virtual resources exceed actual resources, it is rational and feasible. Providing the appearance of unlimited resources to a process greatly simplifies the design and implementation of that process. A programmer need not be concerned with implementing elaborate procedures for determining resource availability. The programmer doesn't need to implement schemes to lock and unlock resources for exclusive use. Programs can be written under the assumption that whatever resources are requested will be provided. The complexity of shifting and substituting real resources to satisfy this assumption is embedded entirely within the operating system. Processes are oblivious to the complex procedures used to satisfy their resource requests.

Providing virtual resources that meet or exceed real resources is accomplished by two means:

► Rapidly shifting resources unused by one process to other processes that need them.

► Substituting one type of resource for another when demand temporarily exceeds supply.

Although each process may "think" that it has control of all hardware resources in the computer system, it is rare for any one process to need them all simultaneously. The operating system shifts resources among processes as demand rises and falls. In this way, the sum of virtual resources apparently available to all active processes can substantially exceed the computer system's real resources.

Certain types of resources can be substituted for one another. The best example of this is storage. Recall that the available storage devices of a computer system form a hierarchy with faster expensive devices (e.g., SRAM cache) at the top and slower inexpensive devices (e.g., magnetic tape) at the bottom. Every process seeks access to a large supply of the fastest storage available. But since the number of running processes is limited to the number of physical CPUs, only one or a few processes need fast storage at any time. Storage needs of inactive or suspended processes (or unused storage of an active process) can be shifted from expensive, high-speed device to less expensive, slower devices. The most common example of this shifting is virtual memory management which is described in more detail in "Virtual Memory Management."

PROCESS MANAGEMENT

A **process** is a basic unit of executing software that is independently identified and managed by the operating system. A process can request and receive hardware resources and operating system services. Processes can be stand-alone entities or part of a group of processes that cooperate to achieve a common purpose. Processes can communicate with other processes executing on the same computer system or with processes executing on other computer systems.

Processes can be created explicitly or implicitly. Implicit process creation is performed by a user through the operating-system command layer. When the user types the name of a program at a command process, an implicit request to create a process has been made. The selection of a program object with a GUI (e.g., double-clicking on a program icon) is also an implicit process-creation request. Implicit requests must be translated into explicit requests by the command layer.

Processes are created explicitly by an application or system program calling the appropriate service layer subroutine. The subroutine is passed a parameter to indicate the location of a file containing the executable code of the process. Other parameters might be passed to the subroutine such as required resources, process owner, and process priority. The operating system creates the process by allocating memory and other resources, loading the process executable code into memory, and creating a data structure to store information about the process.

Process Control Data Structures

The operating system must keep track of a large set of information about active processes, which allows the operating system to perform a number of functions including resource allocation, secure resource access, and protecting active processes from interference by other active processes. The generic name for a data structure that stores information about a single process is a **process control block (PCB)**.

The contents of a PCB vary from one operating system to another. They also may vary within a single operating system depending on whether certain functions (e.g., user resource accounting) are enabled or disabled. The following data items are typically included:

► A unique process identification number.

► The current state of the process (e.g., executing or suspended).

► Events for which the process is waiting.

► Resources allocated exclusively to the process (including memory, files, and I/O devices).

- Machine resources consumed (e.g., CPU seconds consumed and bytes of data transferred to/from disk).
- Process ownership and/or access privileges.
- Scheduling priority or data from which scheduling priority may be determined.

These are only general categories of information; the specific content and format of data within these contents will vary among operating systems.

PCBs are organized into one or more larger data structures, which are often circular linked lists. Various names may be applied to these lists, including **process queue** and **run queue**. These lists may be searched by various resource-allocation routines. For example, the CPU scheduler searches the list looking for processes that are ready to execute. A user control process might search the list for all processes owned by a certain user so that they may be terminated when that user logs off.

Speed of searching and update is an important consideration when designing PCBs and the data structures within which they are embedded. Efficient data structures allow operating-system management and resource-allocation subroutines to execute more quickly, which frees machine resources for use by application programs. The contents of the process queue and individual PCBs are changed frequently as processes are created and terminated and as resources are allocated to and released by individual processes.

Executing processes also may need to create other processes and to communicate with those processes during their execution. The creation of a process by another process is sometimes referred to as **spawning**. The original process is called the **parent process**, and the newly created process is called the **child process**. Parent processes may spawn multiple child processes (i.e., **sibling processes**). Child processes may, in turn, spawn children of their own. A group of processes descended from a common ancestor (including the common ancestor itself) is sometimes called a **process family**.

Process spawning frequently is used by application programs that need to execute general purpose utility programs that are not part of the operating system service layer. A text editor, for example, may allow a user to sort the contents of all or part of the file currently being edited. Rather than implement instructions for sorting within the editor program itself, the editor might request that an external sort utility program be loaded and executed.

Process spawning and/or the subdivision of a large task into independent processes may improve execution speed, which is particularly true if the computer system on which the processes are executed contains multiple CPUs. Independent processes can execute simultaneously on a multiprocessor computer system, which results in a shorter time interval between the start of a complex task and its completion.

Even if only one CPU is available, execution speed can be increased because of better resource allocation by the operating system. For example, consider a word-processor user who wants to edit one document while printing another. If both editing and printing code are part of the same process, then only one of these tasks can be performed at a time. The user must wait for the printing code to complete execution before beginning to edit the next document.

If printing and editing are contained within separate tasks (e.g., if the word processor spawns a separate printing process), then both tasks may execute concurrently. Because the two tasks use different mixes of computer resources (i.e., heavy I/O and light CPU use for the print process and the reverse for the editing process), the operating system can allocate resources efficiently so that both processes move quickly toward completion. In general, breaking large processes up into smaller processes with simpler resource requirements allows the operating system to more efficiently allocate resources, which may improve total system throughput.

Threads

There is a diminishing return associated with subdividing large processes into smaller ones. As the number of active processes increases, the operating system overhead required to track and manage them also increases. Each process needs its own PCB and its own allocation of primary storage and other resources. If the number of active processes becomes too large, the complexity of searching and updating process management and resource-allocation data structures will result in excessive resource consumption by the operating system itself. This will reduce the resources available to execute application processes and, thus, reduce system throughput.

Some operating systems attempt to minimize this extra overhead while still allowing a large degree of process segmentation. To do so, they allow processes to divide themselves into smaller units of execution called threads. A **thread** is a portion of a process that has independently allocated CPU resources. Process threads can execute concurrently on a single processor or simultaneously on multiple processors. Threads share other resources allocated to their parent process, including primary storage, files, and I/O devices.

The advantage of organizing a task into a process/threads family instead of multiple independent processes is that operating system overhead for resource allocation and process management is reduced. Because all threads of a process share storage and I/O resources, the operating can track allocation of those resources on a per-process basis. This reduces the size of the data structures used to track these resources and, thus, makes searching and updating those structures faster. The operating system must keep track of some information about threads so that they can be scheduled independently (i.e., allocated CPU time), but the information required for this purpose is substantially less than the normal content of a PCB.

A process or program that divides itself into multiple threads is said to be **multithreaded**. The division of a program or process into threads is not automatic. The programmer or compiler must specifically identify code segments that can execute concurrently and create a thread for each. Operating systems that support multithreaded processes include Windows NT and newer versions of UNIX.

Interprocess Communication

Cooperation among active processes requires mechanisms for synchronizing their execution and for passing data among them. The nature of this synchronization depends on the nature of the data passed among the processes and the timing relationships among their execution cycles. A number of interprocess relationships are possible, including the following:

- ▶ One process executes another as a subroutine with no data passed in either direction.
- ▶ Each process executes one process cycle in synchronization with the other process.
- ▶ One process executes another as a subroutine with data passed in one or both directions.

Each of these scenarios requires a different approach to interprocess synchronization and data communication.

Interprocess Signals

In the first scenario, support for **interprocess synchronization** is relatively simple. The parent process executes a service call to request that the child process be loaded and executed. Synchronization is implemented by a **signal**, sent from parent to child, indicating the termination of the child process. The use of such a signal requires a mechanism by which the parent process waits for the completion, or termination, of the child process.

The following program statement is an example of a service call to create a child process from within a parent process:

```
create_process("child","/usr/bin",process_id)
```

In this statement, the parent process requests the operating system to create a new process. The name of the process (as known to the file-management system) is "child" and is located in the directory "/usr/bin". The operating system will search for the file, load its contents into memory, and create a process control block. An identification code for the new process is returned to the parent through the process_id parameter.

If the parent needs to wait for the child to complete before continuing its own execution, it must execute a service call such as:

wait(process_id)

immediately after executing the create_process call. The wait function causes the operating system to suspend execution of the parent pending the termination, through an error or normal exit, of the process identified by the process_id parameter of the wait call.

Multiple threads of a single process also may need to use signals. For example, consider the three threads shown in Figure 12-3. Threads A and B execute independently, but thread C cannot execute until both A and B have completed execution. The operating system provides a service call for this purpose. The parameter passed to this service call is the identification number of a thread, not a process. Signals can be passed only among threads with a common parent.

Figure 12-3

A process containing three threads. Execution of thread C cannot begin until threads A and B are complete.

Process (Word Processor)

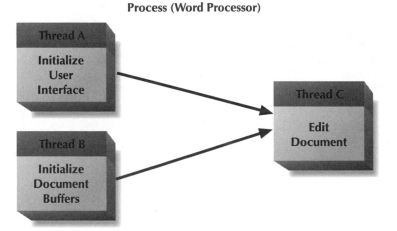

Most operating systems provide a mechanism to return an error or exit status code when a process terminates. These codes are generally implemented as an unsigned integer that is sent to the parent process when the child process terminates. For example, the service call:

wait(process_id,exit_code)

causes the parent process to be suspended pending the completion of the child process. The exit code of the child process is sent to the parent through the exit_code parameter. The parent can then test the value of that code to determine whether the child process executed successfully.

More complex signals are required when processes must coordinate the execution of individual process cycles. Consider, for example, the two cooperating processes shown in Figure 12-4. Process 1 retrieves input data from an I/O or secondary storage device and stores that data in a shared region of memory. Process 2 performs some transformation on the data and sends it to an output device. Each process has a repetitive cycle. Process 1 repetitively reads input records and stores them in memory, and process 2 repetitively transforms and outputs those records.

Figure 12-4

An example of two cooperating processing cycles coordinated through the use of interprocess signals

```
Process 1                      Process 2

read(record)                   signal(process 1)
while not end-of-file          wait(process 1)
    wait(process 2)            retrieve(record)
    store(record)              while (record = 'eof')
    signal(process 2)              signal(process 1)
    read(record)                  transform(record)
end while                          write(record)
store('eof')                      wait(process 1)
                                  retrieve(record)
                               end while
```

Signals are required to coordinate the operations of each of these processes. Process 2 cannot process a record until process 1 has read and stored it. Similarly, process 1 cannot store the next record until process 2 has retrieved the previous record. Coordination between these two processes can be implemented by a pair of signals. Process 1 sends a signal to process 2 after each record is stored in shared memory. Process 2 sends a signal to process 1 each time a record is retrieved from shared memory. These signals prevent process 1 from overwriting shared memory with a new record before process 2 has retrieved the old record.

As in the prior example, the operating system suspends each process pending the arrival of a signal from the other process. Notice that coordination of the processes in the event of an end of file condition also could have been implemented through signals. A separate signal could have been used for this condition or a single signal with multiple values could have been used for all synchronization activities (see Figure 12-5).

```
Process 1                              Process 2

read(record)                           signal(process 1, 'ready')
while not end-of-file                  wait(process 1, condition)
    wait(process 2)                    while (condition = 'ready')
    store(record)                          retrieve(record)
    signal(process 2, 'ready')             signal(process 1, 'ready')
    read(record)                           transform(record)
end while                                   write(process 1, condition)
signal(process 2, 'eof')               end while
```

The implementation of interprocess signals is an important, basic function of most operating systems because signaling is a general method for controlling access to shared resources. In the previous two examples, signals are used not only to synchronize the processes, but also to coordinate their use of a shared resource (i.e., the shared region of memory used for data communication).

The operating system must implement similar controls over access to many different shared resources. Access to the CPU, memory, secondary storage, and I/O devices by multiple processes requires mechanisms to prevent conflict. An operating system implements these mechanisms by the same basic signaling methods used in the previous examples. However, the use of signals for these resource-management functions generally is hidden from executing processes other than the operating system itself.

Interprocess Data Communication

It is unusual for interprocess data communication to be implemented entirely through signals and process-controlled shared memory. In general, the implementation of shared memory shown in the previous example requires both processes to be compiled and/or linked as a unit. When linked, the two processes are in fact a single process for purposes of process management within the operating system. This is the normal situation for threads of a single process, but processes that are truly independent must each occupy a separate region of memory. Thus, direct memory sharing arrangements are difficult, or impossible, to implement.

An operating system provides one or more mechanisms for indirect memory sharing for purposes of data communication among processes. The allocation of the shared memory region and the management of data movement to and from that region are handled directly by the operating system. Independent processes make service calls to create such shared memory regions and to read and write data to/from them. The common term for such a region of shared memory is a **pipe** (i.e., a data pipeline between processes) although some operating systems use different terminology. A conceptual representation of a pipe is shown in Figure 12-6.

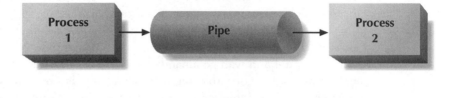

Figure 12-6

A conceptual diagram of a pipe between two processes with one-way data flow

The implementation of a pipe requires mechanisms whereby a process can request the creation of pipe between itself and another process. Mechanisms also are required to allow data to be written into the pipe or read from the pipe. Thus, a set of service calls such as the following is implemented within the operating-system service layer:

► open_pipe(process_id,direction,pipe_id)
► close_pipe(pipe_id)
► read_pipe(pipe_id,data)
► write_pipe(pipe_id,data)

The open_pipe call directs the operating system to establish a region of shared memory for communication between the calling process and the process identified by process_id. Depending on the particular operating system, this shared storage can be implemented in memory or as a temporary file. To allow a process to use multiple pipes simultaneously, the operating system returns pipe_id as an identifier for use in subsequent pipe operations. The argument "direction" in the open_pipe call tells the operating system whether data will flow to or from the calling process. It also may be possible to use the pipe for bidirectional data communication, in which case possible values of direction are "to," "from," or "both." The close_pipe call informs the operating system that the shared storage is no longer needed.

The write_pipe call directs the operating system to move the information stored in the data structure "data" to the storage region allocated to the pipe. The read_pipe call directs the operating system to move data from the storage region to the data structure referenced in the "data" argument. Notice that the location of the storage region, the movement of data into and out of that region, and the data structure(s) used to implement the pipe are entirely under control of the operating system.

In some operating systems (e.g., UNIX), input and output to/from pipes is implemented in a manner similar to input and output to/from files. That is, once created, a process interacts with a pipe in essentially the same manner as it interacts with a file (i.e., it uses file I/O service calls). Thus normal functions such as reading, writing, buffering, and end-of-file conditions are the same for pipes as for files. The advantage to such an approach is that it can be used both for data communication and for synchronization.

Figure 12-7 shows the example from Figure 12-4 modified to use pipes for both data communication and process synchronization. Notice that most of the explicit signal and wait commands have been eliminated. The signal to coordinate read and writing to/from shared memory is completely replaced by I/O to/from the pipe. Attempts by process 1 to write to a full pipe cause the operating system to suspend the process until process 2 has consumed some or all of the data in the pipe. Similarly, an attempt by process 2 to read from an empty pipe causes the operating system to suspend the process until data is written into the pipe by process 1. The end of file condition is handled by explicitly read and writing the end-of-file marker to/from the pipe.

Figure 12-7	Process 1	Process 2

An example of two cooperating processing cycles using a pipe for synchronization and data communication

```
Process 1                             Process 2

open_pipe(process 2,"to",pipe_id)     wait(process 1,pipe_id)
signal(process 2,pipe_id)             read_pipe(pipe_id,record)
read(record)                          while (not eof(pipe_id))
while not end-of-file                      transform(record)
     write_pipe(pipe_id,record)          write(record)
     read(record)                        read_pipe(pipe_id,record)
end while                              end while
close(pipe_id)                        close(pipe_id)
```

CPU ALLOCATION

Allocation of the CPU to active processes is one of the most important activities of the operating system. Processes make progress toward completion of their task only when they have CPU cycles with which to execute their instructions. The large number of processes that are active within a multitasking operating system make the process of CPU allocation extremely complex. The operating system must rapidly make decisions about which process receives access to the CPU and for how long it retains that access. It also must update information in the run queue each time the CPU is reallocated.

No more than one process can control a CPU at any given time. Thus, the number of actively executing processes is limited to the number of CPUs within the computer system. The operating system manages a much larger number of active processes than the number of available CPUs. The operating system executes these processes **concurrently** by rapidly reallocating the CPU(s) among them (see Figure 12-8) in a technique sometimes called **interleaved execution**.

Figure 12-8

Concurrent (interleaved) execution of processes on a single CPU

	Time						
Process 1	Running	Idle	Idle	Idle	Idle	Running	Idle
Process 2	Idle	Running	Idle	Running	Idle	Idle	Idle
Process 3	Idle	Idle	Running	Idle	Running	Idle	Running
CPU	Process 1	Process 2	Process 3	Process 2	Process 3	Process 1	Process 3

Process States

An active process may be in one, and only one, of the following states of operation:

▶ Ready.

▶ Running.

▶ Blocked.

Figure 12-9 depicts these process states and the movement of processes from one state to another. Processes in the **ready state** are waiting for access to a CPU. There are typically a large number of ready processes at any point. When a CPU becomes available, the operating system chooses one ready process to take control of that CPU.

Figure 12-9

The movement of processes among process states

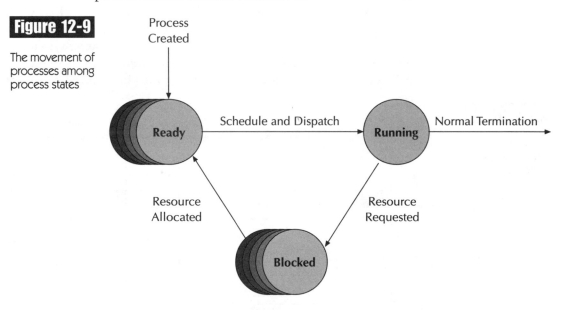

The act of giving control of a CPU to a ready process is called **dispatching**, which is implemented by placing the address of one of the process's instructions in the instruction pointer. On the next fetch cycle, the process takes control of the CPU and its directly related resources (e.g., general purpose registers). Once dispatched, a process has entered the **running state**.

A process in the running state retains control of the CPU until one of the following events:

▶ The process terminates (halts) normally.

▶ An interrupt occurs.

When a process terminates its own execution, control returns to the operating system. The exact mechanisms by which this occurs vary among CPUs and operating systems. The most common implementation is for the process to execute an exit service call. Recall that service calls are generally implemented as process-generated interrupts (i.e., a process explicitly generates an interrupt to call a subroutine in the operating system service layer).

Interrupts can occur for a variety of reasons other than a process executing an exit service call, including:

► Executing some other service call (e.g., request for additional memory).

► A hardware-generated interrupt indicating an error (e.g., overflow) or a critical condition (e.g., power-off alarm from a backup power supply).

► An interrupt generated by a peripheral device (e.g., by a network interface unit in response to the arrival of a data packet).

Recall that any interrupt causes the currently active process to be suspended and pushed onto the stack. The process remains on the stack until processing of the interrupt is completed. During this time, the process is considered to be in a **blocked state**. Whether or not a process remains in the blocked state after an interrupt is processed depends on the type of interrupt and the processing that occurred for that interrupt.

Interrupt Processing

A process in the blocked state is waiting for the occurrence of some event. Typically, that event is either the allocation of a requested resource or the correction of an error condition. Error conditions may or may not be correctable. If they are not correctable (e.g., overflow or memory protection fault), then the process is killed immediately. If the error can be corrected, then the process remains in the blocked state until the error can be corrected.

For example, a process attempting to access a file on diskette will generate an error if the diskette is not in the drive or if the drive bay door is not closed. The operating system typically generates an error message to the user asking that a disk be inserted or the drive bay door be closed. If the user complies, the error condition is cleared and the process leaves the blocked state. If the user does not comply (e.g., clicks an abort button), then the error is not corrected and the requesting process is killed.

Most service calls either directly or indirectly require the allocation of resources. For example, a service call that requests additional memory for a process is a direct request for resource allocation. Service calls for file and/or device I/O represent indirect requests for resource allocation. For example, a request to open a file requires the allocation of memory buffers to hold file data being transferred to/from secondary storage. If the open request is for read/write access, then it also represents a request for exclusive access to the file. A request for file input is an indirect request for access to the secondary storage device, its controller, and the I/O channel(s) that connect it to its allocated memory buffers.

Some resource allocations can be performed relatively quickly (i.e., during the execution of the interrupt handler). Examples include allocation of additional memory to a process (assuming that any is currently available). But many resource allocations require a period of time to complete. For example, file input is not complete until a command has been sent to the disk controller and data has been accessed and moved to a memory buffer. Because secondary storage devices are typically much slower than the CPU, there may be a time lag—from dozens to thousands of CPU cycles—between the time an access command is issued and the time data is available for transfer to memory.

The process that requested input from the file remains in the blocked state during the time lag between a resource request and satisfaction of that request. Thus, for example, a service call for file input results in the following sequence of events:

1. The running process requests file input by issuing an appropriate interrupt.

2. The running process is automatically pushed on the stack and the appropriate interrupt handler (service layer subroutine) is called. The formerly running process is now in the blocked state.

3. The interrupt handler executes. It first checks to see if the requested data is already in a memory buffer. If it is, then the requested resource is already available. The interrupt handler transfers the data from the memory buffer to the data area of the requesting process. The interrupt handler then exits, and the original process is now in the ready state.

4. If the requested data isn't already in a memory buffer, then it must be obtained from a secondary storage device. The interrupt handler generates an appropriate read command and sends it to the secondary storage controller. The interrupt handler then exits without waiting for the data to be returned. The original process stays in the blocked state.

5. When the physical access to data on disk is completed, the secondary storage controller sends an interrupt to the CPU. The interrupt handler transfers the data from the secondary storage controller to a memory buffer over the system bus and then copies it to the data area of the requesting process. The interrupt handler then exits. The requesting process is now in the ready state.

In summary, a process enters the blocked state when it directly or indirectly requests a resource (i.e., memory, creation of a child process, input from a file, or data transferred to/from an I/O device) or generates an error. The process remains in the blocked state until the requested resource is provided or the error is corrected. Once the resource has been provided or the error has been corrected, the process moves to the ready state.

Scheduling

All processes in the ready state compete for access to the CPU. The term **scheduling** refers to the decision to give CPU access to one of the ready processes, and, by extension, not to give CPU access to the other running processes. The term **scheduler** describes that portion of the operating system kernel that makes scheduling decisions. Operating systems vary widely in their approach to implementing the scheduler. Dispatching decisions can be based on any of the following methods:

▶ Preemptive scheduling.

▶ Priority-based scheduling.

▶ Realtime scheduling.

Many operating systems combine aspects of these methods and the exact mechanisms by which they are implemented vary widely.

Preemptive Scheduling

Preemptive scheduling is a generic term that describes any scheduling method that allows a process to be removed from the running state involuntarily. A running process controls the CPU by controlling the content of the instruction pointer. To be involuntarily removed from the running state, there must be some mechanism to place the address of an instruction from another process (preferably, the operating system) in the instruction pointer. The automatic interrupt handling of the CPU is the method by which this occurs. That is, any interrupt received by the CPU results in the running process being pushed on the stack and the starting address of an operating system routine being placed in the instruction pointer.

The portion of the operating system that receives control when an interrupt is detected is sometimes called the **supervisor**. Recall from Chapter 7 that the role of the supervisor is to serve as a master interrupt handler. The supervisor looks up the value of the interrupt register in the interrupt table and executes a call to the associated address (i.e., the address of the corresponding interrupt handler). When the interrupt handler finishes, control is passed back to the supervisor. At this point, the supervisor passes control to the scheduler.

The scheduler has four tasks to perform:

1. Update the status of any process affected by the processing of the last interrupt.

2. Decide which process to dispatch to the CPU.

3. Make appropriate changes to PCBs and the stack to reflect the scheduling decision.

4. Dispatch the selected process.

The processing of an update results in a state change for at least one process. If the interrupt resulted from a request for a resource that couldn't be immediately provided, then the process currently on top of the stack must be moved from the running state to the blocked state. If processing the interrupt cleared an error condition or supplied a resource for which some process was waiting, then that process must be moved from the blocked state to the ready state. The scheduler makes updates to the appropriate process control blocks to reflect these state changes. Figure 12-10 illustrates the processing sequence for both a service request and a delayed satisfaction of that request.

Figure 12-10

Steps of interrupt processing. The processing steps on the left result from an I/O service call by program 1. The processing steps on the right result from the completion of the I/O task by the I/O device.

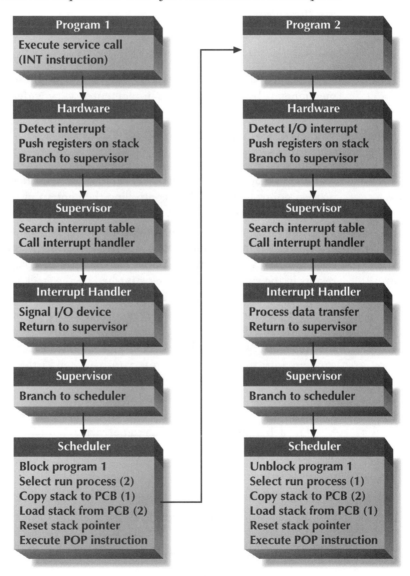

Program 1

Execute service call (INT instruction)

Program 2

Hardware

Detect interrupt
Push registers on stack
Branch to supervisor

Hardware

Detect I/O interrupt
Push registers on stack
Branch to supervisor

Supervisor

Search interrupt table
Call interrupt handler

Supervisor

Search interrupt table
Call interrupt handler

Interrupt Handler

Signal I/O device
Return to supervisor

Interrupt Handler

Process data transfer
Return to supervisor

Supervisor

Branch to scheduler

Supervisor

Branch to scheduler

Scheduler

Block program 1
Select run process (2)
Copy stack to PCB (1)
Load stack from PCB (2)
Reset stack pointer
Execute POP instruction

Scheduler

Unblock program 1
Select run process (1)
Copy stack to PCB (2)
Load stack from PCB (1)
Reset stack pointer
Execute POP instruction

The scheduler then decides which process will be placed in the running state next. It examines the PCBs of all processes to determine which processes are eligible for dispatch (i.e., in the ready state). It chooses one process from among this set based on its predefined scheduling algorithm. Scheduling algorithms are discussed in detail in the next few sections.

Once the scheduler decides which process will run next, it needs to perform any needed updates to the stack and/or process control blocks. This task is trivial if the process that was last interrupted will be returned to the running state. The process control block for that process will already indicate that it is in the running state. The only updates that may need to be made are updates of resource-utilization statistics such as total CPU time consumed. Transfer of control back to the process is also trivial: The scheduler simply issues a POP instruction and the process resumes execution on the next CPU cycle.

Updates of the stack and PCB are much more complex if the next process to execute will not be the process most recently interrupted. In that case, the contents of the stack must be saved to the PCB of the interrupted process. Notice that the entire stack must be saved, not just the topmost element because any elements below the topmost element represent suspended subroutines, or functions, of the interrupted process.

For example, assume that a process (main) begins execution, calls one subroutine (A), that subroutine calls another subroutine (B), and an interrupt is received before any of the subroutines completes execution. After the interrupt is received and control is passed to the supervisor, there will be three process elements on the stack (see Figure 12-11): main (pushed when subroutine A was called), A (pushed when subroutine B was called), and B (pushed when the interrupt was detected). Thus, the state of the process is not simply the topmost element of the stack; rather, it is the entire content of the stack. That entire content must be saved to the process's PCB so that it can be restored to the stack when the process is dispatched at some future time.

Figure 12-11

A sequence of process subroutine calls is shown on the left. On the right are the contents of the stack after an interrupt is received while executing subroutine B . The process state consists of the entire stack contents.

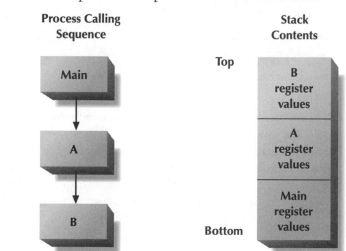

Process Calling Sequence

Stack Contents

The PCB of the process to be dispatched must be changed to indicate it is in the running state. Its most recent stack contents must be copied from its PCB to the stack. The scheduler then transfers control to the process by issuing a POP instruction.

Timer Interrupts

Because the timing of interrupts is unpredictable, the time that a process will remain in the ready state is also unpredictable. Theoretically, it is possible for a process to control the CPU indefinitely if it makes no service calls, generates no error conditions, and no external interrupts are received. Under these conditions a process stuck in an infinite loop could theoretically execute forever.

Most CPUs periodically generate an interrupt simply to give the operating system scheduler an opportunity to suspend the currently executing process if it so chooses. A **timer interrupt** is generated at regular time intervals of between several dozen and several hundred CPU cycles. As with other interrupts, the currently executing process is pushed onto the stack and control is transferred to the supervisor. Because a timer interrupt isn't a "real" interrupt, there is no interrupt handler to call. Instead, the supervisor simply passes control to the scheduler.

Timer interrupts are an important CPU hardware feature for multitasking operating systems because (1) they guarantee that no process can hold the CPU for long periods of time and (2) they provide the scheduler with an opportunity to place other processes in the running state. This ensures that all processes make some progress toward completion at regular intervals.

Priority-Based Scheduling Methods

Various methods are available to determine which ready process should be dispatched to the CPU, including:

► First come first served (FCFS).
► Explicit prioritization.
► Shortest time remaining.

An operating system can implement any or all of these methods, and the exact methods for implementing the second two vary widely.

The simplest scheduling method is **first come first served (FCFS)**. The implementation of this method requires only that the scheduler keep track of when each job most recently entered the ready state. This can be accomplished by recording the time of the state change in the process control block or by maintaining a queue of ready processes in order of arrival. An ordered queue is normally implemented as a circular linked list of PCBs.

Explicit prioritization assigns each process a specific priority level. There must be some mechanism by which priorities are initially assigned to processes, which may be based on the default priority of the process owner (user), a priority stated explicitly in an operating-system command, or any number of other methods. Many operating systems automatically assign higher priority levels to their own component processes than to user processes, with the rationale that operating system processes must be completed quickly to maintain total system throughput.

Priority levels can be used in several ways, including the two most common:

▶ Always dispatch the process of highest priority when the CPU is available.

▶ Assign larger time slices to processes of higher priority.

The first strategy ensures that high-priority processes will always be dispatched before lower-priority processes. It also guarantees that a newly created high-priority process will be immediately dispatched to the CPU. Dispatching based solely on priority levels has a significant negative side-effect. That is, it can result in extremely long or infinite idle time for long jobs of low priority because they are constantly moved to the "back of the queue" and are unable to obtain sufficient CPU time to complete. Many operating systems automatically increase the priority level of "old" processes to compensate for this. Priority-based scheduling using variable length time slices also can address this problem.

Shortest time remaining scheduling chooses the next process to be dispatched based on the expected amount of CPU time needed to complete the process. This can be accomplished directly by dispatching processes based on information about time remaining to completion. It can be accomplished indirectly by increasing the explicit priority of a process as it nears completion. In either case, the scheduler must know how much CPU time will be required for the process to execute completely and how much time has been used already. This implies that the required CPU time was provided to the scheduler when the process was created and that this information was stored in the process control block. It also implies that the amount of CPU time used by a process is stored in its process control block and is updated each time the process leaves the ready state.

Realtime Scheduling

A **realtime scheduling** algorithm guarantees certain minimum levels of CPU time to processes, if requested. Realtime scheduling is employed when one or more processes must be allocated enough resources to complete their function within a specified time interval. Examples of processing environments that require realtime scheduling include some transaction processing applications, data acquisition, and automated process control. In the case of a transaction processing application, a realtime processing requirement might specify a minimum response time for a query against a database. For example, an automated bank teller machine expects a response to a query of a customer's balance within a specified time interval.

Realtime scheduling is often employed for data acquisition and process control applications—similar application types in that data arrives at a constant rate from one or more hardware devices. In a data acquisition process, data inputs are simply copied to a storage device such as tape or disk, and data analysis is performed by a separate process later. This type of processing typifies many scientific environments (e.g., radio astronomy).

A process control program actively processes the data as it arrives. For example, a chemical manufacturer may use a process control program that receives a constant flow of data from sensors within pipelines. This information might represent temperature, flow rate, or other significant data for controlling the manufacturing process. Typically, the data inputs from these sensors are buffered and the process control program extracts input from the buffers. The realtime processing requirement arises from the need to extract and process the data inputs quickly enough to keep the input buffer(s) from overflowing, and thus losing data. Thus, the process control program must be allocated sufficient resources to ensure that it can consume incoming data at least as quickly as it arrives.

In general, realtime processing requirements state a minimum time interval (i.e., elapsed or "wall clock" time) for completing one **program cycle**. A program cycle may involve the processing of a single transaction, the retrieval and storage of data from an I/O device, and/or many other possible actions. The operating system must know both the CPU time required to complete one program cycle and the maximum allowed elapsed time. The scheduler must track the progress of the process by updating CPU time used in the process control block. Furthermore, it must regularly check the status of realtime processes to determine whether or not they need immediate dispatching to meet their response requirements, which usually necessitates a check of the PCB every time a timer interrupt is generated.

Windows NT Scheduling

The Windows NT operating system supports a number of advanced features, including multitasking, multithreaded processes, and preemptive priority-based scheduling and dispatching. Scheduling, dispatching, and interrupt handling are performed by an operating-system component called the microkernel (i.e., the microkernel performs the jobs of the supervisor and scheduler).

Each process managed by the microkernel has a base priority level. Base priority levels are integers with values ranging from 1 to 31. Higher values represent increasing scheduling priority. Priority levels are grouped into four categories called base priority classes: Idle (priority levels 1–6), Normal (priority levels 6–10), High (priority levels 11–15), and Realtime (priority levels 16–31). A process is assigned a base priority class when it is first created. Most user processes start execution in the Normal base priority class. The microkernel generally raises the base priority class to High when a process is moved from the background to the foreground (i.e., when the process is currently receiving input from the keyboard and mouse). It is important to note that the Realtime base priority class does not meet the definition of realtime scheduling given earlier in the chapter; rather, it simply defines the highest group of priority levels in an explicit prioritization scheme.

Process threads have base priority levels that normally vary within the range of priority levels for their parent process's base priority class. Thread priority classes are defined as Idle, Lowest, Below Normal, Normal, Above Normal, Highest, or Time Critical. The precise meaning of these classes within process base priority classes is summarized in Table 12-1.

TABLE 12-1	Base Priority Class	Thread Priority Class	Priority Level
Raw priority levels associated with each combination of process base priority class and thread priority class	Realtime	Time Critical	31
		Highest	26
		Above Normal	25
		Normal	24
		Below Normal	23
		Lowest	22
		Idle	16

TABLE 12-1	Base Priority Class	Thread Priority Class	Priority Level
Continued		Time Critical	15
		Highest	15
		Above Normal	14
	High	Normal	13
		Below Normal	12
		Lowest	11
		Idle	1
		Time Critical	15
		Highest	10
		Above Normal	9
	Normal	Normal	8
		Below Normal	7
		Lowest	6
		Idle	1
		Time Critical	15
		Highest	6
		Above Normal	5
	Idle	Normal	4
		Below Normal	3
		Lowest	2
		Idle	1

The current raw priority level of a thread is called its dynamic priority. The initial value of a thread's dynamic priority level is based on the base priority class of the thread and parent process. The microkernel subsequently may alter dynamic priority to improve system performance and to reflect a thread's high or low demand for system resources. Dynamic priority levels may be adjusted over a wide range of possible values.

Figure 12-12 shows the display of a Windows NT utility called Process Viewer. It shows information about active processes, including the process's base priority level and the priority class and dynamic priority of its threads. This example shows information for the process explorer and its five threads. The base priority class of explorer is Normal. The initial base priority class for the highlighted thread (zero) is Above Normal. Thus, the initial dynamic priority for this thread was 9. Note that the current value (shown at the bottom of the display) is 14. Thus, it is apparent that the microkernel has adjusted the dynamic priority upward from its initial value.

Figure 12-12

Detailed information about a process and its threads, including base priority and thread priority classes

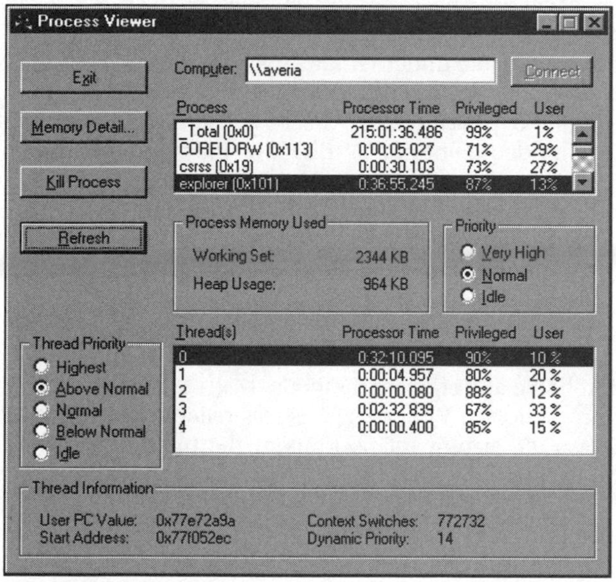

The microkernel implements a strict priority-based scheduling algorithm. A higher-priority thread will always be executed before a lower-priority thread unless the higher-priority thread is in a blocked state. The creation of a new thread with a higher priority level than the currently executed thread will cause the microkernel to preempt the current thread. The current thread will be moved to the ready state and remain there until it all higher-priority threads terminate or become blocked.

Which Windows for the Business Desktop?

Microsoft currently markets two competing versions of Windows. Windows 95, touted as "Windows for the masses," was clearly designed to succeed Windows 3.1. Windows NT, on the other hand, is marketed as an "industrial strength" desktop operating system. It was designed from the ground up to offer maximal reliability, portability, and security. The user interface, or command layer, of both versions is nearly identical, but there are many differences "under the hood."

One of the most significant distinctions is the degree of backward compatibility with programs designed for DOS. Windows 95 was designed for maximal DOS compatibility, and many DOS programs execute under Windows 95. Some programs will run only by exiting Windows 95 to DOS, and Windows 95 automates the operating system exit/reentry for these programs. Many DOS programs will not execute under Windows NT, however. Windows NT routes all access to hardware devices through its own device drivers and prohibits applications from bypassing them. Because many DOS programs access hardware (e.g., video controllers) directly, they are incompatible with Windows NT.

Both Windows versions implement multitasking but do so using different techniques. Applications designed specifically for Windows 95 are multitasked using a generally reliable method, but applications designed for DOS and earlier Windows versions cannot be multitasked by this method. The multitasking method used for these applications allows one application to lock up or crash the entire system. Multitasking in Windows NT is implemented with reliable and efficient methods.

Memory protection in Windows 95 is less reliable than in Windows NT. Under Windows 95, significant portions of the operating-system memory space can be accessed by application programs. Applications designed for Windows 95 are protected from one another but older applications share memory space and, thus, can interfere with one another. Program lockups and system crashes are relatively common with Windows 95. Windows NT applications each run in their own protected memory space. All operating-system code is protected from applications. The result is a highly reliable execution environment in which crashes are rare.

Both operating systems incorporate extensive support for networking. This support is sufficient to meet the vast majority of typical networking needs. The primary difference is in the methods used to share resources among computers. Resource sharing among Windows 95 computers is less secure than resource sharing among Windows NT computers. Network security is also easier to manage centrally with Windows NT than with Windows 95.

Hardware requirements are higher for Windows NT than for Windows 95. A minimal usable configuration for Windows 95 is an 80486 processor, 16 MB of memory, and several hundred MB of disk space. A minimal usable configuration for Windows NT is a Pentium processor, 32 MB of memory and at least 1 GB of disk space. Windows NT can utilize some non-Intel CPUs, including the DEC Alpha. It can also efficiently use multiple CPUs. Memory requirements for Windows NT are generally 30 to 50% greater when utilizing non-Intel CPUs.

Questions:
1. Which operating system would you recommend for a home user?
2. Which operating system would you recommend for a small business user? Assume that the business has 10 PCs connected to file and print servers.
3. Which operating system would you recommend for a local office of a Fortune 500 company? Assume the office has 125 PCS, five network servers, and dedicated high-speed network connections to a nationwide corporate network.

MEMORY MANAGEMENT

The complexity of memory allocation depends on the number of applications that are supported as well as the type of memory addressing and allocation used. For a modern multitasking operating system, the goals of the memory allocation and management function are:

▶ Allow as many processes as possible to be active.

▶ Respond quickly to changes in memory demand by individual processes.

▶ Prevent unauthorized changes to a process's memory region(s).

▶ Implement memory allocation and addressing as efficiently as possible.

Because these goals are somewhat in conflict, achieving them requires a balanced and carefully designed approach to memory management.

Single Tasking Memory Allocation

As with the allocation of other resources, memory allocation in a single tasking operating system is relatively simple. A memory map for a single tasking operating system is shown in Figure 12-13 on the next page. The bulk of the operating system normally occupies the lower addresses within memory and the application program is loaded immediately above it. Addressing individual instructions and data items within the application program is normally accomplished by some form of offset addressing, as discussed in Chapter 6.

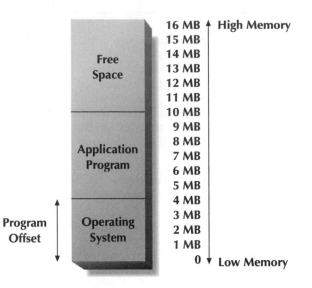

Figure 12-13

Allocation of memory between the operating system and a single application program

Program Offset

Free Space

Application Program

Operating System

16 MB ▲ High Memory
15 MB
14 MB
13 MB
12 MB
11 MB
10 MB
9 MB
8 MB
7 MB
6 MB
5 MB
4 MB
3 MB
2 MB
1 MB
0 ▼ Low Memory

Because many operating systems would consume an excessive amount of storage if entirely loaded into memory, it is common for some operating system routines to be loaded only on demand. If such routines must be loaded during the execution of an application program (e.g., to implement a service call), they will normally be loaded in the highest regions of memory, as depicted in Figure 12-14. This allows free space between the top of the application program and these service routines to be easily allocated to the application program, if needed.

Figure 12-14

Allocation of memory between the base operating system, a single application program, and extended operating system routines

O/S Extensions

Free Space

Application Program

Operating System

16 MB
15 MB
14 MB
13 MB
12 MB
11 MB
10 MB
9 MB
8 MB
7 MB
6 MB
5 MB
4 MB
3 MB
2 MB
1 MB
0

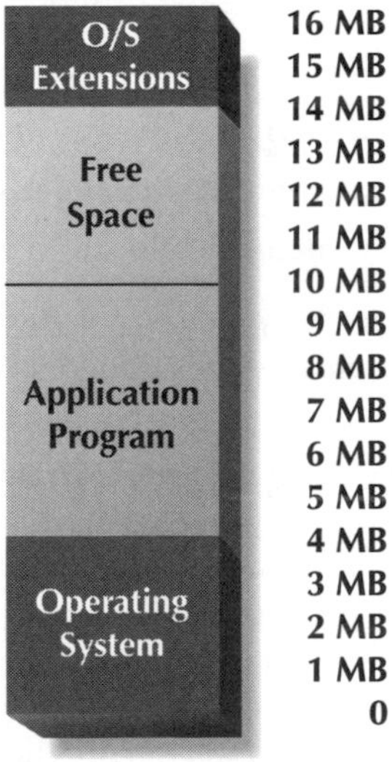

For the application program in Figure 12-14, memory allocation is **contiguous**. That is, all portions of the program are loaded into sequential locations within memory. In contrast, the memory allocation for the operating system shown in Figure 12-14 is **noncontiguous** (or **fragmented**). Although most of the operating system is stored contiguously in low memory, a portion is stored in high memory, with intervening space that is not allocated to the operating system.

As discussed in Chapter 6, most CPUs utilize some form of offset addressing. Offset addressing allows programs to be located within physical memory in locations other than those assumed at the time of compilation and linking. Programs are compiled and linked, assuming that the first instruction will be located at the first address (zero) in physical memory. If offset addressing is to be used, such programs can be referred to as **relocatable** because they can be placed anywhere within physical memory.

A memory reference within a program occurs when a program instruction uses a memory address as an operand. Examples of instructions that contain memory references include load, store, and branch. The process of determining the physical memory address that corresponds to an address reference within a program is called **address mapping** or **address resolution**. The resolution of addresses used within a program is simple when memory is allocated contiguously. When offset addressing is used, each address reference within a program can be mapped to its equivalent address in memory by adding the offset value. For the application program shown in Figure 12-14, all address references can be mapped to physical memory addresses by adding the value 1048577 (4 MB + 1) to the reference.

Just as the operating system may require the allocation of additional memory while it is executing, so too may an application program. As stated earlier, the placement of additional operating system routines in high memory leaves the largest possible amount of contiguous free space above an application program. Thus, if the program requests additional memory, it is simply allocated from the empty region immediately above it in memory.

Because the region above the application may contain additional operating system routines, the operating system must keep track of memory allocated for this purpose so that it is not allocated to the application program. Thus, each time high memory is allocated to the operating system, that fact must be recorded so that a request for additional memory by the application program can be checked against that boundary. If no operating system routines are present in high memory, application program requests still must be checked to ensure that the requested memory is available between the current top of the program and the last address in physical memory.

Contiguous Memory Allocation for Multitasking

Memory allocation is much more complex when multiple programs are allowed to reside in memory simultaneously. Memory must be partitioned for multiple programs that do not coordinate their activities or their demands for memory. Thus, the operating system is responsible for finding free memory regions to load new programs, loading those programs into those regions, and reclaiming those regions when application programs terminate. Multitasking capabilities require a more sophisticated approach to both the procedures for allocating memory to programs and the mechanisms by which free and allocated memory are managed.

Essential to multitasking memory management is the concept of **partitioned memory**. In its simplest sense, this refers to the division of memory into a number of regions, each of which may hold all or part of an application program. Figure 12-15 depicts memory divided into several fixed-size partitions. Each partition may hold either an operating system fragment, an application program fragment, or nothing at all (i.e., free space).

Figure 12-15

Partitioning of main memory into several fixed-sized regions

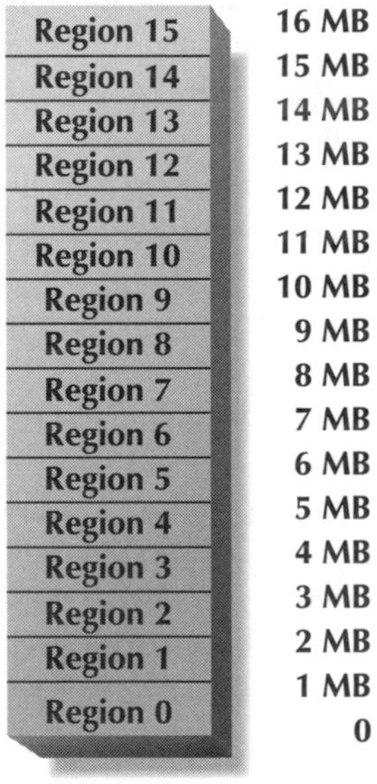

Region 15	16 MB
Region 14	15 MB
Region 13	14 MB
Region 12	13 MB
Region 11	12 MB
Region 10	11 MB
Region 9	10 MB
Region 8	9 MB
Region 7	8 MB
Region 6	7 MB
Region 5	6 MB
Region 4	5 MB
Region 3	4 MB
Region 2	3 MB
Region 1	2 MB
Region 0	1 MB
	0

Figure 12-16 shows three programs and the operating system loaded into fixed-size memory partitions. Program 1 occupies three memory regions, and program 2 occupies two memory regions. Each program fully utilizes all of the memory within its allocated regions. Program 3 occupies four complete regions and part of a fifth. The unused portion of the fifth region is wasted space because it cannot be allocated to another program.

Figure 12-16

Several programs
and the operating
system loaded
into memory that
has been
partitioned into
segments of fixed
size (1 MB)

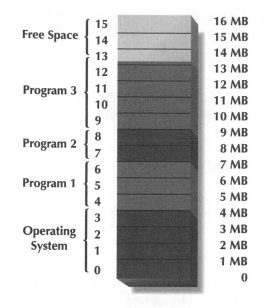

A multitasking operating system that uses partitioned memory must keep track of the status of each partition, which requires a data structure in which to store the status of each memory partition. An example of such a data structure is provided in Table 12-2. The operating system must track each memory partition in terms of its location, status (allocated or free), and the program to which it is allocated, if applicable.

TABLE 12-2	Partition	Starting Address	Status	Allocated To
Data on fixed-sized memory partitions maintained by the operating system	0	0 MB	Allocated	O/S
	1	1 MB	Allocated	O/S
	2	2 MB	Allocated	O/S
	3	3 MB	Allocated	O/S
	4	4 MB	Allocated	Program 1
	5	5 MB	Allocated	Program 1
	6	6 MB	Allocated	Program 1
	7	7 MB	Allocated	Program 2

TABLE 12-2

Continued

Partition	Starting Address	Status	Allocated To
8	8 MB	Allocated	Program 2
9	9 MB	Allocated	Program 3
10	10 MB	Allocated	Program 3
11	11 MB	Allocated	Program 3
12	12 MB	Allocated	Program 3
13	13 MB	Allocated	Program 3
14	14 MB	Free	
15	15 MB	Free	

When a program is ready to be loaded for execution, the operating system must search this table to find a sufficient number of contiguous free partitions to hold the program. If partitions are found, the table is updated with the appropriate data and the program is loaded into the free partition(s). When a program terminates, the table must be updated to show that its memory partition(s) are now free.

Contiguous program loading coupled with fixed-size memory partitions usually results in wasted memory space. As with program 3, the unused portion of the last region allocated to a program is wasted. The total amount of wasted space can be reduced by reducing the size of the partitions. In general, the smaller the size of the partitions, the less wasted space will result. This benefit is offset, however, by the larger number of partitions that must be managed. Small partition size increases the size of the table in which partition allocation data is stored and increases the time necessary to search and update that table.

One way to overcome the limitations of fixed size memory partitioning is to allow memory partitions to vary in size. An example of this type of allocation is shown in Figure 12-17. Notice that each program occupies a memory partition that is exactly the size required by the program. As a result, no memory is wasted initially for unused portions of fixed-size partitions. Variable-sized partitions create additional complexity in the procedures by which the operating system allocates and manages memory. The data structure used to track partitions and their allocation (see Table 12-3) has extra rows of information and its length (i.e., number of rows) varies depending on the number of active programs and memory partitions. Thus, the algorithm used to search and update this table is also more complex.

Figure 12-17

Allocation of
memory to
programs using
variable-sized
partitions

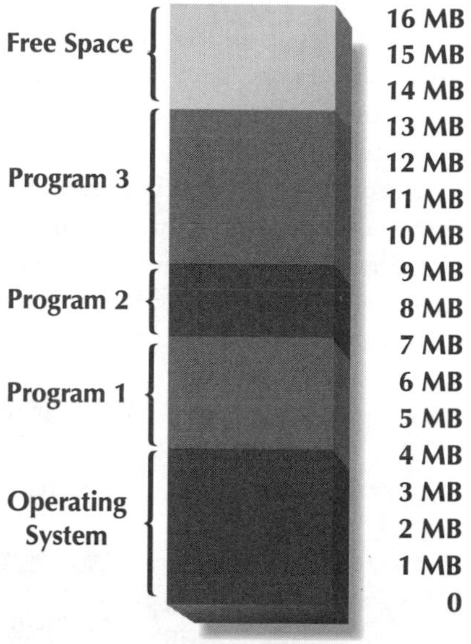

Free Space	16 MB
	15 MB
	14 MB
Program 3	13 MB
	12 MB
	11 MB
	10 MB
Program 2	9 MB
	8 MB
	7 MB
Program 1	6 MB
	5 MB
	4 MB
Operating System	3 MB
	2 MB
	1 MB
	0

TABLE 12-3

Data about
variable-sized
memory parti-
tions maintained
by the operating
system

Partition	Starting Address	Size	Status	Allocated To
1	0 MB	4 MB	Allocated	O/S
2	4 MB	3 MB	Allocated	Program 1
3	7 MB	2 MB	Allocated	Program 2
4	9 MB	4.25 MB	Allocated	Program 3
6	13.25 MB	2.75 MB	Free	

To search for free space for a program, the table must be searched for a free partition of sufficient size. If an exact match is found, then the entry in the table is simply updated to indicate the allocation. If the partition is larger than needed, the corresponding entry in the table must be split. One entry is made for the portion of the partition allocated to the program, and another entry is made for the unallocated (free) portion. In general, the size of the table can grow or shrink as adjacent free partitions are split or combined. In contrast, fixed-size partitions guarantee a table with a fixed number of entries. Updating a fixed-sized table is substantially more efficient than updating a variable-sized table.

Both fixed- and variable-sized memory partitioning suffer problems of memory fragmentation. As programs are created, executed, and terminated, the allocation of memory partitions changes accordingly. In general, the continual allocation and deallocation of memory partitions leads to an increasing number of small free partitions and/or free partitions separated by allocated partitions.

This process is depicted in Figure 12-18. Figure 12-18(b) depicts the allocation of memory after program 3 has terminated and program 5 has been allocated a portion of the newly freed partition. Notice that because program 5 is not as large as program 3, a new partition of free space has been created. In Figure 12-18(c), program 2 has terminated and its former partition is shown as free.

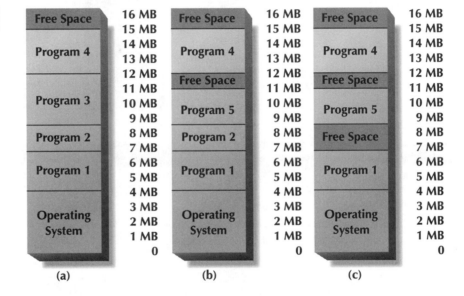

Figure 12-18

Changes over time in the allocation of variable-sized memory partitions as programs execute and terminate

There are now three small partitions of noncontiguous free space. The largest of these is 1.5 MB in size and the total free space is 3.5 MB. Although there may be other programs waiting to execute that require less than 3.5 MB of memory, only those that require 1.5 MB or less can be run due to the fragmentation of available free space. In general, large programs will have difficulty in obtaining required memory allocations due to the necessity to store the entire program in a contiguous block of memory.

One way to address the problem of fragmented free space is to relocate all programs in memory periodically. This process, referred to as **compaction**, results in all free space being collected into a single partition in high memory. Figure 12-19 shows memory contents before and after compaction. Compaction is extremely expensive because the entire contents of programs must be moved within memory, and many entries in the partition allocation table must be updated. The amount of work entailed by these operations is larger than the overhead required to implement a more common strategy—noncontiguous memory allocation to programs, usually coupled with virtual memory techniques.

Figure 12-19

Multiple fragments of free space (a) are collected into a single contiguous fragment (b) by compaction

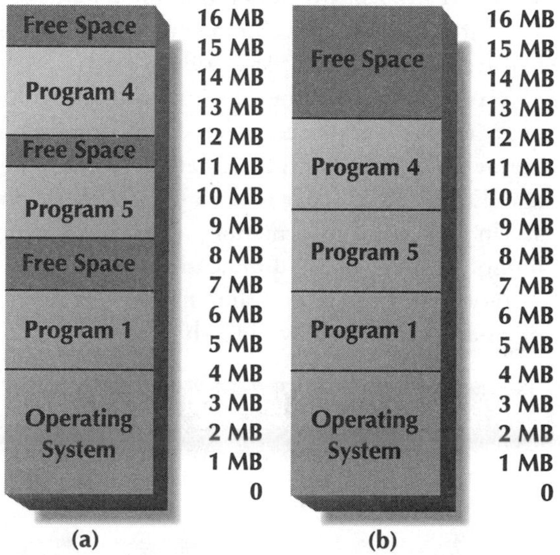

(a)   (b)

Virtual Memory Management

In general, the only portions of a program that must be in memory at any given point during execution are the next instruction to be fetched and any data required to support that instruction (e.g., a memory address referenced by an operand). Thus, at most, only a few bytes of any program must reside in memory at any one time. Although no operating system carries memory-allocation strategies to this logical extreme, several techniques attempt to minimize the amount of program code and data that is stored in memory to relatively small portions. These techniques free large quantities of memory for use by other programs and greatly increase the number of processes that can execute concurrently. However, these techniques also require a considerable amount of operating-system overhead to manage memory allocation and the swapping of program fragments between primary and secondary storage.

Virtual memory management (VMM) divides a program into segments called **pages**. Each page is a relatively small portion of a program (normally between 1 and 4 KB), and page size is a constant. Main memory is also divided into pages of the same size as program pages. Each page of main memory is sometimes referred to as a **page frame**. During program execution, one or more pages are held in memory and the remainder are held in secondary storage. As pages in secondary storage are needed for current processing, they are copied into main memory by the operating system. If necessary, existing pages in main memory may be swapped out to disk to make room for pages being swapped in. Each memory reference made by a program must be checked to see if it refers to a page in memory. References to pages not contained in memory are referred to as **page faults**.

As in simpler memory addressing and allocation techniques, tables—called **page tables**—are used to store information about the allocation of program segments to pages and the location of those pages in primary or secondary storage. Each active process has a page table or portion of a page table dedicated to storing its page information. The information in the table includes the page numbers, a status field indicating whether or not the page is currently held in memory, and the page frame number in main memory or on disk. An example of a page table is shown in Table 12-4.

TABLE 12-4	Page Number	Memory Status	Frame Number	Modification Status
Portion of a program page table. This table is used when resolving memory references made by the program under virtual memory management.	1	In memory	214	No
	2	On disk	101	N/A
	3	On disk	44	N/A
	4	In memory	110	Yes
	5	On disk	252	N/A

Because page size is fixed, memory references relative to address zero can be converted easily to the corresponding page number and offset within the page. The page number can be determined by an integer divide of the memory address by the page size. The remainder of the division is the offset into that page. For example, if page size is 1K, a reference to address 1500 is equivalent to an offset of 476 (1500 − 1024) into page number 2 (1500/1024 + 1). If the table is stored sequentially, the corresponding entry in the process's page table can be computed quickly as an offset into the table. If the reference is to an address in a page held in memory, the corresponding memory address is an offset into the memory page indicated in the table. Using Table 12-4 as an example, a reference to address 700 (offset of 700 into page 1) would translate into an offset of 700 into memory frame 214. Once again, fixed page size allows us to quickly calculate the corresponding memory address as 219836 ($217 \times 1024 + 700$).[1]

An area of secondary storage—the **swap space**—is reserved exclusively for the task of holding pages not held in (or swapped from) main memory. This space is divided into pages in the same manner as main memory. A memory reference to a page held in the swap space results in that page being loaded into a page frame in memory. As with the resolution of addresses in main memory, the location of the page within the swap space can be quickly computed by multiplying the page number by the page size.

If the system is fully loaded (i.e., all page frames are currently in use), a page fault in the currently executing process implies not only a page swapped into memory but an existing page swapped out of memory. The selection of the page to be swapped out—sometimes called the **victim**—can be made on any number of criteria. Some commonly used methods for selecting the victim are:

► Least recently used.

► Least frequently used.

Each of these strategies requires the operating system to maintain information about the utilization of each page in memory. Searching and updating this information is a part of the system overhead associated with VMM.

When a victim has been selected, it may or may not be copied back to the swap space. Some operating systems maintain a copy of all process pages in the swap space. In this case, the contents of the page in memory are simply copied to the corresponding page in the swap space. The sample data in Table 12-4 shows an entry that indicates whether a page has been modified since it was swapped into memory. If a page has not been modified, the copy held in the swap space is identical to the copy in memory, which implies that the contents of the memory page do not have to be copied to the swap space if that page has been selected for replacement.

Memory Protection

Memory protection refers to the protection of memory allocated to one program from unauthorized access by another program, which may apply to interference between programs in a multitasking environment or interference between a program and the operating system in either a single or multitasking environment. A lack of memory protection allows errors in one program to

---

[1] Note that all of these calculations are similar to those used for addressing array contents, as described in Chapter 4.

generate errors in another. If the program being interfered with is the operating system, this may result in the entire computer "locking up."[2]

In its simplest form, memory protection requires that each write to a memory location be checked to ensure that the address being written has been allocated to the program performing the write operation. Complicating factors include the use of various forms of indirect addressing, VMM, and cooperating processes (i.e., two or more processes that "want" to share a memory region). Regardless of the form of addressing used, memory protection adds overhead to each write operation to check ownership and other protection information.

Memory Management Hardware Trends

Early implementations of multitasking, protected memory access, and VMM were based exclusively on modifications to system software. However, implementation through software imposes severe performance penalties on the system as a whole. Consider, for example, the overhead required to map program memory references using VMM. Each reference requires the operating system to search one or more tables to locate the appropriate page and to determine the corresponding memory or disk location of that page. Thus, a memory reference that should consume only one or a few CPU cycles consumes many additional cycles for the paging, swapping, and address mapping functions.

The benefits of advanced memory addressing and allocation schemes are quickly offset when they must be implemented in software. Because of this, one trend in the design of computer systems in general and of the CPU in particular has been to incorporate support for these techniques directly in the hardware. This hardware support has existed in mainframe systems for decades and has been implemented within microprocessors since the late 1980s. All Intel microprocessors since the 80386 have hardware support for VMM and protected memory management.

---

[2] This phenomenon is a frequent occurrence with simple microcomputer operating systems such as MS-DOS or early versions of Windows. The only way to recover from such an error is to reboot the computer.

Intel Pentium Memory Management

The Intel 80X86 family of microprocessors has incorporated extensive hardware support for memory protection (MP) and virtual memory management (VMM) since the 80386 processor. MP in the Pentium family of processors is closely tied to the Intel segmented memory model (first described in Chapter 6 on page 214).

Six registers—named CS, DS, ES, FS, GS, and SS—are dedicated to holding data structures called segment descriptors. Segment descriptors contain various information about a segment, including:

► Physical memory address of the first byte of the segment (i.e., its base address).
► Size of the segment.
► Segment type.
► Access restrictions.
► Privilege level.

A memory address used in an operand consists of a reference to one of the segment registers and an offset value. For example, the operand DS:018C refers to byte 018C (hexadecimal) within the segment pointed to by the DS segment descriptor. The CPU converts this to a physical memory address by adding the segment base address and the offset.

Segments may range in size from 1 byte to 4 GB. Segment types include data segments and code segments (i.e., executable instructions). The data type is further subdivided into "true" data and stack data. The CS register always contains a segment descriptor for a code segment. The DS, ES, FS, and GS registers always contain segment selectors for a data segment, or a null descriptor. The SS register always contains the segment descriptor of a stack.

Various forms of MP are implemented by checking instruction operands that reference segment descriptors. The simplest of these checks is a limit check on the offset value. For each memory reference, the CPU compares the offset value to the size parameter stored in the segment descriptor. If the offset is larger than the segment size, then an error interrupt is generated. This form of MP prevents processes from reading or writing data outside the boundaries of their segments.

Access types must be specified for each segment. Data segments can be marked as read-only (RO) or read-write (RW). Code segments can be marked as execute-only (EO) or execute-read (ER). The access level of a segment may restrict the execution of certain types of instructions with respect to memory locations within the segment. An error interrupt will be generated if an instruction attempts to write to a segment marked RO. An error interrupt also is generated if an instruction attempts to read from a segment marked EO. Write operations to code segments always result in an error interrupt. The CPU also checks segment types when loading segment descriptors into a register. Segment descriptors for data segments cannot be loaded into the CS register. A segment descriptor for a code segment cannot be loaded into the DS, ES, FS, GS, or SS registers.

Two different methods are implemented for preventing any access to code segments not allocated to a particular process. The first of these is implemented by specifying a privilege level for each process and segment. Privilege levels are numbered 0 (zero) through 3, with 0 being the most privileged. Every process is assigned a privilege level, and it cannot access a segment with a privilege level numbered lower than its own privilege level (an error interrupt is generated). This form of MP is used to protect operating system components from interference by application programs. Segments allocated to the kernel typically have a 0 privilege level. Segments allocated to other portions of the operating system typically have a privilege level of 1 or 2. Application processes typically have a privilege level of 3.

The existence of some segments may be hidden from a process. The processor maintains a global descriptor table (GDT) containing descriptors for all segments currently defined. If an application process uses the GDT, then it can "see" all of the defined segments. However, the operating system can issue a specific instruction that causes a process to be allocated a local descriptor table (LDT). The operating system defines the content of the LDT prior to issuing the instruction. Typically, the operating system places segment descriptors for only those segments allocated to a process in its LDT. Segment descriptors for all other segments are omitted from the LDT, which makes it impossible for a process using an LDT to know of any segments other than its own.

Support for VMM in the Pentium processor family is implemented by the definition and automatic use of VMM tables. Page size is normally set to 4 KB. The CPU uses two types of VMM tables—page directories and page tables. A page directory is a table of pointers to page tables. By convention, the entries in the page directory correspond directly to the segment descriptors contained in a process's LDT (see Figure 12-20). Each entry in a page directory points to one entry in a page table, and each entry in a page table contains descriptive information about one 4 KB page.

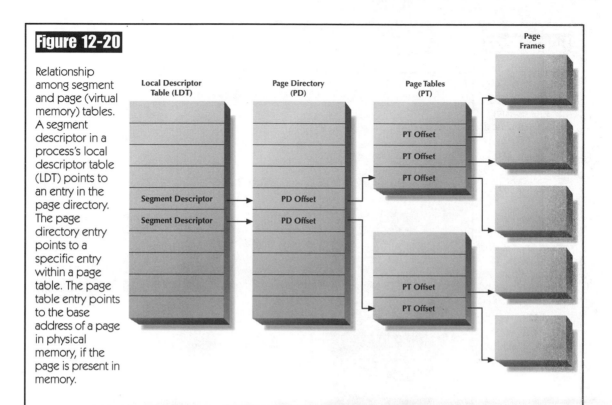

Figure 12-20

Relationship among segment and page (virtual memory) tables. A segment descriptor in a process's local descriptor table (LDT) points to an entry in the page directory. The page directory entry points to a specific entry within a page table. The page table entry points to the base address of a page in physical memory, if the page is present in memory.

Local Descriptor Table (LDT)

Segment Descriptor

Segment Descriptor

Page Directory (PD)

PD Offset

PD Offset

Page Tables (PT)

PT Offset

PT Offset

PT Offset

PT Offset

PT Offset

Page Frames

A page table entry includes the physical memory address of the page if it is loaded into memory as well as a number of flags (bit fields), including a flag to indicate whether the page is in memory, a flag indicating whether or not the page has been accessed since it was loaded, and a flag to indicate whether the content of a loaded page has been written to. A significant omission from a page table entry is the disk location of a swapped out page.

The responsibility for implementing VMM is split between the CPU and the operating system. When VMM is enabled, the processor converts addresses consisting of a segment descriptor and offset into addresses consisting of a page directory offset, a page table offset, and an offset into the 4-KB page. The CPU automatically performs a lookup in the page directory and page table based on these offsets to find the base physical address of the page (see Figure 12-21 on the next page). It then adds the page offset to generate a physical address. If the page table entry indicates that the page is not in memory, then the CPU generates an interrupt. The CPU also sets the access and write flags in the page table page contents are read or written.

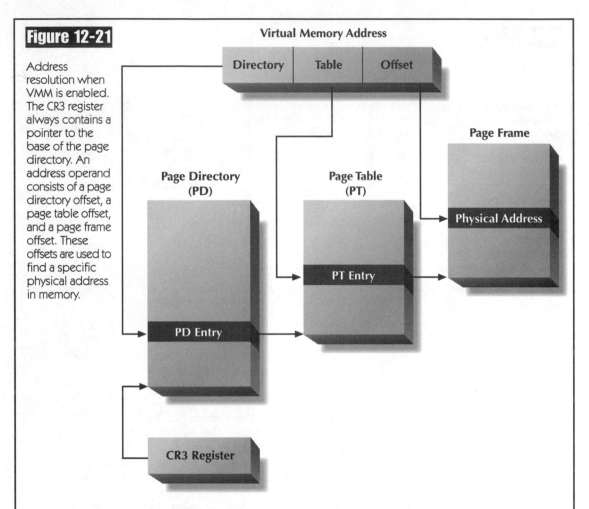

Figure 12-21

Address resolution when VMM is enabled. The CR3 register always contains a pointer to the base of the page directory. An address operand consists of a page directory offset, a page table offset, and a page frame offset. These offsets are used to find a specific physical address in memory.

Virtual Memory Address

Directory | Table | Offset

Page Frame

Page Directory (PD)

Page Table (PT)

Physical Address

PT Entry

PD Entry

CR3 Register

The operating system maintains its own table that maps VMM pages to disk locations in the swap space. It also provides an interrupt handler to implement page swaps when an access to a swapped out page is attempted. The operating system also is responsible for clearing the access and write flags when a page is loaded into memory as well as for setting and clearing the page table entry "in memory" flag each time page load or swap operations are performed.

The performance improvements gained by implementing MP and VMM within the CPU are substantial. A case in point is the UNIX operating system (a multitasking operating system that implements VMM). A few versions of UNIX were developed to run on the Intel 80286 CPU. This processor had little support for MP and no support for VMM. Thus, the operating system had to implement all MP and VMM functions in software. The result was execution so slow as to be unusable.

The 80386 CPU, introduced in 1987, provided MP and VMM hardware support similar to that described earlier. It also increased word size from 16 to 32 bits. When the 80286 UNIX versions were rewritten for the 80386 system, throughput was improved by a factor of 10 to 20 times. The doubling of word size and minor increases in clock speed would have predicted a performance improvement of only two to three times. The extra performance gain realized was attributable to the movement of many MP and VMM functions from software to hardware.

SUMMARY

An operating system is the most complex component of system software. Its primary purpose is to manage hardware resources and to provide support services to users and application programs. In functional terms, the operating system is responsible for the management of the CPU, memory, processes, secondary storage (i.e., files), I/O devices, and users.

From an architectural viewpoint, the operating system consists of the kernel, service layer, and command layer. The kernel performs all resource-allocation functions and directly controls and communicates with hardware devices. The service layer consists of a set of programs that provide basic system services to users or their application programs. These can be classified as process, memory, file, and I/O services. The command layer is the user interface to the operating system. It is used to control the execution of application and utility programs and for direct user control over hardware resources and files.

The operating system is responsible for allocating hardware resources to individual user processes on demand. This is a simple matter in single tasking operating systems because only one application program can be active at a time. Multitasking operating systems must implement substantially more complex procedures for resource allocation. The operating system must mediate contention by multiple processes for access to a limited set of resources while ensuring that each process receives the resources it needs. This mediation is performed primarily through operating system procedures for scheduling, dispatching, and interrupt processing.

The resource-allocation function is complex and requires extensive record-keeping to track current resource commitments and availability. It requires complex procedures to maintain these records and to make allocation decisions. The programs that implement resource allocation consume hardware resources. A key design goal for any operating system is to minimize the hardware resources consumed for resource allocation while ensuring that resource-allocation goals are met.

Application software is simpler to develop if programs are unaware of resource-allocation functions. Operating systems make a large set of virtual resources appear to be available to each application program. The sum of these apparent virtual resources generally exceeds the real resources that exist within the computer system. The operating system implements virtual resources by rapidly reallocating real resources among programs and by substituting one resource for another where appropriate and necessary.

A process is the basic unit of software to which the operating system allocates resources. The operating system stores information about each process in a process control block (PCB). PCB contents are continually updated as resources are allocated to and freed by active processes. Some operating systems allow processes to define executable subunits called threads. Threads, independent software entities for purposes of CPU scheduling, share all other resources with their parent process. Execution speed can be increased by allowing a process to create multiple threads and execute them concurrently or simultaneously. A program that creates and uses multiple threads is said to be multithreaded.

Processes often need to create other processes in a procedure called spawning. Process families may need to synchronize their actions and/or transfer data among themselves. The operating system provides several mechanisms to accomplish this, including signals and pipes. Signals are flags or single-valued data items passed among processes, whereas pipes are data conduits that connect data producers and consumers. Pipes are generally implemented with shared memory and managed directly by the operating system.

An active process is always in one of three states—ready, running, or blocked. In a single-CPU machine, only one process may be in the running state at any one time. Ready processes are waiting only for access to the CPU. Blocked processes are waiting for some event to occur. Such events are normally the completion of a service request or the correction of an error condition. Various methods exist for determining the order and duration of process access to the CPU. Methods for determining order of dispatch include first come first served (FCFS), shortest time remaining, and explicit prioritization. Methods for determining the duration of a process's control of the CPU include priority interrupt and time sharing. Operating systems often implement dispatching and scheduling by a combination of these methods.

Memory is allocated through a partitioning scheme. A memory partition is a fixed or variable portion of available main memory. Processes are allocated one or more memory partitions to store instructions and data. The operating system maintains tables of information concerning the allocation of specific memory partitions to specific processes. Memory references made by a program are mapped to its memory partitions through table lookups and address calculations.

Modern operating systems implement a form of memory allocation and management called virtual memory management (VMM). Portions of processes are held in small partitions of memory called pages. Some pages may be held in main memory, and others may be held in the swap area of secondary storage. Process pages are swapped between main memory and secondary storage as dictated by the needs of individual processes and the availability of main memory.

Complex memory management procedures require a substantial amount of operating system overhead. Techniques such as partitioned, virtual, and protected memory management require address calculations and the maintenance of tables of memory allocation information. To reduce the amount of processor time allocated to such tasks, modern CPUs implement many of these functions within the CPU. This speeds such processing considerably and leaves more CPU cycles free to execute user application processes.

Key Terms

address mapping
address resolution
batch operating system
batch process
blocked state
child process
command layer
compaction
concurrent execution
contiguous memory allocation
dispatching
explicit prioritization
first come first served (FCFS)
fragmented memory allocation
interleaved execution
interprocess synchronization
kernel
multitasking operating system
multithreaded process
multiuser operating system

noncontiguous memory allocation
page
page fault
page frame
page table
parent process
partitioned memory
pipe
preemptive scheduling
process
process control block (PCB)
process family
process queue
program cycle
ready state
real resource
realtime scheduling
relocatable
run queue

running state
scheduler
scheduling
service layer
shortest time remaining
sibling process
signal
single tasking operating system
spawning
supervisor
swap space
system overhead
task
thread
timer interrupt
victim
virtual memory management (VMM)
virtual resource

Vocabulary Exercises

1. A type of processing in which the operating system supports multiple active processes is called _____ processing or _____.

2. Under virtual memory management (VMM), the location of a memory page is determined by searching a(n) _____.

3. A(n) _____ occurs when a program requires a memory page not held in memory.

4. A(n) _____ is a shared memory region that is managed by the operating system and used to send data from one active process to another.

5. Dispatching a process moves it from the _____ to the _____.

6. The CPU periodically generates a(n) _____ to guarantee the scheduler an opportunity to allocate the CPU to another ready process.

7. A process in the _____ requires only access to the CPU to continue execution.

8. In the _____ scheduling method, processes are dispatched in order of their arrival.

9. A(n) _____ process contains subunits that can be executed concurrently or simultaneously.

10. _____ scheduling guarantees that a process will receive sufficient resources to complete one _____ within a stated interval.

11. The hardware resources consumed by the resource allocation functions of the kernel are sometimes referred to as _____.

12. Process control blocks are normally organized into a linked or circular list called the _____ or the _____.

13. Under virtual memory management (VMM), the location of a memory page is determined by searching a(n) _____.

14. _____ scheduling refers to any type of scheduling in which a running process can lose control of the CPU to another process.

15. The act of selecting a running process and loading its register contents is called _____ and is performed by the _____.

16. To achieve efficient use of memory and a large number of concurrently executing processes, most operating systems use _____ memory management.

17. A(n) _____ application program requires no user interaction during its execution.

18. When a process makes an I/O service request, it is placed in the _____ until processing of the request is completed.

19. Memory pages not held in primary storage are held in the _____ of a secondary storage device.

20. On a single-processor computer, multitasking is achieved through _____ execution of processes.

21. Under _____, all portions of a process must be loaded into sequential physical memory locations.

22. The _____, _____, and _____ are the primary layers of an operating system.

23. A(n) _____ operating system can support multiple active processes at any one time.

24. A(n) _____ is the unit of memory read or written to the swap space.

25. A(n) _____ is apparent to a process or user, although it may not exist physically.

26. Under a(n) _____ memory allocation scheme, portions of a single process can be located physically in scattered segments of main memory.

27. A(n) _____ process spawns a(n) _____ process by means of a process-creation service request.

28. Under virtual memory management (VMM), memory references by a process must be converted to an offset within a(n) _____.

29. Information about a process's execution state (e.g., register values, status, etc.) is held in a(n) _____ for use by the _____.

30. Under the _____ scheduling method, processes requiring the least CPU time are dispatched first.

31. The detection of a(n) _____ causes the currently executing process to be _____ and control passed to the _____.

32. The process of converting an address operand into a physical address within a memory partition or page frame is called _____.

33. A(n) _____ can be used for interprocess _____ when no data needs to be passed.

34. A(n) _____ is an executable subunit of a(n) _____ that is scheduled independently but shares memory and I/O resources with its parent.

Review Questions

1. Describe the functions of the kernel, service, and command layers of the operating system.

2. What is the difference between a real resource and a virtual resource?

3. What hardware resources are allocated to processes by the operating system?

4. What are the goals of an operating system resource-allocation function? Describe the conflict among them.

5. How and why does a process move from the ready state to the running state? How and why does a process move from the running state to the blocked state? How and why does a process move from the blocked state to the ready state?

6. What is process control block, and what is its use?

7. What methods does an operating system provide for interprocess synchronization and communication? Under what conditions would each be most applicable?

8. What is a thread? What resources does it share with other threads (within the same process)? What resources doesn't it share?

9. Briefly describe the most common methods for making scheduling decisions.

10. What scheduling complexities are introduced by realtime processing requirements?

11. What are the comparative advantages and disadvantages of memory allocation using fixed and variable size memory regions?

12. Describe the operation of virtual memory management (VMM). Under what assumptions is it more efficient than other methods of memory management?

13. What is memory protection? What capabilities does it imply in the operating system and/or hardware?

Research Problems

1. The RISC approach to processor design (described in Chapter 5) avoids internal CPU complexity to gain raw processor speed. Strict adherence to this design philosophy might be construed to eliminate CPU support for memory protection and virtual memory management (VMM). Investigate the memory management techniques of a RISC CPU such as the DEC Alpha (www.digital.com) or IBM/Motorola PowerPC (www.mot.com). What support (if any) is provided for memory protection and VMM?

2. Investigate the memory protection and multitasking implementation of the Windows NT operating system (www.microsoft.com). How are application programs isolated from the operating system? How are application programs isolated from each other? Is it possible for one application to cause another application or the operating system to crash? How are CPU cycles allocated to application programs? Is it possible to lock up the system (i.e., for a program to be allocated the CPU and never release it)?

Chapter

13

Mass Storage Access and Management

FILE-MANAGEMENT OVERVIEW

A file is the fundamental unit of storage for both data or programs. Files are stored on secondary storage devices, including optical disks, magnetic disks, and magnetic tape. Because of their number and importance, management of files is a critical function of system software. For clarity, this text refers to the entire collection of system software programs that perform file-management and access functions as a **file-management system**. In most systems, the bulk of these programs and functions are provided by the operating system. However, some systems use additional software (e.g., a database management system [DBMS] or a separately purchased backup utility) to extend the file-management capabilities of the operating system.

File Content and Type

To restate the definition given in Chapter 4, a **file** is a collection of related data items. A file has a unique identifier (e.g., a symbolic name) and can be manipulated in various ways as a single entity (e.g., copying, deletion, etc.). Many different types of data may be stored in a file, including:

▶ Character, numeric, and other types of data used by application programs.

▶ Symbolic instructions (i.e., source code).

▶ Machine instructions (i.e., executable code).

Data used by application programs can be of many types. For example, a customer file used by a billing program can contain account numbers, names, addresses, account balances, and other data items. Each of these items may be of a different data type or structure. For example, the account number might be encoded as an integer, the name and address as character arrays, and the account balance as a floating point number.

Symbolic instructions are stored as sequences of characters. Thus, the source code of a COBOL or FORTRAN program is stored as a sequence of characters comprising the program statements. The sequence of characters in the file follows the sequence of the characters as read or written on a printed page. In contrast, machine instructions are stored in a format that can be directly manipulated by the CPU. Thus, an executable program stored in a file consists of a sequence of instructions encoded in the CPU's instruction format.

Because of the wide variety of file contents and methods of accessing those contents, different files may require different methods of storage and manipulation. Because file manipulation is such a pervasive aspect of information processing, it is desirable to implement all file-manipulation functions within system software (e.g., as utility programs and/or as service layer functions). However, it is difficult to design a file-management system that accounts for all possible variations in file content and organization because such a system would be excessively complex and have a very large number of file-oriented commands and service routines.

Most file-management systems strike a compromise in which a limited number of predefined **file types** are provided. For example, a simple file-management system may define the following types of files:

- Executable programs.
- Operating system commands.
- Data.

When file types are defined, every file must be of one, and only one, type. Among other things, the type of a file determines:

- Physical organization of data items and structures within secondary storage.
- Access methods that may be used to read and write file content (e.g., sequential, random, or indexed access).
- Operations that may, or may not, be performed upon the file.
- Restrictions on the symbolic name of the file.

The file-management system uses different physical organizations for different types of files. Thus, the physical storage format of an executable program (i.e., file of machine instructions) differs from the physical structure of a file that contains data. These differences are implemented for the sake of efficiency. For example, files containing machine language programs are organized to minimize the complexity of loading their contents into memory for execution. Operating systems that implement virtual memory management (VMM) may use a physical organization that facilitates paging and swapping operations.

Some file-management systems further differentiate physical structure within file types. For example, different physical storage formats may be used for data files, depending on whether access to those files is sequential or random. In such a case, an application program would typically declare the type of access when interacting with a file through service layer functions. The operating system service layer provides a different set of file I/O service calls for each type of access. A programming language (e.g., BASIC) or application development tool (e.g., a DBMS) may impose additional physical structure on files containing data.

Within many file-management systems, the type of a file determines which operations may be performed upon it. For example, it should not be possible for a user to load and execute any file that does not contain a machine language program. Similarly, it should not be possible to edit a file containing a machine language program with an ordinary word processor or text editor. Implementing access and use restrictions requires a mechanism by which system software can determine the content or structure of a file.

File type is declared at the time a file is created. In some systems, file type is stored as part of a file's directory entry. The file type can be viewed when examining a directory, but is not otherwise obvious. In other systems, the file type may be indicated by a naming convention for the file's symbolic name. A common approach is to require that the filename extension, or appending identifier, match the intended use. The file-management system uses these identifiers to match file type to allowable operations.

For example, consider the naming conventions for executable files used by MS-DOS and Microsoft Windows. Filenames with the extensions .COM and .EXE are assumed to contain executable machine language programs, each stored in a specific format. Files with the extension .DLL are assumed to contain a searchable library of executable subroutines. Filenames with the extension .BAT are assumed to contain the text (ASCII characters) of one or more (batch) commands to be interpreted by the operating system command layer. Many application tools make additional assumptions about filenaming syntax and extensions. Examples include .ASM (macro assembler source code file), .DOC (Microsoft Word document file), .CDR (Corel Draw graphics file), and many others.

The implementation of file types and associated rules and restrictions increases the complexity of the file-management system. Additional programming logic is needed to check file types and implement processing restrictions. A further disadvantage can be that programmers and users will be limited to the set of predefined file types, none of which may be entirely suitable for a specific application.

File-Management System Functions and Components

The primary purpose of the file-management system is to provide facilities for read and writing to/from files. Thus, many file-management service layer functions implement file I/O operations. However, because the number of files in a typical computer system is large, the file-management system also must provide facilities for managing and administering groups of files and secondary storage devices. All of the functions can be loosely grouped into the following categories:

► Creation and manipulation of files.

► File I/O operations.

► Management of secondary storage devices.

► File security.

► File backup and recovery.

These functions are normally implemented within the operating system; however, some systems may use additional system software (e.g., a DBMS) to extend the native capabilities of the operating system.

A file-management system is implemented in layers similar to those of the operating system as a whole. These layers form a hierarchy that spans from the lowest, or device, level to the highest, or user, level. From high to low, layers within a file-management system include:

► Application and command.

► File control.

► Storage I/O control.

► Secondary storage devices.

These levels of the file-management hierarchy are illustrated in Figure 13-1 and are compared to the more general representation of operating system software layers.

Figure 13-1

Layers of the file-management system and their relationship to the more general concept of operating system layers

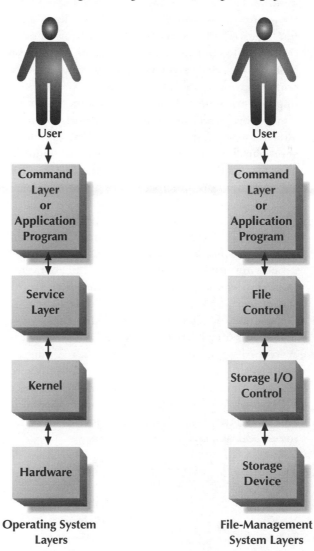

The storage device (or hardware) level includes both storage devices and storage device controllers. Storage devices implement the physical storage of bits, bytes, and blocks within a storage medium. Storage device controllers interact with the bus and with operating system device drivers to direct the actions of storage devices. As discussed in Chapter 7, the device controller presents a logical view of the storage device or media to the device driver. This logical view is a linear sequence of storage locations (i.e., a linear address space).

The **storage I/O control layer** is responsible for implementing access to specific storage locations and for managing the movement of data between secondary and primary storage. Software modules within this layer include device drivers, interrupt handlers, and routines that manage buffers and caches. These modules reside within the kernel of the operating system. Commands to read or write specific secondary storage device locations are received from the file control layer. These are translated into equivalent low-level hardware commands.

The **file control layer** implements basic file-manipulation capabilities (e.g., creating files and read/writing their content). Commands are accepted from the command and application layers by service calls. The file control layer also implements a number of complex functions, including the directory management, storage allocation, and file security. Directory management and storage allocation are closely related tasks. The directory provides a mapping between symbolic filenames and the storage locations that hold file contents. Storage allocation is the process by which storage locations are tracked and allocated to specific files and directories. File security is implemented through access controls based on information contained within the directory.

The file-control layer forms a bridge between the physical file-management system and the logical file-management system. Users and application programs see a file as a collection of data that is structured to satisfy specific information-processing needs. The user's view of file content and organization is called the **logical file structure**, which is different from the underlying **physical file structure**.

Figure 13-2 shows the logical structure of a typical data file. The file is subdivided into multiple **records**, and each record is composed of multiple **fields**. A record typically corresponds with a single person (e.g., a customer or employee), thing (e.g., a product held in inventory), or event (e.g., a transaction). A field usually contains a single data item that describes the subject of the record. For example, a record might contain information about a customer and be composed of individual fields for account number, customer name, customer address, and so forth.

Figure 13-2

Logical structure of a file as a group of related records, each of which is, in turn, a group of related fields

Fields

Records	1	2	3	4	5	6	
1	99635	Smith	George	R	1414 Oak Road	Los Angeles	...
2	04667	Jones	Alice	L	12 Main Street	Buffalo	...
3	52046	Torres	Maria	H	9823 7th Avenue	Chicago	...
4	34421	Miller	Fritz	S	451 Quincy Road	Atlanta	...
5	63752	Chang	Wendy	B	310 G Street	Fort Worth	...

The logical file structure exists independently of the physical devices on which it might be stored. The physical file structure is the manner in which individual bits and bytes of data are represented and organized on a physical storage medium. A number of physical structure characteristics are simplified or ignored in the corresponding logical file structure, including:

- Physical storage allocation.
- Method(s) of physical data access.
- Data encoding.

Ignoring these issues allows the file-management interface provided by the service layer to be simplified. The application or command layer asks for specific contents of a named file. Knowledge of the physical structure that implements the file and communication with the storage device that holds it is not required.

Physical storage allocation considerations include the placement of fields and records within a file and the distribution of file components across storage locations on specific devices. Physical data access considerations include the use of directories (and related structures) and access to nonsequentially stored file contents. Issues of data encoding include the data structures and coding methods used to represent individual fields. Related issues include data encryption and data compression. In sum, the logical view of a file is a sequential set of records and fields that are immediately and exclusively available to a single user or application program. Physical storage and access considerations are simply not relevant to this view.

STORAGE ALLOCATION

Storage allocation refers to the methods by which individual storage locations are correlated with specific files. Like other resource allocation functions, storage allocation functions are implemented by allocation data structures and the algorithms that maintain and use those data structures. Information about the allocation of specific storage locations is recorded within the data structure(s). The file-management system uses the data structure content to make allocation decisions and record those decisions by updating the data structure.

Allocation Units

An **allocation unit**—the smallest unit of storage that can be allocated to a file—cannot be smaller than the unit of data transfer to/from secondary storage devices (i.e., device block size). Secondary storage device block size typically ranges from 512 bytes to 4 KB, in multiples of 512 bytes. Thus, 512 bytes is the smallest possible allocation unit size.

Allocation unit size is determined when an operating system or secondary storage device is installed. Some operating systems (e.g., DOS and early Windows versions) automatically determine unit size whereas others, such as UNIX and Windows NT, allow the installer to select unit size. It is possible to select different unit sizes for each individual storage device or partition. Once unit size is set, it is difficult or impossible to change.

The choice of allocation unit size represents a trade-off among:

► Efficient use of secondary storage space for files.

► Storage space used for storage-allocation data structures.

► Efficiency of storage allocation procedures.

Smaller allocation unit size results in more efficient use of available storage space. For example, a file containing a single byte will be allocated one unit of storage. If allocation unit size is 512 bytes, then 511 bytes store no data. Larger allocation unit size increases the amount of wasted space (e.g., a 1024-byte unit size results in 1023 wasted bytes).

The advantage of larger allocation unit sizes is a reduction in the size of storage-allocation data structures. As allocation unit size decreases, the number of allocation units increases. For example, consider a storage device 1 GB (1024^3

bytes) in size. There are 2,097,152 ($1024^3 \div 512$) allocation units in the device if unit size is set to 512 bytes, whereas there are 262,144 ($1024^3 \div 4096$) allocation units in the device if unit size is set to 4 KB.

A **storage-allocation table** is a record of the allocation of storage locations to individual files. A storage-allocation table contains one entry for each allocation unit. Smaller allocation unit size increases the number of entries in the storage allocation table. Larger table size increases the time required for searches and updates which, in turn, slows down any processing function that changes file size (e.g., an update or delete operation).

The format and content of the storage-allocation table differs for contiguous and noncontiguous storage allocation.[1] As with the allocation of other types of storage (e.g., memory), contiguous allocation is relatively simple and requires little management overhead. Noncontiguous allocation is more complex but inherently more flexible. With either method, updates to the storage-allocation table must be coordinated with updates to the data structures used to store directory information.

Contiguous Storage Allocation

With contiguous allocation, most storage-allocation information is stored in a file's directory entry. This information includes the identifier (number) of the first allocation unit allocated to a file and the number of units allocated to a file. Figure 13-3 (on the next page) shows a set of allocation units and the allocation of those units to three files—the units allocated to each file are similarly shaded. Notice that all units allocated to an individual file are in a contiguous sequence. A simplified directory table that includes storage-allocation information for these files is shown in Table 13-1. A fourth file named SysFree has been added to the table to represent the unused allocation units.

TABLE 13-1	Filename	Owner Name	First Allocation Unit	Length (in Allocation Units)
Simplified directory entries for the files depicted in Figure 13-3	File1	Smith	0	17
	File2	Jones	17	3
	File3	Smith	20	9
	SysFree	System	29	7

---

[1] It may be helpful to review the discussion of memory allocation in Chapter 12 prior to reading the following material.

Figure 13-3

Contiguous
allocation of
storage blocks to
three different
files

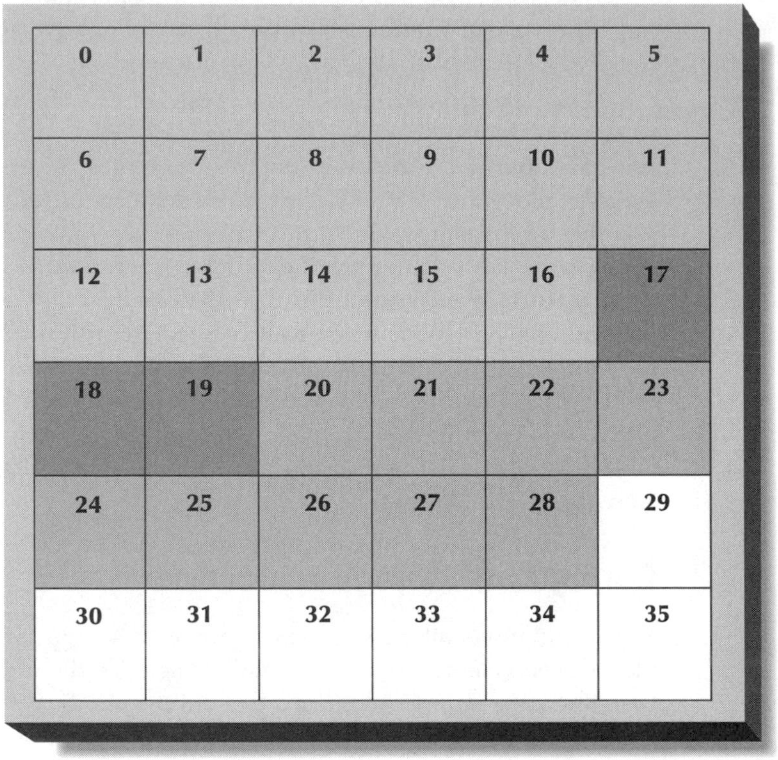

0	1	2	3	4	5
6	7	8	9	10	11
12	13	14	15	16	17
18	19	20	21	22	23
24	25	26	27	28	29
30	31	32	33	34	35

The most significant problem with contiguous storage allocation is the fragmentation of allocation units that occurs after the creation and deletion of many files. Figure 13-4 shows the storage-allocation table from Figure 13-3 after the deletion of File1 and the creation of a new, but smaller, file. There are now two separate regions of unallocated space on the device. Thus, the simple use of the dummy file SysFree is no longer adequate to connect all unallocated storage units. Multiple files—one for each contiguous region of unallocated units—are required. Over a long period, the number of these regions tends to grow and the size of individual regions tends to shrink. Because units allocated to a file must be contiguous, the maximum size of a newly created file is limited to the size of the largest region of unallocated blocks.

One method of counteracting the problem of many small, unallocated storage units is **compaction**, which refers to the collection of multiple groups of unallocated storage units into a single, contiguous region. Compaction is performed by moving all files to the lowest-numbered allocation units possible, which, in effect, "squeezes out" small, unallocated storage regions and moves them to a contiguous storage region. Figure 13-5 (on the next page) shows the results of compacting the files originally shown in Figure 13-4. The process of compaction is relatively expensive because it requires the movement of large numbers of allocation units. In some file-management systems, compaction is performed automatically when needed (e.g., when a user attempts to create a file that exceeds the size of the largest available group of unallocated storage units). In most file-management systems, compaction is performed by a special utility only when the system administrator or user explicitly requests it.

Figure 13-4

Allocation of storage blocks after the deletion of FILE1 and the creation of a new, smaller file

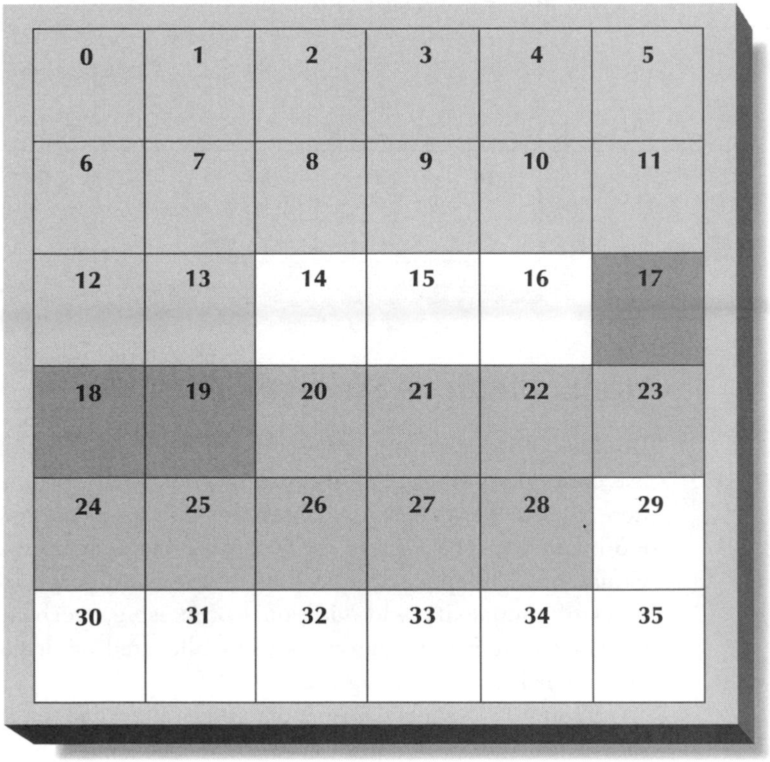

0	1	2	3	4	5
6	7	8	9	10	11
12	13	14	15	16	17
18	19	20	21	22	23
24	25	26	27	28	29
30	31	32	33	34	35

Figure 13-5

Results of
compacting the
files shown in
Figure 13-4

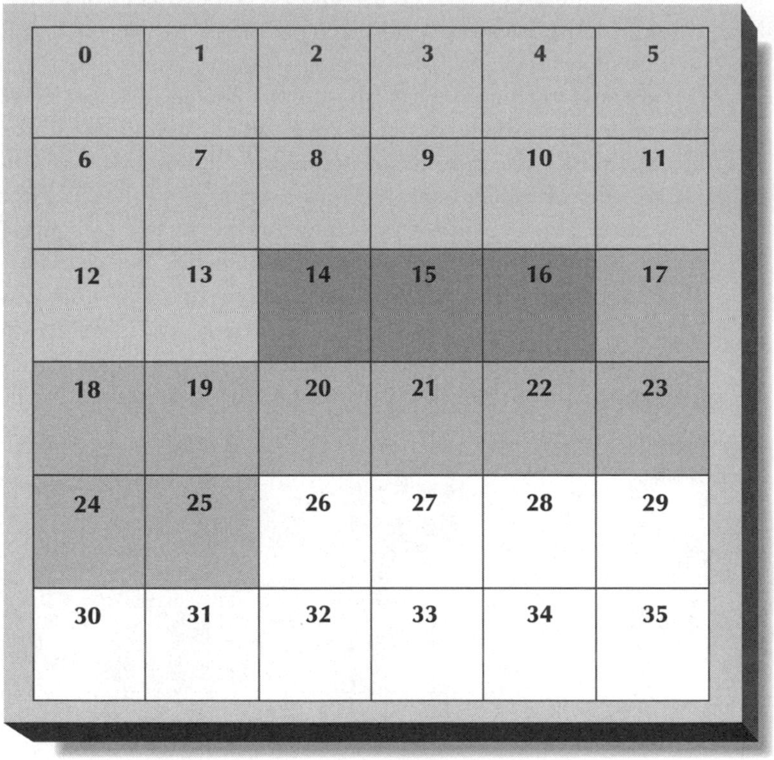

0	1	2	3	4	5
6	7	8	9	10	11
12	13	14	15	16	17
18	19	20	21	22	23
24	25	26	27	28	29
30	31	32	33	34	35

Noncontiguous Storage Allocation

Because of its relatively low degree of flexibility, few file-management systems use contiguous storage allocation. Instead, some form of noncontiguous allocation is used. However, noncontiguous allocation methods require substantially more complex procedures for recording the allocation of storage units to individual files. Keeping track of unallocated units is also more complex. Both types of complexity add additional processing overhead to any file I/O operation that requires storage units to be allocated or deallocated (e.g., appending new records to an existing file).

Figure 13-6 shows a noncontiguous allocation of storage units to the files shown previously in Figure 13-5. Simplified directory entries for these files are shown in Table 13-2; these entries are identical in structure to those shown in Table 13-1. Although the storage-allocation information stored in the directory is the same as under contiguous allocation, it cannot be used in the same manner because the second, third, and subsequent blocks of a file can no longer be assumed to sequentially follow the first allocation unit. Because units allocated to a file may be widely dispersed throughout a storage device, some method (other than the assumption of contiguous allocation) must be used to correlate allocated units.

Figure 13-6

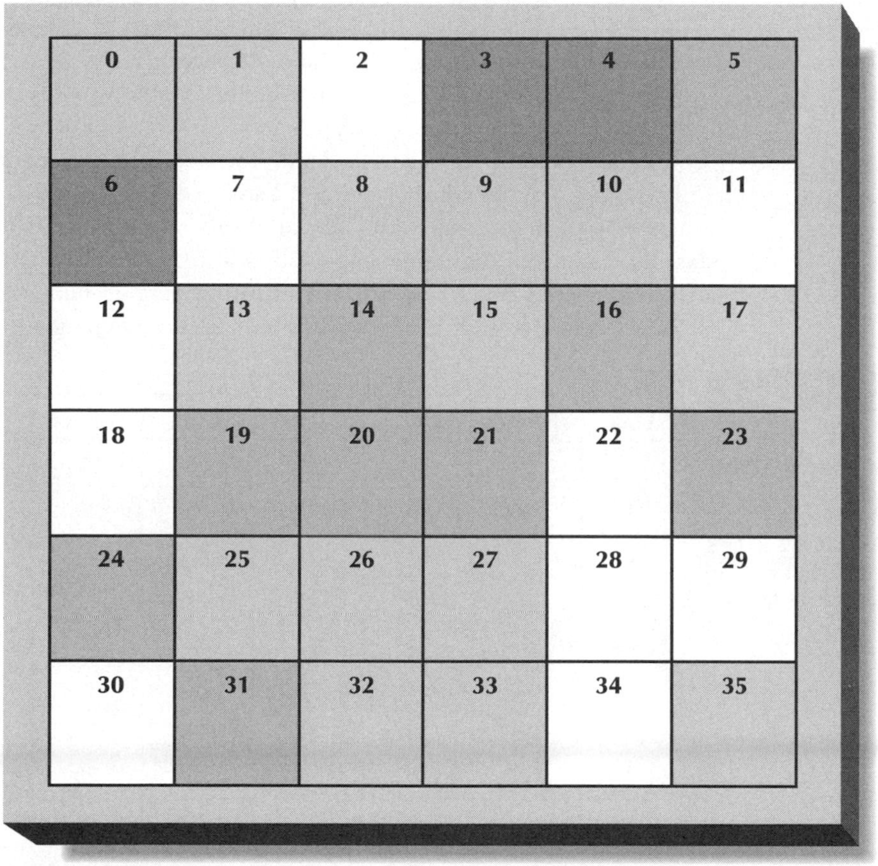

Noncontiguous allocation of storage blocks to three files

TABLE 13-2	Filename	Owner Name	First Allocation Unit	Length (in Allocation Units)
Simplified directory entries for the files depicted in Figure 13-6	File4	Smith	0	14
	File2	Jones	3	3
	File3	Smith	5	9
	SysFree	System	2	10

Noncontiguous storage allocation requires a separate data structure to record the allocation of storage units to individual files. A number of different data structures can be used, including arrays (tables), linked lists, and indices. A commonly used structure is an array containing embedded linked lists—the method used by DOS and some versions of Windows, which is called a **file-allocation table (FAT)**.

Table 13-3 shows a file-allocation table for the storage device depicted in Figure 13-6. The table contains an entry for each allocation unit. All of the entries allocated to a file are "chained" together by a series of pointers. The table entry for each allocated unit contains a pointer to the next allocated unit of the file. Thus, the table entry for unit zero (the first unit allocated to File4) contains a pointer to the table entry for unit 1 (the second unit allocated to File4). The pointers form a linked list, or chain, that ties together the table entries of all the storage units allocated to a specific file. The table entry of the last allocated unit contains a special code to indicate that it is the final block. All unallocated units also are linked into a single chain, which simplifies the task of finding storage units to allocate to new, or expanded, files.

TABLE 13-3

A noncontiguous storage allocation table. Table entries match the allocations to files shown in Figure 13-6.

Unit	Pointer	Unit	Pointer	Unit	Pointer	Unit	Pointer
0	01	9	10	18	22	27	32
1	08	10	13	19	20	28	29
2	7	11	12	20	21	29	30
3	04	12	18	21	23	30	34
4	06	13	15	22	28	31	End
5	14	14	16	23	24	32	33
6	End	15	17	24	31	33	35
7	11	16	19	25	26	34	End
8	09	17	25	26	27	35	End

Sequential access to a file's storage units is relatively efficient when a linked list is used. Random access is much less efficient, particularly if the file is large. Some operating systems define a separate file type for random access files. An index or similar data structure is stored in the file's first allocation unit, or units, to record all of the units allocated to the file. The contents of this index are redundant (i.e., the same storage-allocation information is also contained within the file-allocation table), but the index allows specific units of a file to be efficiently located and accessed.

Blocking and Buffering

At the level of application programs, files are composed of **logical records**. A logical record is a collection of data items, or fields, that is accessed by an application program as a single unit. A **physical record** is the unit of data storage that is accessed by the CPU or storage device controller in a single operation. For disk and other devices that use fixed-size units of data transfer, a physical record is equivalent to a block. For storage devices with variable-sized data transfer units (e.g., tape drives), block size may differ from physical record size or it may be undefined.

If logical record size is less than physical record size, then multiple logical records may be contained within a single physical record (see Figure 13-7[a]). If logical record size is larger than physical record size, then multiple physical records are required to hold a single logical record (see Figure 13-7[b]). The grouping of logical records within physical record storage areas is called **blocking**, which is described by a numeric ratio of logical records to physical records. (This is sometimes called the **blocking factor**.) The blocking factors illustrated in Figures 13-7(a) and 13-7(b) are 4:3 and 2:3, respectively. If a physical record contains just one logical record, the grouping is said to be **unblocked**.

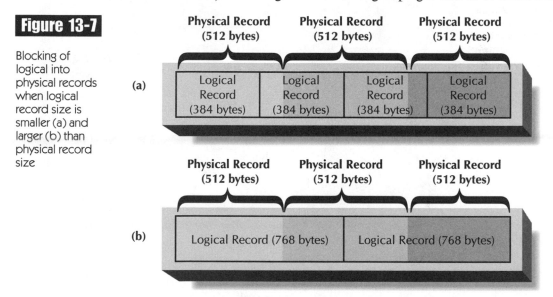

Figure 13-7

Blocking of logical into physical records when logical record size is smaller (a) and larger (b) than physical record size

Read and write operations to/from an unblocked file can be implemented with simple and efficient algorithms because each logical record I/O operation requires the movement of exactly one physical record. File I/O is much more complex when logical and physical records have different sizes. The file-management system must coordinate the movement of physical records to and from memory and must extract logical records on behalf of the requesting program.

The file-management system uses buffers in primary storage to temporarily store data as it moves among programs and secondary storage devices. These buffers are allocated and managed by the operating system on behalf of application programs. Each buffer is the size of a block, and multiple buffers may be used for each open file. As physical records are read from secondary storage, they are transferred to the buffer(s). Logical records are extracted from the buffers and copied to the data area of the application program (see Figure 13-8). The procedure is reversed for write operations.

Figure 13-8

A buffering technique for file-read operations. The file-management system loads a physical record from disk into a buffer area in memory and then extracts the desired logical record and moves it to the program data area.

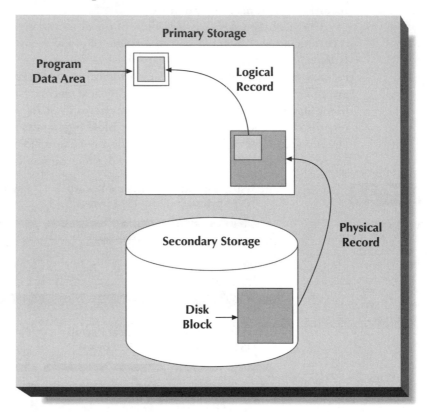

Buffering provides a scratchpad area in memory for mapping logical record accesses into physical record accesses. As discussed in Chapter 7, buffering also improves performance if a sufficient number of buffers are used. For high blocking factors, a relatively small number of buffers can dramatically improve performance. For example, if 10 logical records are contained in each physical record, then reading a single physical record provides sufficient data for 10 logical read operations. Reading the first logical record results in 10 logical records (i.e., one physical record) being copied from secondary storage to a buffer. The next nine read operations—assuming they are sequential—can be satisfied from the buffer content without further input from secondary storage.

Low blocking factors (i.e., larger logical records) require a correspondingly larger number of buffers to achieve significant performance improvements. The file-management system must combine the physical records to reconstruct the required logical record. Typically, the file-management system will use a sufficient number of buffers to hold all of the physical records corresponding to a single logical record. If the size of the logical record exceeds the capacity of allocated buffers, then the logical record must be moved to the program data area in a series of physical read and buffer copying operations.

Example

The following example is typical of storage-allocation and file I/O procedures in simple file-management systems. In fact, the procedures and data structures described are almost identical to those of MS-DOS and early versions of Windows. Assume that the directory entries in Table 13-2 represent actual files stored on a disk drive. The storage-allocation (i.e., block) size used by both the disk drive and the file-management system is 512 bytes. Blocks are allocated to files nonsequentially as shown in Figure 13-6, and the file-allocation table is shown in Table 13-3. Assume that records are the unit of file input and output and that each record is 55 bytes in length. Further assume that logical records are stored sequentially within the blocks allocated to the file.

In response to any read operation performed by the application program, the file-management system must perform the following tasks:

1. Determine which allocation unit contains the requested record.

2. Load that allocation unit into the buffer, if it is not already there.

3. Copy the portion of the allocation unit that contains the desired data to the application program's data area in memory.

4. Increment a pointer to the current position in the file.

The exact implementation of these procedures will vary with the file-access method.

Sequential Access For the first read operation, the first storage unit allocated to the file will always contain the first record, or the first part of that record if it is larger than an allocation unit. The address of the first storage unit allocated to File3 can be found in its directory entry (block 5). Thus, the first byte of the first logical record in File3 is contained within the first byte of allocation unit 5. The file-management system will issue a read request to the disk controller for allocation unit 5 and load that storage unit into the buffer when it has been retrieved from disk. Because the application program is reading the first record in the file, the record will begin in the first byte (byte 0) of the buffer.[2] Thus the file-management system will copy 55 bytes from the buffer to the application program's memory region starting at byte 0. At the conclusion of the operation, the file pointer will be incremented, from byte 0 to byte 55.

Subsequent read operations require calculations to determine which allocation unit contains the record to be read and the byte offset within that allocation unit. These calculations are performed using the logical record size, file pointer value, and the contents of the file-allocation table. For the second record, the calculation is as follows:

$$\frac{(Pointer-1) \times Record.Size}{Block.Size} = \frac{(2-1) \times 55}{512} = 0, remainder\ 55$$

Thus, the next (second) record begins in the first storage unit allocated to the file at byte offset 55. From the directory and file-allocation table, it can be determined that the first block allocated to the file is allocation unit 5. Because that block is already in the buffer, the file-management system copies 55 bytes starting at offset 55 and increments the pointer value to 110—the old value of 55 plus the number of bytes just copied.

Direct Access Now assume that a program requires direct access to be performed. As before, the file-management system will allocate a memory buffer when the File3 is opened. However, no pointer need be initialized for direct access. Assume that the first read operation requests the 37th record of the file. Using the same formula as before, the calculation is:

$$\frac{(Pointer-1) \times Record.Size}{Block.Size} = \frac{(37-1) \times 55}{512} = 3, remainder\ 499$$

Thus, the 37th record begins in the fourth storage unit allocated to File3 at byte offset 499.

To determine the corresponding allocation unit the file-management system follows the chain of entries in the file-allocation table for File3. The first allocated storage unit is unit 5, as recorded in the directory. The second allocated unit is the pointer field contained in entry 5 of the file-allocation table (allocation unit 14). The third allocated unit is the pointer field contained in entry 14 of the file allocation table (allocation unit 16). The fourth allocated unit is the pointer field contained in entry 16 of the file allocation table (allocation unit 19). Thus, the 37th record starts in allocation unit 19 at byte offset 499.

The file-management system loads allocation unit 19 into the buffer and begins the transfer of 55 bytes to the application program memory region starting from byte offset 499. However, all of the record contents are not contained within the buffer. While copying, the file-management system will come to the end of the buffer before it has finished copying 55 bytes. The system must

---

[2] Allocation unit and offset numbering start at zero, not one.

detect this condition, determine the next storage unit allocated to File3, issue a read request for that storage unit, load it into the buffer, and continue the interrupted copy operation from the beginning of the buffer. Thus, the system must follow the chain of pointers in the file-allocation table from allocation unit 19 to the next allocation unit of File3 (allocation unit 20). It must issue a read request for this unit and load it into the buffer. Then, it copies the first 42 bytes of the buffer to complete the read operation.

DIRECTORY CONTENT AND STRUCTURE

A directory contains descriptive data about files contained in a storage device. This data—usually accessible by both the operating system and the user—may include:

► Filename.
► File type.
► Location.
► Size.
► Ownership.
► Access controls.
► Time stamp(s).

Tables are the most common data structures used to implement directories. Directory tables are sometimes supplemented by an index to speed searching operations.

The filename is the means by which users and application programs access a file. If the directory is indexed, then the filename is used as the index key. Operating systems vary in the format required for valid filenames. Many older operating systems restrict both the length of a name and the characters used. For example, MS-DOS uses a two-part name with a maximum of eight characters in the first part, three characters in the second part, and a mandatory '.' symbol separating the parts (this format is sometimes referred to as an 8.3 filename). Embedded space characters are not allowed nor are most nonalphabetic and non-numeric characters. Modern operating systems (e.g., Windows NT) are far less restrictive in their requirements for filenames.

As described earlier, file type may be stored implicitly through a filenaming convention (e.g., .EXE for an executable file). This approach is typical of microcomputer operating systems such as DOS and early versions of Windows. Other operating systems (e.g., UNIX and MVS) store file type within a specially coded field in the directory. The type field can be displayed in a human-readable form when the contents of a directory are listed or otherwise displayed.

The location field of a directory entry stores the linear disk address of the file's first allocation unit. The content of this allocation unit may be the data or a data structure (e.g., an index) that contains pointers to other allocation units of the file. The size of the file can be computed by accessing and counting all of its allocation units. Storage of a size field in the directory is far more efficient and is almost universally implemented in operating systems. Some operating systems store the size of the data actually stored in the file, whereas others compute and store file size based on the number of storage units allocated to the file. These methods will yield different results unless the data content of a file is an exact multiple of allocation-unit size.

File security information includes file ownership and any access controls that may apply. File ownership is the primary basis for implementing file security. Access to a file is restricted to its owner unless other access controls are specified. The number and nature of these access controls varies widely.

One or more time stamp fields are stored in a directory entry. If only one such field is provided, then it records the date and time of the last file write operation. Other time stamps that might be provided include file-creation date/time and date/time of last access. This information is frequently used by file-backup and recovery-utility programs. It also can be used to allow multiple versions of the same file (with identical filenames) to be stored in the same directory.

Directory Structure

Early microcomputers used relatively simple directory structures, which reflected the single-user nature of these systems and their relatively small secondary-storage capacity. Small storage capacity limited the number of files that could be stored on a single storage device (or storage medium) to a few dozen. A single directory was provided for each device or medium that stored information about all files on the medium. Such a directory structure is called a **single-level directory**.

The number of files stored on a modern microcomputer typically numbers in the thousands. Within larger multiuser computer systems, tens or hundreds of thousands of files may be stored. Such a large number of files requires a multilevel system of directories (i.e., a system that uses directories of directories). This is implemented through a **hierarchic directory** (or **tree structure directory**) system.

Tree Structure Directories

The left pane of Figure 13-9 shows a portion of a hierarchic directory structure as displayed by Windows Explorer. A root (highest level) directory exists for each storage device in the computer system (nodes labeled A: through E:). Each root directory can, in turn, contain other directories, files, or a mixture of the two. Directories below the root directory also can contain other directories, files, or a mixture of the two. Theoretically, the number of recursively descending directory levels is unlimited.

Figure 13-9

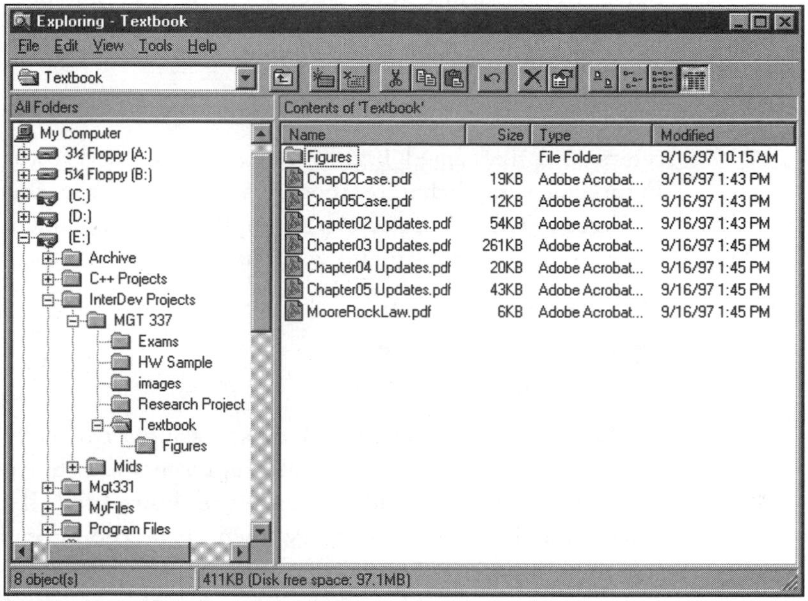

A portion of a hierarchical directory structure. The left pane shows directory nodes in the hierarchy. The right pane shows the contents of the current directory (Textbook).

The right pane of Figure 13-9 shows the content of a single directory (Textbook). It contains a number of files and a single embedded directory named Figures. For each process or user, the operating system maintains a pointer to the directory that is currently being accessed. That directory is called the **current directory** or **working directory** (Textbook is the current directory in Figure 13-9).

In a multiuser operating system, each user has a default working directory (or **home directory***)*. When a user logs in interactively or runs a batch process, the home directory is made the default current directory. The user or process can subsequently change the current directory by issuing a command through the command layer or by making the appropriate service call.

Within the hierarchy of directories, names of access paths can be specified in one of two ways. A **complete path** name begins at the root directory and proceeds through all nodes along a path to the desired file. Such a path is also referred to as a **fully qualified reference** because the entire access path is fully stated (qualified). For example, the name:

e:\InterDev Projects\MGT 337\Textbook\Chap02Case.pdf

is a fully qualified reference to the file Chap02Case.pdf displayed in the right pane of Figure 13-9 (where the symbol '\' is used to separate directory names). A **relative path** name begins at the level of the current directory. In a conventional tree structure, the assumption is that the path proceeds downward through subdirectories that are contained within the current directory. For example, the name:

Figures\Figure01a.jpg

refers to a file named Figure01a.jpg in the subdirectory of Textbook named Figures (assuming that Textbook is the current or working directory).

Tree structure directories simplify the implementation of several file-management system functions, including:

► Directory searching.

► Grouping related files.

► Enforcing access controls.

Searching for a desired file or accessing a particular node is facilitated by the tree structure. An access path represents the shortest route that traverses the network of directories to reach the desired file. If the system were organized as a single level directory, then all preceding entries would have to be searched to find the desired file. Within a tree structure, each branch selection eliminates a significant portion of the available routes at that level and greatly increases the efficiency of the search.

Tree structures are naturally suited to hierarchically grouping related files. Directories near the root directory level are used to represent high-level aggregations of files (e.g., C++ projects and InterDev Projects in Figure 13-9). Lower-level directories implement more specific groups of files and directories. For example, the directory Figures contains figures for a single textbook within a single InterDev project workspace. Hierarchical file and directory groupings are advantageous when related groups of files or directories must be manipulated as a unit (e.g., archived or copied to another location).

In a tree structure, the owner of a directory is considered to be the owner of all files and directories contained within. This rule of ownership extends recursively down through lower levels of the directory structure. Full access permissions are granted to a user's home directory. By extension, they are extended to any files or directories below within that branch of the directory structure. Other branches at the same level and directories and files at higher levels are inaccessible. Thus, the level at which a user's home directory is created can be used to assign the user a level of access authorization, as illustrated in Figure 13-10.

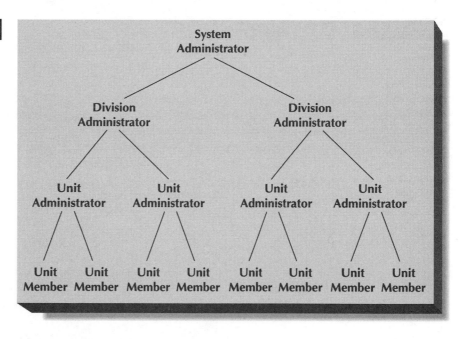

Figure 13-10

An example of hierarchical organization relationships that can be implemented as access controls within a tree structured directory

Graph Directories

A **graph directory structure** (see Figure 13-11 on the next page)—more flexible than a tree structure directory—relaxes two of the restrictions enforced within a tree structure directory:

► Files and directories can be contained within multiple directories.

► Directory entries can form cycles within the directory structure.

The files Syllabus (within MGT 331) and Course List and the directory Assignments are all examples of relaxing the first restriction. Syllabus and Assignments are contained within two separate directories, and Course List is contained within three directories.

An example of relaxing the second restriction is the relationship among the directories Directory Root, WWW Pages, and MGT 337. MGT 337 directly contains Directory Root and Directory Root indirectly contains MGT 337 (through the intermediate directory WWW Pages). These relationships form a cycle (or loop) in the directory graph. Cycles create special problems for programs that recursively search directories. Such programs can become stuck in an endless loop as they traverse the cycle repeatedly. Special processing procedures are required to avoid this situation.

Figure 13-11

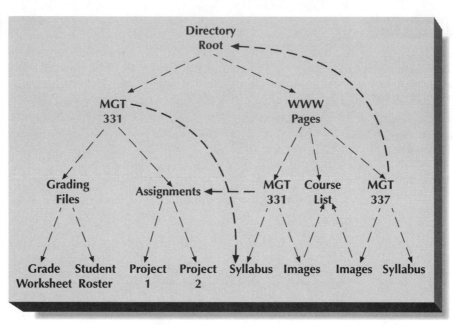

Few file-management systems directly implement graph directories. Most implement them indirectly by using two types of links among directories and their contents. The first type is a "contains" link. The second type is a special link reserved for referencing directories or files already contained within another directory. The name of this type of reference varies among operating systems (e.g., UNIX uses the term "link" and Windows uses the term "shortcut"). Programs that recursively access directory contents can avoid infinite loops by not traversing links or shortcuts.

Directory Operations

The file-management system provides service calls for directory creation and directory deletion. These service calls can be used directly by system or application programs. One example of directory creation by a system program is the creation of a home directory for a newly created user account. Typically, a program is provided for use by system administrators to add, modify, or delete user accounts. Among other functions, this program adds, modifies, or deletes the home directories associated with user accounts. Directory creation/deletion also can be included in a system program that installs (or removes) large software packages. Complex software packages consist of many files, organized in a multilevel directory structure. Thus, installation programs must be capable of creating and manipulating directory structures in addition to files.

Directory creation, modification, and deletion abilities are provided to end users through command layer facilities. Equivalent functions are provided to programs through the service layer. These functions may be complex to implement because operations must be performed not only on a single directory, but on subdirectories as well. For renaming and movement operations, the uniqueness of path names must be guaranteed.

The most commonly used directory operation is the listing or display operation. At minimum, a directory list or display includes the symbolic names of files and subdirectories. In most systems, all other directory information (e.g., size, location, usage data, ownership, type, etc.) also may be listed or displayed. Listing facilities provide a number of alternate presentation formats and options (e.g., column-oriented lists and sorted output). As with other directory operations, these operations are implemented within the service layer and may be used directly by application programs or by end users through command-layer facilities.

FILE ORGANIZATION

File organization refers to the placement of records within the storage space allocated to a file. Notice that this is an entirely separate issue from the allocation of storage units to a file. That is, the storage units allocated to a file should appear as a linear sequence for purposes of accessing file contents. The physical position of those units on a storage device should be unknown, and irrelevant, to file I/O procedures.

Several methods of file organization are possible, including sequential, direct, and indexed. Hybrid methods also can be implemented (e.g., indexed sequential). **Sequential file organization** is the simplest method because it makes no assumptions about file content and requires no special data structures to implement file-access methods. The records of the file are stored in adjacent logical storage locations. Although data or records may be sorted or otherwise ordered within this sequential space, that fact is unknown by, and irrelevant to, the file-management system.

Direct file organization assumes that the storage location of a record depends on the value of some field within that record. In general, it requires the use of a key field that can be mapped into a logical record number within the file. For example, consider a customer file in which each record is identified by a unique field called ACCT#. Assume further that account numbers are always six-digit integers and that the lowest numbered account is 100,000. Direct access to records of this file can be implemented by correlating logical record numbers to account numbers through the formula:

$$\text{ACCT\#} - 100{,}000 = \text{logical\_record\_number}$$

Thus, for example, the customer record corresponding to account number 100,423 is stored in logical record 423 (100,423 – 100,000).

Direct file organization works well under the following assumptions:

▶ The key field is unique and is easily converted into a logical or physical storage address.

▶ Allowable key values fall within a relatively narrow range.

▶ There are few gaps in the sequence of keys currently in use.

The first assumption assures that minimal processing overhead will be consumed by access operations. In general, the more complex the conversion procedure, the more processing overhead will be added to each record access. The last two assumptions are closely related in that both help to ensure little or no wasted storage space. The first assumption guarantees that the fewest number of storage locations must be allocated to the file, whereas the second guarantees that few allocated storage locations will be empty. For example, if the first currently used key value is 100,000 and the next currently used key value is 100,423, then 422 storage locations in the file are currently empty (i.e., those for key values 100,001 through 100,422). Empty storage locations are minimized when keys are assigned in sequence and when old keys are reused as soon as possible (e.g., the account number of a deleted customer is assigned to the next new customer).

Indexed file organization is used frequently when all the assumptions needed to assure successful use of direct organization cannot be met. It also can be used to support both sequential and random access to the same file. Indexed organization uses a table of key values and storage locations that is stored separately from the file's records. The index contains an entry for each currently used key value and correlates it with a logical record number or physical storage location. The entries are sorted by ascending key value, to allow efficient searching of the index. In some file-management systems, the index is stored in an entirely separate file apart from the data records, whereas in others, storage blocks for the index are allocated at the beginning or end of the file.

Indexed file organization requires more storage space than sequential or direct organization because both the index and the data records must be stored. In addition, accesses using the index require processing overhead to search for key values in the index. However, these disadvantages are often offset by the increased flexibility of file access made possible by indexed organization.

Each type of file organization makes certain assumptions about file contents and internal organization. Because the assumptions and structures differ, the method of file organization must be declared when a file is created. This allows the file-management system to initialize the appropriate storage structure for that type of organization. A file's organization method is stored within the directory for use by the operating system when performing file I/O operations.

File-Access Methods

Access to file contents may vary by the units of file I/O and the order (or lack thereof) in which those units are read or written. Virtually all file-management systems support byte-oriented file I/O. In this form of I/O, a file is assumed to consist of a linear sequence of byte. No assumptions are made by the file-management system as to the interpretation of those bytes or their organization into higher-level data structures (e.g., arrays, records, and indices). Some file-management systems also support record-oriented file I/O. In this form of I/O, entire records are read or written as integrated units. For fixed-length records, record boundaries are determined by formulas based on record size. For variable-length records, some method must be provided to mark the start (and end) of records within the file. This is implemented using a control character (e.g., ASCII 30).

A file-management system may support a variety of methods for accessing file contents. Some of the simpler methods that are supported are:

► Sequential (byte- or record-oriented).

► Direct (byte- or record-oriented).

► Indexed (record-oriented only).

Depending on the file-management system, different file organization methods may be used to support different file-access methods. An exact equivalence between file organization and access method is not always required. For example, direct access can be made to a sequential file if the range of keys and storage location addresses is known in advance. Similarly, sequential access to an indexed file can be implemented if the index is stored after the data records or in a separate file. Access to file content is most efficient when the organization and access methods are the same.

Sequential file access—by far the simplest type of access to implement—refers to reading and writing of file contents in the order in which they are stored. With sequential file access, the file-management system keeps track of the current position within the file and maintains an internal pointer for this purpose. When a file is first opened, the pointer is set to the first byte or record in the file. Subsequent read or write operations advance this pointer to subsequent bytes or records. Thus, the next byte or record to be read/written is always represented by the value of the pointer. In general, the pointer is always incremented in a positive direction, and it is not possible to advance it more than one position at a time. Thus, to read the 37th record in a file, the previous 36 records must be read first in order to advance the pointer to the proper position.

Direct file access allows the pointer position to be specified directly by the application program. Thus, to read the 37th record in a file, the application program issues a service call to the file-management system that specifies the position of the desired record. It is not necessary for the application program to directly manipulate any of the first 36 records in the physical sequence. If record-oriented file I/O is not supported, direct access may be implemented on byte positions within the file. In this case, the pointer value represents the current byte within the file instead of the current record.

Direct file access allows certain types of file I/O operations to be implemented much more efficiently than with sequential access. For example, consider a file of 10,000 employee records sorted by ascending social security number (SSN). Assume that an application program accepts a social security number from a user, searches the file for the record with that SSN, and displays it to the user. If sequential file access is used to search the file, then each record must be read sequentially and its SSN compared to the one input by the user. The search stops when a match is found. This method of search is called a **linear search**. On average, half the number of records in the file (5000 in this example) will have to be searched for each SSN input by the user. For large files, the searching overhead can seriously degrade performance due to the large number of extraneous records that must be read.

With direct file access, this same search could be implemented using a **binary search**. In a binary search, the number of records in the file must be known in advance. This information is stored in the directory by file-management systems that support record-oriented I/O. It also can be computed based on the length of the file (in bytes) and the length of a record (in bytes), if record length is fixed. The search begins by directly accessing a record in the middle of the file (record 5000 in this example). The SSN of that record is then compared to the SSN input by the user; if they match, the search is completed.

If the input SSN is greater than the SSN in the record just read, then the search continues within the upper half of the file (i.e., the next record read will be number 7500). If the input SSN is less than the SSN in the record just read, then the search continues within the lower half of the file (i.e., the next record read will be number 2500). The search proceeds iteratively with the number of records currently under consideration cut in half on each iteration. For the file of 10,000 records in this example, a match should be found by executing no more than 14 read operations (computed as n such that $2^n \geq 10,000$ and $2^{n-1} < 10,000$).

Indexed file access is a variation of record-oriented direct access in which a requested record is specified by the value of a key field, rather than by position. In the previous example, the file-management system was unaware of the key value corresponding to each logical record. The key value was compared to the search value by the application program, not by the file-management system. With indexed access, the job of searching and comparing key values is moved from the application program to the file-management system. The application program requests a record by key value, and the file-management system is responsible for whatever processing is necessary to locate the corresponding record.

FILE INPUT AND OUTPUT OPERATIONS

An **open operation** can be performed only on a file that has already been created. Opening a file makes its content available for access by the user or requesting process. Within the operating system, opening a file means that it is allocated logically to a specific process. Typically, the operating system performs the allocation by entering the symbolic name of the file in a table of open files that is maintained for each process. Other "housekeeping" chores such as allocating buffer space in primary storage are performed at this time also. The file-management system creates and maintains a data structure called a **file control block** for each open file. This data structure is used to store information about the file, including current position, the number and location of I/O buffers, and the current state of each buffer (e.g., empty, full, modified, or not modified). This data structure is updated, if necessary, as file I/O operations are performed.

A file **close operation** severs its logical relationship with the current process. Within the file-management system, file closure is accomplished by removing the symbolic name of the file from the file table of the current process. Other resources associated with the file/process connection (e.g., buffers) are released. The file control block is used to determine what these resources are. It also is used to determine if any I/O operations must be completed (e.g., copying modified buffers to disk). By closing a file, the file-management system releases it and its related resources for reallocation to other processes.

Read and Write Operations

The set of available read and write operations and their exact implementation depends on the type of file being read and its method of file organization. In general, sequential access to a sequentially organized file is the simplest and most efficient to implement. Direct access to a sequentially organized file is a bit more complex but still relatively straightforward. Access to indexed storage structures requires the most complex I/O operations.

File I/O operations can be defined as either record-oriented or byte-oriented. In record-oriented I/O, an entire logical record is the unit of data transfer between the file-management system and an application program. In byte-oriented I/O, data is transferred between the file-management system and an application program as a sequence of bytes. In this type of file I/O, the file-management system is unaware of the logical structure of the data. Thus, all conversion between logical record structures and streams of bytes must be performed within the application program.

Sequential Access

A set of file access service calls to support byte-oriented sequential file access is:

► read_n_bytes(file,n,data)

► write_n_bytes(file,n,data)

where "file" is a file identifier (typically, an integer returned by a file open service call), "n" is an integer containing the number of bytes to be read or written, and "data" is the memory address of a data element or structure containing the data after reading or before writing. Data are read or written entirely as sequences of byte values, with no regard to their interpretation or logical structure. For example, the service call:

read_n_bytes(file,4,customer_balance)

simply reads the next four bytes from the file and copies them into memory starting at the address of the data item customer_balance. A field such as customer balance is probably a real number (4 bytes in length, in this case), but that fact is unknown by, and irrelevant to, the file-management system.

With all methods of sequential file access, the file-management system must keep track of the current position within the file. For byte-oriented data transfer, this position is the sequential number of the most recently read (or written) byte in the file. This pointer is initialized when the file is opened and incremented immediately after data is read or written. Subsequent read and write operations use the pointer to determine where within the file to read or write data.

A simple set of service calls for record-oriented file I/O is:

► read_record(file,data)

► write_record(file,data)

where "data" is the starting address in memory of a data structure comprising a record. In file-management systems that support record-oriented I/O, some method must be provided for marking the end of one record, and the beginning of another, within a file. A control character (e.g., ASCII 30_{10}) is placed after each group of bytes that comprises a record. A read_record service call causes the file-management system to sequentially read all bytes up to and including the end of record control character. These bytes are copied sequentially to the programs data area. The write_record service call copies bytes from memory to a file up to and

---

Mass Storage Access and Management

including the end-of-record control character. As with byte-oriented I/O, a pointer is maintained to indicate the current position (record) within the file.

Sequential file I/O, the most basic method of file access, is useful in a wide variety of situations. Many batch application programs process input records sequentially and generate sequential output files. The operating system itself uses sequential I/O to perform routine functions such as loading the executable image of a program from a file into memory. However, many types of information processing require nonsequential file-access methods.

Direct Access

Direct access byte-oriented file I/O refers to the ability to read or write any byte in the file without accessing any other bytes other bytes. Direct access in record-oriented file I/O refers to the ability to read or write any record in the file without accessing any other records. With such access methods, the concept of current position (or a file pointer) is irrelevant. Read and write operations are not restricted to the record or byte at the position following the current pointer. They can operate on any portion of the file's content.

A typical set of service calls to support direct file I/O is:

► read_n_bytes(file,position,n,data)

► read_record(file,position,data)

► write_n_bytes(file,position,n,data)

► write_record(file,position,data)

where the parameters "file," "n," and "data" are the same as previously defined for sequential I/O. The parameter position is an integer representing the position within the file where reading or writing is to occur. This integer may represent a sequential byte or record position, depending on the type of file I/O operation (byte- or record-oriented). For example, the service call:

read_record(File3,37,RECORD)

will cause the 37th record of File3 to be read and copied to the data structure RECORD within the application program's data area.

Indexed Access

Indexed access is similar to direct access except that records are specified by key value instead of position. A typical set of service calls to support indexed file I/O is:

► read_n_bytes(file,key,n,data)

► read_record(file,key,data)

►write_n_bytes(file,key,n,data)

►write_record(file,key,data)

where the parameters "file," "n," and "data" are the same as previously defined for sequential I/O. The parameter "key" is a data item that matches the index key used to access file contents. A key field is an integer or character array. For a read operation, the system searches the index to find the corresponding record position and then issues another read command using direct access. Write operations require that the index be updated, as discussed next.

Edit Operations

Several kinds of edit operations are available for modifying data within files. Specific edit operations include:

►Update.

►Insert.

►Append.

►Delete.

Typically, a copy of a record in main memory is modified and then written to secondary storage. Depending on the physical organization of the file or storage device, the record can overwrite the prior version. Alternatively, the edited version may be written to a new location and the prior version retained as part of a backup system.

In an **update operation**, data within a field or record is replaced with new content. Updating may be performed in place or by rewriting the record or file to another device. In an **insert operation**, a new record is added within the existing sequence of records. In an **append operation**, data is added to the end of the existing file. This operation moves the physical end-of-file marker, or indicator, to the first available record position after the appended portion. In a **delete operation**, a record is logically or physically removed from a file.

Edits of Sequential Files

Updating an existing record in a sequentially organized file is relatively straightforward. The record to be modified is first located and read into memory. The contents of the record (in memory) are altered and then written back to the same storage location in an operation commonly called an **update in place**.

Append and insert operations on sequential files are similar to those same operations on arrays (see Chapter 4). For an append operation, a new storage location large enough to hold the appended data must be allocated to the end of the file. In many file-management systems, a special type of file access called

append mode is declared when a file is opened. The file-management system opens the file and positions the pointer at the very end. Subsequent write commands automatically cause storage to be allocated at the end of the file and new data to be written there.

Insertion is substantially more difficult to implement. For example, consider the insertion of a new record after the 500th record in a file of 1000 records. The following procedure is required:

▶ Allocate additional storage space at the end of the file to hold a new record.

▶ Starting at record 1000 and working backward to record 501, copy each record to the next logical position (e.g., record 1000 is copied to the 1001th position).

▶ Write the new record at logical position 501.

This procedure is extremely inefficient, especially with large files. For this reason, few file-management systems directly support the insertion of new data into sequentially organized files. For similar reasons, deletion of records is not supported. Implementation of these operations may be performed in application software by copying a file into memory as a linked or indexed data structure, performing the insert and delete operations, and then writing the contents of memory to a newly created file.

Edits of Nonsequential Files

Update operations on direct files are implemented in essentially the same manner as for sequential files. That is, the record to be updated is located, read into memory, altered (in memory), and written back to the same location in the file. Append operations also are similar to those for sequential files. For both direct and indexed files, new space is allocated at the end-of-record storage area and the new record is written to that location. For indexed files, an entry for the new record must be added to the index, possibly necessitating the allocation of additional space to the index itself.

Insert operations are not supported for direct files for reasons similar to those of sequential files. If insertion were supported, the correspondence between key values and record positions would be altered. This is not a problem for indexed files because any newly created record can be placed at the end of the record storage area of the file. Thus, in an indexed file, there is little logical difference between an append operation and an insert operation. The differences that do exist are confined to the update of the index (as opposed to the records of the file). For an append operation, a new index entry is created at the end of the index. For an insert operation, the index entry must be created in the interior of the index. If the index is organized as an array (or table), then

---

the index must be updated in essentially the same fashion as described earlier for record insertion in sequential files.

Delete Operations

A **logical delete operation** is performed by marking a record as deleted. The data content of the record is not removed physically from secondary storage. A logical delete can be implemented by overwriting the beginning of the record with a predefined control character sequence. Subsequent read or edit operations must detect these control characters and alter their behavior accordingly. If an attempt is made to read a deleted record directly or by an index, the file-management system should return an error message indicating that no such record exists. For sequential read operations, the file-management system should skip over the logically deleted record as if it didn't exist.

Over time, many deleted records can accumulate within a file resulting in a substantial amount of wasted storage. To counteract this problem, logically deleted records must periodically be physically deleted. The process of physically deleting logically deleted records is called **file compaction**, or **file packing**. The file is said to be compacted because the empty spaces (occupied by logically deleted records) within the file are "squeezed out," resulting in a physically smaller file.

In a **physical delete operation**, compaction is performed immediately. Storage space within the file is reorganized so that the deleted record is either overwritten by another record or the space that it occupied is deallocated from the file. Logical deletes with periodic file compaction (as opposed to physical deletes with immediate file compaction) are used to improve the efficiency of edit operations. Compaction requires accessing and manipulating every record in a file. Thus, it consumes a substantial amount of CPU and I/O resources. It is impractical to perform compaction every time a record is deleted, particularly if the file is large or if delete operations are numerous and frequent.

FILE MANIPULATION

File creation by a program is made directly through the service layer routine. Such a routine creates a directory entry, allocates one unit of storage, and positions the file pointer at the beginning of the file. In some file-management systems, a call to a file open service routine with a nonexistent file reference results in the creation of that file. In others, such a call returns an error code.

File creation by end users is typically accomplished using the facilities of a utility program called an **editor**. Editors are used to create files containing source code, job control language commands, and sometimes to create data files for application programs. Editors for such files are called **text editors**. Other editors (called **binary editors**) may be provided to allow a user to make changes in other types of files (e.g., executable files).

Under most file-management systems, deletion of a file is performed by removing its name or identifier from the directory of the corresponding storage device or volume. The delete operation merely makes the file's storage area available for reallocation, which is similar to the logical delete of a record. The space formerly occupied by the file is reallocated as other files are created, and the data content of the previous file is destroyed only when the reallocated space is overwritten.

Because the file content is not destroyed immediately by a delete operation, it is possible to restore the file provided the space has not been reallocated. Data content can be recovered by recreating the filename in the directory and reestablishing the pointer values that associate it with the storage area. This process is sometimes referred to as an **undelete operation**.

Deletion is implemented this way primarily for the sake of speed. However, security may be sacrificed because unauthorized persons who have sufficient technical knowledge may be able to gain access to a deleted file by rebuilding its entry in the directory.[3] To prevent this, some file-management systems include an optional **file destruct** capability, which removes the filename from the directory and overwrites the storage space with null values. The trade-off for greater security is the increased time required to perform the erasure.

Renaming a file is a simple operation that only requires changing the name field in the directory. All other data about the file remains unchanged. Moving a file also may be implemented simply if both the target and destination directory are within the same secondary storage device. In that case, only the directory entry need be copied from the old entry to the new. If a file is moved from one storage device to another, then both its directory entry and its data content must be copied to the new location. The implementation of a file move in this manner can consume a considerable amount of system resources if the file is large. Every record of the original file must be both read and written.

---

[3] An interesting example of this security problem occurred in the early 1990s when a large number of computers owned by the U.S. Justice Department were sold to a computer dealer. The computers were confiscated a short time later when it was discovered that the contents of the disks has not been wiped clean. The contents included a considerable amount of data (e.g., names of suspects and informants) about ongoing criminal investigations.

FILE SECURITY AND INTEGRITY

Computerized information systems can be viewed as tools for creating and maintaining large collections of data. Because data is an important and valuable organizational resource, a file-management system must provide facilities to prevent the loss, corruption, and unauthorized access to data stored in files. These protections must also extend to files containing programs because they are both a resource and the means by which data is stored, modified, and viewed. Access to files by multiple users also raises issues of security and accidental corruption of data. Thus, complete file protection in a multiuser system must provide facilities for file sharing while preventing data corruption.

Ownership and Access Controls

In a multiuser computer system, ownership of system resources is an important feature that must be supported by the operating system. Within the file-management system, the concept of ownership extends to files, directories, and, possibly, to entire secondary storage devices. Access controls for files (and other resources) must be based on a separate system of user identification and authorization. This is provided through the operating system by a system of user accounts, login, and logout processing. Each user must have an account, and each account has a unique identifier (e.g., a user name or account number). Initial access to the system is controlled by login processing. A user must enter an account identifier and a password to establish and authenticate his or her identity. Once established, user identity is made available to the file-management system for access-control purposes.

Many different schemes of access control and user privilege may be implemented. A relatively simple system, such as that used by the UNIX operating system, separates access controls into four levels. From highest to lowest, these levels are system administrator (superuser), user (owner), group, and public. System-level access is restricted to a specific user account assigned to the system administrator. This user has unrestricted access to all system resources including files, directories, and secondary storage devices.

By default, individual users are deemed to be the owner of files that they create. As owners, they have the ability to grant or deny file access privileges to other users. Groups—predefined sets of user accounts—may represent project work groups within an organization, all of the members of an organizational unit, or any other grouping criteria implemented by the system administrator. In UNIX, each user must belong to one and only one group. In other systems that implement group identification (e.g., Novell Netware and Windows NT), users may belong to many different groups. The public is simply a term referring to all known users or accounts.

An individual user can control the level of access provided to files he or she owns. In UNIX, three types of file-access privileges are defined—read, write, and execute access. Read access allows a user or process to view the contents of a file. Write access allows a user or process to alter the contents of a file or to delete it altogether. Execute access allows a user or process to execute a file, assuming that the file contains an executable program or job control stream. A file owner may reserve any of these access privileges to himself, thereby denying those access privileges to all other users except the system administrator. If the user desires, any access privilege (or set of privileges) may be extended to other members of the group or to the public. Thus, for example, a user might grant read and write access to the group but only read access to the public. The user may even deny certain access privileges to him- or herself. For example, a user might deny him- or herself write access to prevent accidental deletion of an important file. Of course, the user can alter those access controls at any time.

In addition to providing a system of access controls for files, most file-management systems also provide similar access controls for directories. Users are deemed to be the owner of their home directories and any directories that exist below it in a hierarchical directory structure. Access privileges for reading (listing directory contents) and writing (altering directory contents) are defined. As with file-access privileges, these may be individually granted or denied to a group or to the public. Directory privileges may be used to provide additional security for individual files. Read permission to a directory is denied to all other users, thus making it difficult or impossible for others to even know of the existence of a file.

Access controls are enforced within the service routines that access and manipulate files and directories. For example, a request to open a file will first check the access controls that exist for the file and then determine if the requesting user has sufficient privilege to read the file. Service routines for record editing, record deletion, and file deletions also implement access controls. For purposes of verifying access privilege, a process executed by a user generally inherits that user's access privileges (or lack thereof). Although access controls are a necessary part of file manipulation and I/O in most systems, they impose additional processing overhead on many file-manipulation operations. In some file-management systems, the system administrator may choose among several different levels (i.e., degrees of enforcement) for file-access controls. These allow an administrator to balance overall file-management system performance against the relative need for file security.

A file-management system must also provide additional controls to ensure that normal file-management system controls are not circumvented. These controls are placed on secondary storage devices and on the data structures directly related to them (e.g., storage-allocation tables and master directories). Such controls prevent unauthorized access to data and programs through direct access to the secondary storage devices on which they are stored. A file-management system may also implement a system of logging access to files. Individual access to files (e.g., file opens, writes, creations, and deletions) may be automatically recorded in a log file that the system administrator can view later. This provides an audit trail to identify unauthorized access to files and directories. As with other access controls, logging imposes a performance penalty on all file accesses.

Multiuser File I/O

A file-management system that allows multiple users or processes to simultaneously access a single file must provide additional access controls to prevent incorrect (and unintentional) corruption of file contents. This need arises from the use of memory buffers and the timing problems inherent in multiuser file access. As an example, consider the following sequence of operations:

file_open(file)

read_record(file,record)

/* change record contents */

write(file,record)

file_close(file)

When the file_open call is executed, a buffer will be allocated for file I/O. The read_record call then causes a record to be copied from secondary storage to the buffer and, from there, to the program's data area. At this point, the program alters the contents of the record (i.e., changes the value of one or more fields). The program then executes a write_record call, which causes the modified record contents to copied from the program's data area to the file buffer. The buffer will be flushed (i.e., copied to disk) when the file_close call is executed.

By itself, this file-manipulation sequence causes no inherent processing problems. However, when multiple programs execute this same sequence within the same time frame, unintended results may occur. For example, consider the same programming sequence executed by two different programs as depicted in Figure 13-12. The execution of these commands results in a problem commonly called the **lost update**, which arises from the fact that each process modifies a copy of the record within its own data area. At time 2, process 1 reads a record, which is ultimately copied to its own data area. At time 3, process 2 reads that same record, which ultimately is copied to its own data area. Both processes modify the record copy in their own data area and then use a write_record call to copy the record back to the file. However, neither process coordinates its activity with the other. As a result, the modifications made by process 1 are lost because the write_record and file_close calls issued by process 2 (at times 5 and 6) causes the modifications made by process 1 to be overwritten.

Figure 13-12

A simple file update procedure executed by two different processes with a time lag. The update made by process 1 is lost.

Time	Process 1	Process 2
1	file_open(file)	
2	read_record(file,record)	file_open(file)
3	/* alter record */	read_record(file,record)
4	write_record(file,record)	/* alter record */
5	file_close(file)	write_record(file,record)
6		file_close(file)

To prevent lost updates, a file-management system must provide a mechanism to prevent multiple processes from attempting to update the same record at the same time. The general means of implementing such a control is through use of locks. In essence, a lock prevents access to a resource until the lock is removed. Locks may be applied on an entire file or on individual records within a file. If **file locking** is used, the file-management system locks a file when a process issues a file_open call. The process that issued the call is allowed access to the file, attempts by other processes to open that same file are denied. The lock is released when the process that opened the file issues a file_close call.

File locking is relatively simple to implement in a file-management system. However, its use can impose severe performance penalties, particularly when many different processes require access to the same file. File locking allows only one process at a time to access a file, thus drastically reducing the amount of I/O that can be performed on any one file.

In **record locking**, the file-management system allows multiple processes to access the same file. However, it prevents multiple processes from simultaneously modifying the same record by applying locks to individual records. Because of the use of memory buffers for file I/O, the unit of locking is a physical record (i.e., allocation unit) rather than a logical record. Record locking allows many programs to access the same file, as long as they access different records. However, record locking is substantially more complex to implement because the file system must track every file access — not just opens and closes — and maintain a potentially large set of locks.

As discussed thus far, locking is a process implemented without the direct knowledge of application programs. That is, application programs are unaware of the locking process. To support this independence, the file-management system must provide a mechanism to delay the execution of an access to a locked resource. This can be implemented by queuing access requests to locked processes and blocking the requesting processes until the lock is released. In this way, application programs are unaware of the locking mechanism, although the user may be aware of the delay.

A simpler option is to simply fail the access request: That is, return an error code to an application program when it requests access to a locked resource. This method transfers much of the processing overhead of locks from the file-management system to application programs. Typically, an application program will test each access for failure due to a lock and implement procedures to retry the access later. Because of the complexity required of application programs, it is generally preferable to implement access spooling and/or retries within the file-management system.

FILE SYSTEM ADMINISTRATION

The oversight and management of a collection of data files is a major function of **file system administration**. Specific administration functions supported by most large-scale file-management systems include:

► File migration (version control).

► Automatic and manual file backup.

► File recovery.

Smaller-scale file-management systems (e.g., those in LAN and PC operating systems) typically do not support file migration but do support backup and recovery.

Windows NTFS

Early Microsoft operating systems including MS-DOS, Windows 3.x, and Windows 95 used a file system called the FAT (file-allocation table) file system. OS/2, which was a joint effort between Microsoft and IBM during its early years, used a file system called HPFS (high-performance file system). In the late 1980s, Microsoft considered using these file systems for a developmental operating system that later became Windows NT.

Windows NT was targeted to high-performance and "mission critical" application software. Microsoft considered a number of file-system characteristics to be crucial in such environments, including:

▶ High-speed directory and file operations.

▶ Ability to handle large disks, files, and directories.

▶ Secure file and disk content.

▶ Reliability and fault tolerance.

FAT and HPFS lacked many of these characteristics and, thus, Microsoft decided to develop a new file system called NTFS (NT file system).

NTFS organizes secondary storage as a set of volumes. A volume normally corresponds to one partition of a disk drive but may span multiple partitions and drives. Volumes consist of a collection of storage allocation units called clusters. Cluster size can be 512, 1,024, 2,048, or 4,096 bytes. Each cluster is identified by a 64-bit logical cluster number (LCN) within a linear address space (i.e., LCNs are numbered consecutively starting with zero to a maximum of $2^{64} - 1$). Thus, a volume can be as large as 4×2^{64} KB.

A volume's master directory is stored in a data structure called the master file table (MFT). The MFT contains a sequential set of file records, one for each file on the volume. All volume contents are stored as files including all the user files, the MFT itself, and other volume-management files (e.g., the root directory, storage allocation table, bootstrap program, bad cluster table). The first 16 MFT entries (numbered 0 through 15) are reserved for the MFT and volume-management files. All subsequent MFT entries (numbered 16 and higher) store records about user files.

Conceptually, a file is an object with a collection of attributes, including the filename, access restrictions (e.g., read only), and security descriptor (i.e., owner and owner-defined access controls). The data content of a file is just another attribute—though it is usually much larger than the other attributes. Each attribute type is assigned a numeric code, and file attributes are stored in ascending order of their code within the file's MFT record (see Figure 13-13a on the next page). MFT record size is either 1, 2, or 4 KB. MFT record size is determined by the operating system when a volume is formatted.

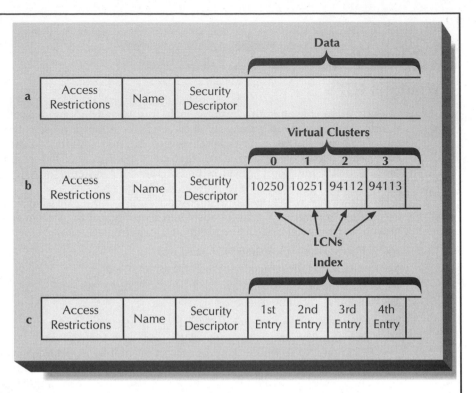

Figure 13-13

Master file table (MFT) records for (a) a small file, (b) a large file, and (c) a small directory

Each file attribute consists of a header and a data value. The header contains the attribute name, a flag that indicates whether or not the attribute value immediately follows the header, the length of the header, and the length of the attribute value. Attributes can be either resident or nonresident. A resident attribute is stored within an MFT record immediately following its header. Most resident attributes have "short" values (e.g., filename and access restrictions).

The file system allocates clusters to store nonresident attributes and records the cluster addresses within the MFT record. Data attributes are the most common type of nonresident attribute, though other attributes (e.g., the security descriptor) may grow too large to fit within an MFT record. The clusters assigned to a nonresident attribute are called virtual cluster numbers (VCNs). The area immediately following the header of a nonresident attribute stores a sequence of LCNs (see Figure 13-13b). The VCN corresponds to the position of an LCN within this sequence.

The volume root directory and all user-defined directories are stored in the same manner as files. That is, they are stored as sequences of attributes within an MFT record. A directory's data attribute contains an index of files within the directory (see Figure 13-13c). The index is sorted by filename and also contains the file number (equivalent to the MFT record number), time stamps, and size. This information is redundant with the contents of the MFT records of individual files in the directory. Duplicating this information within the directory index allows directories to be listed more quickly. The index of small directories is stored sequentially within the an MFT record. Larger directories are stored as b+ trees.

File security is implemented through the object manager facilities of the operating system. Files, I/O devices, and many system services are all managed as objects by the Windows NT operating system. Accessing an object automatically invokes a security subsystem that compares the object's security descriptor against the security descriptor of the accessing entity (e.g., a process or interactive user). The file security descriptor stored within the MFT is equivalent in structure and content to the security descriptors of other object types.

Fault tolerance is implemented in several ways within NTFS, including redundant storage of critical volume information, bad cluster mapping, logging of disk changes, and optional disk mirroring and striping. The MFT is always stored at the beginning of a volume, but a partial second copy is maintained in the middle of the disk in case a block assigned to the primary MFT becomes corrupted or unreadable. The file-management system tracks unreadable blocks during formatting and subsequent read and write operations. Clusters containing bad blocks are marked as unreadable in a separate bad cluster file.

NTFS uses a delayed (or lazy) write protocol. Disk blocks are cached in virtual memory, and write operations are made to the cache and immediately confirmed to the requesting process. Cache flushing is performed by a background process. Write operations that affect volume structure (e.g., file creation, file deletion, and directory modification) are written to the cache and also written immediately to a log file stored on disk. In the event of a system crash, the contents of the log file are always current and can be used to restore the volume structure to a consistent state.

File Migration

When a user alters the contents of a file, the unaltered (original) version of the file is overwritten by the new version. However, it is often desirable to maintain the original version as a record of the previous file state. This record may be used as a backup copy or to "undo" the most recent set of changes to the file. Many commonly used application tools such as word processors and text editors automatically generate such copies. For example, the standard MS-DOS text editor (Edit) will automatically rename the original version of a file to a new name that has the same prefix, followed by the filename extension ".BAK."

Many application programs also are written to preserve original versions of files. For example, the original version of a bank account master file may be copied prior to processing daily transactions (e.g., checks, deposits, etc.). The application program then uses a separate file of transactions to update the copy of the account master file (see Figure 13-14). The original account master is commonly called the **father** and the copy that has been updated to reflect new transactions is called the **son**. After another set of transactions is processed, the father becomes the **grandfather**, the son becomes the father, and the new copy of the account master file becomes the new son.

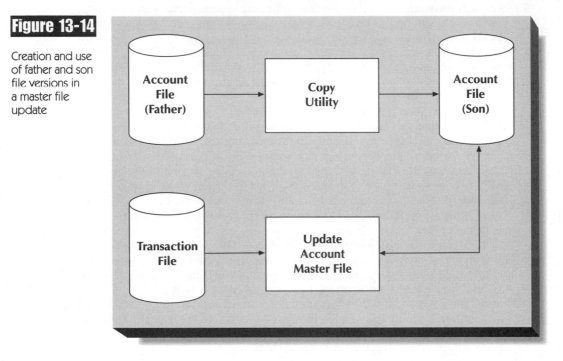

Figure 13-14

Creation and use of father and son file versions in a master file update

In a large-scale file-management system, the process of naming and storing original versions of altered files may be automated. That is, the original version of a file is automatically archived by the file-management system whenever the file is modified. In such systems, it is common to attach a version number to each filename. When a file is created originally, it is assigned a version number of 1. The first time it is altered, the altered copy is assigned a version number of 2, and so forth. The file-management system automatically generates and stores copies and tracks version numbers. A user or application program may access an older version of a file by explicitly referring to its version number.

As files are modified, older versions will accumulate on secondary storage. This can quickly swamp the secondary storage capacity of a computer system, especially when files are large and/or altered frequently. To compensate, a file-management system can implement a system of **file migration**. File migration is a file-management technique that balances the storage cost/performance of each file version with anticipated user demand for that version. As a file version becomes increasingly outdated, the probability of an access to it will decrease. The file-management system responds by migrating the file version from online to off-line storage and eventually to an archive. Ideally, a file version should reside at a point in the storage hierarchy at which an expected frequency of access can be met at the lowest possible storage cost.

The migration of file versions is logged so that they can be located and recovered. File migration typically is a semi-automated process. The generation and tracking of file versions is performed automatically by the file-management system. The timing of the movement of older versions to off-line storage also may be determined automatically, subject to some control by the system administrator. However, the physical movement of old versions to off-line storage is integrated with normal manual backup procedures.

File Backup

The file-management system provides general purpose utilities to generate backup copies of current versions of files. These utilities consist of a set of programs that may be run manually or scheduled to automatically execute periodically. Backups are needed in case of loss or damage to the original files. Loss or damage may result from a number of causes, including accidental alteration or erasure, malicious alteration or erasure (e.g., due to a virus), and the partial or total physical failure of a secondary storage device (e.g., a "head crash"). Backups protect against loss or damage to files themselves and to the directory and storage-allocation data structures needed to access them.

Automated backups performed by a file-management system and/or system administrator usually include the following options:

► Full backup.
► Incremental backup.
► Differential backup.

When a **full backup** operation is performed, the file-management system copies, or dumps, all current file content to auxiliary storage devices (e.g., magnetic tape). This strategy, which assures that protection copies exist for all active files, consumes a significant amount of time because of the large number of files and the relatively slow write speeds of auxiliary storage devices. For this reason, full backups usually are performed at relatively long intervals (e.g., weekly) and during off-peak hours.

An **incremental backup** operation archives only those files that have been modified since the previous incremental or full backup. To make incremental backup operations possible, the file-management system must keep track of when backups were performed and when files are modified. The system maintains a record of the date and time of the most recent backup in a log file.

File-directory entries are used to store the date and time of the most recent modification to each file.

Most large-scale file-management systems use both full and incremental backups. For example, incremental backups might be done at the end of each business day, with full backups run each weekend. The process can be automated if the operating system supports delayed scheduling of programs. Full automation requires auxiliary storage devices with large-capacity media. Otherwise, media must be manually swapped during or between backup operations.

A **differential backup** is similar to an incremental backup in that only files changed since the most recent backup are archived. It differs from an incremental backup in the contents of the archive. Instead of archiving the entire file, only the portions of the file that have been changed since the last backup are archived.

Differential backup utilities are complex because they require the file-management system to log changes to files at the logical or physical record level. The differential backup utility uses this log to determine which records to write to the archive. Recovery operations also are more complex. Changed records stored in an incremental backup archive must be merged with the contents of a complete file backup (e.g., from an earlier incremental or full backup operation). This process is far more complex than restoring a complete file from an incremental or full backup.

Differential backups are seldom used as standalone file-system administration utilities. However, they are sometimes implemented as part of an automatic (i.e., operating system controlled) file-system recovery operation that is invoked when an operating system is first booted after a file system crash.

Backup Methods

File backup operations may be performed in physical or logical order. If copying is in physical order, the entire content of a storage device is copied block by block (or track by track). This includes not only the files, but also the storage allocation and directory data structures. Thus, the physical organization of the backup will be identical to the physical organization of the source. Such a copy is called a **mirror image backup**.

If the order of the dump proceeds logically, symbolic filenames are passed to the copying routine in the order in which they appear in the directory. Thus, copying is said to proceed file-by-file. In performing a **file-by-file backup**, the copying routine must create each file on the destination device before the file can be written. These actions reallocate the physical organization of the files on the destination device and create new storage allocation and directory data structures. The resulting copy may be more efficiently organized than the original. For example, blocks of a single file that were randomly scattered over a storage media may be written to the copy in sequential order. In addition, logically deleted records may be deleted physically during the backup process.

Backup copies should be made to a different storage device or volume from the original. For example, it makes little sense to backup files on one partition of a hard disk to another partition of the same disk drive. Both the original and the backup would be lost in the event of total failure of the drive. For backups made to removable media (e.g., tape or removable disk), the copies should be stored in a separate physical location. Large computer centers store backup copies in a separate building or site to minimize the probability of a disaster destroying both copies (e.g., a fire that destroys an entire building). With the advent of large-scale computer networking, backup copies are frequently transmitted directly to remote storage facilities by data communication lines, which eliminates the cost, delay, and danger inherent in the physical transportation of storage media.

Transaction Logging

Transaction logging (also called **journaling**) is a form of automated file backup. The use of the term *transaction* in this context should not be confused with the more generic meaning of the term (e.g., a business transaction). In this context, a transaction is any single change to the content of a file (e.g., a newly added record, a modified field, etc.). In a file-management system that supports transaction logging, all changes to file content are recorded automatically in a separate storage area. Thus, for example, the execution of a write_record service call causes the file buffer to be modified and that modification is also written to a log file. To maximize protection, the log entries should be immediately or frequently written to a physical storage device, as opposed to a buffer.

Transaction logging provides a high degree of protection against loss of data due to program or hardware failure. When an entire computer system fails (e.g., due to a power failure or fatal system software error), the contents of buffers for open files are lost. Because these buffers may not have been written to physical storage prior to the failure, the content of those files becomes corrupted. In addition, any content changes within the buffers are lost. Transaction logging provides the ability to recover most or all of the lost changes and to repair corrupted files. When the system is restarted, the contents of the transaction log are reviewed and compared to the file content on disk. Lost updates are identified and written to the files.

Transaction logging is used commonly in large-scale online transaction processing systems. It is not used in other situations because of its substantial processing and storage overhead.

File Recovery

Backup procedures and utility programs must be supplemented by a reliable set of recovery procedures to form a complete file-protection mechanism. Recovery procedures consist of both automated and manual components. For example, the replay of a transaction log and subsequent repair of a damaged file is an example of an entirely automated recovery procedure. Recovery procedures utilizing full or incremental backups stored on removable media must rely on manual procedures at least to some degree.

The file-management system maintains logs of backups to aid in locating backup copies of lost or damaged files. Utility programs may be used to search these logs for particular files or groups of files. These logs identify the storage device on which the backup copies are located. At the time of the backup, some form of device identification should have been written to the backup medium itself and on a manually prepared backup medium label. The system administrator is responsible for locating the appropriate backup medium and mounting it in the appropriate device. The system reads the automated label to verify that the correct medium has been mounted before beginning recovery operations.

Recovery procedures for a crashed system or physically damaged storage device are more sophisticated and highly automated. In these cases, damage may not only have occurred to files, but also to directories and storage allocation structures. An automated facility is provided to reconstruct as much of the directory and storage allocation data structures as possible. A consistency check must be made to ensure that:

► All storage locations appear within the storage allocation data structure.
► All files have correct directory entries.
► All storage locations of a file can be accessed through the storage-allocation data structure.
► All storage locations can be read and/or written.

Performing the consistency checks and repair procedures consumes time (anywhere from a few minutes to several hours), but it mitigates to the need to do large amounts of data recovery from backup copies and thus minimizes the amount of current data that is lost. It also may prevent the need to reinstall system and application software.

Fault Tolerance

As applied to file-management systems, the term **fault tolerance** describes methods of securing file content against hardware failure. Hardware failure of secondary storage devices is the primary are of concern. Magnetic and optical disk drives are highly complex devices. To gain greater performance, disk designers have employed high rates of spin and small distances between read/write heads and recording media. The result is a device that provides high performance at the cost of occasional catastrophic failure.

Common causes of disk failure include burned-out motors and bearings and head crashes—contact between a read/write head and a spinning platter.

Repair of a failed disk drive is impossible or prohibitively expensive because of the nature of modern manufacturing methods. The mean time between failure of a modern magnetic disk drive is approximately 10 years, assuming continuous use, but the large number of disks in use in modern LANs, WANs, and mainframe computers guarantees that some failures will occur.

The file backup/recovery and transaction logging methods previously discussed are forms of protection against disk failure, but both methods require time to implement recovery procedures. File backups must be loaded onto a new disk from archival media—typically tape. Transaction logs must be reapplied to bring the files up to their state just prior to the failure. Recovery procedures require minutes or hours to perform. During recovery, the computer system and the data within its files is unavailable to users.

In many processing environments, an occasional down-time period for file recovery is acceptable, but in other processing environments such periods are unacceptable. Examples of such processing environments include banking, retail sales, and production monitoring and control. In general, any business or organization that performs continuous updates and queries against files and databases is candidate for advanced methods of fault tolerance.

Mirroring

Disk mirroring is a relatively simple method of protecting data against disk failure. All files are maintained on two separate storage devices. In some cases, the two devices may be located physically in different cabinets, rooms, or buildings. All updates to file and database content are simultaneously, or concurrently, written to both storage devices. Thus, if one device fails, the other device contains a duplicate of the data. Data is available continuously because both devices are continually available to respond to read and write requests.

Disk mirroring can be implemented through software. That is, the file-management system can be configured to perform duplicate writes to duplicate storage devices. But the complexity of managing duplicate storage devices can substantially reduce system performance. Software-based mirroring is required if duplicate disks are not attached to the same disk controller. When duplicate disks are located within the same cabinet, it is far more common to implement mirroring within hardware, which reduces CPU and system bus overhead.

Hardware-based disk mirroring is implemented within a disk controller (e.g., a SCSI controller). Multiple disk drives are attached to the controller and write operations are duplicated automatically by the controller to redundant drives. Read operations may be split among duplicate drives to improve performance. These processing operations take place without any intervention by system software. Special utility programs are required to configure the disk controller for mirroring and to initialize a new duplicate drive if a failure occurs.

Disk mirroring provides a high degree of protection against data loss. Hardware-based mirroring adds little or no processing overhead to the CPU or primary I/O channels. The primary disadvantages of mirroring are the cost of redundant disk drives and the higher cost of disk controllers that implement mirroring. Mirroring at least doubles the cost of data storage.

Technology Focus

RAID

Redundant array of inexpensive disks (RAID) is a disk-storage technique originally developed at the University of California Berkeley in the late 1980s; the original technique is now commonly known as RAID level zero. RAID was conceived as a means of improving disk performance and reducing storage cost. Early implementations of RAID were primarily found in large mainframe computer systems and were usually custom-designed for a particular computer installation.

RAID has evolved considerably since that time, and a large number of products are now available commercially. The original purposes of RAID have been expanded to include fault tolerance. The flurry of RAID development in the early 1990s resulted in many incompatible approaches and products. In 1992, the RAID Advisory Board (RAB, http://www.raid-advisory.com) was formed to define standard methods of implementing RAID including levels one through five (see Table 13-4). Level one is disk mirroring, as described in the previous section.

TABLE 13-4	Level	Description
RAID levels defined by the RAID Advisory Board	0	Data striping without redundancy.
	1	Mirroring.
	2	Data bit striping with multiple error checksums.
	3	Data byte striping with parity check data stored on a separate disk.
	4	Data block striping with parity check data stored on a separate disk.
	5	Data block striping with parity check data stored on a multiple disk.

All RAID levels except level one use some form of data striping. Data striping is a technique for breaking a unit of data up into smaller segments and storing those segments on multiple disks. For example, a 16-KB block of data could be divided into four 4-KB segments and each segment written to a separate disk (see Figure 13-15). A read or write operation of the original 16-KB

---

block would access all four disks simultaneously. The size of a block and of the segments can be varied to suit the characteristics of the disk drives and of accesses to the data.

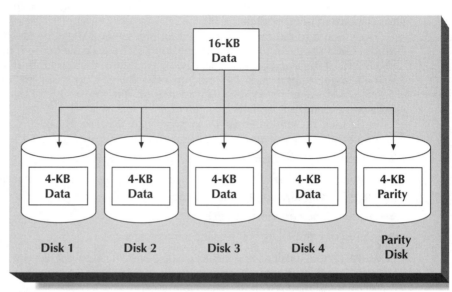

Read performance improves as a result of data striping. A large read operation is broken into multiple smaller read operations, each of which is executed in parallel on a different disk. The elapsed time to perform the entire read operation is reduced because a single disk can perform a small read operation faster than a large read operation. Write performance slows only slightly as a result of data striping if segments are at least as large as the disk block size. The penalty arises from the overhead of generated multiple write requests and the fact that all disks are busy at the same time—thus postponing the processing of any subsequent requests. Write operations to individual disks occur more quickly because of smaller data size, but most operating systems don't wait for a write confirmation so this is of no real benefit.

Fault tolerance is achieved through the generation and storage of redundant data. The redundant data is typically a set of parity bits; one parity bit is generated for each bit position within a block segment. The failure of any single drive does not result in a loss of data. The missing bit—stored on the failed drive—can be reconstructed from the remaining data bits and the parity bit.

For example, assume the use of even parity and the storage of 4 one-valued bits in equivalent bit positions of four different disks. Assume further that a fifth disk holds the parity bit. The parity bit generated for the data bits is a one-valued bit (four is an even number of one-valued bits). If any of the disks fails, there is a single missing bit. That bit must have held either a one or zero value. The value it held is determined by comparing the parity bit to the remaining bit values. Because there are 3 one-valued bits and the parity bit is one-valued, then the missing bit must be one-valued. The ability to reconstruct missing data depends on the assumption that only one disk might fail at a time; thus, there is never more than one bit value to reconstruct per stored parity bit.

The amount of redundancy required to implement RAID depends on the number of disks used. In the previous example, five disks were used, resulting in 20% of the available disk space being used for redundant information. Using a larger number of disks would decrease the portion of space used for parity bits at the expense of a slight increase in the probability of data loss because of multiple drive failures.

RAID may be implemented in software or hardware. Hardware implementations are fairly common for a number of reasons, including:

► The ability to configure all RAID components in a single cabinet.

► Extended hardware fault tolerance through the use of redundant power supplies and disk controllers.

► Reduced load on the host CPU.

► Reduced complexity of file-management software.

A RAID storage device appears as a single large disk drive to an operating system. A dedicated controller performs all RAID-related processing, including segmentation of read and write operations, generation of parity data, and reconstruction of missing data when necessary. Hardware-based RAID systems for LAN and small WAN servers are based on the SCSI bus. RAID systems for larger computer systems use fibre-channel or other high-capacity communication channels.

SUMMARY

The file-management system is responsible for managing and controlling access to files and secondary storage devices. A file-management system is implemented in layers similar to those of the operating system. The file-control layer accepts commands from the command and application layers through service calls. It implements basic file-manipulation capabilities (e.g., creating files and read/writing their content). The storage I/O control layer is responsible for implementing access to secondary storage devices and for managing the movement of data between secondary and primary storage.

The file-control layer forms a bridge between the physical and logical file-management systems. The logical file structure subdivides a file into records and fields. The physical file structure is the manner in which individual bits and bytes of data are physically stored. Physical characteristics that are simplified or ignored in the logical file structure include storage allocation, data access methods, and data encoding. Ignoring these issues allows the file-management interface provided by the service layer to be simplified.

A file is the fundamental unit of storage for both data or programs. Many different types of data may be stored in a file, including data used by application programs, source code, and executable code. Different files may require different methods of storage and manipulation. The file-management system defines a limited number of file types. A file's type determines its physical organization, allowable access methods, allowable operations, and restrictions on its name.

Storage allocation refers to the methods by which individual storage locations are assigned to specific files. An allocation unit is the smallest unit of storage that can be allocated to a file. Allocation unit size represents a trade-off among efficient use of secondary storage space, size of storage-allocation data structures, and efficiency of storage-allocation procedures. A storage-allocation table records the allocation of storage locations to specific files.

Storage allocation can be either contiguous or noncontiguous. With contiguous allocation, a file's directory entry contains the address of the first allocation unit and the number of units allocated. The most significant problem with contiguous storage allocation is the fragmentation of unallocated, or free, space. The problem can be counteracted by implementing periodic file compaction. Most file-management systems use noncontiguous storage allocation because of its inherent flexibility. However, it requires a separate data structure to record storage-allocation information and more complex allocation procedures.

Primary storage buffers are used to improve the performance of file I/O operations. In a read operation, disk blocks (or physical records) are moved from secondary storage to primary storage. Logical records are then extracted and copied to a program's data area. A write operation reverses this process. Buffering adds processing complexity to implement the procedures and calculations required to locate and access logical records within physical records.

Directories are used to store information about files and to organize files into related groups. File information stored in a directory includes name, type, location, size, ownership, access controls, and time stamps. Various directory structures may be employed, including single-level directories, tree structure directories, and graph directories. Single-level directories provide a single directory for an entire storage device. Tree structure directories allow multiple directories per storage device with a strictly hierarchical relationship among directories and files. Graph directories are less restrictive and allow multiple references to a single file or directory.

File organization refers to the placement of records within the storage space allocated to a file. Several methods of file organization can be implemented, including sequential, direct, and indexed, and hybrid methods. Sequential file organization is the simplest method because it makes no assumptions about file content and requires no special data structures to implement file-access methods. Direct file organization computes the storage location of a record based on the value of some field within that record. Indexed organization uses a table of key values and storage locations that is stored separately from the records of the file.

File-access methods may be sequential, direct, or indexed. Access is most efficient when file organization is the same as file access method. Sequential access refers to reading and writing of file contents in the order in which they are stored. Direct file access allows the pointer position to be specified directly by the application program. Indexed file access is a variation of direct access in which a requested record is specified by the value of a key field rather than by position.

The file-management system allocates buffer space and creates a file control block when a file open operation is performed. The file control block stores information about the file, including current position, the number and location of I/O buffers, and the current state of each buffer. The file control block is updated as file I/O operations are performed. Closing a file flushes its buffers to secondary storage, releases its buffer space, and deletes its file control block.

Read and write operation service calls are provided for each method of file access. Required parameters and implementation vary by access method. Edit operations include update, insert, append, and delete. An update operation replaces data within a field or a record is replaced with new content. An insert operation adds a new record within the existing sequence of records. An append operation adds a record to the end of the existing file. A delete operation logically or physically removes a record from a file. A logical delete operation marks a record as deleted but does not overwrite or reallocate its storage. A physical delete operation reallocates and/or overwrites a record's allocated storage.

A file-management system provides facilities to prevent the loss, corruption, and unauthorized access to data stored in files. These include ownership and access controls and multiuser file access controls. File ownership allows the creator of a file to control access to a file by other users. Access controls include restrictions on the ability to read, write, or execute file contents. Multiuser access controls prevent lost updates resulting from simultaneous buffered file updates. Access controls can be implemented by file and/or record locking.

File-system administration functions include file migration (version control), file backup/recovery, and transaction logging. File migration tracks changes to files and maintains previous versions of the file in online or archival storage. File backup utilities provide a means to copy files, directories, and entire storage devices to archival storage. File backups can be either complete (full) or incremental. File recovery utilities provide a means of copying files from archives back to online storage. Transaction logging records changes made to files in a separate log file. If the system crashes, files are updated from the log contents during the next system boot up.

Fault tolerance describes means of protecting file content against secondary storage device failure. Methods include disk mirroring and redundant arrays of inexpensive disks (RAID). Both methods employ hardware redundancy and automatically generate protection copies as file updates occur. RAID methods are relatively sophisticated and are implemented using special purpose hardware devices.

Key Terms

allocation unit
append mode
append operation
binary editor
binary search
blocking
blocking factor
close operation
compaction
complete path
current directory
delete operation
differential backup
direct file access
direct file organization
disk mirroring
editor
father
fault tolerance
field
file
file-allocation table (FAT)
file-by-file backup
file compaction
file control block

file control layer
file destruct
file locking
file-management system
file migration
file packing
file system administration
file type
full backup
fully qualified reference
grandfather
graph directory structure
hierarchic directory
home directory
incremental backup
indexed file access
indexed file organization
insert operation
journaling
linear search
logical delete operation
logical file structure
logical record
lost update
mirror image backup

open operation
physical delete operation
physical file structure
physical record
record
record locking
redundant array of inexpensive
    disks (RAID)
relative path
sequential file access
sequential file organization
single-level directory
son
storage-allocation table
storage I/O control layer
text editor
transaction logging
tree structure directory
unblocked
undelete operation
update in place
update operation
working directory

Vocabulary Exercises

1. A(n) _____ is the unit of file I/O to/from an application program, whereas a(n) _____ is the unit of file I/O to/from a secondary storage device.

2. _____ describes the number of _____ contained within a single _____.

3. The storage-allocation table records the assignment of _____ to specific _____.

4. A file _____ releases allocated buffers and flushes their content to secondary storage.

5. _____ eliminates logically deleted records and reallocates their associated storage units.

6. A(n) _____ operation allocates buffers for file I/O and creates a(n) _____ to record information about an active file.

7. A user owns and sets access controls for all files and directories contained in his/her _____.

8. The content of a logically, but not physically, deleted file may be recovered in a(n) _____ operation.

9. File and record locking are used to prevent the _____ problem, which may occur when two programs attempt to update a buffered file at the same time.

10. _____ describes the tracking of old file versions and their movement to off-line and archival storage devices.

11. Under _____, changes to files are written to a log file as they are made.

12. The _____ layer presents a service layer interface to application programs and the command layer. The _____ layer manages the movement of data between secondary and primary storage.

13. A(n) _____ specifies a storage device and all directories leading to a specific file. A(n) _____ specifies file location with respect to the current or working directory.

14. Under _____ file organization, the physical location of a record within the record sequence is computer-based on the value of a key field.

15. In a(n) _____ directory, a file may be located within no more than one directory. This restriction does not apply in a(n) _____ directory structure.

16. _____ is the simplest but least flexible method of file access.

17. A(n) _____ consists of a master (root) directory, one or more subdirectories, and a filename.

18. _____ allows records to be directly retrieved by specifying the value of a key field.

19. A(n) _____ marks a record as removed without physically removing it from a file.

20. _____ directory structures allow directories to be located within other directories.

21. A(n) _____ deletes directory and other references to a file and releases allocated storage locations. A(n) _____ also does this and also destroys the data content of all storage locations that were allocated to the file.

22. In a(n) _____, all files within a directory or storage device are copied to backup storage. In a(n) _____, only those files altered since the last backup are copied to backup storage.

23. A(n) _____ writes a new record at the end of an existing file.

24. A(n) _____ backup stores files, directories, and storage allocation data structures exactly as they appear on a storage device. A(n) _____ backup operation reallocates storage as it copies file contents.

25. DOS and some Windows versions record storage allocation information in a(n) _____.

26. When a file is _____, logically deleted records are physically removed from storage.

27. A file-management system may implement _____ through disk mirroring or _____.

28. A(n) _____ records the allocation of storage locations to specific files.

29. Under _____ file access, a key field is used to search a table of record locations.

30. When old versions of master files are saved, the current version may be called the _____, the previous version the _____, and the version before that the _____.

Review Questions

1. List the layers of a file-management system and describe their functions.

2. What is the difference between the logical and physical structure of a file? What advantages are realized by not allowing an application program to interact directly with a physical file structure?

3. What file types are normally implemented by a file-management system? How is file type information used?

4. What is an allocation unit? What are the advantages of using relatively small allocation units? What are the disadvantages?

5. By what methods may storage device blocks be allocated to files? What are the comparative advantages and disadvantages of each?

6. Describe the use of buffers in file I/O operations. When are buffers allocated? When are they released?

7. Describe the structure of a hierarchical directory. What are its advantages and disadvantages as compared to graph structure directories?

8.   Describe the various types of file organization. Describe the various method of file access. Must file access method always be the same as file-organization type?

9.   Describe the various types of file-update operations and the methods of implementing each type.

10.   How is file deletion accomplished? What security problems might arise from this method?

11.   What levels of access privilege may exist for a file?

12.   What is the lost update problem? How can it be prevented?

13.   What is the difference between a full dump and a selective (or incremental) dump?

14.   What is transaction logging (journaling)? Describe the performance penalty that it imposes on file-update operations.

15.   Describe various methods of implementing hardware fault tolerance. What are their comparative advantages and disadvantages?

Research Problems

1.   To increase efficiency of application program execution, some file-management systems use a large number of file organization and access methods, each optimized to a specific type of processing. The MVS (IBM mainframe) operating system contains such a file-management system. Investigate this system to determine the various methods of file organization and access that are supported. For what type(s) of application program file manipulation is each method intended?

2.   The UNIX operating system, and its embedded file-management system, were not originally designed to support large-scale transaction processing systems. For this and other reasons, UNIX lacks many aspects of file management that would support such an environment (e.g., transaction logging, indexed files, file migration, etc.). Identify the specific deficiencies of the UNIX operating system with respect to large-scale online transaction processing. How do/might application developers overcome these deficiencies?

Chapter

14

Operating System
Input/Output

Chapter Goals

► To describe service-layer functions that support character-, full-screen-, graphic-, and windows-based I/O.

► To describe methods of interactive command and control, including command line, form-based, window-based, and object-oriented interfaces.

► To describe batch control of software with job control languages (JCLs).

► To describe software components used to access networks and distributed resources.

INPUT/OUTPUT SERVICE FUNCTIONS

The service layer provides a set of subroutines that allow processes to interact with input/output (I/O) devices and, therefore, directly with users. I/O service layer capabilities—driven by the introduction of new I/O devices and new methods of human-computer interaction—have changed dramatically over the last three decades. The variety of I/O devices and methods supported by operating systems results in a large, diverse set of supporting service layer functions.

Character-Oriented I/O

Early operating systems implemented I/O as streams of individual characters, which mirrored the capabilities of early I/O devices (e.g., line printers, card readers, and teletypes). Very few control functions were required by early I/O devices and those that were implemented were based on transmitting nonprinting characters (e.g., form feed, carriage return, line feed, etc.) to the device.

The service routines required to support character-based I/O are relatively simple:

▶ read_character(device_id,character)

▶ write_character(device_id,character)

Both of these routines interact with an I/O device, or device controller, through a specific I/O port identified by the device_id argument. The parameter "character" is usually a memory address where the character to be read or written is stored, or should be stored once input.

Because characters are read and written in groups (e.g., words, fields, lines of a report, etc.), service calls also are provided to read and write entire strings. The following service calls provide these additional capabilities:

▶ read_string(device_id,string)

▶ write_string(device_id,string)

As with single-character I/O calls, device_id identifies a specific I/O device. The parameter "string" is the starting address of a character array in memory.

Notice that string input requires some mechanism for identifying the end of one string and the start of another. In most operating systems, a special character is placed at the end of each string array to signify its end. UNIX, for example, terminates character arrays with an ASCII zero (NULL) character. Terminating strings with control characters allows string I/O commands to be implemented as a loop containing character I/O commands. For example, the call write_string might be implemented as shown below.

```
procedure write_string(device_id,string_start_address)
begin
        i=0
        character=string_start_address
        while (character ≠ 0)
                write_character(device_id,character)
                i=i+1
                character=string_start_address+1
        end while
        write_character(device_id,character)
end
```

A similar strategy can be used for the input and output of printed text lines. In most operating systems, text lines are character strings containing a special "end-of-line" character or character sequence (e.g., line feed in UNIX and line feed followed by carriage return in DOS and Windows). Thus, a service call to output a text line can be implemented by a subroutine similar to the one shown above. The test for a null character is replaced by a test for the end-of-line character(s).

Full-Screen I/O

I/O service calls for single characters, strings, and text lines are cumbersome to use with video display terminals (VDTs) and programs that emulate VDT input/output to or from a computer monitor (e.g., telnet executing on a micro-computer). Output to a VDT may not flow in a linear sequence from left to right and top to bottom. VDT input is often form-based, which requires printing a blank form on the screen and then allowing the user to move (i.e, tab) from field to field to input individual data items. Altering individual fields requires the ability to move the input cursor directly to a particular position on the screen, without overwriting other characters already printed there.

Additional complexities of VDT I/O include special display features such as underlining, inverse-video, bold-intensity, half-intensity, and invisible characters. Such features require complex control capabilities, and the specific control methods used to enable and disable those features varies from one VDT to another. Most VDTs activate and deactivate special display features when sequences of control characters are received. An ANSI (American National Standards Institute) standard exists for some of these control character sequences (see Table 14-1 for a partial listing).[1] However, not all manufacturers adhere to the ANSI standard, and many VDTs implement commands and display features not included in the standard.

TABLE 14-1	Feature or Command	Control Character Sequence
A subset of the ANSI standard control character sequences for full-screen character-based displays. The string <escape> stands for ASCII control character #27.	position cursor	<escape>[#;#H
	underline on	<escape>[4m
	inverse video on	<escape>[7m
	bold on	<escape>[1m
	all attributes off	<escape>[0m
	erase to end of line	<escape>[K
	erase entire screen	<escape>[2J
	cursor up	<escape>[#A
	cursor down	<escape>[#B
	cursor left	<escape>[#D
	cursor right	<escape>[#C

Many operating systems provide a set of I/O service calls to perform full-screen I/O functions such as cursor positioning, screen clearing, and activating and deactivating special display functions. Such operating systems also provide a mechanism to vary the implementation of these services to match the capabilities and requirements of specific VDTs. Implementation methods include:

► Asking the user to provide device type or identifier during login processing.

► Querying the I/O device directly; most full-screen terminals transmit an identification code in response to the proper control sequence.

► Specifying a default device in a file of initialization information and allowing the user to alter this information, if needed.

---

[1] Loading the device driver ANSI.SYS under MS-DOS or Windows enables character-based screens and windows to respond correctly to ANSI full-screen control character sequences.

Once the VDT device type is determined, that information is stored for later use by full-screen I/O service routines. Whenever a process calls a full-screen I/O service function, the function tests the I/O device type to determine the proper control sequence to implement that function on that particular device. To improve efficiency, a table of control sequences is sometimes created and stored in memory when the I/O device is first identified. All full-screen service routines then use the content of that table instead of testing the device identifier each time they are called. For example, UNIX uses a shell variable called TERMCAP (terminal capability) to store a list of full-screen capabilities and control sequences for use by full-screen I/O service functions.

Operating system service layers support character-based printer output in essentially the same manner they support output to VDTs. That is, a set of service calls is provided for basic character-output functions (e.g., writing characters, strings, and text lines), and an additional set of service functions are provided to control special display functions (e.g., underlining, font selection, and print head positioning). As with VDTs, the operating system must provide a mechanism to customize the implementation of these service calls to the capabilities and control procedures of a specific printer.

Graphic I/O

Graphic I/O devices (e.g., laser printers and microcomputer monitors) provide a much richer set of I/O capabilities than VDTs and character-based printers, and their additional capabilities require a larger and more complex set of I/O service calls. Graphic output devices generate images in terms of pixels rather than characters. Individual pixels of the display or page can be manipulated in terms of binary values (e.g., black or white), grey scales, or color codes (e.g., RGB intensities). As discussed in Chapter 10, communication with graphic output devices may use bit maps, vector drawing commands, a display language, or some combination of these. As with VDTs, there is considerable variation among devices and manufacturers in display capabilities and the methods of controlling those capabilities.

Although there is no standard method of controlling graphic monitors, there is some trend toward standardization of display languages for printers. The Postscript display language is a clear leader in the competition for a standard graphics display language. However, there has been no acceptance of Postscript by any standards organization (e.g., ANSI and ISO), and there are significant competitors (e.g., Hewlett-Packard Graphics Language).

Operating systems rely on a layered set of software components to deal with the variation and lack of standardization in graphic I/O devices (see Figure 14-1 on the next page). The key component of a layer set is a generic graphics device driver. Service layer functions interact with the generic device which, in turn, interacts with the device driver of a specific I/O device. Commands sent from the service layer to a generic device driver are called **device-independent commands**. Specific device drivers accept device-independent commands as input and translate them into commands recognized by a specific graphic I/O device.

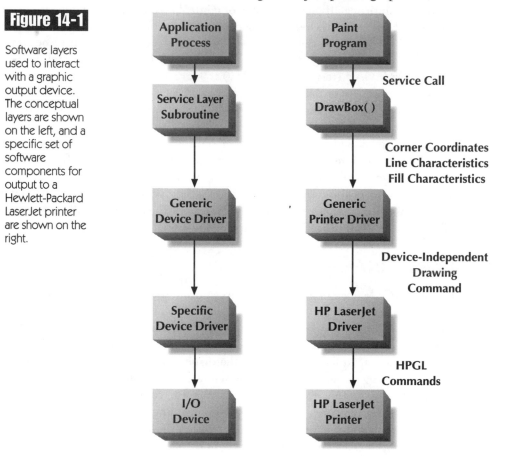

Figure 14-1

Software layers used to interact with a graphic output device. The conceptual layers are shown on the left, and a specific set of software components for output to a Hewlett-Packard LaserJet printer are shown on the right.

The layered approach to graphic I/O provides several advantages to designers of operating systems and graphic I/O devices. Operating system service routines need not be designed to handle variation among I/O devices. All service routines are designed to interact with the generic device driver, thus greatly simplifying the task of developing service layer routines. I/O hardware designers and device driver programmers don't need detailed knowledge of how service layer functions are implemented; instead, they only need detailed knowledge of the device-independent commands generated by the generic device driver. This simplifies the task of developing specific device drivers and defines a common pool of knowledge used by all device driver programmers.

The device-independent command language is a critical component of the layered approach to graphic I/O. The language must implement all of the control capabilities of the specific graphic I/O devices that might be used. The set of control capabilities that can be expressed in the device-independent command language should be the set of graphical output capabilities for all possible I/O devices. The language syntax and semantics should be designed to simplify the process of translating the language into device-specific commands. Designing a powerful, expressive, and easy-to-translate language is not a simple task.

Device-independent command languages must be updated regularly to account for new features in the latest graphic I/O devices. These updates should retain backward compatibility with earlier versions to minimize the need to rewrite specific device drivers and/or operating system service layer routines.

Window-Based I/O

Most operating systems provide service layer functions that support interactive window-based interfaces. Figure 14-2 shows an example of a window-based display. **Window-based I/O** extends graphic I/O in a number of ways, including:

► Division of a display into multiple output regions (i.e., windows).

► Use of colors and background patterns.

► Ability to use various font styles and sizes.

► Provision of standard facilities for menu display and control (e.g., pop-up menus, push buttons, scroll bars, etc.).

Figure 14-2

A sample window-based graphic display

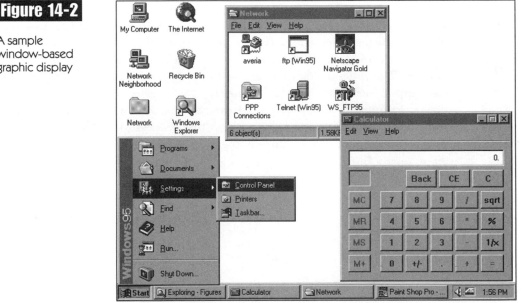

A number of different window-based interfaces are common, including Presentation Manager (used in OS/2), X Windows (used in many UNIX versions), and Explorer (used in Windows). These interfaces provide a set of service routines that implement:

► Definition of window size and display characteristics (e.g., background color, scrolling capabilities, font style and size, available control functions, etc.).

► Definition and display of menus and the input of user menu selections.

► Display of graphic images based on bitmaps, vector commands, or a display language.

► Standardized facilities for window control, including resizing, movement, overlap with other windows, and so forth.

► Copying of text and/or images between windows (i.e., "cut and paste" capabilities).

► User input by keyboard or pointing device (e.g., mouse, trackball, or digitizing tablet).

The number of service calls needed to implement window-based I/O is large (typically, more than 100). As with graphic I/O, the operating system must provide a mechanism to translate service calls into specific commands to a graphic display device. Window-based interfaces consume a substantial amount of hardware resources (e.g., CPU cycles for translation, memory for display images, and disk space for service routines and fonts).

The organization of software components within a typical window-based interface is shown in Figure 14-3. Notice that the command layer and the application programs don't interact directly with I/O services. Instead, all communication is routed through the **window manager**, which acts as a switch to route messages from one software layer to another. The window manager ensures that only one interactive program at a time receives access to the I/O devices. The program that has access to the input devices currently is said to have **input focus**. The window manager usually gives input focus to the window containing the mouse pointer. Keystroke sequences are sometimes defined to force input focus to move from one active process to another (e.g., the Alt-Tab keystroke combination in Windows).

Figure 14-3

Software
components
of typical
window-based
interface

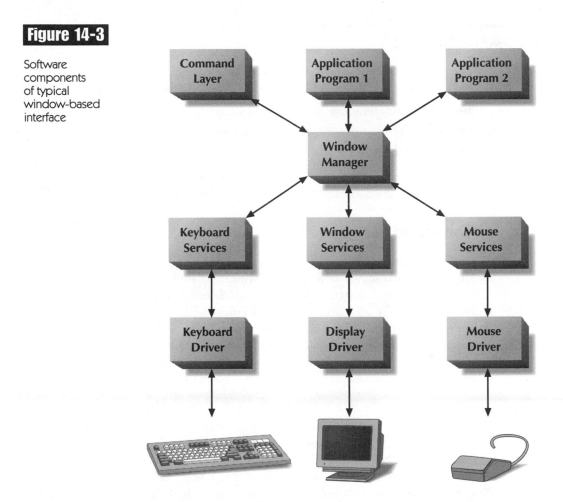

The window manager has other responsibilities, including maintaining the content of screen areas not occupied by active programs (i.e., the background or desktop) and implementing full-window management functions (e.g., minimize, move, and resize). The window manager keeps track of which display areas are occupied by which windows. As windows are created, destroyed, moved, or resized, the window manager sends appropriate messages to programs to update their portion of the display. For example, if the Calculator program in Figure 14-2 is terminated, the window manager instructs the program that owns the window titled "Network" to redraw itself, thus updating the lower-right corner. The window manager will also redraw the background color or pattern underneath the remainder of the display area formerly occupied by the calculator window.

COMMAND LAYER

The command layer allows an end user to directly control hardware and software resources. Specific tasks performed through the command layer include:

► Batch and interactive program execution.

► Control of multiple programs or processes.

► File and directory management.

As with other aspects of operating system architecture, the exact features provided and the details of their implementation vary considerably.

Command Languages

Command languages are the oldest form of command layer interface. Processing requests are expressed in a language with rigid syntactic and semantic requirements, similar to a programming language. Commands are accepted from a keyboard or text file and are processed in much the same way as statements of an interpreted programming language. Command layers that require command language input are sometimes said to have **command-line interfaces**.

A typical command in a command language is a line of text (i.e., one or more characters terminated by a carriage return) of the form:

COMMAND PARAMETERS OPTIONS

The COMMAND field states the name of the operation to be performed. PARAMETERS refer to objects to be manipulated by the command (e.g., files). OPTIONS provide a mechanism for altering the normal (default) execution of the command. Some command languages require that all options be specified. In others, defaults values are assumed if their corresponding options are not present on the command line. In some operating systems, options may precede arguments, or they may be interspersed with arguments in the command line.

The organization of the command layer with a command-line interface is shown in Figure 14-4. The **command interpreter** accepts text input and checks it for valid content by verifying correct syntax of the entire statement and its parameters (e.g., pathnames). Errors are immediately reported back to the user. The first word of a valid command line is checked against a table of internally implemented utility programs. If the command name is found in the table, the corresponding program is executed immediately by the internal command processor. Many operating systems implement simple command utilities (e.g., file renaming and directory listing within the internal command processor).

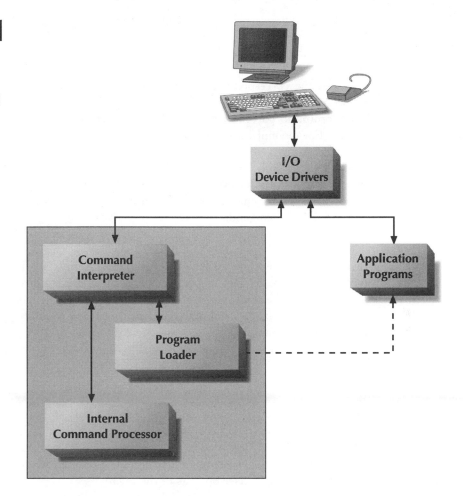

Figure 14-4

Components of the command layer (shown within the shaded box) and their relationship to other software components

If the command is not found within the table of internal commands, then the command name is assumed to be that of an executable program stored on disk. The command interpreter searches for a file that matches the command name. If found, the program loader is called to load the file into memory and begin execution.

The user can exercise some control over the search process for executable files. The command interpreter maintains a list of devices and/or directories to be searched for executable files. This list is usually defined in an initialization file that is read by the command interpreter when the user logs on. The user or system administrator can modify the contents of this file to include additional devices or directories or to exclude one or more default devices or directories. The list of devices and directories is called a **path**, or **search path**.

Users can override the search process by specifying the device and/or directory containing an executable program within the command line. Typically, the device and/or directory are placed prior to the command name on the command line. The command interpreter detects this information and searches the corresponding device/directory for a filename matching the command name.

Command parameters (or arguments) are references to data to be manipulated. For example, the argument C:\ in the MS-DOS command:

DIR C:\

instructs the operating system to display a directory listing of the root directory (\) of device C:. Some commands can have multiple arguments: for example, the MS-DOS command:

COPY C:\FILE1.TXT D:\

instructs the operating system to locate FILE1.TXT in device/directory C:\, make a copy of the file, and store the copy in device/directory D:\ with the same name as the original.

Many commands recognize **command options** to control various processing features. Command options are typically preceded by a special symbol to distinguish them from command names and parameters. For example, the MS-DOS command:

DIR C:\ /P

produces a directory listing and will pause after each screen of text is displayed. The option "P" modifies the default behavior of the DIR command. Options are always preceded by the symbol "/" in MS-DOS.

Other operating systems use different syntactic conventions for options. UNIX, for example, uses options of the form:

-option value

where the "-" symbol indicates that an option identifier follows, "option" is a character or string representing the option, and "value" is a character or string containing the value of the option. For example, the UNIX command:

lpr -d printer2 file3.txt

instructs the operating system to print the file file3.txt. The option "-d printer2" alters the default destination for the printed output to printer2.

In MVS, options are specified in the form:

option1=value, option2=value, option3=value

where "option" is a keyword indicating the option being set and "value" is the value of the option. The "=" symbol separates the keyword from the value, and the "," symbol separates multiple options on the same command line.

Statements of a command language can be stored in a text file for later use. Such files are useful for two purposes. First, they allow users to interactively execute a series of commands without typing each individual command line at the keyboard. Most operating systems allow text files containing command language statements to be marked as executable files. The name of such a file can be typed as part of a command line, and the command layer searches for it as it would search for a file containing executable code. When found, the contents of the file are accepted as input by the command interpreter as if an interactive user had typed them at the keyboard.

Batch Service

Files of command lines are useful for controlling the execution of batch application programs. Examples of application programs that normally run in batch mode include periodic reporting, file-maintenance activities, and transaction processing activities that do not use online input (e.g., payroll and accounts payable systems). A command language used to control batch execution, called a **job control language (JCL)**, may be a superset of an interactive command language or a completely different language. Figure 14-5 on the next page shows a portion of an IBM MVS JCL program.

Figure 14-5

Portion of an MVS JCL program. The program segment compiles, links, and executes a COBOL program.

```
//IDMSPRE.SYSIPT   DD   *
//COB        EXEC PGM=IKFCBL00,
//                 PARM=(NOLIB,'BUF=30504',NOADV,NODYNAM,SOURCE,APOST)
//STEPLIB  DD   DSN=SYS1.CO24.VSCOLIB,DISP=SHR
//SYSPRINT DD   SYSOUT=*
//SYSLIN   DD   DSN=&OBJ,SPACE=(3120,(40,40),,,ROUND),UNIT=VIO,
//                 DISP=(MOD,PASS),
//                 DCB=(BLKSIZE=3120,LRECL=80,RECFM=FBS,BUFNO=1)
//SYSUT1   DD   DSN=&SYSUT1,SPACE=(1024,(120,120),,,ROUND),UNIT=VIO,
//                 DCB=NCP=1
//SYSUT3   DD   DSN=&SYSUT3,SPACE=(1024,(120,120),,,ROUND),UNIT=VIO,
//                 DCB=NCP=1
//SYSUT2   DD   DSN=&SYSUT2,SPACE=(1024,(120,120),,,ROUND),UNIT=VTO,
//                 DCB=NCP=1
//SYSUT4   DD   DSN=&SYSUT4,SPACE=(1024,(120,120),,,ROUND),UNIT=VIO,
//                 DCB=NCP=1
//SYSUT5   DD   DSN=&SYSUT5,SPACE=(1024,(120,120),,,ROUND),UNIT=VIO,
//                 DCB=NCP=1
//SYSLIB   DD   DUMMY
//SYSIN    DD   DSN=&&DMLOUT,DISP=(OLD,DELETE)
***
********************************************************************
***
//LKED       EXEC PGM=IEWL,COND=(5,LT,COB),PARM='LIST,XREF,LET,MAP'
//SYSPRINT DD   SYSOUT=*
//SYSLMOD  DD   DSN=&&LNKLIB(X),DISP=(,PASS),UNIT=SYSW,
//                 SPACE=(TRK,(1,1,1)),DCB=(RECFM=U,BLKSIZE=6144)
//SYSUT1   DD   DSN=&SYSUT1,SPACE=(1024,(120,120),,,ROUND),UNIT=VIO,
//                 DCB=BUFNO=1
//SYSLIB   DD   DSN=&COBLIB,DISP=(SHR,PASS)
//         DD   DSN=USER.T87250.IDMS.LOADLIB,DISP=SHR
//         DD   DSN=USER.T87250.IDMS.PGMLIB,DISP=SHR
//LIB      DD   DSN=USER.T87250.IDMS.LOADLIB,DISP=SHR
//         DD   DSN=USER.T87250.IDMS.PGMLIB,DISP=SHR
//SYSLIN   DD   DSN=&OBJ,DISP=(OLD,DELETE)
//         DD   DSN=USER.T87250.IDMS.SRCLIB(I102LKED),DISP=SHR
***
********************************************************************
***
//GO         EXEC PGM=X,COND=(1,LT,LKED)
//STEPLIB  DD   DSN=&&LNKLIB,DISP=(OLD,DELETE)
//         DD   DSN=USER.T87250.IDMS.LOADLIB,DISP=SHR
//         DD   DSN=USER.T87250.IDMS.PGMLIB,DISP=SHR
//         DD   DSN=SYS1.CO24.VSCLLIB,DISP=SHR
//SYSOUT   DD   SYSOUT=*
//SYSDBOUT DD   SYSOUT=*
//SYSJRNL  DD   DUMMY
//DBFILE1  DD   DSN=USER.T&FILE#..DBFILE1,DISP=SHR
//DBFILE2  DD   DSN=USER.T&FILE#..DBFILE2,DISP=SHR
//GO.RPTOUT   DD   SYSOUT=*,DCB=(RECFM=FBA,LRECL=133,BLKSIZE=133)
//GO.TRANIN   DD   *
```

Many large application systems are partitioned into multiple programs or processes. A collection of related programs that are run within the same time frame (e.g., the set of programs executed for monthly payroll processing) is called a **job**. Each individual program, or step, in a job is called a **job step**. A set of JCL commands that execute a series of job steps is called a **job stream**.

A JCL provides control mechanisms to execute individual job steps, coordinate multiple programs, and define recovery procedures for likely error conditions. In many operating systems, the complexity of the job control language used to control batch program execution is as great as the complexity of the programming language statements used to define individual programs. Programming language features such as control structures and variable declarations are a required feature of any JCL.

Control Structures

The control structures common to all procedural programming and command languages are:

► Sequence.
► Selection.
► Repetition.

Sequence control refers to the sequential execution of multiple programs. Most programming and JCLs express sequence by placing commands on consecutive lines or in consecutive statements. Figure 14-6 on the next page shows an example of a batch transaction processing system organized as a sequence of job steps. Figure 14-7 on the next page shows a JCL program that executes those job steps sequentially.

Figure 14-6

A batch
transaction
processing
system consisting
of threee
sequential
processes

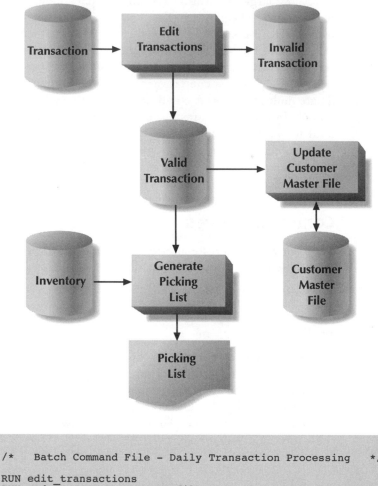

Figure 14-7

Example of a
batch command
file. Each program
is executed, one
after the other,
without user
intervention.

```
/*    Batch Command File - Daily Transaction Processing    */

RUN edit_transactions
RUN update_customer_master_file
RUN generate_picking_list
```

Selection refers to the ability to select alternate processing paths, depending on stated conditions and/or the results of previous processing steps. An **if statement** or **case statement** implements selection in a JCL or programming language. Figure 14-8 shows an example of conditional processing within a JCL program. In this example, the system date is tested to determine if the current date is the last day of the year. If the test is successful, an additional job step is executed in addition to the normally executed job step.

Figure 14-8

An example of a
batch command
file with
conditional
execution of
end_of_year_
program

```
/*   Batch Command File - Daily Transaction Processing   */

RUN edit_transactions
RUN update_customer_master_file
RUN generate_picking_list
if (sys_date = 'December 31') then
    RUN end_of_year_program
endif
```

Selection can be used to detect processing errors. As described in Chapter 12, most operating systems report an exit or status code when a process terminates execution. An example of a JCL program that tests program exit codes is shown in Figure 14-9. The exit status for the most recently terminated process is assumed to be stored in the variable exit_code by the operating system. The symbol "$" is used within the JCL command to denote a variable name. The sample program tests the value of exit_code after the first processing step to determine whether remaining processing steps should be executed. (The file valid-trans may not be correct if the program edit_transactions fails.) In the event of failure in any process, an error message is generated to identify the failed process and its exit code.

Figure 14-9

Example of a
batch command
file with nested
conditional
executions. The
exit status of each
program is tested
to determine
whether to
execute the next
step or print an
error message.

```
/*   Batch Command File - Daily Transaction Processing   */

RUN edit_transactions
if ($exit_code = 0) then
   RUN update_customer_master_file
   if ($exit_code = 0) then
      echo "failure in process update_customer_master_file
      echo "exit_code=" $exit_code
   endif
   RUN generate_picking_list
   if ($exit_code = 0) then
      echo "failure in process generate_picking_list"
      echo "exit_code=" $exit_code
   endif
else
   echo "failure in process edit_transactions, exit code=" $exit_code
   echo "remaining processing steps aborted"
endif
```

Repetition control structures allow one or more processing steps to be executed multiple times. Various forms of the repetition control structure include:

► for each (list of items) do (processing)

► while (condition) do (processing)

► repeat (processing) until (condition)

As with other control structures, the exact syntax and semantics varies widely between JCLs. Figure 14-10 shows the example from Figure 14-9, modified to use a for-each control structure.

Example of a batch command file with nested conditional executions. Sequential execution and testing of exit code is implemented with a for-each control structure.

```
/*   Batch Command File - Daily Transaction Processing   */

$LIST=(edit_transactions,update_customer_master_file,generate_picking_list)
for each $PROGRAM in $LIST do
    RUN $PROGRAM
    if ($exit _code   0) then
        echo "failure in process" $PROGRAM
        echo "exit_code=" $exit_code
        halt
    endif
end for
```

I/O Re-direction

It is often necessary to alter the source and/or destination of a program's I/O data. The command interpreter itself is an example of this. When executing interactively, input is received from a keyboard or other interactive I/O device; when executing in batch mode, input is received from a file of previously stored commands.

The ability to change the source of input data or the destination of output data—referred to as **I/O redirection**—requires a mechanism for stating alternate data sources/destinations on the command line. It also demands that processes be written in such a way that they can accept input from—or generate output to—a variety of sources/destinations.

To support I/O re-direction, programs are designed to write to "standard" input and output devices. The operating system controls these standard devices and can route I/O to or from standard devices to any physical device or file. By default, interactive processes use a VDT or monitor/keyboard for the standard input and standard output devices.

A specific command line syntax is used to indicate I/O re-direction, as in:

RUN application_program < input_file

In this example, the symbol "<" directs the operating system to perform input re-direction. The operating system loads and executes application_program and will assign the file input_file as the standard input device. Similarly, the command:

RUN application_program > printer1

loads and executes the application_program with output to the standard output device re-directed to printer1.

I/O re-direction is often combined with the operating system's interprocess data communication (e.g., piping) facilities. For example, the command:

RUN program1 < file1 | program2 > printer2

instructs the operating system to load and execute program1 using the contents of file1 as the standard input and to route standard output into a pipe, as represented by the "|" symbol. The pipe is the standard input for program2 (executing concurrently), which, in turn, routes its output to printer2. Neither program explicitly references a pipe. Thus, the programs contain no explicit calls to create or use pipes. In essence, the operating system "fools" the programs into thinking that they are communicating with physical I/O devices.

Resource Declarations

Some operating systems require JCL programs to specify the resources (e.g., memory and disk space) required by a job stream. Any such statement of resource needs in a job stream is called a **resource declaration**. Resource declarations are commonly used in large-scale batch processing systems. Data-intensive programs that require up to several hours to execute on a mainframe computer are typical in such systems. Allowing large programs to fail during execution because of lack of resources results in wasted computer resources and creates scheduling problems for downstream processing tasks.

Resource declarations provide two important benefits to the operating system. First, they allow the operating system to determine whether sufficient resources exist to successfully complete a job prior to starting execution. The operating system examines the resource declarations and attempts to allocate those resources before the first job step is executed, which prevents failure of a later jobs step because of the lack of a required resource.

The other benefit of explicit resource declarations is the additional information provided to the operating system for resource allocation. Resource declarations allow the operating system to determine whether sufficient resources exist to execute all pending jobs. If resources are insufficient, the operating system can attempt corrective actions, such as delaying the execution of one or more jobs or shifting resources from one use to another.

Dynamic resource allocation does not require JCL programs to state their resource requirements explicitly. Instead, the operating system responds to resource demands as they occur (i.e., as programs are loaded and executed). For example, an operating system using dynamic resource allocation doesn't attempt to determine the maximum memory that will be used during a job; rather, it simply waits for programs to request memory and then attempts to fill those requests. Dynamic resource allocation allows jobs to fail because of unavailability of required resources. Such failures are unlikely in environments where computing resources are not severely constrained.

Form-Based Command Interfaces

The complexity of interactive command languages often leads to errors in user input. The user must memorize the names of valid commands and the type and syntax of arguments and options for each command. In essence, the user speaks to the computer in an artificial control language. Form-based command interfaces simplify the user interface. A form-based interface also relies on a command language; that is, a user still must memorize a basic set of command names. However, the form-based interface provides a much simpler means to specify command arguments and options.

Once a command name has been entered at the command prompt, the command interpreter places a form, or menu, on the screen. In the case of a form, the user is presented with a set of blank fields for argument names and option values. Each of these blank fields is preceded by the name of the argument or option. Thus, the user need not be concerned with the placement, or syntax, of these fields within a command line. In the event of errors, an appropriate error message can be displayed (typically at the bottom of the screen) and the cursor moved automatically to the fields that need to be changed. An example of a form-based interface is provided in Figure 14-11.

Figure 14-11

An example of form-based interface. The interface is for a file-management utility in IBM Structured Programming Facility (SPF). Command names, parameters, and options are entered to the right of the ===> prompt string. The user moves among fields with tab or arrow keys.

```
-------------------------- EDIT  ENTRY PANEL --------------------------
COMMAND ===>

ISPF LIBRARY:
    PROJECT ===> ABC.USER12
    GROUP   ===> XYZ         ===>            ===>            ===>
    TYPE    ===> SOURCE
    MEMBER  ===>                    (Blank or pattern for member selection list)

OTHER PARTITIONED OR SEQUENTIAL DATA SET:
    DATA SET NAME  ===>
    VOLUME SERIAL  ===>             (If not cataloged)

DATA SET PASSWORD ===>             (If password protected)

PROFILE NAME      ===>             (Blank defaults to data set type)

INITIAL MACRO     ===>        LOCK       ===> YES    (YES, NO or NEVER)

FORMAT NAME       ===>        MIXED MODE ===> NO     (YES or NO)
```

A form-based interface can be supplemented with a menu, or list, of argument and option values. For example, a user that executes a command to list a directory might be provided with a list of directories from which to choose. Also, the user might receive a list of options or option values (e.g., sorted listing, short or long display, etc.).

Window-Based Command Interfaces

A window-based command interface is easier to learn and use than command line- or form-based interfaces because of the extensive use of menus and graphic images. Most or all of the available commands are provided in menus. The large number of commands may necessitate a hierarchical layering of menus (see Figure 14-12 on the next page). Application and system-utility programs provide menu bars and pull-down menus for commands and some options and parameters (see Figure 14-13 on the next page). Pop-up windows also may be used to query the user for required parameters and other information.

Figure 14-12

Example of
layered menus

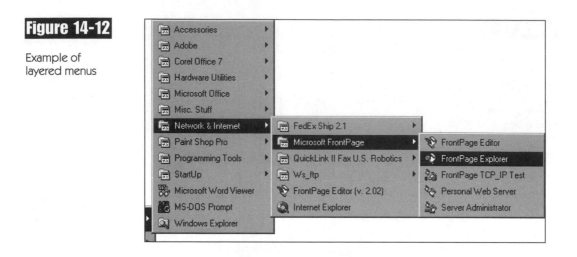

Figure 14-13

Example of a
menu bar and
pull-down
menus

Graphic images often are used to supplement commands listed in menus. Pictorial metaphors (called **icons**) can be used to represent certain commands or objects to be manipulated. For example, directories can be represented as pictures of file folders, and documents within them can be shown as pictures of printed paper (see Figure 14-14). Common command actions such as moving, deleting, and printing documents can be implemented by mouse movements with the icons (e.g., dragging and dropping). A command interface that implements such features is said to be **object-oriented**.

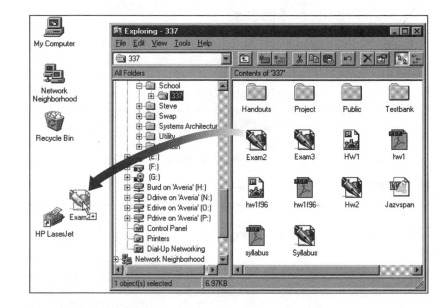

Figure 14-14

Example of object-oriented drag and drop. The document object Exam2 is being dragged to the printer object HP LaserJet.

Object-oriented command interfaces are used widely for interactive command and control, but there are limits to the visual metaphors that they employ. For example, there are no obvious drag-and-drop metaphors for common actions such as scanning a disk for errors and installing new application software. Thus, object-oriented command interfaces must provide alternative command access through menus or forms for some command tasks. Command and control of noninteractive processing is clearly not suited to interactive object-oriented interfaces and, thus, JCLs remain in wide use.

The boundary between the command layer and other system software components is not well defined with a window-based or object-oriented command interface. Command and control features such as a program or system settings menu usually are integrated with the window manager. A command processor may be loosely integrated with the window manager as a cooperating process, or the two components may be tightly bound within a single software component.

NETWORK I/O

Network I/O services have grown in complexity and breadth over the last decade. The days of centralized hardware, software, and software resource management have long passed. Thus, an operating system must provide users and programs with access to resources and services on remote servers.

Protocol Stacks

An operating system implements network I/O and services as a complex set of software layers. The **Open Systems Integration (OSI) network model** was an early attempt to standardize the number and function of these software layers.[2] The lower levels of the OSI model are commonly called the network **protocol stack**. Figure 14-5 shows two example protocol stacks. Stack one is typical of the protocols used by a workstation client (e.g., a microcomputer) to access a LAN server running Novell NetWare. Stack two typifies an Internet Web server (e.g., a minicomputer running UNIX).

Figure 14-15	Layer Description	Stack 1	Stack 2
Two sample network protocol stacks	OSI Session Layer	Transport Layer Interface	Sockets
	OSI Transport Layer	Sequenced Packet Exchange	Transmission Control Protocol
	OSI Network Layer	Internet Packet Exchange	Internet Protocol
	Logical Network Interface Driver	Open Data Interface	Network Driver Interface Specification
	OSI Data Link Layer (Physical Network Interface Driver)	10 Mbps Ethernet	Asynchronous Transfer Mode
	OSI Physical Layer	Twisted-Pair Copper Wire	Fiberoptic Cable

The **physical layer** contains the transmission lines and network interface devices used to send and receive network messages. The **data link layer** contains the physical device drivers for the network interface units. The **logical network interface driver** is similar in function to the generic printer driver described earlier—that is, a switch connecting the layers above to the physical network interface device drivers. The logical network interface driver defines a generic network interface device and translates commands for that device into commands specific to the physical device driver. **Open Data Interface (ODI)** and **Network Driver Interface Specification (NDIS)** are two widely implemented logical network driver standards.

---

[2] You may want to review that section of Chapter 9 before reading the remainder of this chapter.

The **network layer** divides outgoing network messages into packets and reassembles incoming packets into messages to upper layers. The **transport layer** implements error checking and manages quality of service. Quality of service is an especially important issue in the asynchronous transfer mode (ATM) networks. **Transmission Control Protocol (TCP)** and **Internet Protocol (IP)** are widely implemented protocols that form the "language" of the Internet; **Sequenced Packet Exchange (SPX)** and **Internet Packet Exchange (IPX)** are their equivalents within Novell NetWare LANs.

The **session layer** defines a high-level interface used for client-server interaction. A concept borrowed from the UNIX operating system—**sockets**—has become a standard method of interfacing clients and servers over the Internet. **Transport Layer Interface (TLI)** provides equivalent services in Novell NetWare LANs.

Commonly, multiple protocols stacks exist simultaneously on a single computer system. A typical arrangement for a workstation client is shown in Figure 14-16. Two different network, transport, and session layer stacks share a single physical device driver and physical network layer. The dual stack allows the client to access Novell NetWare LAN servers (through TLI, SPX, and IPX) and Internet servers (through sockets, TCP, and IP). The ODI layer is the "glue" that allows both upper-level protocol stacks to share a common physical network interface.

Figure 14-16	**Layer Description**	**Stack 1**	**Stack 2**
Example of a complex network stack with two upper-level protocol stacks sharing a single network interface	OSI Session Layer	Transport Layer Interface	Sockets
	OSI Transport Layer	Sequenced Packet Exchange	Transmission Control Protocol
	OSI Network Layer	Internet Packet Exchange	Internet Protocol
	Logical Network Interface Driver	Open Data Interface	
	OSI Data Link Layer (Physical Network Interface Driver)	10-Mbps Ethernet	
	OSI Physical Layer	Twisted-Pair Copper Wire	

Protocol stacks provide several advantages for implementing network I/O and services. They divide the task of network interaction into several well-defined pieces that can be implemented and installed separately. They provide the flexibility needed to keep pace with the rapid evolution of protocol standards.

Also, they insulate application programs and many portions of the operating system from details of low-level protocols and physical network implementation, which ensures portability of software across a wide range of network protocols and implementations.

Service Layer Functions

A modern operating system implements a set of service layer subroutines to access remote resources through one or more session layer protocols. The number of routines provided ranges from a few dozen to a few hundred, depending on the number and complexity of supported session layer protocols. The nature of network service I/O layer is evolving continuously. Early approaches treated network interfaces similarly to other I/O devices. Later approaches have incorporated the client-server model of computing directly into the design of the operating system, which has resulted in a fundamental redefinition of the role of the operating system service layer and the mechanisms by which access to local hardware resources is implemented and controlled.

Resource Access

An important function of the operating system network interface is to hide the location of resources from users and application programs. This operating system characteristic is often called **location transparency**, or **network transparency**. The goal is for users and programs to interact with resources and services using identical methods, regardless of whether those resources and services are provided by the local operating system or obtained from a remote server. Thus, for example, a word processor should use the same method to access a document file stored on a local hard drive and a document file stored on a server disk. Similarly, a Web browser should access Web services provided by the native operating system in the same manner it accesses Web services from a computer system located on another continent.

An operating system can use one or more **re-directors** to implement location transparency. A re-director is a software component that translates service calls that access remote resources into an appropriate set of network messages to implement the access. An operating system implements a separate re-director for each class of resource (e.g., one for printer I/O and another for file I/O). The re-director maintains a table of resource names and their corresponding local or remote resources. The interaction between the re-director and other software components is illustrated in Figure 14-17.

The re-director intercepts every service call and determines, whether the resource being accessed is local or remote. Local accesses are passed along to "normal" access routines (e.g., service routines that provide access to a local printer through a local device driver). Accesses to remote resources are routed through the network service layer to the appropriate server. The re-director establishes a connection with the remote server, if necessary, and sends a request to the remote server. When the resource is delivered to the local machine, the re-director passes it back to the requestor in the same format as a locally accessed resource.

Figure 14-17

Interaction between the re-director and other software components

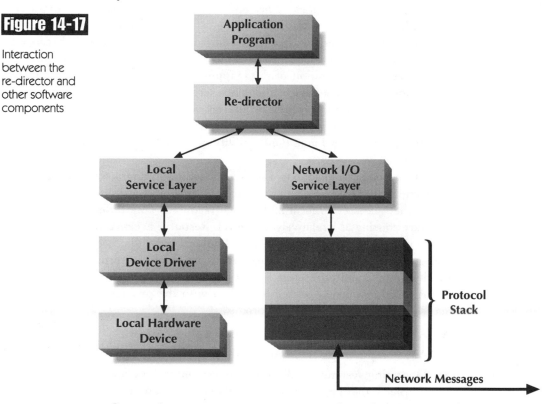

Connections to remote resources can be either static or dynamic. **Static connections** are initialized by the user or system administrator prior to accessing remote resources. The remote resources are mapped to local object, resource, or service names. For example, the disk drive names H:, N:, O:, and P: in Figure 14-14 are local names for directories or volumes named Burd, Ddrive, Edrive, and Pdrive stored on a remote server named averia. Once the connections to those resources have been established, their contents are available to local processes through the locally defined name. Thus, for example, a text editor could open a file named P:FILE.TXT using the same file I/O service call that it would use to open a file stored on a local disk drive.

Re-directors and static connections are relatively old approaches to accessing remote resources. They were commonly employed in the 1980s to make operating systems such as MS-DOS, UNIX, and Windows 3.1 "network aware." In essence, they are remote resource access mechanisms that

---

have been grafted onto existing operating system service layers. This approach allowed network capabilities to be quickly added to existing operating systems with minimal reimplementation of existing code.

There are two primary disadvantages to using re-directors and static network connections. The first is the difficulty inherent in initializing and maintaining static connections. The operating system must be configured to establish a connection each time it boots up and to map the remote resource to a local name. Location transparency is implemented only partially because configuring static connections requires explicit knowledge of remote resource locations. The initial definition of the mapping between the remote resource and local name must come from either the user or system administrator. Once the connection is created, it cannot be altered easily. For example, assigning a new name to a remote network resource requires reconfiguring all connections to that resource, which limits the flexibility of system administrators to reconfigure network resources.

The second disadvantage of static connections and redirectors is the underlying complexity of the operating system service layer. Return to Figure 14-17 and notice that accesses to local and remote resources take different paths through software starting at a relatively high level. Resource access differs fundamentally, depending on resource location. Those fundamental differences result in complex and somewhat redundant software functions which, in turn, increase the complexity of the operating system and reduce its flexibility and efficiency.

A more modern approach to dealing with the problem of remote resource access is based on the following premises:

► All resources are potentially shared across a network.

► Any computer system is potentially both a client and a server.

► Operating systems, application program interfaces, and user interfaces are simpler if there is no distinction between local and remote resource access.

The first two premises go hand in hand. That is, if every local resource should be available to other computer systems, then every computer system is both a client and a server. If all resources are potentially shared, then all resource management software must incorporate "server-like" functions. The key design feature that follows from these premises is that all resource access software should be implemented as if the access were coming from a remote machine. Operating systems with this design feature are said to implement **service-oriented resource access**.

Figure 14-18 illustrates the arrangement of software components that support service-oriented resource access. Notice the differences between this figure and Figure 14-17. The re-director has been replaced by a **resource locator**, and its function has been moved to a lower software layer. There is no longer a separate network I/O service layer because all I/O resource accesses are routed through the local service layer. The resource locator functions as a switch in the same fashion as a re-director. The key differences in its function are the types of messages being routed and the mechanism used to locate resources.

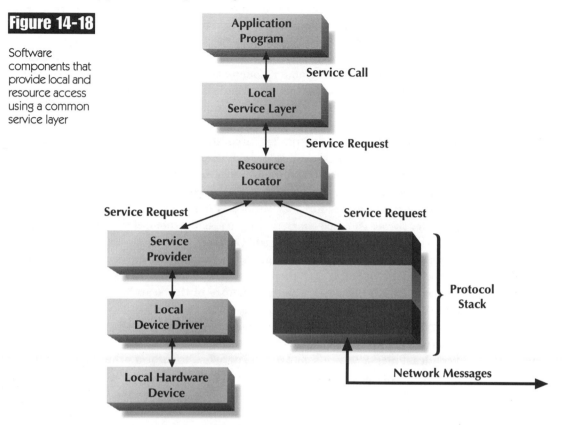

Figure 14-18

Software components that provide local and resource access using a common service layer

The function of the local service layer is changed fundamentally under service-oriented resource access. The local service layer no longer interacts directly with local device drivers to access resources. Instead, it performs the much simpler task of translating a local service call into a network service request to which either a local or remote server process can respond. Thus, the language used to access local resources is identical to the language used to access remote resources. The job of interacting with local device drivers to respond to a network service request is moved to a new software component—the service provider process.

At first glance, it may appear that the former job of the local service layer has been divided into two more complex pieces (i.e., the new local service layer and the service provider process). This is true if the only concern is local processes accessing local resources. Every resource managed by the local machine, though, potentially is shared through the network with other computer systems. Remote access to local resources requires a local server process for

---

each type of resource. The service provider accesses local resources on behalf of requesting processes in an identical fashion—whether the request comes from a locally executing application program or from a process executing on a remote machine. Thus, locally executing applications simply are clients that happen to be executing on the same machine as the service provider processes.

In contrast to a re-director, the resource locator does not need to translate service calls that access remote resources into network service requests because the inputs to the resource locator already are formatted as network requests. The primary tasks of the resource locator are (1) to determine where the requested resources reside and (2) to route the service requests to that location. To accomplish these tasks, the resource locator must know what resources are available on the local machine and on the network.

The resource locator maintains a **resource registry** containing the names and locations of known resources and services. The format of the names is the same, whether the resource or service is local or remote. When local service provider processes are started, they register themselves with the resource provider, which updates the registry accordingly. Resource requests from the local service layer are checked against the registry. If the requested resource or service name is not found, then the resource locator initiates a search for the resource on the network. The exact nature of this search depends on the high-level network service protocols in use.

Most network service protocols use a distributed directory of services and resources. As services and resources are registered with a resource locator, broadcast messages are sent over the network to inform other resource locators of their existence. Some machines on the network may be designated as primary repositories of this information. When a resource locator receives a request for a resource not listed in its registry, it queries the closest primary repository to find the service. That repository may, in turn, query other repositories until the requested resource is located.

The interaction between a resource locator and a primary repository of resource registrations is inherently dynamic. Thus, connections established through those interactions are **dynamic connections**. A resource locator usually updates its own registry based on query responses, which improves the efficiency of future accesses to a previously accessed resource. But the registry contents are in a constant state of flux as local and remote server processes are started, stopped, moved, or reconfigured. Thus, the resource locator provides a dynamic mechanism for reconfiguring connections to services and resources.

The resource locator also acts as a router for resource access requests arriving from remote machines. Such requests are received by the local physical network layer and passed up through the protocol stack to the resource locator. Then the resource locator passes the request to the appropriate local service provider process and passes the response back to the requestor through the protocol stack.

Distributed Computer Environment

The Distributed Computing Environment (DCE) is a standard for distributed operating system services defined by the Open Group (formerly known as the Open Software Foundation). DCE is a wide-ranging standard covering network directory services, file services, remote procedure calls, remote thread execution, system security, and distributed resource management. Its primary goal is to promote interoperability of distributed software across operating systems and related products. A number of operating systems partially comply with DCE, including Windows NT, OS/2, many versions of UNIX, and OS/390 (MVS). IBM, DEC, and HP are the principal supporters of and contributors to the standard.

DCE does not outline a standard for a fully functional operating system; rather, it defines a standard for a subset of operating system services and a uniform means to access those services. Thus, DCE software must be incorporated into an operating system or provided as an optional component (see Figure 14-19). In theory, DCE services can replace their counterparts in an existing operating system. In practice, however, DCE-compliant services are provided as add-ons to an operating system that translates DCE-compatible interfaces into native operating system calls. This may change as newer software versions are released.

Figure 14-19

Software layers used to implement Distributed Computing Environment (DCE) services

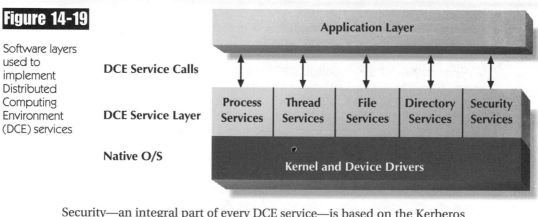

Security—an integral part of every DCE service—is based on the Kerberos security model, which defines a set of interactions between clients, services, and a trusted security service. Clients and servers authenticate one another by asking the security server to authenticate the other party. The security server provides security "tickets" to each party, which they exchange to verify their identity.

The security database also maintains an access control list for each service and resource. Once a client has been authenticated, its identity is checked against the access control list for the resource it is attempting to access. If the client is part of the access control list, then it is issued a ticket that it presents to the server to gain access. Tickets, passwords, and other security mechanisms are all encrypted during network transmission. Tickets are also time-stamped and expire within a relatively short period (e.g., minutes or hours).

DCE directory services are based on the 1998 version of the X.500 standard. Two different levels of resource and service naming—cell directory services and global directory services—are provided. A cell is an independent unit of directory management similar to an Internet domain. All the resources of a single organizational entity (e.g., a department, organization, or enterprise) normally belong to a single cell. Each cell has a primary directory server that provides resource and name-lookup services for the cell. The global directory service—a superset of all cell directory services—acts a master name server for all DCE-compliant directories. The entire set of directory names forms a hierarchy similar to a tree structured file-system directory.

DCE file services are based on a fully distributed file and directory structure. The file system name space sits below the cell name space maintained by the directory services. Thus, every file within a DCE file system has a unique name, which consists of its cell name, a file-system resource name within the cell, and its pathname within the file-system resource. File systems can be distributed or replicated across multiple servers. DCE file services implement transaction logging for all updates to enable rapid recovery in the event of a crash.

Remote procedure calls (RPCs) form the heart of any DCE implementation. Clients, servers, and all DCE services interact through remote procedure calls. A remote procedure call is a network- and operating-system independent message that requests the execution of a specific procedure on a specific server. Parameters are passed as part of the request message and returned as part of the response. RPC messages are formatted according to an Interface Definition Language (IDL). DCE-compliant client and server programs are compiled and linked with a set of DCE library routines that implement the IDL. These routines implement the low-level aspects of message passing, parameter exchange, data-format conversions, and interaction with DCE security services.

The DCE standard appears to be growing in vendor and user acceptance. It is mostly widely implemented in UNIX operating system variants, but it also has deeply penetrated other operating systems. There are competing standards—most notably, the Common Object Request Broker Architecture (CORBA). CORBA standards are increasingly accepted and are based on a more modern object-oriented model, compared to DCE's procedural model. However, DCE standards are more rigorously specified at the client and service interface level, which provides a higher degree of interoperability than CORBA can currently provide. Thus, implementers of distributed applications that need guaranteed interoperability have little choice but to embrace DCE.

SUMMARY

The service layer provides a set of subroutines that allow processes to interact with I/O devices and, therefore, directly with users. Early operating systems implemented I/O as streams of individual characters. A relatively simple set of service layer routines allowed reading and writing of characters and strings.

VDT input is often full-screen or form-based. Also, VDT I/O includes special display features such as underlining, inverse video, bold-intensity, half-intensity, and invisible characters. Many operating systems provide a set of full-screen I/O service calls to control functions such as cursor positioning, screen clearing, and activating and deactivating special display functions. To improve efficiency, a table of control sequences sometimes is created and stored in memory when the I/O device is first identified. All full-screen service routines then use the content of that table instead of testing the device identifier each time they are called.

Graphic I/O devices (e.g., laser printers and microcomputer monitors) provide a much richer set of I/O capabilities than VDTs and character-based printers. Their additional capabilities require a larger and more complex set of I/O service calls. There is no standard method of controlling graphic monitors. Operating systems use a generic device driver to deal with graphic I/O devices' variation and lack of standardization. Graphic I/O device drivers accept device-independent commands as input and translate them into commands that are recognized by a specific graphic I/O device.

A device-independent command language implements all of the control capabilities of the possible physical graphic I/O devices. Language syntax and semantics are designed to simplify the process of translating the language into device-specific commands. Device-independent command languages must be updated regularly to account for new features in the latest graphic I/O devices.

Window-based I/O extends graphic I/O in a number of ways, including division of a display into multiple output regions (i.e., windows), use of colors and background patterns, ability to use various font styles and sizes, and provision of standard facilities for menu display and control. Window-based I/O service layers provide a set of service routines that implement definition of window size and display characteristics, definition and display of menus and the input of user menu selections, display of graphic images, facilities for window control, cut-and-paste capabilities, and user input by keyboard or pointing device.

A window manager acts as a switch to route messages from the application layer to the service layer. The window manager usually gives input focus to the window containing the mouse pointer. In addition, the window manager maintains the content of screen areas not occupied by active programs and implements full-window management functions.

A command language expresses processing requests with rigid syntactic and semantic requirements that are similar to a programming language. A command interpreter accepts command input and checks it for valid syntax and semantics. Command names may correspond to internal commands or stored executable programs. The command interpreter uses a previously defined path to search for executable files.

Files of command lines are useful for controlling the execution of batch application programs. A command language used to control batch execution—often called a job control language (JCL)—provides control mechanisms to execute individual job steps, coordinate multiple programs, and define recovery procedures for likely error conditions. JCLs implement control structures similar to those found in programming languages, including sequence, selection, and repetition.

The ability to change the source of input data or the destination of output data is referred to as I/O re-direction. To support I/O, re-direction programs are designed to write to standard input and output devices. The operating system controls these standard devices and can route I/O to or from standard devices to any physical device or file.

Some operating systems require JCL programs to specify the required resources in resource declarations. Resource declarations allow the operating system to determine whether sufficient resources exist to complete a job successfully and provide the operating system with additional resource-allocation information. Operating systems that implement dynamic resource allocation respond to resource demands as they occur. Dynamic resource allocation allows jobs to fail because of unavailability of required resources, but such failures are unlikely unless resources are severely constrained.

Form-based command interfaces provide a relatively simple means to specify command arguments and options. Once a command name has been entered at the command prompt, the command interpreter places a form or menu on the screen and the user is presented with a set of named blank fields for argument names and option values. A form-based command interface can be supplemented with a menu of argument and option values.

A window-based command interface is easier to learn and use than command line- or form-based interfaces because of the extensive use of menus and graphic images. Most or all of the available commands are provided in menus. Graphic images are used often to supplement commands listed in menus. Icons can be used to represent certain commands or objects to be manipulated. Object-oriented command interfaces are used widely for interactive command and control. Command and control features such as a program or system settings menu usually are integrated with the window manager.

An operating system implements network I/O and services as a complex set of software layers called a protocol stack. Protocol stacks provide several advantages, including dividing the task of network interaction into independent components and insulating application programs and many portions of the operating system from details of low-level protocols and physical network implementation.

The operating system implements location or network transparency for remote resource access. The goal is for users and programs to interact with resources and services using identical methods, regardless of whether those resources and services are provided by the local operating system or obtained from a remote server. An operating system can use one or more re-directors to implement location transparency. A re-director is a software component that translates service calls that access remote resources into an appropriate set of network messages to implement the access. A re-director intercepts every service call, routes accesses to local resources to the local service layer, and routes accesses to remote resources through the network. A re-director uses static connections to map remote resources to locally defined names.

Static connections are difficult to initialize and maintain. Using them violates location transparency because knowledge of resource locations is required to initialize or update connections. Static connections and re-directors also increase the complexity of the operating system service layer because of the use of multiple processing paths—one for local resources and the other for remote resources.

Modern operating systems assume that all resources are potentially shared across a network and that any computer system is potentially both a client and a server. Operating systems implement service-oriented resource access methods and dynamic connections. Local resources are managed by independent server processes. A resource locator establishes dynamic resource connections, tracks them in a resource registry, and uses this information to route service requests from local or remote client processes to appropriate server processes. The format of service requests is the same for both local and remote resources.

Key Terms

case statement
command interpreter
command language
command option
command parameter
command-line interface
data link layer
device-independent commands
dynamic connections
dynamic resource allocation
icon
if statement
input focus
Internet Packet Exchange (IPX)
Internet Protocol (IP)
I/O re-direction
job
job control language (JCL)
job step
job stream
location transparency
logical network interface driver
Network Driver Interface Specification (NDIS)
network layer
network transparency

object-oriented command interface
Open Data Interface (ODI)
Open Systems Integration (OSI) network model
path
physical layer
protocol stack
re-director
repetition
resource declaration
resource locator
resource registry
search path
selection
sequence
Sequenced Packet Exchange (SPX)
service-oriented resource access
session layer
socket
static connection
Transmission Control Protocol (TCP)
transport layer
Transport Layer Interface (TLI)
window manager
window-based I/O

Vocabulary Exercises

1. _____ describes the use of operating-system facilities to alter the source and/or destination of data to or from a process.

2. A physical network interface unit device driver occupies the _____ layer of the OSI model.

3. A window-based command interface may use _____ to represent files, programs, or other objects that the user can manipulate.

4. A command language uses specific syntactic conventions such as a "-" or "/" to designate _____.

5. A(n) _____ is a single processing step (usually a complete program) within a(n) _____.

6. Some job control languages (JCLs) require explicit _____, thus allowing the operating system to allocate needed resources before job execution begins.

7. The ODI and NDIS protocols define standards for implementing a(n) _____ driver.

8. The _____ and _____ protocols are the standard transport and network layer protocols of the Internet.

9. The control structures supported by most job control languages (JCLs) include _____, _____, and _____.

10. Client access to a Novell NetWare LAN server uses the _____, _____, and _____ protocols to implement the network, transport, and session layers, respectively.

11. The _____ routes keyboard input to the window or program that has _____ (i.e., the window currently containing the mouse pointer).

12. The set of software layers that implement access to remote resources through a LAN or WAN is called a(n) _____.

13. A(n) _____ interface requires the user to type the command, parameters, and options in response to a prompt.

14. A(n) _____ command interface allows users to perform some command actions by directly manipulating icons that represent files, directories, and hardware devices.

15. A command interpreter uses a predefined _____ to help it locate files containing executable programs.

16. An operating system that uses _____ does not require resource declarations in its job control language (JCL).

17. An operating system that implements _____ hides the location of resources and services from users and application programs.

18. _____ to remote resources must be initialized by the user or a system administrator.

19. A(n) _____ intercepts service calls that access remote resources and translates them into network messages (i.e., service requests).

20. A(n) _____ establishes dynamic connections to local and remote services and resources and records those connections in a(n) _____.

21. A language used to control the execution of complex batch processes is called a(n) _____.

1. How is full-screen I/O more complex than character-based I/O?

2. How is graphic I/O more complex than full-screen I/O?

3. How does the operating system service layer deal with variation in the methods used to control special display functions on VDTs?

4. How does an object-oriented command interface differ from a window-based command interface?

5. What command language features/capabilities are required to support batch job execution?

6. What is a search path?

7. What are standard input and output devices? What is I/O re-direction?

8. What are job control language (JCL) resource declarations? How are they used?

9. What is a device-independent I/O command language? Why are such languages commonly used?

10. Describe the switching/routine functions of a window manager. Why is this function needed?

11. What is a network protocol? What specific stack layers are implemented? What are the advantages of using protocol stacks?

12. What is network or location transparency? How is it implemented by a modern operating system?

13. What are static remote resource connections? Why are they difficult to initialize and maintain?

14. A modern operating system acts as both client and server. How are software components organized to accomplish both functions at the same time? How and why does this organization improve location transparency?

1. Most operating systems provide a rich set of service layer functions for printer output. These functions support text formatting (e.g., font selection and formatting), graphic object display (e.g., boxes, circles, and line drawing), and bitmap displays (e.g., backgrounds and images). Investigate the printer service layer routines of a desktop operating system (e.g., Microsoft Windows, Macintosh System 8, or IBM OS/2). What specific capabilities are provided? How easy are they to use?

2. Input/output re-direction and pipes are provided by many different operating systems (e.g., UNIX, MS-DOS, and Windows). However, the methods by which I/O re-direction is implemented vary. Investigate the implementation of pipes and standard I/O device re-direction in UNIX, MS-DOS, and Windows. Consider a sequence of processes, I/O re-directions, and pipes such as:

   input_file < program_1 | program_2 | program_3 > output file

   Compare and contrast the methods by which each operating system implements this sequence. Which method of implementation is more efficient and why is it more efficient?

3. Windows 95 and Windows NT implement access to network resources in different ways. Windows 95 primarily relies on static connections and a re-director. Windows95 can share its resources with other computers using a method originally implemented in Windows 3.11. Windows NT takes a modern client/server approach to accessing and sharing resources. It implements dynamic resource connections, a resource locator, and a directory of network resources. Investigate each operating system's network capabilities. How have the additional capabilities of Windows NT changed the internal arrangement of its service layer functions as compared to Windows 95? What are the resulting costs and benefits of those changes?

Chapter

15

System Administration

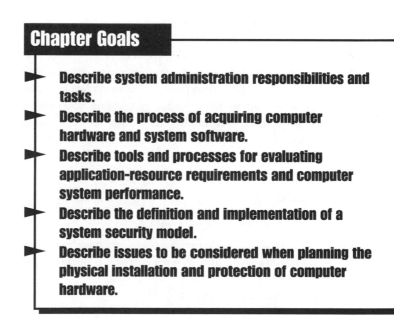

Chapter Goals

► Describe system administration responsibilities and tasks.

► Describe the process of acquiring computer hardware and system software.

► Describe tools and processes for evaluating application-resource requirements and computer system performance.

► Describe the definition and implementation of a system security model.

► Describe issues to be considered when planning the physical installation and protection of computer hardware.

SYSTEM ADMINISTRATION

Within the context of information systems, the term **system administration** covers a range of activities and responsibilities. The primary responsibility of a system administrator is to ensure efficient and reliable delivery of information system (IS) services. There are many specific tasks that a system administrator performs to meet these broad responsibilities, grouped into the following broad categories:

► Acquiring new IS resources.
► Maintaining existing IS resources.
► Designing and implementing an IS security policy.

There is considerable variation among organizations in the assignment of these responsibilities to specific individuals. In small organizations (e.g., a dozen employees or less), the user or organizational administrator may assume these responsibilities directly. Large organizations (e.g., more than 500 employees) partition the responsibilities. Medium-sized organizations may have a single technical specialist assume all system administration responsibilities.

A variety of specific system administration tasks are performed within the activity categories outlined above. Some of these tasks are relatively mundane and address a short time frame (e.g., maintaining user accounts and performing file-system backups), whereas others require the exercise of many different skills and address a relatively long time frame (e.g., IS planning and resource acquisition).

Strategic Planning

The acquisition and deployment of IS resources should occur only in the context of a well-defined **strategic plan** for the organization as a whole. For the purposes of this chapter, a strategic plan is defined as a set of long-range goals and a plan to attain those goals. The planning horizon is typically three years and longer. Minimally, the goals must include services to be provided and resources needed to provide the services. The strategic plan addresses the following issues with respect to achieving stated goals:

► Strategies for developing services and a market for them.
► Strategies for acquiring sufficient resources for operations and growth.
► Organizational structure and control.

In all cases, strategic plans must address the basic question of, "How do we get there from here?" with respect to the strategic goals.

The IS component of an organization is only one part of the entire organization. Thus, its strategic plan is only one part of the organization's strategic plan. All portions of its plan must be evaluated in concert with those of other organizational units. The need for coordination is driven largely by the service nature of ISs in most organizations. That is, ISs are normally a support service for other organizational units and functions (e.g., customer services, accounting, manufacturing, etc.). Thus, the strategic plan for ISs tends to follow, rather than lead, the strategic plans of other organizational units.

Hardware/Software as Infrastructure

The resources devoted to most organizational activities can be roughly classified into two categories—**capital expenditures** and **operating expenditures**. Capital expenditures are used to purchase **capital resources**, or assets, which are expected to provide benefits for more than one operating period (i.e., beyond the current fiscal year). Examples of capital expenditures in an organization include buildings, land, equipment, and research and development (R&D) costs. Although the expected useful lifetime of computer hardware and software has decreased in recent years because of rapid technological change, it remains long enough to be considered a capital resource.

Many capital resources provide benefits to a wide range of organizational units and functions. An office building, for example, provides benefit to all of the units and functions that are housed within it. Such resources, referred to as **infrastructure**, have the following characteristics:

▶ Service to a large and diverse set of users.

▶ Difficulty in allocating costs to individual users.

▶ Recurring need for new capital expenditures.

▶ Significant operating costs for maintenance.

Computer hardware and system software used to provide information systems services are infrastructure. This is obvious in companies that rely extensively on large computer systems used by many units within the organization. It is less obvious—but no less true—in organizations that have highly decentralized hardware and system software systems. The strategic planning issues in ISs thus are similar to those in many infrastructure-based service organizations.

Examples of infrastructure-based organizations include those that provide services such as communications, electrical power, and water. In general, the strategic issues that must be addressed in such an environment are:

▶ What services will be provided?

▶ How will service users be charged?

► What infrastructure is required to provide the services?

► How can the infrastructure be operated, maintained, and improved at minimal cost?

Consider, for example, the provision of communication services by local telephone companies. The primary strategic question to be addressed is what types of services should be provided (e.g., basic telephone only, expanded telephone services, computer communication services, mobile services, information storage and retrieval, etc.). The answer to that question leads to decisions regarding the nature of the required infrastructure and its associated capital and operating costs (e.g., cellular transmission facilities, fiber-optic transmission, total communication capacity, devices required/available for user interaction with the network, etc.).

Standards

Providing infrastructure-based services to a wide variety of users requires the adoption of **service standards**. Standardization, however, can stifle innovation and produce suboptimal solutions. Once again, consider the local telephone system as an example. All users agree on and abide by a number of telephone-service standards, including allowable user devices, basic service availability, standards for interacting with the infrastructure (e.g., line voltage, signal encoding, etc.), and others. The nature of infrastructure requires standardization in order to provide service at reasonable cost.

Standardization often causes problems for some users, especially those who demand services at or near the leading edge of technology. For example, there are still some telephone lines in this country that are not capable of touch-tone dialing, which limits the range of services available to users of those lines. Consider also the use of fiber-optic communication lines. The lack of fiber-optic connections in local telephone grids limits the use of high-speed computer communication, which, in turn, slows the trend toward employees working at home. An infrastructure-based service provider must constantly balance the benefits of standardization (e.g., reduced costs and simplified service) against its costs (e.g., stifled innovation and failure to meet some user's needs).

It is particularly complex to manage the standardization issue when considering computer hardware and system software because of the large number of choices in both categories. In addition, although there has been some progress in hardware and software compatibility, it remains the exception. A further complication is the diverse set of components required for information processing within even a modest-sized organization.

Competitive Advantage

Discussion of computer hardware and system software strictly as infrastructure ignores certain types of opportunities. Infrastructure management concentrates on the provision of short-term services at minimal cost. Such an outlook tends to preclude major technical innovations as well as radical re-definition of services to be provided.

The term **competitive advantage** describes a state of affairs in which one organization employs resources so as to give it a significant advantage over its competitors. This can take a number of forms, including:

▶ Provision of services that others are unable to provide.

▶ Provision of unusually high-quality services.

▶ Provision of services at unusually low price.

▶ Generation of services at unusually low cost.

Computer hardware and system software can be applied to achieve competitive advantage in any or all of these areas. Examples of applications include automated bank tellers, scanning grocery checkouts, computer integrated manufacturing, and many others.

Notice that each of these examples was considered a competitive advantage at one time but is now commonplace. Unfortunately, this is the nature of applying technology for competitive advantage. Rapid technology changes and adoption by competitors severely restrict the useful life of most technology-based competitive advantages. Furthermore, there are substantial risks in pursuing competitive advantage through new technology. For example, the costs of technology decrease rapidly after introduction, especially with computer hardware and software. Early adopters also face the inefficiency of starting at the beginning of a learning curve. These factors often combine to create high costs with limited benefit. Late adopters may incur substantially lower costs while realizing most of the benefits.

A Standard Hardware Platform?

Cooper State University (CSU) is a large school, with over 1,000 faculty members and a full-time enrollment of 20,000 students. Academic programs span a wide range of fields, including hard sciences, humanities, engineering, education, management, law, and medicine. CSU offers many graduate degree programs and is a highly ranked university in terms of externally funded research.

CSU has a campus-wide computing organization called Computer Information Services (CIS). CIS is responsible for supporting the computing and information processing needs of both academic and administrative users. It supports administrative users through the operation of a large mainframe computer, several local area networks (LANs), and many administrative applications (e.g., payroll, accounts payable, class scheduling, class registration, and student academic records). It supports academic users through shared minicomputers, shared file servers, and a large number of microcomputers in dedicated classrooms and general purpose facilities. CIS also operates the campus network, which links most offices, classrooms, and buildings in a two-tiered campus network.

CIS has been under considerable budgetary pressure for several years. Demand for computing services has grown rapidly, but funding has not kept pace. Although declining hardware costs have been a budgetary bright spot, they have been more than offset by increases in the cost of supporting a wide variety of hardware platforms, software packages, and administrative applications. Integrating this disparate collection of hardware and software with the growing campus network has been a particularly difficult and costly support task.

Recently, CIS has proposed a method of minimizing some of these costs. They propose to standardize the hardware platforms and operating systems used for microcomputers and minicomputers. Also, they intend to contract with a single vendor for each class of machines and to negotiate volume pricing on hardware and operating software and a hardware-maintenance agreement as part of the contract. All CSU purchases of microcomputers and minicomputers (including those made with non-CIS funds) would be required to comply with the contract provisions.

Reactions to the proposal have ranged from indifference to near revolt. Most administrative departments and many academic departments are supportive, provided that CIS can assure them that their computing and information-processing needs will continue to be met. In general, these departments rely on CIS for most or all of their computing and information-processing needs. Some academic departments that fund some of their own computer purchases are concerned with loss of control over their own acquisition process.

The computer science, engineering, IS, and physics departments—vehemently opposed to the proposal—argue that their computing needs are unique and require hardware and software at the cutting edge of technology. Furthermore, they claim that standardized platforms would lag the cutting edge of technology and, thus, interfere with their teaching and research missions. They point out that most of their computing needs are met with funds derived from external sources (e.g., research grants and contracts). As a result, they vigorously oppose restrictions on expenditures.

Questions:

1. Are the benefits that CIS anticipates likely to be realized?

2. Are the concerns of the engineering, physics, and chemistry departments valid? Should any other departments share these concerns?

3. Should the proposal be adopted as is? Should the dissenting departments be allowed to opt out of the contract?

ACQUISITION PROCESS

The process of acquiring computer hardware and system software is ongoing in most organizations. New hardware and software may be acquired to:

► Support entirely new applications.

► Increase the capability to support existing applications.

► Reduce the cost of supporting existing applications.

The exact nature of the acquisition process depends on which of these cases (or a combination thereof) motivates the new acquisition. It also depends on a number of other factors, including:

► The mix of applications that the hardware/software will support.

► Existing plans for upgrade or change in those applications.

► Requirements for compatibility with existing hardware and software.

► Existing technical capabilities.

Notice that applications to be supported figure prominently in both the motivation for new acquisitions as well as factors for choosing among them. Remember that hardware and software exist merely to support present and future applications. Therefore, planning for acquisition is little more than guesswork if a thorough understanding of present and anticipated application needs is lacking.

The acquisition process consists of the following steps:

1. Determine the applications that the hardware/software will support.
2. Specify detailed requirements in terms of hardware and software capability and capacity.
3. Draft a request for proposals and circulate it to potential vendors.
4. Evaluate the responses to the request for proposals.
5. Contract with a vendor (or vendors) for purchase, installation, and/or maintenance.

The acquisition process is initiated by new system implementation projects or major upgrades of an existing system. In either case, hardware and system software requirements are determined early in the design phase of the systems-development life cycle. The analysis phase provides detailed estimates of system activity and data store content (e.g, transaction volume and number of records in a database). Decisions made early in the design phase (e.g., the automation boundary, form of user interfaces, and selection of development tools) allow the analysis-phase activity estimates to be translated into detailed hardware requirements.

Determining and Stating Requirements

As first discussed in Chapter 2, computer system performance is measured in terms of application tasks that can be performed within a given time frame. This may be measured in terms of throughput, response time, or some combination of the two. Thus, the first step in stating hardware/software requirements is a statement of application tasks to be performed, which is supplemented by stated performance requirements for application tasks.

Depending on the motivation for the proposed acquisition (new applications, growth of old applications, etc.), different techniques may be applicable to requirements determination. For existing applications, hardware and system software requirements may be based on measurements of existing performance and resource consumption. Requirements for new applications are more difficult to derive. A number of different techniques that may be used to determine requirements are discussed in later sections of this chapter.

Although application requirements are the primary basis for hardware and system software requirements, other factors also must be considered, including:

► Integration with existing hardware/software.
► Availability of maintenance services.
► Availability of training.

► Physical parameters such as size, cooling requirements, disk space required for system software, and so forth.

► Availability of upgrades.

These factors form an integral part of the overall requirements statement. Some of these are essential requirements (e.g., physical parameters), whereas others are less important bases for differentiating among potential vendors.

Request for Proposal

A **request for proposals (RFP)**—a formal document sent to vendors—is intended to state requirements and solicit proposals to meet those requirements. Often, particularly in governmental purchasing, it is a legal document as well. Vendors rely on information and procedures specified in the RFP. That is, they invest resources in the response process with the expectation that stated procedures will be followed consistently and completely. Thus, an RFP is considered to be a contract offer, and a vendor's response represents an acceptance of that offer. Problems such as erroneous or incomplete information, failure to enforce deadlines, failure to apply procedures equally to all respondents, and failure to state all relevant requirements and procedures can lead to litigation.

The general outline of an RFP is as follows:

1. Identification of requestor.

2. Format, content, and timing requirements for responses.

3. Requirements.

4. Evaluation criteria.

The identification section describes the organization that is requesting proposals. Identification should state the name of a person to whom questions may be addressed; addresses, phone numbers, FAX numbers, and so forth should be included also.

The RFP should clearly state the procedural requirements for submitting a valid proposal. Where possible, an outline of a valid proposal should be given with a statement of the contents of each section. In addition, deadlines for questions, proposal delivery, and other important events should be stated clearly.

The requirements statement comprises the majority of the RFP. Requirements should be organized by type and listed completely, into the following relevant categories:

► Hardware/software capability.

► Related services (e.g., installation and maintenance).

► Warranties and guarantees.

► Financial considerations (e.g., acquisition cost, maintenance cost, lease cost, and payment options).

Requirements should be separated into those that are essential and those that are optional or subject to negotiation. For example, minimum hardware capacity generally is stated as an absolute requirement, whereas some related services may merely be desirable.

Evaluation criteria are stated with as much specificity as possible. A point system or weighing scheme often will be used for the optional, or desirable, requirements. Weight also may be given to factors that are not stated as part of the hardware or software requirements. Such factors might include the financial stability of the vendor and good, or bad, previous experiences with the vendor.

Evaluating Proposals

Proposal evaluation is a multistep process, as follows:

1. Determine minimal acceptability of each proposal.
2. Rank acceptable proposals.
3. Validate high-ranking proposals.

Each proposal must be evaluated to determine if it meets minimal criteria of acceptability: meeting essential (or absolute) requirements, financial requirements, and deadlines. Proposals that fail to satisfy minimal criteria in any of these categories are eliminated from further consideration.

The remaining proposals must be ranked by evaluating the extent to which they exceed minimal requirements, including the provision of excess capability or capacity, satisfaction of optional requirements, and other factors. Measurements of subjective criteria such as compatibility, technical competence, and vendor stability also are considered at this stage.

A subset of the highly ranked proposals is chosen for validation. The subset should be relatively small because of the length and expense of validation procedures. To validate a proposal, the evaluator must determine the correctness of vendor claims and the ability of the vendor to meet commitments in the proposal. Notice that the ranking process relies primarily on the vendor's assertions in the proposal. The validation stage is where the accuracy of those assertions is determined.

Various methods and sources of information are applicable to proposal validation. The most reliable of these is a **benchmark** of the proposed system with actual applications. A benchmark is a performance evaluation of application software, or test programs, using actual hardware and systems software under realistic processing conditions. In the past, benchmarking often was difficult to perform because of the expense of the hardware/software configurations and the length and cost of installation. Currently, this is less problematic thanks to cheaper hardware, streamlined installation procedures, and fierce competition among vendors. For all but the largest systems, it is now common practice for a vendor to deliver and install hardware and system software for customer evaluation.

Alternatives to on-site benchmarking with actual application software include:

► Benchmarking at alternative sites with actual applications.

► Benchmarking of test applications.

► Validation through published evaluations.

With systems that are difficult to install, it may be possible to test applications at an alternative site (e.g., on the premises of the vendor or of another customer).

It may not be possible to test configurations with actual applications if those applications are large, difficult to install, or not yet developed. In such cases, it may be possible to use standard benchmarking software or to construct benchmarking software that simulates the execution of the actual application. Many standard benchmarks are available, primarily for smaller computer systems and for scientific applications.

Care must be taken when using such benchmarks, especially if they are derived in isolation (i.e., if only one benchmark, testing one type of performance, is executed at a time). The overall performance of systems of mixed application demand can vary substantially from individually derived benchmarks results. For standard configurations, benchmark results may be obtained from trade publications, periodicals, or independent testing agencies.

REQUIREMENTS DETERMINATION AND PERFORMANCE EVALUATION

ISs are a combination of computer hardware, systems software, and application software. One of the most difficult system administration tasks is determining hardware requirements for a specific set of application software because:

► Computer systems are complex combinations of interdependent components.

► The configuration of operating and other system software can significantly affect raw hardware performance.

► Application software's use of hardware and system software resources cannot always be precisely predicted.

The starting point for determining hardware requirements is the application software that will run on the hardware platform. The nature and volume of application software inputs must be described to determine what application code will be executed and how often it will be executed. This, in turn, determines the demands that application software will place on system software (e.g., number and type of service calls executed) and on computer hardware (e.g., number of CPU instructions executed).

If the application software has been developed already, then its hardware and system-software resource consumption can be measured. A number of software and hardware tools can be used to monitor resource utilization while application programs are executing. Executing application programs with "live" data inputs can generate precise measurements of resource requirements. These measurements are data inputs to the process of acquiring and configuring hardware and system software to support application software.

Determining resource requirements is much more complex when application software has not yet been developed. If the process of software development were completely mechanical, then it would be possible to accurately determine resource requirements prior to software construction. Application development is not, though, a mechanized process; rather, it has many characteristics that make accurately determining resource requirements a difficult or impossible task.

The application-development process doesn't start with a fully specified set of application requirements; instead, it starts with a general statement of user needs. That general statement is expanded and refined during the analysis and design phases of the systems development life cycle (SDLC). User requirements are not defined fully until the conclusion of the analysis phase. Final decisions about the scope of the automated portion of the system are not made until early in the design phase. Detailed construction plans for application software are not developed until late in the design phase.

The precision of estimated hardware and system software requirements mirrors the precision and level of detail of application-software requirements. That is, general statements of user and software requirements can generate only rough estimates of hardware requirements. More accurate hardware requirements can be determined only based on more detailed application software requirements. But these details are not available until late in the design phase. Thus, accurate hardware requirements cannot be determined until late in the design phase.

Unfortunately, application developers seldom have the luxury of waiting until the end of the design phase to acquire hardware, which is especially true when a large system is being implemented. Sufficient lead time is required to implement the acquisition process described earlier in this chapter. That process is typically started soon after the scope of the automated system has been determined. Thus, hardware and system software requirements must be determined based on the detailed analysis-phase descriptions of processing requirements.

Relatively informal methods of estimating requirements based on analysis phase outputs can sometimes be employed. Such methods employ comparisons to similar application software and work volumes. For example, a developer of an online transaction processing system might attempt to identify a similar transaction processing system within or outside the organization. The system is examined to determine the adequacy of its existing hardware and system software. If current hardware resources match current application demand, then those hardware resources can be used as a baseline for estimating the hardware requirements of the yet-to-be developed system. Requirements can be adjusted to account for differences in transaction-volume and application-processing characteristics.

This basic approach to requirements estimation is widely and successfully used. It is particularly apt for smaller hardware platforms with "garden variety" application software. The risk of inaccurate estimates is minimized for smaller hardware platforms because they can be modified relatively quickly and inexpensively.

Larger hardware platforms and unique large-scale application software requires more formal methods of requirements estimation. Lead times for hardware acquisition and configuration are substantially longer for mainframes and large minicomputers than for smaller hardware classes. Dollar values of hardware and system software components are large, thus increasing the economic risks of inaccurate estimates. Requirements estimation in this environment typically employs highly formal methods based on mathematical models.

Modeling Concepts

A **model** is an abstract representation of physical reality. Models can be of many types, including graphic representations, physical representations, mathematical representations, and combinations thereof. An artist's rendition, or drawing, of the facade of a building is a purely graphical model. On the other hand, a blueprint of a building is a combination of a graphic and mathematical model. A set of simultaneous equations is a purely mathematical model. Finally, a scale model of a building or sculpture is a physical model.

All models limit the features of reality that are represented within the model. The process of choosing features to include in the model and of choosing an appropriate representation for those features is called **abstraction**. Different types of models abstract different features. For example, a drawing of a building facade abstracts visual features while ignoring many others such as construction materials and hidden support structure. A blueprint models physical placement of construction materials in great detail while ignoring many external-appearance issues. The validity of a model depends on the relative importance of features abstracted and ignored and the accuracy of representation. That is, a valid model accurately models all important features and ignores unimportant ones.

Models are used to make a wide variety of decisions primarily because of the expense of building and testing actual systems. To be useful, a model must be easier to build, understand, and manipulate than the corresponding reality. For example, a blueprint is a useful model of a building because it is far easier to create and modify than is an actual building. A model also must be valid for its intended purpose: that is, it must accurately represent all features that are important to the decisions that it will support.

A primary use of a model is as a prediction tool, as shown in Figure 15-1. When used for prediction, a model may be manipulated by modifying one or more model parameters. In a mathematical model, modification can be accomplished by changing the numerical values, or parameters, assigned to one or more model features, or variables. For example, an architect might change assumptions about the materials used to construct a certain part of a building. The new materials have a different mass, or weight, than the old. The mass of the new materials can be used as a new parameter value for various predictions (e.g., weight of a floor or required support strength). A model also may be used for prediction by changing model components, as opposed to parameter values. For example, an architect might add or delete a wall and assess, or predict, the effect of the change on the usability or structural integrity of the building.

Figure 15-1

Process of model-based prediction. A model is created by abstracting relevant features of a system. Model parameters or components are modified, and the results are projected back to the system as a prediction.

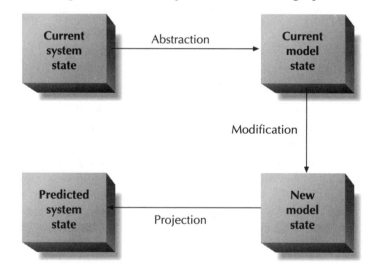

Predictions based on a model are made by projecting the results of model manipulation back onto the system originally abstracted. In other words, it is assumed that modification of model parameters and components will produce results that mirror the effects of those same modifications to an actual system. Every projection from a model to reality carries a risk of incorrect prediction.

The projection may be invalid because of any number of factors, including incomplete abstraction, improper representation of features, and incorrect model manipulation.

Computer System Models

With respect to computer hardware and system software, both performance evaluation and requirements determination can be addressed through the use of predictive models. Models of computer hardware and software can be used for a number of specific purposes, including:

▶ Determining requirements for new hardware and system software to support a new application.

▶ Evaluating the ability of an existing configuration to handle a new application or additional workload from an existing application.

▶ Determining the change in application performance resulting from changes in hardware and/or system software.

▶ Formulating long-range plans for expanding or acquiring new computer capacity.

Variations of a generic modeling strategy can be applied to each of these scenarios. This strategy is outlined in Figure 15-2 and explained in detail next.

Figure 15-2

Generic steps for analytical modeling of application demand and resource delivery

```
              Model Development and Evaluation

1. List each distinct type of application processing task and its
   stated performance requirements.

2. Determine the hardware and system software resource requirements
   for each task.

3. Determine the frequency and distribution of each task.

4. Construct a composite model of application demand.

5. Construct a model of hardware-resource availability for each
   configuration of hardware and system software under consideration.

6. Test each configuration model against the application demand model.
```

Application Demand Models

An **application demand model** (or **workload model**) describes the relationship between user-level processing activities and demands for hardware and system software resources. Examples of user-level processing tasks include transaction updates, database queries, generation of reports, and batch file updates. Each of these processing tasks places different demands on hardware resources, and thus each requires a separate model of resource utilization. For example, a database query requires many accesses to secondary storage with little use of the CPU. A batch update may require many storage accesses (e.g., for a large file) and/or many CPU cycles (e.g., because of complex calculations).

Resource demand by application tasks can be measured in many different ways. These may range from very high-level measurements such as average number of CPU instructions executed to detailed low-level measurements of individual machine actions and operating system service requests. A frequently used set of demand variables is:

► Average number of CPU instructions executed.

► Bytes read from and/or written to secondary storage.

► Bytes read from and/or written to specific I/O devices.

► Bytes read from and/or written to servers or clients over a network.

CPU instructions include only those instructions executed to perform computational and logical processing functions. In practice, this is a difficult number to estimate because it is affected by the type of processing performed within the application, the efficiency of executable or interpreted code, among other factors. These factors are difficult to estimate for applications that have yet to be developed or tested. Estimates can be derived for existing applications through the use of program profiling.

Bytes of secondary storage I/O represent demand for access to mass storage devices and channels as well as the CPU (e.g., operating system service layer) overhead necessary to implement those accesses. Similarly, communication to and from I/O devices and network nodes includes the use of channel capacity and related CPU overhead. For these reasons, I/O to devices or files frequently is measured in terms of specific operating system service calls. For existing systems, these calls can be profiled in much the same manner as application code, in order to derive accurate estimates of CPU overhead.

A composite model of application-resource demand is a combination of workload models for individual processing tasks and estimates of the frequency, or volume, and distribution of those tasks over time. Processing volume can be measured in transactions per minute or hour, reports per day or week, queries per minute, and many other units of measurement. Distribution measures show the expected norms and variations in frequency. For example, the volume of sales transactions can average 200 per hour in a normal week, but typical volume can vary by time of day, day of week, time of year, or other factors. Because of these sources of variation, estimates of processing volume rarely can be stated accurately as a single number. Instead, they require that the statistical distribution of processing demand be known.

Resource Availability Models

A **resource availability model** can describe either a hardware system or a combination of hardware and system software. The combination approach is more accurate, although it introduces additional complexities and possibilities for error. The raw capabilities of hardware represent only a potential. For most applications, access to that capability is obtained exclusively through the system software, and, most importantly, the operating system. Thus, the efficiency with which system software manages and delivers hardware resources to applications is a primary factor determining resource availability.

Resource availability for a configuration of hardware and system software must be expressed in the same terms as for application tasks. Thus, as discussed earlier, resource availability generally is measured in terms of CPU instructions and the ability to move data to and from secondary storage, I/O, and network devices. Time intervals and units of measure for resource demand must match those used in the application demand model.

CPU capability is measured in millions of instructions per second (MIPS) or millions of floating point operations per second (MFLOPS). Several problems may arise from the use of these measures, including:

▶ The execution of long instructions by an application.

▶ Wait states because of memory access bottlenecks.

▶ Incompatibility between the instruction sets of competing hardware platforms.

To simplify the model, application demand for CPU instructions is standardized to a least common denominator (e.g., 32-bit integer or floating point instructions). Configuration capabilities also are standardized to these "generic" instructions and wait states are assumed to be nonexistent or a fixed percentage of instructions executed.

For access to secondary storage and I/O devices, resource availability is measured in terms of sustainable data transfer rates. These rates depend on the characteristics of the devices, device controllers, and communication channels in use. Difficulties can be encountered when substantial differences exist between burst mode data transfer and sustainable data transfer rates. For example, the use of a caching disk controller may allow short bursts of extremely fast data transfer limited only by the speed of cache access.

However, for large transfers that exceed cache capacity, the sustainable data transfer rate would be limited by the physical characteristics of the storage device. An average rate sometimes is used if it can be derived through performance testing.

Model Manipulation and Analysis

The use of mathematical models for application demand and resource availability allows well-established computational techniques to be applied to model evaluation. Computational approaches applicable to computer performance modeling are of two types—static and dynamic analysis methods. Each can be implemented by several specific computational techniques.

Static analysis refers to the use of computational techniques that assume that demand for resources, and the ability to supply them, are constant. Thus, in static analysis, variation over time in application demand or in the ability of a configuration to supply resources is ignored. In contrast, **dynamic analysis** explicitly accounts for variation over time in the values of model parameters. However, computational techniques for dynamic analysis are substantially more complex than those for static analysis.

Static Analysis

When static analysis is used to evaluate models, all parameter values are assumed to be constant. Thus, application tasks are considered to have a constant frequency and resource demand. Configurations of hardware and system software are assumed to have a constant ability to supply computing resources. These assumptions are usually valid for resource delivery, but they are often unrealistic for application demand.

Average resource demands for each application processing task are summed over an interval. For example, assume an application that processes two types of user tasks (*A* and *B*). Assume further that resource demand for each transaction can be expressed by the following formula:

$$A: 15000X_c + 300X_s + 200X_i$$
$$B: 10000X_c + 750X_s + 500X_i$$

where:

$$X_c = \text{thousands of CPU cycles}$$
$$X_s = \text{KB mass storage I/O}$$
$$X_i = \text{KB display device I/O}$$

The composite resource demand model is simply the sum of these two formulas, weighted by their expected volumes over a stated time interval. Thus, if peak volumes for tasks A and B are 100 and 250 per hour, respectively, the peak composite resource demand is:

$$A: 100(15000X_c + 300X_s + 200X_i) = 1500000X_c + 30000X_s + 20000X_i$$
$$B: 250(10000X_c + 650X_s + 500X_i) = 2500000X_c + 162500X_s + 125000X_i$$
$$\text{Composite:} = 4000000X_c + 192500X_s + 145000X_i$$

For each configuration under consideration, a resource availability model is constructed using the same variables, units of measure, and timing conventions. Thus, the capacity of each configuration is expressed as a set of maximum values for each resource class during a one-hour period.

Notice that the maximum capacities of resource classes are interdependent because of trade-offs between various types of processing. System overhead associated with mass storage and I/O accesses consumes CPU cycles and bus capacity. A model of resource availability must account for these complexities. For example, resource availability for a configuration might be stated as:

$$X_c < 5000000000$$
$$X_b < 25000000$$
$$X_s < 8000000$$
$$X_i < 1000000$$
$$X_s = X_b + X_s + 200X_c$$
$$X_i = X_b + X_i + 150X_c$$

where:

$$X_b = \text{KB bus I/O}$$
$$X_s = \text{KB disk device I/O}$$
$$X_i = \text{KB display device I/O}$$

The intermediate variable X_b represents the bus capacity consumed. Resource utilization for each unit of application I/O to disk or display device is expressed as a formula describing its use of CPU, bus, and device resources. The composite variables X_s and X_i represent the sum of all resources consumed for each byte of disk and display device I/O. These composite variables would be substituted for X_s and X_i in the application demand model.

The two models can be analyzed by a number of techniques. For example, a linear program can be constructed to test the feasibility of meeting application demand with a given configuration. More complex formulations can be used to determine detailed configuration requirements (e.g., number of disk drives or I/O channels) or to determine a least-cost configuration among several competing configurations.

Dynamic Analysis

Notice that parameter values in the static analysis example were all single-valued. In dynamic analysis, the distribution of values is modeled explicitly, which implies the use of more model information than with static analysis. In theory, this additional information will yield more accurate results, but this additional information must be accurate to guarantee valid predictions. Thus, although dynamic analysis promises more accurate predictions of system behavior, that accuracy is obtained at the expense of substantially more complex model inputs.

Dynamic analysis of a computer system is implemented by **simulation**. Simulation is, in turn, based on **queuing theory** and **statistics**. A **simulation model** is composed of processing tasks, processing elements, and a set of interconnections between them. Computer hardware and system software is modeled as a linked set of server processes, as shown in Figure 15-3. Each server is dedicated to a specific processing function (e.g., CPU, disk I/O, and display device I/O). Each server accepts input from a queue of pending requests, as shown in Figure 15-4. Output from the server may leave the system entirely, or it may be directed to the input queue of another server.

Figure 15-3

A network diagram of a simulation model. The system is modeled as a set of server processes and input queues to those processes. Circles represent processes, and arrows represent input and output queues.

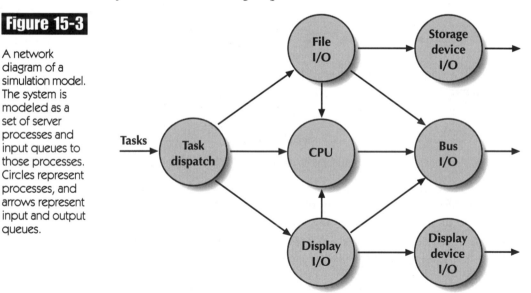

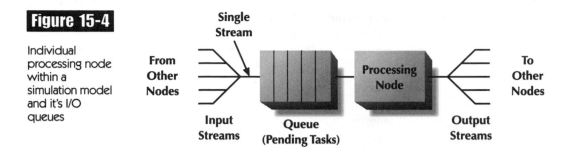

Figure 15-4

Individual processing node within a simulation model and it's I/O queues

Single Stream

From Other Nodes

Input Streams

Queue (Pending Tasks)

Processing Node

To Other Nodes

Output Streams

The arrival rate of processing tasks to initial system nodes (the dispatcher, in this example) is stated as a statistical distribution. This distribution—expressed as a distribution type (e.g., normal or Poisson), mean, and standard deviation—is used to determine the arrival rate for each processing task. Queuing theory may be used to determine the composition of and arrival times for combined queues. Each processing node accepts requests from the input queue and satisfies those requests. The amount of time needed to satisfy a request depends on the nature of the request. The time needed by a processor to satisfy each type of request is stated as part of the model inputs, and it is often stated as a statistical distribution as well.

A number of software products are available to implement and analyze simulation models. Simulation software accepts a description of the modeled system as input. The operation of the system over time is simulated by advancing a computer-generated clock. Each time the clock is advanced, the program generates a set of task arrivals and hardware actions based on the statistical distributions of task arrival and hardware operation. The current state of the system (i.e., hardware states and queue sizes) is updated, based on these calculations. Iterations continue until the simulation model reaches steady-state operation, which simulates the normal operation of the modeled system.

Reports can be produced to show many kinds of system information, including:

▶ Average queue size for each processor.

▶ Average waiting time for each processing request.

▶ Average processing time for each processing request.

▶ Proportion of idle time for each processor.

This information can be used to evaluate the performance of the system and/or to identify performance bottlenecks.

Measuring Resource Demand and Utilization

Both dynamic and static models require accurate data inputs. Accurate model analysis depends on the quality of application demand and resource availability estimates. Some specific types of automated tools are available to measure resource demand and utilization, including:

► Hardware monitors.
► Software monitors.
► Program profilers.

These tools can be used to generate parameter values for application demand models. They also may be used to generate data for evaluating upgrades or modifications to existing applications or hardware/software configurations. The information generated by these monitors describes the behavior of specific devices, resources, or subsystems over some period.

A **monitor** is a program or hardware device that detects and reports processing or I/O activity. A **hardware monitor** is a device attached directly to the communication link between two hardware devices. It monitors the communication activity between the two devices and stores communication statistics or summaries. This data can be retrieved and printed in a report. Hardware monitors often are used to monitor the use of communication channels, disk drives, and network traffic in mainframe computers.

A **software monitor** is a program that detects and reports processing activity or requests. Software monitors are included within operating system service routines and may be activated or deactivated by a system administrator. They can be used to monitor high-level processing requests (e.g., file open, read, write, and close) or low-level kernel routines (e.g., flushing of file or I/O buffers and virtual memory paging). When activated, a software monitor generates statistics of service utilization or processing activity. These statistics can be displayed in realtime or stored in a file for later analysis.

Software monitors can alter the activity being measured. For example, a software monitor that measures systemwide CPU activity generates inflated measurements because the software monitor itself requires CPU cycles to execute. Similar problems can exist for display I/O (e.g., if the monitor displays a continuously updated activity graph) and disk I/O (e.g., if the monitor logs activity measures to a file). Resources consumed by the software monitor must be estimated or measured and subtracted from the overall measurements to determine correct measurements for processes other than the monitor.

Monitors can operate either continuously or intermittently. A continuously operating monitor records all activity as it occurs, whereas a sampling monitor checks for activity periodically (e.g., 20 times per second). The advantage of a sampling monitor is that fewer computer resources are expended executing the monitor process itself. Another advantage is that less data is accumulated in output files, which can grow very large. Continuous utilization statistics can be estimated based on the sampled activity. Continuous monitors provide complete information on activity, but their operation may consume excessive amounts of system resources.

Monitors can be used to identify performance bottlenecks as a precursor to configuring hardware and/or system software for maximal performance. For example, monitoring I/O activity on all secondary storage channels may indicate that some disks are used continuously and others much less so. Based on this information, a system administrator might decide to move highly active files from heavily utilized disks to less-utilized disks to reallocate disks among controllers or I/O channels. An entire system can be tested at full load to determine its maximum sustainable resource delivery.

A **program profiler** is used to determine the resource or service utilization of an application program during execution. Typically, a set of monitor subroutines is added to the program's executable image during link editing. As the program executes, these subroutines record system service requests in a file. They also may record other statistics such as elapsed (i.e., wall clock) time to complete each service request, CPU time consumed by service calls, and CPU time consumed by program subroutines. This information can be used to derive a resource demand model for the application or to identify segments of the program that may be inefficiently implemented.

Monitors and profilers are commonly used tools for performance evaluation and requirements determination. By providing accurate data on resource demands and utilization, they allow accurate modeling of computer system configurations. This, in turn, improves the accuracy of predictions based on those models.

Windows NT Performance Monitoring

The Windows NT operating system is widely used for both workstations and servers. In workstation environments, it is used for applications that demand high performance and reliability. In server environments, it is used to provide network access to distributed file system, database, and World Wide Web (WWW) services. Both environments are characterized by a need to optimize the use of hardware and software resources to intended applications. To address this need, Windows NT provides a sophisticated set of tools to monitor hardware and operating system performance.

The primary tool Windows NT provides is a utility program called performance monitor (PM), which provides a system administrator with the ability to monitor a number of hardware and software resources in realtime. PM defines a number of system objects about which performance and utilization data may be captured. A partial display of these objects is shown in Figure 15-5. Monitoring objects may be individual hardware devices (e.g., the CPU), groups of hardware devices (e.g., all physical disks), operating system services (e.g., a network name service), or resource-management data structures (e.g., a service queue or a virtual memory paging file).

Figure 15-5

Partial listing of the system objects that can be monitored by the Windows NT performance monitor (PM) utility

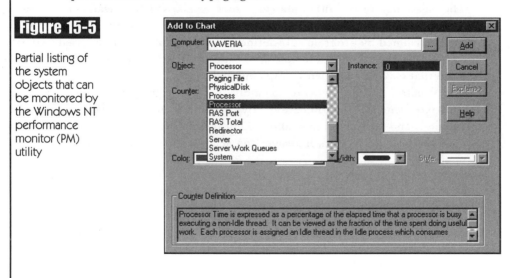

Each object has a set of data items called counters, which monitor specific performance or resource utilization aspects of an object. For example, there are 16 counters for the object "PhysicalDisk," including the current queue length, average seconds per disk read and/or write operation, number of disk read and/or write operations per second, and number of bytes read and/or written per second. The large set of counters defined for each object allows monitoring of both aggregate and highly specific aspects of performance.

Resource utilization and other system activity statistics can be displayed in various formats. Figure 15-6 shows a line graph displaying total CPU utilization (i.e., thin line) and virtual memory page faults per second (i.e., thick line). Statistics (e.g., mean, minimum, and maximum) also are displayed for the currently selected counter (Page Faults/sec, in this example). The display can be configured for any number of counters, and each counter is assigned a specific color and type of line. A similar set of display options are available for other output formats such as bar graphs and text reports. Performance data also may be captured in a file for analysis at a later time (e.g., using a statistical or spreadsheet utility).

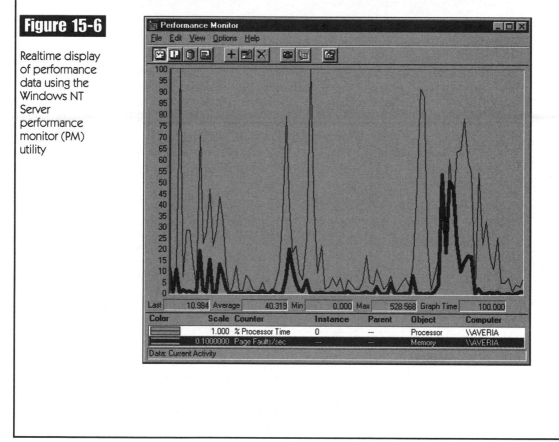

Figure 15-6

Realtime display of performance data using the Windows NT Server performance monitor (PM) utility

A system administrator needs accurate data to make system configuration decisions. For example, optimal decisions about changes to the size and/or location of virtual memory page files require accurate data concerning current virtual memory utilization. Optimal decisions about upgrading CPU capacity require data describing current CPU utilization and the utilization of related resources (e.g., memory and bus I/O). PM provides a flexible and powerful means of acquiring such data.

Accurate data is not, though, the only requirement for good configuration decisions. Interpreting that data requires a solid understanding of the internal functioning of operating system and hardware components, which allows the system administrator to associate cause and effect. It also permits the system administrator to understand the trade-offs among various types of system performance. Without such an understanding, performance assessment is little more than guesswork and configuration decisions may produce suboptimal, or detrimental, effects on overall system performance.

SECURITY

The information-processing resources of many organizations represent a sizeable investment. Some of the resources such as specific items of hardware and software are tangible and have well-defined dollar values. Others, such as databases, user skill, and reliable operating procedures are less tangible but nonetheless of considerable value. As applied to ISs, the term **security** describes all measures by which the value of these investments are protected. These measures include physical protection against loss or damage and economic protection against loss of value through unauthorized disclosure.

A well-integrated approach to system security includes many disparate components including measures to:

► Protect physical resources against accidental loss or damage (e.g., from fire or flood).

► Protect data and software resources against accidental loss or damage (e.g., because of storage-device failure).

► Protect all resources against malicious tampering (e.g., vandalism).

► Protect sensitive software and data resources against unauthorized access and accidental disclosure (e.g., industrial espionage).

Physical protection of resources against accidental loss or damage is discussed later in this chapter. Protection of data and software against accidental loss or damage is primarily implemented by file-system backup and recovery procedures as described in Chapter 13. Controls over malicious tampering and unauthorized access are implemented through a combination of physical and software access controls. Physical access controls are beyond the scope of this text. Software access controls are discussed in the following sections.

Access Controls

Multiuser and network operating systems require users to establish their identity prior to accessing resources in a process called **user authentication**. A user enters a user name or other identifier and a password to prove identity. The operating system verifies the user name and password against a security database. The security database may be maintained by the operating system on the local machine or located on a network server and accessed by network service calls.

An authenticated user name or identification is the basis for determining that user's ability, or lack thereof, to log on and to access resources managed by the operating system. Specific access controls may include restrictions on:

► Locations from which a user may log on to the system or access its resources.

► Days/times when access is allowed or denied.

► Ability to read, write, create, and delete directories and files.

► Ability to execute programs.

► Ability to access hardware resources (e.g., printers and communication devices).

The system administrator is responsible for establishing these access controls to implement the organization's security policies.

Early access-control systems were implemented primarily through login and file-system controls based on individual user accounts, but these methods have proven inadequate as information processing resources have grown more numerous and varied. Modern access-control systems are generally based on a complex directory structure that is a superset of the file management system directory structure. Figure 15-7 on the next page shows an example of a hierarchical directory structure for Novell NetWare version 4.11. Many types of objects are represented, including organizational roles (e.g., Chairman), user groups (e.g., Faculty), organizational units (e.g., ISs), server computers (e.g., Server1), file systems (e.g., Applications), printers and printer input queues (e.g., LaserJet 4Si and PrimaryQueue), and client computers (e.g., ASM2104).

Figure 15-7

NetWare
Directory Services
tree

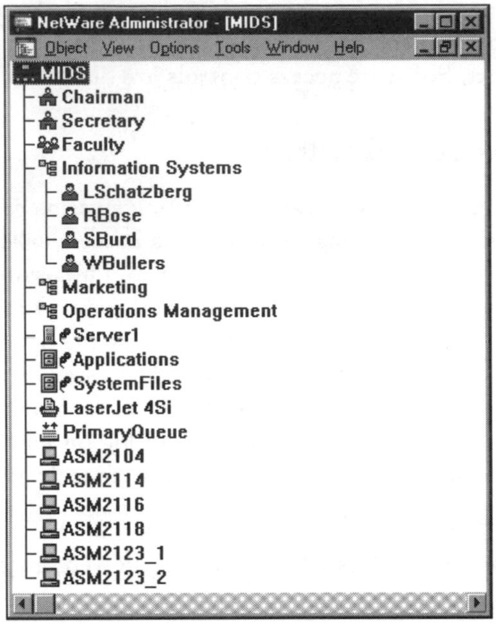

```
NetWare Administrator - [MIDS]                    _ □ ✕
  Object  View  Options  Tools  Window  Help       _ 🗗 ✕
  ⬛ MIDS
   ├ 🏛 Chairman
   ├ 🏛 Secretary
   ├ 👥 Faculty
   ├ ◰ Information Systems
   │  ├ 👤 LSchatzberg
   │  ├ 👤 RBose
   │  ├ 👤 SBurd
   │  └ 👤 WBullers
   ├ ◰ Marketing
   ├ ◰ Operations Management
   ├ 🖳 Server1
   ├ 🖿 Applications
   ├ 🖿 SystemFiles
   ├ 🖨 LaserJet 4Si
   ├ ⬚ PrimaryQueue
   ├ 🖥 ASM2104
   ├ 🖥 ASM2114
   ├ 🖥 ASM2116
   ├ 🖥 ASM2118
   ├ 🖥 ASM2123_1
   └ 🖥 ASM2123_2
◀                                                        ▶
```

Access controls can be defined among directory objects or groups of objects. For example, users can be assigned specific access rights to a server, file system, print queue, or workstation. Access controls can be indirectly assigned to users by their group membership, organizational role, or organizational unit. The variety of objects in the directory structure allows a variety of complex and/or overlapping security policies to be implemented.

Notice that the objects modeled within the sample directory structure extend well beyond the boundaries of a single computer or operating system. The directory incorporates all of the objects within a departmental network. The widespread use of LANs and WANs has complicated the task of implementing security policies. Security controls implemented within a single operating system are no longer sufficient because resources and objects may be spread across many computer and operating systems. Thus, security policies and the means by which they are implemented (e.g., directory structures and access controls) must transcend the boundaries of a single computer system.

Password Controls and Security

User-authentication procedures are a critical component of any access control system. A failure in these procedures effectively disables most software-based security measures. The authentication process must be highly reliable. That is, it must ensure that only authorized users are able to log on and/or gain access to system resources and services. Maximal reliability requires that passwords be securely stored and difficult to guess.

Modern operating systems employ a number of methods to enhance password security, including:

▶ Restrictions on the length and composition of valid passwords.

▶ Requirements that passwords be changed periodically.

▶ Analysis of password content to identify passwords that are guessed easily.

▶ Encryption of passwords in files and during transmission over a network.

▶ Restrictions on locations from which a user may access the system.

Most operating systems allow the system administrator to create and enforce password policies on a per-user, per-user-group, or per-system basis.

Figure 15-8 shows the systemwide password controls that can be set in Windows NT, including restrictions on password length, age, and uniqueness. A related set of policies can be set to deal with failed login attempts. Windows NT allows accounts to be deactivated after a specific number of failed logins. The account can be reactivated manually by the system administrator or automatically after a specified interval. Some of the password controls (e.g., age restrictions) can be overridden for specific accounts.

Figure 15-8

Windows NT account policy utility

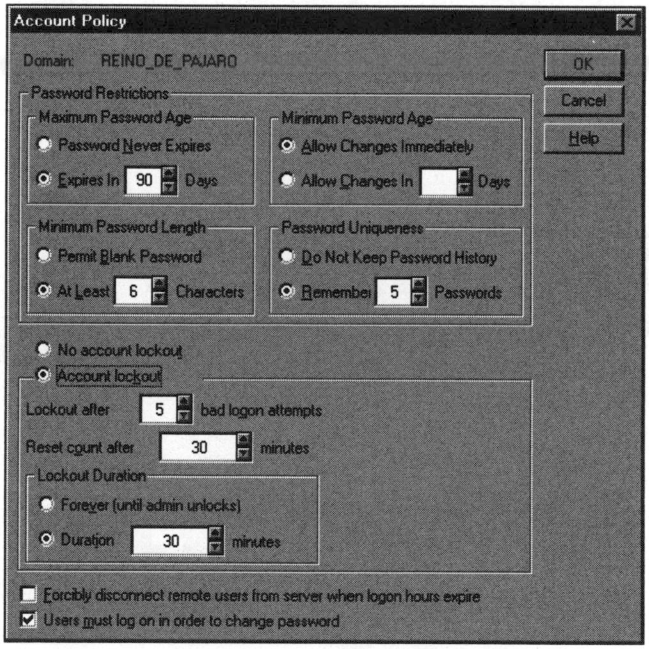

Locking out accounts after a specified number of failed logins is implemented to prevent unauthorized users from repeatedly attempting to guess correct passwords for valid user accounts. Software programs are sometimes used to attempt remote login with words and/or names taken from a dictionary. Because many users choose common words or names as their passwords, it is possible for such programs to eventually find a correct password simply by trying every word in their dictionary. Automatic account deactivation coupled with delayed reactivation severely limits the number of guesses that can be made in any period.

Some operating systems employ similar "password guessing" programs against their own password files on a periodic or continuous basis. When passwords are guessed successfully, the operating system forces the user to change their password at the next login. Such programs also can be employed by the program used by users to change their passwords. An attempt to change a password to an easily guessed value is denied.

System security may be compromised if a copy of the password file or database is somehow distributed to, or stolen by, an unauthorized individual. The password file is normally heavily protected by access controls, but some degree of accessibility must be provided to allow normal login processing. Most operating systems further protect the password file by encrypting part or all of its content. However, a stolen password file can be decrypted successfully if sufficient computing resources are employed to crack the encryption key.

Many operating systems provide further control over unauthorized access by allowing the system administrator to restrict the times and locations from which a user may log in. Figure 15-9 shows the display of a Windows NT utility to set allowable times for user login and access. User logins and system access can be restricted to hours when they are expected to be at work. The system administrator also can restrict the locations (e.g., specific workstations or groups of workstations) from which a user can login.

Figure 15-9

Windows NT login time restrictions

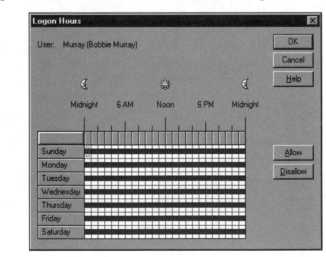

System Administration

PHYSICAL ENVIRONMENT

The installation of computer hardware requires special attention to many aspects of its physical environment. Some of these are a matter of convenience, and others are a matter of protection for the equipment itself. Particular issues to be addressed include:

► Electrical power.

► Heat dissipation.

► Moisture.

► Cable routing.

► Fire protection.

► Disaster planning and recovery.

Each of these is discussed in detail next.

Electrical Power

Computer hardware is very sensitive to fluctuations in power levels. Processing circuitry (e.g., the CPU, device controllers, etc.) is designed to operate at a constant low-power level. Fluctuations may cause momentary loss of operation if power levels drop or damage to electrical circuits if power levels rise. Fluctuations can be of several types, including:

► Momentary power surges.

► Momentary power sags.

► Long-term voltage sags.

► Total loss of power.

Power surges can be caused by a number of events. Lightning strikes in power-generation or transmission facilities tend to cause the most dangerous types of power surges. Similarly, dangerous spikes may be caused by the failure of power transformers or other transmission equipment, which lead to brief surges of very high intensity. Because the surges are brief, they may not engage standard protection devices (e.g., fuses or circuit breakers) before significant damage has occurred. Thus, standard fuses or breakers (e.g., those controlling the distribution of power to a floor or entire building) do not provide adequate protection for computer equipment.

Power sags normally occur when a device that requires a large amount of power is started. This can be seen in home environments when devices such as air conditioners, refrigerators, and electric dryers are started and cause a momentary dimming of lights. Small power sags are almost always present when multiple devices share a single electrical circuit. Large power sags are a symptom of overloaded circuits. Mixing equipment with large variations in power requirements on a single circuit is particularly troublesome.

Longer-term power sags are often caused by the power provider itself. The common term for this event is a **brownout**. Brownouts occur when the demand for electricity exceeds the generation and transmission capabilities of the provider. This commonly occurs during peak demand periods such as hot summer days. The power provider temporarily reduces the voltage level on a systemwide basis to spread the available power evenly.

Most computer equipment is designed to operate reliably over a range of voltage levels. In the United States, transformers within smaller computer systems are designed for 110-volt alternating current power input. Larger computer systems in the United States may use 220- or 440-volt power inputs. Computer systems in other countries use different voltage levels. Computer system power transformers are constructed to tolerate input voltage variations of up to 10%. This characteristic provides some protection against surges and sags because of the startup/shutdown of other equipment as well as brownouts.

Other provisions must be made to protect equipment against more severe variations, including mechanisms to deal with powerful surges and total power loss. Equipment may be protected against high power surges by the use of a **surge protector**, which detects incoming power surges and quickly (e.g., within nanoseconds) diverts them to ground. Surge protectors differ in the speed at which they react, the intensity of the surge that can be grounded, and whether the device can be reused after a surge. Surge protectors can be purchased as separate devices or may be a component of an integrated device (e.g., a line conditioner).

Total power loss alone rarely causes damage to computer hardware. However, a tripped circuit breaker or blackout is often accompanied by one or more power surges. The primary problem with power loss lies in the loss of data. Data held in RAM, including process data areas, secondary storage buffers, communication buffers, and so forth is lost when power is interrupted.

Protection against power loss requires the use of an auxiliary power source, which may take several forms, including secondary circuits, auxiliary generators, battery backup, or some combination of these. An **uninterruptable power supply (UPS)** is a device (usually battery-based) that provides power to attached devices in the event of external power failure. These devices vary in their power delivery capacity, switching time, and duration of operation between power loss and power restoration. Normally, surge protection is incorporated into a UPS because switching between external and internal power supplies may introduce surges.

Some UPSs, particularly those with short delivery times, are designed to work in concert with a computer's operating system. They are attached to a computer-system communication port (e.g., a PC serial port) or directly to the system bus through an interface card. When a power failure is detected, an interrupt is generated, which informs the operating system of the power failure and provides a mechanism for initiating protective actions prior to a total loss of power. The operating system initiates a normal or emergency shutdown procedure when such an interrupt is detected.

Heat Dissipation

All electrical devices generate heat as a byproduct of normal operation. Excessive heat can cause intermittent or total failure of electrical circuits. Thus, all computer equipment requires some means of **heat dissipation**. In equipment that generates little heat, vents in the equipment cabinet are normally sufficient to allow heat dissipation. Care must be taken to ensure that vents do not become blocked, so as to allow a free movement of air through the cabinet.

Many hardware devices supplement venting with the forced movement of air through the unit through the use of one or more fans, which either force cool exterior air into the cabinet or draw hot interior air out. Either method requires at least two vents and a clear pathway for air movement. Vents are positioned at opposite corners of the cabinet to ensure that all components receive adequate cooling. Forced air cooling also requires some filtering to remove dust and other contaminants.

When heat is dissipated from an equipment cabinet, it collects in the room in which the cabinet is located. Some mechanism must be provided to dissipate heat from the room as well as the cabinets, which is especially true when many hardware devices are situated in a relatively small space. Normal room or building air conditioning may be sufficient, or supplemental cooling capacity—which more effectively counteracts heat buildup and provides additional protection if the primary (building) cooling system fails—may be required.

In extreme cases, auxiliary cooling may be provided within an individual equipment cabinet, which may take the form of a refrigerant-based heat exchanger, a liquid cooling system, or even a liquid nitrogen system. Such measures are often used in equipment where semiconductor devices are operated at extremely high clock rates or access speeds (e.g., supercomputers). Liquid cooling has been used in some mainframe computer systems as well.

Moisture

Excessive moisture is an enemy of electrical circuitry because of the danger of short circuits (e.g., water conducting electricity between two otherwise unconnected conductors). Short circuits can lead to circuit damage and are also a fire hazard. Water can damage computer equipment even when power is turned off. Impurities in the water are left on computer components as the water evaporates and can corrode exposed electrical contacts or other hardware components (e.g., disk platters, printed circuit boards, metal cabinets, etc.). These impurities also can form a short circuit.

Well-designed cabinets are one defense against the dangers of moisture, but they protect only against overt spills and leaks. Another protective measure is to mount cabinets (or devices within cabinets) above floor level, which minimizes the danger from roof leaks, broken pipes, and similar problems that can lead to standing water.

Protection also must be provided against condensation because of excessive humidity. Thus, excessive humidity levels must be avoided through direct control (e.g., a dehumidifier). On the other hand, low humidity is also problematic because static electricity can build up as a result. This increases the likelihood of circuit damage because of inadvertent static discharges. In general, the humidity level of a room containing computer equipment should be near 50%.

Cable Routing

Computer facilities must be designed to provide protection for data-communication lines. Because the configuration of these lines changes frequently, ease of access is also key. Computer facilities deal with this problem in two ways—raised floors and dedicated cable conduits.

A raised floor, used in a room that contains multiple hardware cabinets (e.g., a single mainframe computer system or multiple minicomputers), serves several purposes. The primary purpose is to provide an accessible location for cables connecting different devices. The floor consists of a set of load-bearing supports upon which a grid of panel supports is laid. The surface of the floor consists of solid panels that may be installed easily or removed from the grid. Cables are routed in a straight line under walkway areas (i.e., between hardware cabinets). Thus, cabling is easily accessible by removing floor panels.

Secondary reasons for raised floors include protection from standing water and movement of chilled air. Several inches of water may accumulate without reaching the level of equipment cabinets. Moisture sensors are placed below the floor panels to detect the buildup of standing water. The space between the actual floor and the floor panels also may be used as a conduit for chilled air.

When used in this manner, equipment cabinets are vented at the bottom and top. Chilled air is thus forced through the floor into the bottom of the cabinets, and heat is dissipated through the top.

Dedicated cabling conduits are used to provide cable access between rooms or floors of a building. To prevent electromagnetic interference, these conduits should not be used to route both electrical power and electrical data communication lines. (Fiber-optic lines are immune to interference from electrical power lines.) In addition, access panels should be provided at regular intervals to allow the addition, removal, or rerouting of cables. Conduits should be shielded to limit external electromagnetic interference.

Fire Protection

Fire protection is an important consideration both for safety of personnel and protection of expensive computer hardware. As with cooling, the normal fire-protection mechanisms incorporated within buildings are inadequate for rooms that house computer hardware. In fact, such measures actually increase the danger to both personnel and equipment, primarily because of the extensive use of water (i.e., automatic sprinklers).

Carbon dioxide, fire-retardant foams and powders, and various gaseous compounds are alternative methods of fire protection. Generally, carbon dioxide is unacceptable because it is a hazard to humans (e.g., suffocation) and because it promotes condensation within computer equipment. Fire-retardant foams and powders are unacceptable because of their moisture content, and powders have corrosive properties.

Many large computer facilities use halon 1301 gas, which does not promote condensation and also does not displace oxygen to the extent of carbon dioxide. This allows personnel adequate time to evacuate a room or floor. Unfortunately, halon gas is a chlorofluorocarbon (CFC), a class of gases with ozone-depleting properties. Most industrialized countries have signed the Montreal Protocol, which phases out CFC production and bans importation of newly produced CFCs. Various alternatives are currently under study, including halocarbon compounds and inert gas mixtures, but none has yet emerged as a clear successor to halon.

Fire detection also is a special problem within computer facilities. Electrical fires often do not generate heat or smoke as quickly as do conventional fires. Thus, normal detection equipment may be slow to react. Fast detection is an economic necessity. Fires within one item of computer equipment can quickly spread and or/cause damage to attached equipment through power surges.

Typically, fire-detection equipment is supplemented within a computer room. Additional smoke detectors generally are placed near large concentrations of equipment. Smoke detectors also should be placed below raised floors to quickly detect fires in cabling.

Disaster Planning and Recovery

Because disasters (e.g., fire, flood, and earthquake) cannot be avoided, plans must be made to recover from them. Disaster planning is especially critical in online systems and systems where extended downtime will cause extreme economic impact. A number of typical measures include:

▶ Periodic data backup and storage of backups at alternative sites.

▶ Backup and storage of critical software at alternative sites.

▶ Provision of duplicate or supplementary equipment at alternative sites.

▶ Arrangements for leasing existing equipment at alternative sites (e.g., with another company or a service bureau).

The exact measures that are appropriate for any given installation heavily depend on local characteristics.

SUMMARY

The primary responsibility of a system administrator is to ensure efficient and reliable delivery of IS services. Broad categories of system administration tasks include acquiring new IS resources, maintaining existing IS resources, and designing and implementing an IS security policy. There is considerable variation among organization as to the assignment of specific tasks to specific individuals.

The acquisition and deployment of IS resources should only occur in the context of a well-defined strategic plan for the organization as a whole. The IS strategic plan is only one part of the organization's strategic plan. Because IS is normally a support organization, its strategic plan tends to follow (rather than lead) the strategic plans of other units in the organization. IS resources can be considered organizational infrastructure. Many of its resources also are capital assets because they are used during multiple years. Strategic issues relevant to capital-intensive infrastructure include services to be provided, charging for services, infrastructure composition, and infrastructure improvement and maintenance.

Providing infrastructure-based services to a wide variety of users requires the adoption of service standards. However, standardization tends to stifle innovation and to produce suboptimal solutions. Standardization often causes problems for users who need services at or near the leading edge of technology.

Managing IS resources as infrastructure may result in ignoring opportunities to employ IS resources for competitive advantage.

Planning for IS resource acquisition requires a thorough understanding of present and anticipated application needs. Acquisition proceeds by determining the applications that the hardware/software will support, specifying detailed hardware/software requirements, drafting and circulating a request for proposals (RFP), evaluating RFP responses, and negotiating a purchase and/or support contract. An RFP is a formal document sent to vendors that states requirements and solicits proposals to meet those requirements. RFP responses are evaluated by determining minimal acceptability of each proposal, ranking acceptable proposals, and validating high-ranking proposals.

Determining hardware requirements is one of the most difficult system administration tasks. Hardware requirements depend on the hardware and system software resources required by application software. If the application software has already been developed, then its hardware and system software-resource consumption can be measured. Determining resource requirements is much more complex when application software has not yet been developed. In this case, hardware and system software requirements must be determined based on the detailed analysis-phase descriptions of processing requirements. Estimates can be based on comparisons to similar application software or on detailed modeling of application-resource requirements.

Computer hardware and software performance evaluation and requirements determination can be performed with mathematical models. Two separate models are required—an application demand model and a resource availability model. An application demand model describes the relationship between user-level processing activities and demands for hardware and system software resources. A resource availability model describes the ability of hardware system and system software to supply computing resources to application programs.

Computational approaches applicable to computer performance and application requirements modeling are of two types—static and dynamic analysis methods. Static analysis methods assume that all parameter values are constant. This assumption usually is valid for resource delivery but not for application demand. Dynamic analysis explicitly models the variability of model parameters. Dynamic analysis uses simulation and promises more accurate predictions of system behavior at the cost of substantially more complex model inputs and analysis.

Automated tools such as hardware monitors, software monitors, and program profilers are available to measure resource demand and utilization. A hardware monitor is a device that is attached directly to the communication link between two hardware devices to report the communication activity between the devices. A software monitor is a program that detects and reports processing activity or requests. A program profiler is used to determine the resource or service utilization of an application program during execution.

The information-processing resources of many organizations represent a sizeable investment. A well-integrated approach to system security protects these resources against accidental loss or damage, malicious tampering, unauthorized access, and accidental disclosure. System security is enforced through the operating system by access controls and user-authentication procedures. Specific access controls include restrictions on locations from which a user may log on to the system or access its resources; days/times when access is allowed or denied; the ability to read, write, create, and delete directories and files; the ability to execute programs; and, finally, the ability to access hardware resources. Modern access control systems are based on a complex directory structure that encompasses all computing, data, and software resources.

User-authentication procedures are a critical component of any access control system. Passwords are the key component of a user-authentication system. Operating systems employ a number of methods to enhance password security, including restrictions on password length and composition, requirements that passwords be changed periodically, periodic or continuous attempts to guess passwords, encryption of passwords in files and during transmission, and restrictions on locations from which a user may access the system.

The installation of computer hardware requires special attention to many aspects of its physical environment. Particular issues to be addressed include electrical power, heat dissipation, moisture, cable routing, and fire protection. Computer equipment is sensitive to fluctuations in power, including power sags, power surges, brownouts, and power failure. Surge protectors protect computer equipment against brief amounts of excessive power transmission. Uninterruptable power supplies (UPSs) protect against problems due to power failure. Protection against power sags is built into most computer equipment. Power sags can be minimized by careful attention to power-circuit design and to the allocation of electrical devices to individual power circuits.

All electrical equipment produces heat as a result of internal resistance. Within computing equipment, heat also may be generated by motors, display devices, and other equipment. Heat must be dissipated to protect electrical components. Most computer equipment contains some heat-dissipation capability in the form of vented enclosures and fans. For concentrations of computing equipment, additional cooling measures must be used.

Moisture is dangerous to computer equipment because of the possibility of short circuits and component damage. In larger computer installations, equipment is placed on a raised floor to protect against water leaks. Moisture sensors can be used below the floor to detect standing water. The raised floor also provides a convenient place to route communication cables. Humidity levels must be controlled as well. Low humidity increases the buildup of static electrical charges, and high humidity increases condensation.

Fire protection must be provided to protect both personnel and equipment. Standard methods of file protection are inadequate or dangerous in computer installations. Sprinklers cannot be used because of potential water damage and the negative interactions between electricity and water. Carbon dioxide cannot be used because of its danger to personnel and the danger of water condensation. Fire-retardant foams and powders contain chemicals that are harmful to electrical equipment. Halon gas is a commonly used protection method. However, its production has been eliminated in most of the world, and no clear successor has yet been identified.

Key Terms

abstraction
application demand model
benchmark
brownout
capital expenditures
capital resources
competitive advantage
dynamic analysis
hardware monitor
heat dissipation
infrastructure
model
monitor
operating expenditures
power sag
power surge

program profiler
queuing theory
request for proposals (RFP)
resource availability model
service standards
simulation
simulation model
software monitor
static analysis
statistics
strategic plan
surge protector
system administration
uninterruptable power supply (UPS)
user authentication
workload model

1. An estimate of resource delivery by a hardware/software configuration can be obtained by a tested or published _____.

2. _____ are expected to provide service over a period of years.

3. Integrated analysis of the operation of application software, system software, and computer hardware requires a(n) _____ model and a(n) _____ model.

4. _____ uses _____ theory to determine the arrival of service requests.

5. Features and capabilities of an application or computer system are _____ to create a model.

6. _____ analysis explicitly considers variations in model parameters over time whereas _____ analysis ignores those variations.

7. Because IS resources can be considered _____, issues of service standards, cost, and cost recovery are important components of the IS strategic plan.

8. A(n) _____ is a formal (i.e., legal) document that solicits proposals from hardware and/or software vendors.

9. _____ based on passwords is the most critical process in a security system.

10. A(n) _____ detects and reports hardware or software actions.

11. Opportunities to exploit IS resources for _____ may be missed if those resources are managed only as infrastructure.

12. A(n) _____ provides auxiliary power during blackouts and can notify the operating system when it is activated.

13. Provision must be made to protect computer hardware against _____ and _____ in electrical power.

14. _____ is the process of ensuring efficient and reliable delivery of IS services.

15. A(n) _____ is an abstract representation of reality.

16. Long-range acquisition of computer hardware and software should be made in the context of a(n) _____ for the entire organization.

17. The resource demands of an existing application can be measured with a(n) _____.

1. What is infrastructure? In what ways do computer hardware and system software qualify as infrastructure?

2. What are the basic strategic planning questions that are addressed with respect to infrastructure?

3. What are the advantages and disadvantages of standardization in computer hardware and systems software?

4. What is a request for a proposal? How are responses to a request for a proposal evaluated?

5. What problems are encountered when attempting to determine hardware and system software requirements for application software that has not been fully designed or developed?

6. How is computer performance evaluation related to applications?

7. What are some purposes of performance evaluation?

8. What are the essential elements of a simplified performance model, and how may these elements be organized?

9. What is the difference between static analysis and dynamic analysis? What are the comparative advantages and disadvantages of each?

10. What steps are used in model-based requirement determination and performance evaluation?

11. What parameters might be included in a measurement of application resource demand?

12. What is a monitor? List the various types of monitors and the information that they provide.

13. What is user authentication? Why is user authentication required to implement access controls?

14. List the types of access controls implemented by a typical operating system. How is access-control information maintained by the operating system and system administrator?

15. Why are conventional methods of fire protection inadequate or dangerous for computer equipment?

16. What problems associated with electrical power must be considered in planning the physical environment of computer hardware?

1. Some manufacturers and vendors of large minicomputers and mainframes have developed documentation and software to help engineers, account representatives, and purchasers to configure computer hardware and system software. Some of the software can help users to match software demands against the capabilities of particular hardware configurations. Investigate the online offerings of several large computer companies (e.g., www.ibm.com, www.digital.com, www.unisys.com, and www.fujitsu.com).

2. Halon use extends well beyond computer rooms and includes manufacturers or electrical equipment and the U.S. Navy. All these users have large investments in halon fire-protection equipment. Thus, there is a significant economic incentive to developing halon alternatives that work with existing equipment. Investigate the current state of halon replacements with particular attention to cost, availability, fire-protection ability, protection of computer equipment, and compatibility with existing fire-protection equipment. Organizations that monitor halon replacements include the Halon Alternatives Research Corporation (www.harc.org) and the U.S. Navy CFC & Halon Clearinghouse (home.navisoft.com).

Units of Measurement

Many units of measure are employed to describe the capabilities of computer systems and related hardware. Their names and abbreviations are a common source of confusion when reading about hardware capabilities or examining hardware specifications. This appendix defines commonly used measurement units and describes common mistakes in their use.

TIME UNITS

Time intervals (e.g., cycle or access times) are expressed in fractions of a second, represented by the following abbreviations and terms:

Abbreviation	Term	Fraction
ms	millisecond	10^{-3} (one thousandth)
ms	microsecond	10^{-6} (one millionth)
ns	nanosecond	10^{-9} (one billionth)
ps	picosecond	10^{-12} (one trillionth)

Notice the potential source of confusion in the use of the abbreviation "ms" for both millisecond and microsecond. Microseconds are rarely used when describing computer hardware, so a reader can generally assume that ms refers to milliseconds.

CAPACITY UNITS

The basic units of data storage and transmission are bits and bytes. A bit is a single binary digit containing either the value 0 or 1. A byte contains 8 bits. The terms **bit** and **byte** are often quantified with a prefix to indicate larger magnitudes. These prefixes are all based on the value 2^{10} (1024_{10}) because it approximates traditional units based on powers of 10 (e.g., thousands, millions, and billions).

The following abbreviations and prefixes are used in capacity measurements:

Abbreviation	Prefix	Value
K	kilo	1024
M	mega	1024^2 (1,048,576)
G	giga	1024^3 (1,073,741,824)
T	tera	1024^4 (1,099,511,627,776)

Notice that the values represented by these prefixes are all slightly larger than their decimal counterparts: thousand, million, billion, and trillion. This sometimes results in confusion when an abbreviation based on 1024 is used to describe a quantity based on 1000.

For example, consider the capacity of a "standard" 3.5-inch PC floppy diskette, which is frequently (and incorrectly) described with a capacity of 1.44 megabytes. The actual capacity is 1,474,560 bytes (2 sides $\times$ 80 tracks/side $\times$ 18 sectors/track $\times$ 512 bytes/sector), 1,440 kilobytes (1,474,560 ÷ 1024), or 1.40625 megabytes (1,474,560 ÷ 1024^2). Hard disk capacities often are stated with similarly flawed units of measure.

Another source of confusion arises from the application of prefixes to the terms bit and byte. For example, the terms **megabit** and **megabyte** can both be abbreviated as MB. A common method of distinguishing those terms is to abbreviate the term bit using a lowercase letter and the term byte using an uppercase letter. Thus, the abbreviation KB would be interpreted as kilobyte or kilobytes, and the abbreviation Gb as gigabit or gigabits. Unfortunately, this convention is not universally followed, which results in much confusion and possible misinterpretation. Any abbreviation containing the letters b and B deserves careful scrutiny to determine whether the term bit or byte is intended.

DATA TRANSFER RATES

Data transfer rates are used to describe the capacity of communication channels and the throughput rates of some I/O devices. A data transfer rate is always expressed as a quantity of data per some time interval. For example, a LAN might be described as having 100 megabits per second (Mbps) of data transmission capacity. A printer might be described as having an output speed of four pages per minute (ppm).

Communication channel capacities usually are expressed in bits or bytes per second. Magnitude prefixes such as kilo and mega can be used (e.g., 500 kilobits per second and 100 megabytes per second). Bit and byte capacity measures are abbreviated as described in the previous section and the letters "ps" (i.e., per second) are appended. Thus, the unit kilobits per second is abbreviated Kbps, and the unit megabytes per second is abbreviated MBps. As with capacity measures, any abbreviation containing the letters "b" and "B" deserves careful scrutiny to determine whether the term bit or byte is intended.

Printer output rates generally are expressed in characters per second (abbreviated cps) or pages per minute (abbreviated ppm). Notice that a character generally is assumed to be 8 bits, and the number of bytes in a page varies widely depending on the page content and the data-encoding method used. Video output rates may be expressed in frames per second (abbreviated fps). The number of bytes per frame depends on frame size, image content, and data encoding method.

GLOSSARY

100Base-AnyLAN – Synonym of 100BaseVG.

100BaseVG – IEEE standard for 100 Mbps Ethernet.

10BaseT – Type of twisted pair wiring commonly used to implement local area network (LAN) connections. It contains four wire strands and uses RJ45 connectors.

24-bit color – Representation of color using three eight bit unsigned integers to represent intensities of the primary colors red, blue, and green.

3GL – Acronym of Third Generation Language.

4GAD – Acronym of Fourth Generation Application Development.

4GL – Acronym of Fourth Generation Language.

A/D – Abbreviation of analog to digital.

absolute address – Memory reference within a program or CPU circuitry that refers to a specific physical memory location.

absolute addressing – Assumption that all memory references within a program correspond to actual physical memory locations.

absolute value – Value of a positive number or the product of (result of multiplying) a negative number and –1.

abstraction – (1) With respect to software, the conceptual difference between the physical reality of computer hardware and the abstract or logical view of that same hardware provided by system software and/or a programming language. (2) With respect to computer system modeling, the process of selecting and representing computing resources and application demands within a model.

access arm – Device within a disk drive containing one or more read/write heads mounted at one end. The other end is attached to a motor that allows the read/write head(s) to be positioned over a single track of the disk platter.

access time – Elapsed time between the receipt and completion of a read or write command by a storage device.

ACK – Mnemonic for acknowledge.

acknowledge (ACK) – ASCII control character sent from receiver to sender to indicate successful receipt of transmitted data.

ACM – Acronym of the Association for Computing Machinery.

ADC – Acronym of Analog to Digital Converter.

ADD operation – (1) Boolean operation that causes the arithmetic sum of two numbers to be computed. (2) CPU instruction that causes the arithmetic sum of two numbers to be computed and stored in a register or memory location

additive colors – Red, green, and blue.

address – Physical location of a data item (or the first element of a set of data items) within a storage or output device.

address bus – Subset of bus lines or wires used to transmit the address of a storage location to be read from or written to.

address mapping – Synonym of address resolution.

address resolution – As performed by the control unit, the conversion of relative or segmented address references into a corresponding physical memory address.

addressable memory – Maximum amount of memory that can be addressed physically by the CPU.

AITP – Acronym of the Association for Information Technology Professionals.

algorithm – (1) Series of processing steps that describe the solution to a problem. (2) The sequence of instructions within a computer program that implements a complex processing operation.

allocation unit – Smallest unit of storage that can be allocated to an object such as a file or a program.

ALU – Acronym of arithmetic logic unit.

AM – Acronym of amplitude modulation.

American National Standards Institute (ANSI) – Governmental body that promulgates many types of standards, including those for computer-related technology (e.g., programming languages).

American Standard Code for Information Interchange (ASCII) – Standard 7-bit coding scheme used to represent character data and a limited set of I/O device control functions.

amplifier – Hardware device used to increase the strength of an analog signal in a communication channel.

amplitude – Magnitude (i.e., highest value) of wave peaks of an analog waveform.

amplitude modulation (AM) – Data-transmission method that encodes data values as variations in the amplitude of an analog carrier wave.

amplitude shift keying (ASK) – Synonym of amplitude modulation

analog signal – (1) Signal in which any value (of an infinite set) can be encoded by modulating one or more carrier wave parameters in direct proportion to the encoded value. (2) Signal that varies continuously in one or more carrier wave parameters (e.g., frequency or amplitude).

Analog-to-digital (A/D) conversion – Translation of an analog signal into an equivalent stream of digital data.

Analog-to-digital converter (ADC) – Processing device that performs analog to digital conversion.

analysis model – (1) Model of user requirements for information processing, often in the form of data flow diagrams and supporting specifications. (2) Output of the analysis phase of the structured systems development life cycle (SDLC).

AND operation – (1) Boolean operation that generates the value False unless both input values are True. (2) CPU instruction that generates the value False unless both input values are True and stores the result in a register or memory location.

ANSI – Acronym of American National Standards Institute.

append mode – Mode of output in which all write operations are implemented as append operations.

append operation – Output operation that appends a data item (e.g., a record) to the end of a set of data items (e.g., a file).

applet – Program written in the Java programming language that is executed by another program such as a Web browser.

application demand model – Model of the demand for computer resources by one or more application programs.

application generator – Program that generates application program source or executable code based on inputs from a system model.

application layer – Application programs (i.e., software), as compared to systems software and computer hardware.

application program – Program that addresses a single specific need of a user or a narrowly defined class of information processing tasks.

application programmer – Person who creates and/or maintains application software.

application software – Class of software consisting of all application programs.

application tool – Program designed for use by end users that addresses one or more general types of information processing (e.g., a spreadsheet or word-processing program).

arithmetic logic unit (ALU) – Circuitry within a central processing unit (CPU) that performs computation, comparison, and Boolean logic operations.

arithmetic shift operation – Shift operation used to perform division or multiplication.

array – Set of data items stored in consecutive storage areas identified by a common symbolic name.

ASCII – Acronym of American Standard Code for Information Interchange.

assembler – Program that translates an assembly language program into object code or CPU instructions.

assembly language – Programming language that uses mnemonics to represent CPU instructions and storage address operands.

ASK – Acronym of Amplitude Shift Keying.

Association for Computing Machinery (ACM) – Professional organization for computer scientists, programmers, and engineers.

Association for Information Technology Professionals (AITP) – Professional organization for information systems (IS) managers and specialists.

asynchronous transmission – Any transmission method in which sender and receiver do not use synchronized clocks.

attenuation – Reduction in signal power or strength that occurs as the signal travels through a transmission medium.

audio response unit – Device that provides natural language instructions and/or responses to input from the keypad of a telephone or other device.

average access time – Statistical average (or mean) elapsed time required by a storage device to respond to a read or write command.

back-end computer assisted software engineering tool – (1) Synonym of code generator. (2) Software tool that creates program instructions from a system model.

band – Portion of the message carrying capacity of a carrier wave or transmission medium, typically expressed as a range of frequencies (e.g., the band between 200 and 1000 Hz).

bandwidth – Range of carrier wave frequencies that can be transmitted over a transmission medium.

bar code scanner – Device that optically scans surfaces for numbers encoded in sequences of dark bands on a light background.

base address – First physical storage location in which a group of data items or instructions are stored.

base register – Register in which a base address is stored, as typically used by the control unit when performing address resolution.

baseband – (1) Transmission method that uses the entire bandwidth of a communication channel to carry one signal or data stream. (2) Local area network (LAN) transmission method in which multiple digital signals are carried over a single communication channel through time-division multiplexing.

batch – Collection of records (or other input data) assembled into a group of manageable size for processing purposes.

batch processing – Mode of program execution in which all data inputs are processed without interruption or intervention by an interactive user.

BCC – Acronym of block check character.

BCD – Acronym of binary coded decimal.

benchmark – Program or program fragment and related data inputs that are executed to test the performance of a hardware component, a group of hardware and/or software components, or an entire computer system or network.

big endian – Term describing the storage of a multibyte data item with the most significant byte in the storage location with the lowest-numbered address.

binary coded decimal (BCD) – Antiquated 6-bit character coding method used by early IBM computer systems.

binary compatibility – Characteristic of two CPUs or computer systems that allows machine language programs to execute on both with equivalent behavior.

binary editor – Editor that allows the addition, deletion, or modification of bit patterns within a file.

binary search – Efficient search method for sorted data sets that repeatedly divides the search space in half until the desired data item is located.

binary signal – Signal in which one of two values (i.e., zero and one) can be encoded by modulating a carrier wave parameter.

binding – Synonym of link editing.

bit – (1) Value represented in one position of a binary number. (2) Number that can have a value of zero or one. (3) Abbreviation of Binary digIT.

bitmap – Image representation in which pixel content is represented by one or more bits with a set of bits comprising all the pixels of the image.

bit position – Position or location of a single bit within a byte, word, or bit string.

bit string – (1) Sequence of binary digits that forms the encoded representation of a single data item value. (2) Group of binary digits in which each digit is a separate boolean data item (or flag).

bit time – Time interval during which the value of a single bit is present within a carrier wave.

bitwise – Adjective applied to a Boolean processing operation (e.g., bitwise OR) that indicates that the operation should be applied to each bit position as if it were a separate Boolean data value.

block – (1) Series of logical records grouped on a storage device for efficient processing, storage, or transport. (2) Unit of data transfer between a storage device and other computer hardware device. (3) Portion of a program that is always executed as a unit.

block check character (BCC) – Extra 8-bit value containing error-detection information that is appended to the end of a set (or block) of transmitted data items.

block size – Number of bits, bytes, or records within a single storage block.

blocked state – State of a process that is waiting for an event such as completion of another process, arrival of data input, or correction of an error condition.

blocking – Storage of multiple data items or data item fragments within a single storage block or allocation unit.

blocking factor – The number of bits, bytes, or logical records grouped within a single physical record, storage block, or allocation unit.

Boolean (data type) – Data item that may have only the values True or False.

boot up – Initialization sequence for a hardware device including configuration, error checking, and loading and configuration of devices drivers and operating software.

branch operation – (1) Processor operation that causes an instruction other than the next sequential instruction to be fetched and executed. (2) Processor instruction that implements a branch operation by overwriting the current value of the instruction pointer.

branching – The act of or ability to execute a branch instruction.

bridge – Device that copies packets from one network segment to another.

broadband – (1) Communication channel with a wide frequency range and high data-transmission capacity. (2) Network transmission method in which multiple digital signals are carried over a single channel through frequency division multiplexing.

broadcast mode – Transmission method that sends a message from a sender to all devices on the communication channel.

brownout – Reduction in voltage by an electrical power provider, usually because of temporary excess of power demand over power supply.

buffer – Primary storage used to temporarily store data in transit from a sender to a receiver.

buffer overflow – Receipt of more data than can be held in a buffer.

bus – Communication channel shared by multiple devices within a computer system or network.

bus arbitration unit – Processor that mediates competing demands for control of or access to a bus.

bus clock – Clock circuit that generates timing pulses (or ticks), which are transmitted to all devices attached to a bus.

bus clock rate – Rate at which a bus clock generates timing pulses.

bus cycle – (1) Period of time required to perform one data transfer operation on a bus. (2) Elapsed time between two consecutive pulses of a bus clock.

bus line – Single transmission line within a bus.

bus port – Single connection point on a bus for a peripheral device or device controller.

bus master – (1) Any device attached to a bus that is capable of initiating a data-transfer operation or sending a command to another device. (2) Device that controls a data-transfer operation during a specific bus clock cycle.

bus protocol – Communication protocol used on a bus.

bus slave – (1) Recipient of a command or data from a bus master. (2) Device that is incapable of acting as a bus master.

bus topology – With respect to data communication networks, a network implementation that uses a bus as a communication channel.

byte – String of 8 bits.

cache – Area of high-speed memory that holds portions of data also held within another storage device. A cache is used to improve the average speed of read and write operations to/from its associated storage device.

cache controller – Hardware device that manages the contents of a cache.

cache hit – Access to data that resides in a storage device and in its cache.

cache miss – Access to data that resides in a storage device but not in its cache.

cache swap – Operation that copies data from a cache to its associated storage device and then overwrites the data copy in the cache with new data.

capital expenditures – Funds expended to obtain capital resources.

capital resources – Resources expected to provide benefits beyond the current operating period (e.g., fiscal year).

carrier wave – Energy wave (e.g., electricity or light) that carries encoded messages within a communication channel.

Carrier Sense Multiple Access Collision Detection (CSMA/CD) – Media access control (MAC) and transmission protocol that allows message collisions to occur but defines a method of collision recovery.

CASE – Acronym of Computer Assisted Software Engineering.

case statement – Control structure in a high-level programming or job control language (JCL) that chooses among several alternate processing paths, depending on the value of a variable or expression.

cathode ray tube (CRT) – (1) Specific technology for implementing video display using beams of electrons that excite phosphor(s) on the inner surface of a vacuum tube. (2) Generic term for a video display terminal (VDT) or monitor.

CD – Short acronym of Compact Disc read only memory.

CDROM – Acronym of Compact Disc Read Only Memory.

central processing unit (CPU) – Hardware component of a computer system that executes program instructions.

channel – (1) Synonym of input/output (I/O) channel. (2) Shortened form of communication channel.

character – (1) Primitive and indivisible component of a written language. (2) One byte of data.

character framing – Serial data communication technique that groups bit patterns into bytes (coded characters) for purposes of transmission and error detection.

chief information officer – The manager charged with responsibility for planning, maintenance, and operation of all information-processing resources within an organization.

child process – Process created and controlled by another executing process (i.e., the parent).

chromatic depth – Synonym of chromatic resolution.

chromatic resolution – (1) Number of bits used to describe the color of each pixel in a graphic display. (2) Number of different colors that may be assigned as the value of any single pixel.

circuit switching – Technique for allocating circuits to senders and receivers whereby a sender and receiver are granted exclusive use of a communication channel for a time period by means of switching technology.

CISC – Acronym of Complex Instruction Set Computing.

class library – Software library composed of object-oriented classes that may be used directly or from which new classes may be derived.

client – Program or computer that requests and receives services from another program or computer through network communication.

Client/server architecture – Organization of software and hardware components as clients (service requesters) and servers (service providers) that interact through a communication network.

clock rate – (1) Rate at which clock pulses (or clock ticks) are generated by clock circuitry, stated in Hertz. (2) With respect to a central processing unit (CPU), the time required to fetch and execute the fastest instruction.

close operation – Operation that severs the relationship between a file and a process, flushes file I/O buffers, and deletes operating system data structures associated with the file.

CMOS – Acronym of Complementary Metal Oxide Semiconductor.

CMY – Acronym of Cyan Magenta Yellow.

CMYK – Acronym of Cyan Magenta Yellow blacK.

coaxial cable – Transmission medium composed of a single strand of wire (i.e., the signal carrier) surrounded by an insulator, a braided return wire, and a tough plastic outer coating.

COBOL – (1) Third generation programming language used primarily for writing business application software. (2) Acronym of COmmon Business Oriented Language.

code – (1) Synonym of instructions in a program. (2) Representation of a value by symbol(s), signal(s), or a bit string. (3) Act of creating program instructions.

code checker – Software tool that searches for and reports errors in program source code.

code generator – Program that produces source, object, or executable code based on a general description of processing requirements.

coercivity – Ability of an element or compound to accept and hold magnetic charge.

collating sequence – Order of symbols produced by sorting them according to a numeric interpretation of their coded values.

collision – Result of simultaneous transmission by two senders on a shared communication channel.

command interpreter – Component of system software that interactively accepts user commands and executes corresponding commands and processes.

command language – A written language used to direct an operating system's actions.

command layer – Component of system software that accepts user commands and executes corresponding processes, encompassing both batch and interactive methods of program and system control.

command option – Synonym of command parameter.

command parameter – Parameter (i.e., input data or instruction) to an operating system command.

command-line interface – Operating system command layer user interface based on a command language.

communication channel – Combination of a transmission medium and communication protocol.

communication protocol – Set of communication rules encompassing data and command representation, bit encoding and transmission, transmission media access, clock synchronization, error detection and correction, and message routing.

compact disc (CD) – Synonym of compact disc read only memory.

compact disc read only memory (CDROM) – Standardized format for optical data storage on a single-sided disk.

compaction – Process of reallocating storage locations assigned to a file to eliminate empty storage locations.

compensating errors – Errors within the same message that counterbalance one another, thus causing a failure of an error-detection method.

competitive advantage – Use of resources to provide better or cheaper services than others can provide with those same resources.

compiler – Program that translates high-level programming language source code instructions into object code.

compiler library – Set of object-code modules intended to be combined with object-code modules produced by a compiler.

complement – Number substituted for another number to which it has an inverse relationship. For example, in the binary number system, zero is the complement of one.

complementary arithmetic – Implementation of arithmetic operations using complementary values of one or more operands.

complementary metal oxide semiconductors (CMOS) – (1) Class of semiconductors implementing using two or more metal oxides to form a transistor. (2) Synonym of flash ROM or battery-backed RAM used to store hardware configuration information in an IBM-compatible personal computer (PC).

complete path – (1) File or object name that includes all associated directory and subdirectory names within a device or global directory. (2) Named path from the root node of a directory to a specific file through intervening layers of a hierarchical directory structure.

complex instruction set computing (CISC) – (1) Use of a processor instruction set that defines some instructions that are combinations of other processor instructions. (2) Processor architecture that uses complex instructions that embed many primitive processing operations in contrast to reduced instruction set computing.

complex machine instruction – Processor instruction that results in the execution of multiple processor instructions or multiple primitive processing operations.

compression – Reduction in data size using a compression algorithm.

compression algorithm – Algorithm by which data inputs may be translated into equivalent outputs of smaller size.

compression ratio – Ratio of data size before and after applying a compression algorithm.

computer assisted software engineering (CASE) – Process of analyzing, designing, and/or developing computer software with the assistance of automated tools.

computer assisted software engineering tool – Program or integrated set of programs used for computer assisted software engineering.

computer engineering – Field of study encompassing the design and implementation of computer hardware.

computer operations manager – Manager responsible for the operation and maintenance of a large data-processing center or information system (IS).

computer operator – Person responsible for controlling computer operations and responding to processing errors, typically in a batch processing environment.

computer science – Study of the implementation, organization, and application of computer software and hardware resources.

computer system – Integrated set of computer hardware devices and associated system software.

concurrent execution – Execution of multiple processes or threads within the same immediate time frame through interleaved execution on a single processor.

conditional branch operation – (1) Branch operation that is performed only if a boolean variable is True. (2) Processor instruction that implements a conditional branch operation.

conductivity – Ability to serve as a transmission path for an energy wave (e.g., electricity or light).

conductor – Element or compound that exhibits conductivity.

connection-oriented protocol – A communication protocol that explicitly requires a connection to be established between sender and receiver prior to transmitting any data.

connectionless protocol – Communication protocol that does not require a connection to be established between sender and receiver prior to transmission of any data.

constant angular velocity – Constant disk rotation speed as measured in revolutions per time interval.

constant linear velocity – Variable disk rotation speed that maintains a constant rate of linear recording surface passing beneath the read/write head.

contiguous memory allocation – Storage of a set of data elements in sequential (i.e., contiguous) physical storage locations.

control bus – Portion of a bus used to transmit command and status signals among devices.

control character – Command, coded in a manner similar to printable characters, sent to a storage or I/O device (e.g., ASCII character codes 0 through 31).

control structure – High-level programming or job control language (JCL) statement that describes the selection or repetition of other program statements (e.g., an If-Then or While-Do statement).

control unit – Component of a central processing unit (CPU) that is responsible for data movement, instruction fetching and decoding, and control of the arithmetic logic unit and system bus.

core memory – Form of primary storage implemented as a lattice of wires with iron rings wrapped around each wire junction point.

CPU – Acronym of Central Processing Unit.

CRC – Acronym of Cyclical Redundancy Check.

CRT – Acronym of Cathode Ray Tube.

CSMA/CD – Acronym of Carrier Sense Multiple Access Collision Detection.

current directory – Directory that serves as the origin for file and directory search operations.

cursor – Visual symbol on a video display (e.g., a colored box or underline) that indicates the current input or output position.

cyan magenta yellow black (CMYK) – Primary subtractive colors, plus black, as used in color printing.

cyan magenta yellow (CMY) – Primary subtractive colors, as used in color printing.

cycle – Single operation, function, or transfer within a series of similar operations, functions, or transfers, as performed by a processor, storage device, or communication channel.

cycle time – Duration of one cycle.

cyclic redundancy checking (CRC) – Technique for detecting errors in data transmission based on redundant data computed with a complex algorithm and appended to the original message.

D/A – Abbreviation of Digital to Analog.

DAC – Acronym of Digital to Analog Converter.

data bus – Portion of a bus used to transmit data among devices.

data declaration – High-level programming language statement that declares the existence, name, type, and other characteristics of a data element or structure.

data dictionary – Stored set of data declarations or definitions within a database management system (DBMS) or as part of a system model.

data encoding – (1) Representation of data values as a bit string. (2) Process of embedding data or bit values within a carrier wave.

data flow diagram (DFD) – (1) Graphic model of information-processing requirements composed of data sources, data sinks, processes, files (i.e., data stores) and data flows.

data link layer (DLL) – Layer of network system software responsible for media access and bit transmission.

data manipulation tool – Approximate synonym of a database management system (DBMS) or one of its component tools.

data operations – Any processing operation that uses or alters data content.

data structure – Data item composed of multiple primitive data elements (e.g., arrays, linked lists, and records).

data transfer rate – Rate at which data is transmitted over a transmission medium or communication channel, as measured in data units per time interval.

database – Stored set of data elements, organized to facilitate access and manipulation by multiple users and/or application programs, generally through the facilities of a database-management system.

database administration – Process of ensuring data integrity, reliability, security, and availability within a database.

database management system (DBMS) – Set of system software programs that provides access to and administration of a database.

datagram – Smallest unit of data transfer defined by the Transmission Control Protocol.

dB – Abbreviation of deciBel.

DBMS – Acronym of DataBase Management System.

debugger – Software tool (or set of tools) that assist a programmer in locating and correcting program errors through actual or simulated execution.

debugging version – Program that contains data and instructions to aid in locating and correcting errors.

deciBel – Unit of acoustic or carrier wave amplitude measured on a logarithmic scale.

decoding – (1) Process of separating an instruction into its component parts (i.e., op code and operands) and routing data and control signals to their appropriate destinations (e.g., registers and the ALU). (2) Process of extracting data content from a carrier wave.

decryption – Opposite of encryption.

dedicated mode – Mode of operation in which a device controller interacts with a single high-speed storage or I/O device.

delete operation – (1) Removal of a data element from within a data structure. (2) Removal of a record in a file. (3) Removal of a file by deleting its directory entry and deallocating its storage.

device controller – Special purpose processor that controls the physical actions of one or more storage or I/O devices.

device driver – Component of the operating system kernel dedicated to control of and communication with a single hardware device.

device status – Command or signal describing the ability of a device to respond to service requests (e.g., ready and busy).

DFD – Acronym of Data Flow Diagram.

dialogue – Pattern of interactive communication between a user and computer software.

differential backup – Redundant data store containing copies of changes to data content that have occurred since the most recent incremental or full backup.

digital audio tape (DAT) – Data storage format that encodes audio data as digital information on magnetic tape.

digital computer – Computer that represents and manipulates data as digital (usually binary) signals.

digital processing – Manipulation of data values represented as digital signals.

digital signal – (1) Signal in which one of a discrete (countable) set of values can be encoded by modulating one or more carrier wave parameters. (2) Synonym of binary signal.

digital signal processor (DSP) – Processor that manipulates continuous streams of digital data (e.g., audio or video data streams).

digital to analog (D/A) conversion – Translation of a continuous stream of digital data (e.g., audio or video) into an equivalent analog signal.

digital to analog converter (DAC) – Processing device that performs digital to analog conversion.

digitizer – (1) Device that captures the position (row and column) of a pointing device as input data. (2) Synonym of optical scanner.

DIMM – Acronym of Double In-line Memory Module.

direct access – Ability of a storage device to access storage locations directly, or in any desired order.

direct file access – Direct access to records or bytes of a file.

direct file organization – Method of physically organizing file contents such that the value of a field determines the physical location of the record containing that field value.

direct memory access (DMA) – Method of data transfer that allows direct data movement between main memory and a storage or I/O device, bypassing the central processing unit (CPU).

direct memory access controller – Dedicated processor that handles the processing overhead of data transfers during direct memory access (DMA) operations.

discrete signal – Synonym of digital signal.

disk defragmentation – Reorganization of data on disk drive so that directory and file content occupies contiguous storage locations.

disk mirroring – Technique that ensures the integrity of disk content by storing a redundant copy of all data on a second disk drive.

diskette – Small, removable magnetic disk storage medium encased in a protective cover.

dispatching – (1) Act of giving a process or thread control of the central processing unit (CPU). (2) Act of moving a process or thread into the running state.

displacement – Synonym of offset.

display attribute – Term describing a special feature of printed or displayed output (e.g., background color and boldface type).

display list – Group of high-level commands that describe graphic and/or textual images (e.g., a Postscript program).

display object – Primitive graphic object, as described and/or manipulated within a display list.

distortion – The undesirable alteration of one or more carrier wave characteristics resulting from interaction between signal energy and the transmission-medium or signal-propagation equipment.

distributed file system – Set of files distributed among multiple computer systems on a network and the system software by which file content is accessed.

distributed processing – Distribution and execution of processes or threads over multiple nodes of a computer network.

distributed system – (1) An information system (IS) consisting of multiple software and hardware components distributed across a computer network. (2) A set of computer hardware devices distributed across a computer network that can cooperate to execute a single program or support a single IS.

distribution version – Synonym of production version.

dithering – Simulation of continuous color tones using patterns of dots containing a finite set of colors.

DMA – Acronym of Direct Memory Access.

dot matrix printer – Impact printer that generates printed characters using a matrix of print pins and an inked ribbon.

dot pitch – Height and/or width of printed characters as measured in dots or pixels per inch.

dots per inch (dpi) – Measure of print or display resolution (pixel density) for a printing or video display device.

double in-line memory module (DIMM) – Unit of primary storage consisting of a small printed circuit board with memory chips and electrical contacts on both sides.

double precision – Representation of a numeric value with twice the usual number of bit positions for greater accuracy or numeric range.

doubly linked list – Set of stored data items in which each element contains pointers to both the previous and next list elements.

dpi – Acronym of dots per inch.

DRAM – Acronym of Dynamic Random Access Memory.

drive array – Set of disk drives that are managed as if they were a single storage device.

DSP – Acronym of Digital Signal Processor.

dyadic instruction – Instruction that uses two operands as input.

dynamic analysis – Use of models and/or model analysis techniques that incorporate parameter variability and random events in the representation of system behavior.

dynamic connection – Connection between a client and a server or remote resource that is not established until the client accesses the server or resource.

dynamic random access memory (DRAM) – Type of random access memory that implements bit storage with a combination of transistors and capacitors.

dynamic resource allocation – Allocation of resources to processes as the resources are requested or accessed.

EBCDIC – Acronym of Extended Binary Coded Decimal Interchange Code.

edit operation – Any operation in which the content of a stored data item or structure is altered.

editor – Software utility program used to create or modify the contents of files.

EDO (RAM) – Acronym of Extended Data Out Random Access Memory.

EEPROM – Acronym of Electronically Erasable Programmable Read Only Memory.

effective data transfer rate – Error-free throughput of a communication channel under normal conditions with a specific communication protocol.

electroluminescent display – Display device that generates light by the application of electrical power to solid-state elements coated with phosphors.

electromagnetic interference (EMI) – Alteration of the characteristics of an electrical carrier wave caused by external electrical or magnetic phenomena (e.g., electric motors or sun spots).

electronically erasable programmable read only memory (EEPROM) – Read only memory device that can be erased and written by sending appropriate control signals.

EMI – Acronym of ElectroMagnetic Interference.

encoding – Process of embedding data content in a carrier wave.

encryption – Altering data content to prevent unauthorized access, generally implemented as an algorithmic transformation of data content based on an encryption key.

encryption key – Data value used as input to an encryption algorithm.

EPROM – Acronym of Erasable Programmable Read Only Memory.

erasable programmable read only memory (EPROM) – Read only memory device that can be rewritten by first exposing its circuitry to ultraviolet light.

escape sequence – Command to a storage or I/O device consisting of multiple characters, the first of which is an escape (ASCII 27) character.

Ethernet – Specific type of local area network that employs twisted pair or coaxial cables and the carrier sense multiple access collision detection media access protocol.

even parity – Error-detection method that appends a parity bit to each transmitted or stored character. The value of the parity bit is one if the number of one valued bits in a character is even; it is zero otherwise.

excess notation – System of coding integers as bit strings such that the value zero (the midpoint of the numeric range) is represented by all zeros except for the most significant bit (which contains a one).

exclusive OR (XOR) operation – (1) Boolean operation that generates the value True if only one of its inputs is True. (2) CPU instruction that generates the value True if only one of its inputs is True.

executable code – Program that consists entirely of CPU instructions that are ready to be loaded and executed.

execution – (1) Process of carrying out an instruction in the central processing unit (CPU) and completing the operation it specifies. (2) Same process described in (1) as applied to all instructions in a program.

execution cycle – Portion of a central processing unit (CPU) cycle in which an instruction is executed and the processing result is stored in a register.

explicit file transfer – Transfer of files between machines by a communication network as a result of an explicit user request.

explicit prioritization – Process scheduling through explicitly stated priority levels and a corresponding scheduling algorithm.

exponent – (1) Portion of the internal representation of a real number. (2) Numeric value representing the power (or exponentiation) of another value.

expression – Within a program, a formula consisting of constants and/or variable names.

extended data out (EDO) random access memory – Form of dynamic random access memory that improves read access time by fetching and storing the data word following the most recently read data word.

Extended Binary Coded Decimal Interchange Code (EBCDIC) – Standard coding system for representing character data in an 8-bit byte format, most commonly used within IBM mainframe computers.

fast page mode (FPM) RAM – Form of dynamic random access memory that improves read access time by activating the read circuitry for the next word as soon as the preceding word has been read.

father – Original version of a file, after updates have been applied to generate a new version (the son).

fault tolerance – Characteristic of a computer system or file-management system that allows rapid recovery from the failure of a hardware component without data loss.

FCFS – Acronym of First Come First Served.

FDM – Acronym of Frequency Division Multiplexing.

ferroelectric RAM – Type of nonvolatile random access memory that uses iron to store bit values.

fetch cycle – Portion of a central processing unit (CPU) cycle in which an instruction is loaded into a register and decoded.

fiber optic cable – Transmission medium for optical-signal propagation, generally consisting of one or more plastic or glass fibers sheathed in a protective plastic coating.

field – Component data items of a record.

FIFO – Acronym of First In First Out.

file – (1) Collection of related records. (2) Fundamental unit of data storage on secondary storage devices.

file allocation table – Index that records the allocation of storage device locations to files.

file backup – (1) Copy of a file made to allow recovery in case of loss of or damage to the original. (2) Act of creating the copy described in (1).

file control block – Data structure created by the operating system when a file is opened that is used to store information about the file.

file control layer – System software layer that accepts file-manipulation requests from application programs and translates them into corresponding low-level processing commands to the CPU and secondary storage device drivers.

file create operation – Act of creating a directory entry for a file and allocating storage space to it.

file delete operation – Act of deleting a directory entry for a file and deallocating its storage space.

file destruct operation – File delete operation immediately followed by overwriting the deallocated storage locations with null values.

file lock – Operation that allocates a file to a single process and prevents access by any other process.

file management system – Combination of the file control layer and any file or secondary storage-management utility programs provided by the operating system.

file migration – Management technique for secondary storage in which older versions of a file are automatically moved to less costly storage media (e.g., magnetic tape).

file packing – Process of compacting a file's contents.

file specification – Written description of the structure and content of a file within a system model.

file type – Variable that describes the content of a file (e.g., text, binary, indexed, or executable).

file undelete operation – Process of recovering a previously deleted file by reinitializing its directory entry and reclaiming its previously allocated storage locations (and their contents).

file-by-file dump – Operation that creates a full backup by individually copying each directory and file, resulting in a copy with compacted and defragmented storage allocation.

firmware – Software that has been permanently stored on read only memory devices.

first come first served (FCFS) – Scheduling method that services processes in order of their request arrival.

first in first out (FIFO) – (1) Processing or scheduling order equivalent to first come first served. (2) Mode of data storage in which items are read in the same order they were written.

fixed length instruction – One member of an instruction set in which all instructions contain the same number of bits.

flash memory – Modern form of electronically erasable read only memory that requires relatively little time to update memory contents.

flash random access memory – Synonym of flash memory.

flat memory model – Memory organization and access method in which memory locations are assigned a single unsigned integer address corresponding to their linear positions.

flat panel display – Display device that is thin, flat, and lightweight.

flip flop circuit – (1) Electrical circuit that can be switched between two states, thus representing the binary values zero and one. (2) Basic component of static random access memory and processor registers.

floating point notation – Method of encoding real numbers in a bit string consisting of two parts—a mantissa and exponent.

floating point operation – Any processor instruction that uses floating point (i.e., real) numbers as data inputs.

floppy disk – Synonym of diskette.

FM – Acronym of Frequency Modulation.

font – Named set of display formats for printable characters and/or symbols with a similar appearance or style.

form based interface – Style of user interface in which the user is prompted interactively for input by the display of a blank form (or template) on a video display device.

FORTRAN – (1) Third generation programming language used primarily to write scientific and mathematical application software. (2) Acronym of FORmula TRANslator.

formulaic problem – Processing problem that can be represented formulaically and solved without the use of branching operations or control structures.

fourth generation application development (4GAD) – Integrated set of tools for developing and testing programs written in a fourth generation (programming) language.

fourth generation language (4GL) – Programming language that supports nonprocedural programming, database manipulation, and advanced I/O capabilities.

FPM (RAM) – Acronym of Fast Page Mode Random Access Memory.

fragmentation – File or storage device characteristic in which storage locations allocated to one or more files are scattered throughout a storage device in noncontiguous locations.

frame – (1) Synonym of packet. (2) Single video image from a set of images comprising a motion video sequence.

frequency division multiplexing (FDM) – Communication channel sharing technique in which a single broadband channel is partitioned into multiple subchannels (frequency bands), each of which can carry a separate data stream.

frequency modulation (FM) – Data transmission method that encodes data values as variations in the frequency of an analog carrier wave.

frequency shift keying (FSK) – Synonym of frequency division multiplexing.

frequency – Number of complete waveform transitions (i.e., change from positive to negative to positive energy peak) that occur in one second.

front end computer assisted software engineering tool – Program that supports the creation and modification of system models.

FSK – Acronym of Frequency Shift Keying.

full backup – Data store containing copies of all data in a directory or storage device.

full duplex – Type of communication channel composed of two transmission media that allows simultaneous communication in both directions.

fully qualified file reference – Synonym of complete path with respect to a specific file.

function – Named segment of a high-level language program that is always executed as a unit.

gallium arsenide – Compound used to implement devices that have both electrical and optical properties.

gas plasma display – Display technology that generates light by applying electricity to trapped bubbles of neon gas.

gate – Processing device that implements a primitive Boolean operation or processing function by physically transforming one or more input signals.

gateway – Computer system or hardware device that is connected to two network segments and forwards packets from one segment to another.

general purpose processor – Processor with capabilities that can be applied to perform a wide variety of processing tasks.

gigabyte – 1024^3 bytes.

grandfather – Version of a file prior to the father version.

graph directory structure – (1) Directory structure in which allowable relationships between directories can be represented as a graph. (2) Directory structure in which a file or subdirectory can be contained within two or more directories.

grayscale value – Numeric value representing the level of monochrome (i.e., white) brightness for each pixel.

Grosch's Law – Outdated statement of the mathematical relationship between computer size and cost per unit of instruction execution that states that cost per executed instruction decreases as computer system size increases.

half duplex – Type of communication channel that uses only a single transmission medium where sender and receiver take turns transmitting data to one another.

halt instruction – Instruction that terminates the current sequence of execution and suspends the execution of any further instructions by the central processing unit (CPU).

hard disk – (1) Storage medium consisting of a rigid (usually metal) platter coated with a metallic oxide on which data are recorded as patterns of magnetic charge. (2) Synonym of hard disk drive.

hard disk drive – Storage device incorporating one or more hard disk platters, read/write heads, and low-level control circuitry.

hard drive – Synonym of hard disk drive.

hardware independence – Independence of a program or processing method from the physical details of computer system hardware.

hardware monitor – Program or device that records and reports processing and/or communication activity within or between hardware devices.

head to head switching time – Time interval required to switch shared read/write circuitry between two adjacent read/write heads.

heat dissipation – Act of conducting heat away from a device, thus reducing its temperature.

heat sink – Thermal mass placed in direct contact with a processing or other device to improve heat dissipation.

Hertz (Hz) – Unit of measure for signal frequency defined as one cycle per second.

Hewlett Packard Graphics Language (HPGL) – Specific display list language used to describe printed output for Hewlett Packard laser printers and plotters.

hexadecimal notation – Numbering system with a base value of 16, which uses digit values ranging from 0 to 9 and from A to F (corresponding to decimal values of 0 to 15).

hierarchic directory structure – Multilevel system of directories in which directories and files may be related to one another in a hierarchy or inverted tree.

High-level programming language – Programming language that uses symbolic statements to represent computer operations and memory addresses and in which a single instruction generates multiple machine instructions.

high order bit – Synonym of most significant digit.

high sierra format – Standard format for storing a file system on a compact disc.

home directory – Primary directory associated with (and owned by) a single user.

HPGL – Acronym of Hewlett Packard Graphics Language.

Hz – Abbreviation of Hertz.

I/O – Acronym of Input/Output.

IA5 – Acronym of International Alphabet 5.

IC – An acronym of Integrated Circuit.

ICMP – An acronym of Internet Control Message Protocol.

icon – Visual (graphic) representation of an object or command as displayed on a video display device.

Institute of Electrical and Electronics Engineers 802 standards – Group of standards for data communication networks.

IEEE – Acronym of Institute of Electrical and Electronics Engineers.

if statement – High-level programming language statement that executes a set of instructions only if a stated condition is true.

image description language – Language that symbolically describes the content of a printed or displayed image (e.g., Postscript).

implementation model – System model that describes a specific automated method of satisfying information-processing requirements.

in-line function call – Call to a function that is replaced by the executable instructions of the called function.

inclusive OR operation – Synonym of OR operation.

incremental backup – Data store containing copies of data sets (e.g., files) that have been altered since the most recent incremental or full backup.

index – Stored set of paired data items. The first data item in each pair is a key value, and the second is a pointer to the location of the data item possessing (or corresponding to) that key value.

indexed file access – Method of accessing records in a file by searching an index for the value of a key field.

indexed file organization – Method of file organization in which both records and an index to those records are maintained as a single unit.

indirect addressing – Any addressing method where memory references made within a program do not necessarily correspond to physical memory storage locations.

infrared light – (1) Electromagnetic energy waves approximately within the frequency range 1 to 100 teraHertz. (2) Heat.

infrastructure – Facilities or resources that provide services that are pervasively available and commonly used by a large numbers of users.

initiator – Device connected to a SCSI bus that is currently designated as the bus master.

Ink-jet printer – Printer that produces printed images by spraying small drops of liquid ink onto paper.

input focus – Condition of a process or window by which it receives all input from a computer system keyboard and/or pointing device.

input/output (I/O) – Actions and mechanisms by which data and commands are communicated between a user and software or between computer systems and/or their components.

input/output channel – (1) Device controller dedicated to a mainframe bus port that allows many devices to share access to (and the capacity of) the port. (2) Specific hardware component of an IBM mainframe computer system.

input/output device – Any hardware device that implements or facilitates communication between computer systems, computer system components, or between users and software.

input/output port – (1) Synonym of bus port. (2) I/O device connection port on an I/O channel.

input/output re-direction – Act of rerouting process input or output to alternate devices, files, or storage locations.

input/output wait state – An idle processor cycle consumed while waiting for data transmission from a secondary storage or input/output (I/O) device.

input/output unit – Synonym of input/output (I/O) device.

insert operation – Addition of a data item or record in a storage location or ordinal position other than last.

Institute of Electrical and Electronics Engineers (IEEE) – Professional organization that promulgates many standards for computer hardware and system software.

instruction – Bit string containing an operation code and one or more operands that cause a processor to perform a processing action.

instruction code – Synonym of operation code.

instruction cycle – Synonym of fetch cycle.

instruction explosion – Correspondence of a single high-level programming language statement to multiple central processing unit (CPU) instructions.

instruction format – Length and order of the operation code and operands within a machine language instruction.

instruction pointer – Register that stores the address of the next instruction to be fetched from primary storage.

instruction register – Register that holds an instruction prior to decoding.

instruction set – Set of all machine language instructions that can be executed by a central processing unit (CPU).

integer – Whole number, or a value that does not have a fractional part.

integer arithmetic – Arithmetic operations that use integer inputs and generate integer outputs.

integrated circuit (IC) – Semiconductor device, manufactured as a single unit, that incorporates one or more gates.

interactive processing – Mode of program operation in which a computer prompts a user for input and suspends execution until the user supplies that input.

interleaved execution – Scheduling technique by which a processor alternates instruction execution among multiple active processes.

International Alphabet Number 5 (IA5) – International equivalent of ASCII.

International Standards Organization (ISO) – International body with functions similar to those of the American National Standards Institute.

Internet Control Message Protocol (ICMP) – Internet protocol for the exchange of messages among gateways and routers.

Internet Packet Exchange (IPX) – Novell Netware protocol for packet switching and routing.

Internet Protocol (IP) – Internet protocol for packet switching and routing.

Internet site name – Symbolic name that uniquely identifies an Internet node (e.g., www.unm.edu).

interpretation – Process for translating and executing a high-level language program that interleaves statement translation and execution.

interpreter – Program that performs interpretation.

interprocess communication – Communication of data values or status information among multiple active processes.

interprocess synchronization – Coordination of two or more active processes through the exchange of data and/or status information.

interrupt – Signal to the CPU that some event requires its attention.

interrupt code – Numerically coded value of an interrupt, indicating the type of event that has occurred.

interrupt handler – Program or subroutine that is executed in response to an interrupt.

interrupt register – Register in the control unit that stores an interrupt code received over the bus or generated by the CPU itself.

interrupt table – Index of interrupt codes and interrupt handler memory addresses.

IP – Acronym of Internet Protocol.

IPX – Acronym of Internet Packet Exchange.

ISO – Acronym of International Standards Organization.

Java – Object-oriented programming language used to construct programs that can be executed on different hardware and system software platforms.

Java virtual machine (JVM) – Interpreter that translates and executes Java programs.

JCL – Acronym of Job Control Language.

job – Group of programs and job control language (JCL) statements that are executed as a unit.

job control language (JCL) – Control language used to direct the batch execution of programs or groups of programs.

job queue – Synonym of run queue.

job step – Well-defined segment of a job, usually consisting of one program and directly related job control language (JCL) statements.

job stream – Processing sequence specified in a job control language (JCL).

Joint Photographic Experts Group (JPEG) – (1) Organization that promulgates image representation standards. (2) Specific method of compressed image representation.

journaling – Method of file system update in which all changes to file and directory content are also written immediately to a separate log (or journal), which is used to recover lost data in the event of a system crash.

JPEG – Acronym of Joint Photographic Experts Group.

jump – Synonym of branch operation.

JVM – Acronym of Java Virtual Machine.

kernel – Portion of the operating system that manages resources and directly interacts with computer hardware.

key – (1) Access control field that uniquely identifies a record or classifies it as a member of a category of records within a file. (2) In cryptography, a sequence of symbols that control the operations of encryption algorithm.

kilobyte – 1024 bytes.

LAN – Acronym of Local Area Network.

land – Flat bit area on an optical disk that is highly reflective.

Large-scale integration (LSI) – Semiconductor devices (chips) containing thousands of transistors or other primitive processing circuits.

laser – (1) Acronym of Light Amplification by Stimulated Emission of Radiation. (2) Beam of strong coherent light.

laser printer – Printer that uses a laser to charge areas of a photoconductive drum and paper.

last in first out (LIFO) – (1) Processing order in which the most recent arrival is served or processed first. (2) Mode of data storage in which items are read in the same order they were written.

Latin-1 – Standard character coding table containing the ASCII character set in the first 128 table entries and most of the additional characters used by Western European languages in the upper 128 table entries.

least significant byte – Byte within a multiple byte data item that contains the digits of least (smallest) magnitude.

least significant digit – Bit position within a bit string of the least (smallest) magnitude.

level of abstraction – One of a series of levels within a top-down process that breaks a problem or processing task into increasingly more detailed subproblems.

level one (L1) cache – Primary storage cache implemented within the same chip as a processor.

level two (L2) cache – When two levels of primary storage cache are in use, the cache implemented outside of the microprocessor.

library call – (1) Service request by an application program that causes the execution of an operating system program (library program). (2) Programming language statement that causes the execution of a compiler or interpreter library program.

LIFO – Acronym of Last in First Out.

line conditioner – (1) Hardware device that ensures constant voltage and amperage in an electrical power circuit. (2) Device that ensures a stable and error-free signal in a communication network.

line turnaround – Signal sent between a sender and receiver that causes them to reverse roles (i.e., the sender becomes the receiver and vice versa).

linear address space – Logical view of a secondary storage or I/O device as a sequentially numbered set of storage locations.

linear list – Set of data elements or structures stored in sequential storage locations.

linear search – Method of searching a series of data items in which items are sequentially read until the desired item is located.

link – Synonym of pointer.

link editing – Process of combining multiple object code modules into an integrated set of executable code with a consistent scheme of memory addresses and references.

link editor – Program that performs link editing.

link map – Listing of module and data memory addresses produced by a link editor.

linkage editor – Synonym of link editor.

linked allocation – Method of noncontiguous allocation within which physically separated blocks are correlated by a set of links or pointers to form a single set of storage locations.

linked list – Set of stored data items in which each element contains a pointer to the previous or next list element.

linker – Synonym of link editor.

linking – Synonym of link editing.

liquid crystal display (LCD) – Display device that uses liquid crystals that can be changed from transparent to opaque.

little endian – Term describing the storage of a multibyte data item with the least significant byte in the storage location with the lowest-numbered address.

load balancing – Movement of suspended or pending processes from one CPU or computer system to another to balance demand for CPU resources.

load operation – Act of copying a word from primary storage to a register.

load sharing – Synonym of load balancing.

loader – Program that copies a program (executable code) from secondary storage into primary storage and then transfers control to the first instruction.

loading – Act of copying an application program from secondary storage into main memory for execution.

local area network (LAN) – Network that spans a limited area such as a single building or office floor.

location transparency – Characteristic of software such that resource access is implemented identically whether the resource is part of the local computer system or located within a remote computer system.

logic circuit – Synonym of gate.

logic instruction – Instruction that implements a boolean operation or comparison operation.

logical – Characteristics of objects and methods that are apparent to users and/or application programs, as opposed to the physical objects and methods upon which they are based.

logical access – Access to a storage location expressed in terms of a linear address space.

logical delete operation – Deletion of a data item by deleting address references (e.g., pointers) to it without actually deallocating or overwriting the corresponding storage location.

logical file structure – Internal organization of a file, as seen and manipulated by a user and/or application program, but as distinguished from the physical organization of file contents.

logical file view – Synonym of logical file structure.

logical model – Model of user-processing requirements that is independent of physical methods of satisfying those requirements, as used in the structured system development life cycle (SDLC).

logical network interface driver – Driver that implements a generic interface to a network interface unit.

logical record – Record formatted and described in terms of a logical file structure.

long integer – Double-precision representation of an integer.

longitudinal redundancy check (LRC) – Scheme of parity checking where parity bits are determined based on equivalent bit positions in a group of characters.

lost update – Modification to a file that is never applied to physical storage because of interference caused by another (simultaneous) update and the use of buffers.

low order bit – Synonym of least significant digit.

LRC – Acronym of Longitudinal Redundancy Check.

LSI – Acronym of Large Scale Integration.

MAC – Acronym of Media Access Control.

machine code – Synonym of executable code.

machine independence – Synonym of hardware independence.

machine language – Language consisting solely of instructions from the instruction set of a specific central processing unit (CPU).

machine primitive – One action from the set of most basic actions of which a processor or hardware component is capable.

machine state – Current processing state of the central processing unit (CPU), as represented by the values currently held in its registers.

magnetic decay – Loss in strength of a stored magnetic charge because of passage of time.

magnetic leakage – Reduction in strength of a stored magnetic charge because of interference from one or more adjacent magnetic charges of opposite polarity.

magnetic tape – Ribbon of acetate or polymer coated with a metallic compound used to store data.

magneto-optical disk – Secondary storage device that reads and writes data bits using a combination of magnetic and optical methods.

main memory – (1) Synonym of primary storage. (2) Set of devices that implement primary storage excluding cache.

mainframe – High-capacity computer system designed to simultaneously support hundreds of interactive users and processes.

mainframe channel – Dedicated I/O processor attached to a bus port of a mainframe computer system.

Manchester coding – Bit encoding method for digital transmission that represents bit values as transitions between voltage states.

mantissa – Portion of the encoded value of a real number that is multiplied by a scaling factor to derive the intended data value.

mark sensor – Input device that recognizes printed marks (e.g., bars or shapes) at predetermined locations on an input document.

MAU – Acronym of Media Access Unit.

media access control (MAC) protocol – Set of rules that regulate access to a transmission medium.

media access unit (MAU) – Control device for token ring networks that serves as a central wiring point, provides network initialization, and reconfigures network connections in the event of a failed network node.

medium scale integration (MSI) – Semiconductor devices (chips) containing hundreds of transistors or other primitive processing circuits.

megabyte – $1,024^2$ bytes.

megaHertz (MHz) – Measurement of wave or clock frequency, one million cycles per second.

memory – (1) Synonym of primary storage. (2) Random access memory devices used anywhere within a computer system, peripheral device, or device controller.

memory allocation – Allocation of primary storage resources to multiple active processes.

memory map – Synonym of link map.

memory-mapped I/O – Method of data transfer to input/output ports in which communication with I/O devices is routed through designated primary storage locations.

memory/storage hierarchy – Set of possible data-storage devices in a computer system and the cost/performance relationships among those devices.

message passing – (1) Method of interprocess coordination and communication used between objects. (2) Method by which a client interacts with a server.

message – (1) With respect to communication networks, a command, request, or response sent from one network node to another. (2) Request sent from one object to another.

method – Process that is embedded within an object and responds to messages sent to it by other objects.

MFLOPS – Acronym of Millions of FLoating point Operations Per Second.

MHz – Acronym of MegaHertz.

microchip – Semiconductor device that implements integrated electronic components in a single unit.

microcomputer – Computer designed to meet low-intensity processing needs of a single user.

microprocessor – Microchip on which all of the components of a central processing unit (CPU) are implemented.

microsecond – One-millionth of a second.

MIDI – Acronym of Musical Information Digital Interchange.

millions of floating point operations per second (MFLOPS) – Measure of processor or computer system speed in terms of the number of floating point computation operations executed per second.

millions of instructions per second (MIPS) – Measure of processor or computer system speed in terms of the number of CPU instructions executed per second.

millisecond – One-thousandth of a second.

minicomputer – Computer system designed to meet the processing needs of a small- to medium-sized group of interactive users (e.g., several dozen).

MIPS – Acronym of Millions of Instructions Per Second.

mirror image backup – File, directory, or storage device copy that is identical in structure and content to the original.

mnemonic – Shortened or abbreviated character name, usually written in uppercase letters.

model – Abstract representation of a system or other physical reality.

modem – (1) Contracted form of the term MOdulator-DEModulator. (2) Device that translates analog signals into digital signals (and vice versa), allowing computer hardware to use voice-grade telephone lines for data communication.

modulator-demodulator – See modem.

modulation – Act of varying the amplitude, frequency, or phase of a carrier wave.

module – Set of instructions in a program that are executed as a unit (e.g., a subroutine).

monadic instruction – Instruction that transforms a single data input.

monitor – (1) Video display device. (2) Hardware or software element that monitors and reports processing or communication activity.

monophonic – Capable of generating only one acoustic frequency and/or timbre (voice) at a time.

Moore's Law – Statement that the transistor density—and therefore power—of microprocessors doubles every 18 to 24 months at no additional cost per unit.

most significant byte – Byte within a multiple byte data item that contains the digits of most (largest) magnitude.

most significant digit – Bit position within a bit string of the most (largest) magnitude.

Motion Pictures Experts Group (MPEG) – (1) Organization that promulgates motion video-image representation standards. (2) Specific method of compressed motion video-image representation.

mouse – Interactive pointing device used to control cursor position and to select objects on a video display.

move operation – Act of copying the contents of one register to another register.

MSI – Acronym of Medium Scale Integration.

multichannel – Any transmission mode that uses two or more communication channels.

multilevel coding – Any data-encoding scheme that encodes multiple-bit values in a single signal event.

multimedia controller – Device controller for multiple audio and/or visual I/O devices.

multimode PCM – Digital transmission method that combines pulse-code modulation with frequency division multiplexing.

multinational character – Character similar to an English language character used only in Western European languages other than English (e.g., ñ and é).

multiple master bus – Bus in which more than one attached device may control access to the bus (i.e., become a bus master).

multiplexing – Merging two or more data streams for transmission over a single communication channel.

multitasking – Ability of an operating system to support multiple active processes.

multithreaded process – Process divided into two or more threads, each of which may be scheduled and executed independently.

Musical Information Digital Interchange (MIDI) – Standard method of encoding control information for musical instruments and synthesizers.

NAK – Acronym of Negative AcKnowledgment.

name server – Network node that responds to requests to supply the physical or logical network address corresponding to the symbolic name of another network node.

NAND gate – Gate combining the primitive functions of NOT and AND.

nanosecond – One-billionth of a second.

NDIS – Acronym of Network Driver Interface Specification.

needs analysis – Synonym of systems analysis.

negative acknowledgment (NAK) – ASCII control character used by receiver to indicate the unsuccessful receipt of transmitted data.

network – Group of computers, peripherals, and office machines that communicate and share resources through one or more communication channels and related software.

network administrator – Manager charged with operating and maintaining a local (LAN) or wide area network (WAN).

network computer – Personal computer (PC) or workstation with no locally stored operating software, application software, or configuration information.

Network Driver Interface Specification (NDIS) – Specific standard for a logical network interface driver.

network hub – Network interface device that serves as the connection point of many computer systems to a network.

network interface card (NIC) – Synonym of network interface unit, particularly with respect to a microcomputer, network computer, or workstation.

network interface unit (NIU) – Physical network interface device used by a single computer system.

network layer – Layer of network system software responsible for routing packets to their destination.

network message – Command, signal, or data item, communicated between two processes by network facilities.

network transparency – Synonym of location transparency.

NIC – Acronym of Network Interface Card.

NIU – Acronym of Network Interface Unit.

noise – Combination of electromagnetic interference and distortion as added to a carrier wave during transmission.

noncontiguous allocation – Any storage allocation scheme under which multiple storage locations allocated to a file or program need not be physically adjacent to one another.

nonmaskable interrupt – Interrupt signal that demands immediate access to the processor and arises from an emergency condition such as a loss of power or failure of a hardware device.

nonvolatile storage – Term describing storage devices that retain their contents indefinitely (e.g., magnetic or optical disk).

NOT operation – (1) Boolean operation that generates the value False if its input is True and True if its input is False. (2) CPU instruction that implements the Boolean NOT operation and stores the result in a register or memory location.

numeric range – Set of all data values that can be represented by a specific data-encoding method.

object – Grouping of data and related processes into a single software unit.

object code – Output of an assembler or compiler that contains machine instructions and unresolved references to external library routines.

object-oriented command interface – Interactive command layer interface with which the user manipulates visual representations of files and programs to specify commands.

object-oriented programming (OOP) – Software construction method or paradigm that configures software as cooperating objects.

object-oriented system development – Any system-development method that models information-processing requirements using objects, classes, and messages.

OCR – Acronym of optical Character Recognition.

octal notation – Numbering system with a base value of 8 that uses digit values ranging from 0 to 7.

odd parity – Error-detection method that appends a parity bit to each transmitted or stored character. The value of the parity bit is one if the number of one valued bits in a character is odd, and zero otherwise.

ODI – Acronym of Open Data Interface.

off-line – Device or storage medium that is not currently active or connected to a computer system or network.

offset – Constant that must be added to every program memory reference to derive the corresponding physical memory address.

offset register – Register in which an offset is stored, as typically used by the control unit when performing address resolution.

online – (1) Mode of software execution in which user input is repetitively accepted and processed for an extended or indefinite period of time. (2) Device or storage medium that is not currently active or connected to a computer system or network.

OO – Acronym of Object Oriented.

OOP – Acronym of Object Oriented Programming.

op code – Synonym of operation code.

open operation – Process of associating a file with an active process by allocating buffers and creating a file control block (FCB).

open systems integration (OSI) model – Standardized architecture for computer network software and hardware consisting of seven predefined layers.

Open Data Interface (ODI) – Specific standard for a logical network interface driver.

operand – Component field of an instruction containing a data value or address used as input to or output from a specific processing operation.

operating expenditures – Funds expended during the current operating period (e.g., fiscal year) to support normal operations.

operating system – Set of software programs that manage and control access to computer resources (e.g., hardware, software, and data).

operation – Synonym of instruction.

operation (op) code – Coded value or bit string representing the function or operation to be performed by a processor.

optical character recognition (OCR) – Methods, programs, and devices by which printed characters are recognized as computer system input or data.

optical disk – Secondary storage device that uses rotating disks of reflective material that can be read by a reflected laser beam.

optical scanner – Device that can scan printed graphic inputs and convert them to computer system input or data.

optical sensor – Device that senses optical input, usually by a reflected laser.

OR operation – (1) Boolean operation that generates the value True if either or both of its inputs are True. (2) CPU instruction that generates the value True if either or both of its inputs are True.

OSI – Acronym of Open Systems Integration.

overflow – Error condition that occurs when the output bit string of a processing operation is too large to fit in the designated register.

packet – Fundamental unit of data communication in a computer network.

packet header – Packet component that provides information for packet routing and for the reassembly of multiple packet messages by the receiver.

packet switching – Form of communication channel sharing in which multiple message streams are broken into packets and packets are individually transmitted and routed to their destination.

packing – Synonym of compaction.

page – Small fixed-size portion of a program, swapped between primary and secondary storage under virtual memory management.

page fault – Condition that occurs under virtual memory management when a program references a memory location in a page not currently held in primary storage.

page frame – Portion of primary storage designated to hold a page under virtual memory management.

page table – Table of pages and information about them maintained by an operating system that implements virtual memory management.

palette – Set (or table) of colors used by a display controller or device.

parallel access – Simultaneous access to multiple portions of a data item through multiple communication channels or transmission media.

parallel transmission – Transmission of multiple signals simultaneously over multiple transmission media or channels.

parallelism – Ability to implement multiple simultaneous processing or I/O operations.

parent process – Process that initiates and controls the execution of another (i.e., child) process.

parity bit – Bit appended to a small data unit (e.g., an ASCII character) that stores redundant information used for error checking. See even parity and odd parity.

parity checking – Act of validating data by recomputing the value of a parity bit.

partitioned memory – Division of primary storage into multiple segments, usually for the purpose of efficient memory allocation to multiple active processes.

password – Protected word or string of characters that identifies or authenticates a user, a specific resource, or an access type.

path – Statement of the location of a file or object within a directory structure.

path name – Synonym of path.

PC – Acronym of Personal Computer.

PCB – Acronym of Process Control Block.

PCM – Acronym of Pulse Code Modulation.

peripheral processing unit – Synonym of I/O channel, particularly as applied to minicomputers and mainframes not manufactured by IBM.

personal computer (PC) – Synonym of microcomputer.

phase – Characteristic of an analog wave that indicates its current cycle position (in degrees) with respect to the cycle origin.

phase difference – Timing difference between two identical waveforms.

phase shift keying (PSK) – Synonym of phase shift modulation.

phase shift modulation – Data-transmission method that encodes data values as variations in the phase of an analog carrier wave.

phoneme – Individual vocal sound comprising a primitive component of human speech.

photoelectric cell – Device that generates electrical current in response to light energy.

physical – Characteristics of objects and methods that are implemented within hardware, as distinguished from the corresponding logical objects and methods that are apparent to users and/or application programs.

physical access – Physical actions required to implement an access to/from a storage location or I/O device.

physical delete operation – Act of physically deleting data from a storage device, usually by overwriting it with null values or another data item.

physical file structure – Physical organization of data in a file on a storage device.

physical layer – Layer of network system software responsible for physical transmission of signals representing data.

physical memory – Physical primary storage capacity in a computer system, in contrast to virtual memory and addressable memory.

physical model – Model of information-processing requirements containing information about specific hardware and software components that meet those processing requirements.

physical record – Unit of physical data transfer to or from a storage device.

picosecond – One-trillionth of a second.

pipe – Software mechanism by which the data output of one process is routed to another process as its input.

pit – Rough or concave bit area on an optical disk that is poorly reflective.

pixel – (1) Abbreviation of the term **picture element**. (2) Single unit of data in a graphic image. (3) Single point on a display surface.

platter – Disk within a disk drive, the surface or surfaces of which are used to store data.

plotter – Device that generates printed output through the movement of paper and one or more pens.

point – (1) ½ inch. (2) Unit of measurement for font size.

pointer – (1) Data element that contains the address (location in a storage device) of another data element. (2) Device used to input positional data or control the location of a cursor.

polarization – (1) Alignment of positive and negative poles of a magnetic charge or field. (2) Parallel alignment of photon paths within a beam of light.

polyphonic – Capable of generating many acoustic frequencies and/or timbres (i.e., voices) simultaneously.

pop operation – Process of removing an item from the top of a stack and copying its content to processor registers.

port – Physical connection point on a bus, communication channel, or input/output device.

positional number system – System in which numeric values are represented as multiples of one another according to the placement of digits within a digit string.

Postscript – Specific language for representing and transmitting complex images and display contents.

power sag – Momentary reduction in the voltage or amperage of electrical power.

power surge – Momentary increase in the voltage or amperage of electrical power.

preemptive scheduling – Scheduling method that allows a higher-priority process to interrupt and suspend a lower-priority process.

presentation layer – Layer of network system software responsible for input and output to I/O devices (e.g., video display terminals [VDTs]).

primary storage – High-speed storage within a computer system, accessed directly by the central processing unit (CPU), used to hold currently active programs and data immediately needed by those programs.

printed circuit board (PCB) – Electrical component containing many electrical devices on a board in which intercircuit wiring is embedded.

procedural knowledge representation – Representation of processing knowledge within a specific algorithm or procedure by which that knowledge is applied to data inputs.

procedural programming – Programming paradigm based on procedural knowledge representation.

procedure – Group of instructions that are always executed as a unit (similar to a function or subroutine).

process – (1) Program or program fragment that is separately managed and scheduled by the operating system. (2) To transform input data by the application of processing operations. (3) Representation of data or information processing on a data flow diagram.

process control block (PCB) – Data structure maintained by an operating system that contains information about a currently active process.

process family – Parent process and all its descendants.

process queue – Synonym of run queue.

process specification – Written or graphical description of a process on a data flow diagram.

processor – Any device capable of performing data transformation operations.

production version – Program that contains no debugging components and has been optimized for minimal consumption of hardware resources.

program – Sequence of processing instructions.

program counter – Synonym of instruction pointer.

program cycle – Set of program instructions that can be executed repetitively.

program development tool – Class of programs (e.g., text editors, compilers, interpreters, and debuggers) used to assist in the development of other programs.

program flowchart – Graphic representation of the flow of control among program components.

program profiler – Software utility that monitors and reports the activities and resource utilization of another program during execution.

program specification – (1) Written description of a program on a system flowchart. (2) Written description of processing requirements to be implemented in a program.

program status word (PSW) – Bit string held in a register within the control unit that stores status information in single bits.

program translator – Program that translates instructions in one programming language or instruction set into equivalent instructions in another programming language or instruction set (e.g., a compiler, interpreter, or assembler)

program verifier – Synonym of debugger or code checker.

programmable read only memory (PROM) – Read only memory device that is manufactured blank and may be written to only once.

programmer – Person who creates and/or maintains programs.

programmer/analyst – Person who performs the duties of a programmer and a systems analyst.

programmers' workbench – Integrated set of programs that support program creation, editing, translation, and debugging.

programming language – Any language in which computer-processing functions or instructions can be expressed.

PROM – Acronym of Programmable Read Only Memory.

protocol – Formal set of rules that govern the exchange of data and commands over a communication channel.

protocol stack – Set of software programs, each of which implements one protocol in a layered set of protocols.

PSK – Acronym of Phase Shift Keying.

PSW – Acronym of Program Status Word.

pulse code modulation (PCM) – Signal coding method under which bit values are represented as bursts of light.

push operation – Process of copying register values to the top of a stack.

QIC – Acronym of Quarter Inch Committee.

Quarter Inch Committee (QIC) – Committee that promulgates standards for cartridge magnetic tapes.

queue – (1) List of data items that is updated on a first in/first out basis. (2) Waiting line. (3) Synonym of run queue.

queuing theory – Body of mathematical knowledge that provides a basis for simulating computer-system behavior.

radio frequency (RF) – Electromagnetic radiation with a frequency less than one teraHertz propagated through space.

radix – Base of the number system (e.g., 2 for the binary numbering system and 10 for the decimal numbering system).

radix point – Symbol usually referred to as the decimal point (.) in the decimal number system, but that marks the boundary between whole and fractional quantities in a positional numbering system of any base.

RAID – Acronym of Redundant Array of Inexpensive Disks.

RAM – Acronym of Random Access Memory.

random access – Synonym of direct access.

random access memory (RAM) – (1) Generic description of semiconductor devices used to implement primary storage. (2) Device used to implement primary storage that provides direct access to stored data.

rapid prototyping – System-development methodology in which programs are rapidly developed and iteratively refined.

raster display – Device that produces video or printed output from sequences of horizontal scan lines.

raw data transfer rate – Data transfer rate of a communication channel exclusive of capacity used for header information, error detection data, and retransmission of incorrectly received data.

read only memory (ROM) – Primary storage device that can be read but not written.

read operation – Access operation that retrieves or copies data from an input or storage device.

read/write head – Mechanism within tape and disk drives that reads and writes data to/from the storage medium.

ready state – State of an active process that is waiting only for access to the central processing unit (CPU).

real resource – Hardware or software resource that physically exists.

realtime scheduling – Any scheduling method that guarantees the execution of a program or program cycle within a stated time interval.

record – (1) Data structure composed of data items relating to a single entity such as a person or transaction. (2) Unit of data transfer. (3) Primary component data structure of a file.

record locking – Method used to prevent a lost update by preventing multiple users or processes from simultaneously updating the same record.

recording density – Closeness or spacing of bit positions on a storage medium, typically as measured in bytes or tracks per inch.

red green blue (RGB) – Primary (additive) colors, used to produce color output on video display devices.

re-director – System software component that intercepts service calls that accesses a remote resource and interacts with a remote server to implement the access.

reduced instruction set computing (RISC) – (1) Use of a small set of simple processor instructions that cannot be decomposed into other processor instructions. (2) Processor architecture that emphasizes short, primitive instructions that reference operands contained in registers in contrast to complex instruction set computing.

redundant array of inexpensive disks (RAID) – Set of techniques for ensuring data integrity using redundant disk storage.

refresh cycle – Period during a dynamic random access memory refresh operation when the storage device is unable to respond to a read or write request.

refresh operation – (1) Operation that recharges the capacitors of dynamic random access memory. (2) Operation that redraws the display surface of a video display device.

refresh rate – Number of times per second a refresh operation occurs on a video display device.

register – High-speed storage location within a central processing unit (CPU) that can hold a single word.

relative addressing – Specific form of indirect addressing in which a program memory reference is interpreted as an offset from the content of a base register.

relative path – Path that begins at the level of the current directory.

relocatable code – (1) Program instructions with address values that are changeable, or not absolute. (2) Synonym of object code.

relocation – Act of moving instructions and data from one memory region to another and readjusting all program memory references to reflect the new location.

relocator – Program that performs relocation.

remote job entry (RJE) – Form of batch processing under which batches of data are captured and stored at remote points and then transmitted to a central processing site at scheduled times.

remote login – Ability of a user to start an interactive session on a remote computer using a program executing on a local computer.

remote process execution – Execution of a process on a remote computer through a request from another computer transmitted over a communication network.

removable media – Storage medium that can be removed physically from its storage device (e.g., a diskette).

repeater – Device that relays signals from one network segment to another.

repetition – (1) Repetitive execution of a set of program instructions. (2) Programming language statement or construct that causes repetitive execution of a set of program instructions.

request for proposals (RFP) – Formal document stating hardware and/or software requirements and requesting proposals from vendors to meet those requirements.

requirements analysis – Synonym of system analysis.

resistance – Phenomenon in electrical transmission whereby electrical energy is converted to heat, thus reducing signal strength.

resolution – Synonym of address resolution.

resolving the reference – As performed by a link editor, the act of replacing a library call in object code with the object code of the library routine.

resource availability model – A model of the computing resources that can be provided by a given combination of computer hardware and system software.

resource declaration – Within a job control language (JCL), a statement of the resources that will be required to execute a job or job step.

resource locator – System software component that searches for remote resources on a local (LAN) or wide are network (WAN).

resource registry – Database of servers and resources available within a local (LAN) or wide area network (WAN).

response time – Delay between a processing request from a user or application program and the receipt of a processing result.

return wire – Path in a communication channel that completes an electrical circuit between the sending and the receiving device.

reverse engineering – Process of duplicating the function of an existing program where the new program differs in programming language, operating system, and/or hardware environment.

revolutions per minute (RPM) – Measure of the rotational speed of a circular device (e.g., a disk drive platter).

RF – Acronym of Radio Frequency.

RFP – Acronym of Request for Proposals.

RGB – Acronym of Red Green Blue.

ring topology – Network configuration in which each network node is connected to two other network nodes, with the entire set of connections and nodes forming a ring.

RIP – Acronym of Routing Information Protocol.

RISC – Acronym of Reduced Instruction Set Computing.

RJE – Acronym of Remote Job Entry.

Rock's Law – Observation that states that the cost of fabrication facilities for the latest generation of semiconductor devices doubles every four years.

ROM – Acronym of Read Only Memory.

root directory – (1) Topmost directory (or entry point) in a hierarchical directory structure. (2) Topmost directory for a storage medium, file system, or resource registry.

rotate operation – Modified form of the shift operation in which bit values shifted beyond one end of a bit string are placed at the opposite end of the bit string.

rotational delay – Waiting time for the desired sector of a disk to rotate beneath a read/write head.

router – Device that examines packet destination addresses and forwards them to another network segment, if necessary.

routing protocol – Set of rules governing the routing of packets and the exchange of routing information among network nodes.

Routing Information Protocol (RIP) – Internet protocol for the exchange of routing information among network nodes.

RPM – Acronym of Revolutions Per Minute.

run queue – Ordered set of programs or processes awaiting execution.

run time library – Set of library programs used by an interpreter during program translation and execution.

running state – State of a process that is currently executing within the central processing unit (CPU).

sampling – (1) Process of measuring and digitally encoding one or more parameters of an analog signal at regular time intervals. (2) Process of digitally encoding sound waves.

sandbox – Protected area of a program within which a Java applet is executed.

scan code – Coded output generated by a keyboard for interpretation by a keyboard controller or a processor.

scanning laser – Laser that is automatically swept back and forth over a predefined viewing area (e.g., as used in a bar-code reader)

scanner – Synonym of optical scanner.

scheduler – Program within the operating system that controls process states and access to the CPU.

scheduling – Process of determining and implementing process priorities for access to hardware resources.

SDLC – Acronym of System(s) Development Life Cycle.

search path – Set of directories that are searched automatically by the command layer for executable files.

secondary storage – Set of computer system devices that implement large-capacity long-term data storage.

sector – (1) Smallest accessible unit of a disk drive. (2) For many disk drives, 512 bytes.

segment register – Register that holds the base address of a memory segment, as used in a central processing unit (CPU) that uses a segmented memory model.

segmented memory – Memory allocation and paretitioning based on equal sized segments and the use of segmented memory addresses.

selection – Program or job control language (JCL) control structure that chooses processing statements for execution on the basis of a condition test.

semiconductor – Material with resistance properties that can be tailored between those of a conductor and an insulator by adding chemical impurities.

sequence – Implicit control structure in which processes or statements are executed in sequential order, with each process or statement awaiting the completion of the previous process prior to commencing execution.

Sequenced Packet Exchange (SPX) – Novell Netware protocol for network message transport.

sequential access time – Time required to access the second of a stored sequential pair of data items.

sequential access – Access technique whereby data items are read or written in an order corresponding to their position within allocated storage locations.

sequential execution – Execution of program instructions in the order in which they are stored.

sequential file organization – Method of file organization in which contents are stored sequentially and contiguously in allocated storage locations.

serial access – Synonym of sequential access.

serial transmission – Method of transmission in which individual bits are transmitted sequentially over a single communication channel.

server – Network node that makes hardware or software resources available to other network nodes.

service call – Request for an operating system service by an application program.

service layer – Portion of the operating system that accepts service calls and translates them into low level requests to the kernel.

service standards – Characteristics of services and hardware devices that are associated with an infrastructure.

session layer – Layer of network system software responsible for establishing connections with server processes on other network nodes.

shell – User interface (or command layer) of an operating system.

shift operation – Monadic operation whereby individual bits of a bits string are moved left or right by a stated number of positions, empty positions are filled with zeros, and bit values that shift beyond the bounds of the bit string are truncated (i.e., lost).

shortest time remaining – Scheduling method that assigns highest priority to processes or requests with shortest time remaining until completion.

sibling process – Another child process created, or spawned, by the same parent.

sign bit – Bit (typically, the high-order bit) used to indicate the sign (positive or negative) of a numeric value.

sign magnitude notation – Data representation that appends a sign bit to a bit string containing an ordinary binary number.

signal – (1) Carrier wave with data embedded as one or more modulated wave parameters. (2) Message sent from one active process to another.

signal event – Period of time during which a carrier wave parameter is modulated to encode one data unit (e.g., a bit).

signal-to-noise (S/N) ratio – Mathematical relationship between the power of a carrier signal and the power of the noise in the communication channel, measured in deciBels.

signal wire – Component of an electrical communication channel (i.e., circuit) used to transmit a carrier wave.

signed integer – Integer with a sign bit (i.e., it can have a negative or positive value).

SIMM – Acronym of Single In-line Memory Module.

simple (machine) instruction – Processor instruction that results in the execution of only one primitive processor action (e.g., load, store, XOR).

simplex – Type of communication channel composed of a single transmission medium that allows transmission in only one direction.

simulation – (1) Process of examining the behavior of a system by creating and manipulating a dynamic model of the system. (2) Technique of mathematical analysis based on statistics and queuing theory.

simulation model – System model designed for manipulation by simulation techniques.

sine wave – Analog waveform that varies continuously between positive and negative states.

single in-line memory module (SIMM) – Unit of primary storage consisting of a small printed circuit board with memory chips on one or both sides and electrical contacts on one side.

single level directory – Directory structure in which only one directory is used for each storage medium or device.

single mode pulse code modulation – Modulation scheme used if an optical fiber carries only one signal.

single tasking operating system – Operating system in which only one process may be active at a time.

singly linked list – Set of stored data items in which each element contains a pointer to the next list element.

skew – Timing differences in the arrival of signals transmitted simultaneously on parallel communication channels.

S/N – Acronym of signal-to-noise.

socket – Internet standard for service identification that combines an Internet address with a service port number (e.g., 129.24.8.4:80).

software monitor – Program that monitors and reports the utilization of a software resource.

son – File containing the results of updating a master file (i.e., the father) based on the contents of a transaction file.

sorted list – List in which the component data elements are ordered based on their value or upon the value(s) of one or more component data items.

sound board – Synonym of sound card.

sound card – Hardware device containing components for sound input and output.

source code – Instructions in a high-level programming language.

spawn – Creation of a child process by a parent process.

speaker dependent – Speech recognition system that recognizes input from only one user and must be trained to accurately recognize that user's specific speech patterns.

special purpose processor – (1) Processor with a limited set of processing functions. (2) Processor capable of executing only a single program.

speech recognition – Process of recognizing human speech as computer system input.

speech synthesis – Process of generating human speech based on character or textual input.

spooler – (1) Acronym of Shared Peripheral Operation On Line. (2) Software utility that allows multiple processes to access a single output device through the use of storage buffers.

SPX – Acronym of Sequenced Packet Exchange.

SQL – Acronym of Structured Query Language.

square wave – Digital waveform with two discrete states and no continuous transitions between those states.

SRAM – Acronym of Static Random Access Memory.

stack – (1) List of data items maintained in last in/first out order. (2) Set of registers or memory locations used to store the register values of temporarily suspended processes.

stack overflow – Condition that occurs when an attempt is made to add data to a stack that is already at its maximum capacity.

stack pointer – Register containing the primary storage address of the last (most recently added) stack element.

star topology – Physical network connection pattern in which all nodes are attached to a central node.

start bit – Bit that is always zero used at the beginning of each character in asynchronous serial data transmission.

static analysis – Use of models and/or model analysis techniques that do not incorporate parameter variability and random events in the representation of system behavior.

static connection – Mapping between a local resource name and a remote resource that does not change once initialized and must be explicitly initialized prior to use.

static random access memory (SRAM) – Type of random access memory that implements bit storage with a flip-flop circuit (e.g., two or four transistors).

statistics – Branch of mathematics concerned with the description of phenomena that exhibit variable and/or random behavior.

status code – Signal sent from a peripheral device to a processor to indicate its ability or readiness to respond to a command.

stop bit – Bit appended to the end of a byte in asynchronous serial transmission.

storage allocation table – Table that correlates storage-device allocation units with specific files, directories, or unallocated space.

storage I/O control layer – Layer of system software responsible for controlling the input and output to/from storage devices.

storage medium – Fixed or removable component of a storage device that serves as the physical repository for data (e.g., a platter of a hard disk drive).

store operation – Operation in which the contents of a register are copied to a memory location.

strategic plan – Long-range plan stating the services to be provided by an organization and the means for obtaining and using needed resources.

string – Ordered set of related data elements, usually stored as a list or array.

structure chart – Graphic representation of the hierarchical organization of processing functions (i.e., modules) in a program or system.

structured query language (SQL) – Standardized language for data manipulation operations in a relational database management system (DBMS).

structured systems development – Systems-development method that follows the structured systems development life cycle.

structured systems development life cycle – Particular form of system development life cycle in which analysis is clearly separated from design and in which specific system-modeling tools are used.

subcarrier – Analog signal of a specific frequency that is transmitted to represent a specific data value.

subroutine – Named segment of a high-level language program that is always executed as a unit.

subtractive colors – Cyan (absence of red), magenta (absence of green), and yellow (absence of blue).

supercomputer – Computer designed for very fast processing of real numbers, typically implemented with some form of parallel processing.

superconductor – Conductive material that exhibits little or no resistance to the flow of electrons.

supervisor – Operating system program that serves as the master interrupt handler.

surge protector – Hardware device that eliminates electrical power surges.

sustainable data transfer rate – Maximum data transfer rate that can be sustained by a device or communication channel during large or lengthy data-transfer operations.

swap space – Area of secondary storage used to hold virtual memory pages that cannot fit into primary storage.

switch – (1) Device that exists in one of two states and can be instructed to alternate between them. (2) Network node that quickly routes packets among network segments with switching technology.

symbol table – List of program and data names and their allocated primary storage locations.

symbolic debugger – Debugger that can report program execution progress and errors in terms of symbolic module and data names used within program source code.

synchronous – Precise (timed) coordination of two independent devices or programs through the use of a shared clock signal.

synchronous idle message – Message without data content sent continuously between sender and receiver to keep their internal clocks synchronized.

synchronous transmission – Reliable high-speed method of data transmission in which data are sent in packets and in timed sequences.

syntax – Set of construction rules for a programming or other language.

system administration – Managerial actions required to ensure efficient, reliable, and secure computer system operation.

system bus – Bus shared by most or all devices in a computer system.

system call – Synonym of service call.

system clock – Digital circuit that generates timing pulses and transmits them to other devices over the system bus.

system model – Formal symbolic representation of processing requirements, software elements, and/or hardware elements (e.g., a system flowchart or data flow diagram).

system overhead – Resource consumed for resource-allocation and translation functions.

systems analysis – Process of ascertaining and documenting information processing requirements.

systems analyst – Person who performs system analysis.

systems design – Process of selecting and organizing hardware and software elements to satisfy stated processing requirements.

systems designer – Person who performs system design.

systems development life cycle (SDLC) – Formal process for system analysis, design, implementation, maintenance, and evaluation.

systems development tool – (1) Software tool designed to aid in the creation of groups of programs comprising an entire system. (2) Approximate synonym of computer assisted software engineering (CASE) tool.

systems flowchart – Graphic model of manual processes, software elements, files, and the flow of data and control among them.

systems implementation – Process of software construction, installation, and testing on a given hardware configuration.

systems programmer – Person who creates and/or maintains system software.

systems software – Software programs that perform hardware-interface, resource-management, and/or application-support functions.

systems survey – Process of initially ascertaining user processing requirements and determining the feasibility of meeting those requirements with computer-based solutions.

T connector – Specific type of tap used to connect network nodes to a communication bus or network.

tap – General term describing a connection to a bus communication channel.

task – Process or thread that is independently managed by the operating system.

TCP – Acronym of Transmission Control Protocol.

TCP/IP – Acronym of Transmission Control Protocol/Internet Protocol.

TDM – Acronym of Time Division Multiplexing.

teletype – I/O device consisting of a keyboard and a character printer.

terabyte – $1,024^4$ bytes.

terminal – Synonym of video display terminal (VDT).

text editor – Program that allows a user to create and/or modify files containing character data.

third generation language (3GL) – High-level programming language that does not possess advanced capabilities for interactive input/output, database processing, or nonprocedural programming.

thread – Subcomponent of a process that can be independently scheduled and executed.

threshold – Value of a carrier wave parameter at which the interpretation of its data content changes value.

throughput – Volume or amount of work that a computer system can perform within a given time interval.

time division multiplexing (TDM) – Technique that subdivides the capacity of a communication channel into discrete time slices, each of which may be allocated to a separate sender and receiver.

timer interrupt – Interrupt automatically generated by the CPU in response to the passage of a specific interval.

TLI – Acronym of Transport Layer Interface.

token – Control packet used to regulate access to a communication channel.

token passing – Media access protocol that uses tokens passed between network nodes.

track – Set of sectors on one side of a disk platter that form a concentric circle.

track-to-track seek time – Time required to move a disk read/write head between two adjacent tracks.

transaction – (1) Event or system input that initiates the execution of one or more program segments. (2) Real-world event from which data input is derived (e.g., a sale). (3) A general term describing any change to stored data (e.g., the addition of a record to a file).

transaction logging – Synonym of journaling.

transceiver – Device that detects signals passing through a transmission medium and generates an equivalent signal on another transmission medium.

transistor – Solid state electrical switch that forms the basic component of most computer processing circuitry.

transistor-to-transistor (TTL) logic – Digital signaling method based on square waves.

transmission control protocol (TCP) – Internet protocol for translating messages into packets and guaranteeing their delivery.

transmission medium – Physical communication path through which a carrier wave is propagated.

transparent file access – Access to files stored on a remote machine over a communication network, without the explicit knowledge or control of a user or application program.

transparent file transfer – Transfer of a file between computers over a communication network, without the explicit knowledge or control of a user or application program.

transport layer – Layer of network system software responsible for converting network messages into packets or other data structures suitable for transmission.

Transport Layer Interface (TLI) – Novell Netware standard for addressing remote resources and services.

tree structure directory – Synonym of hierarchic directory structure.

truncation – Act of deleting bits that will not fit within a storage location.

TTL – Acronym of Transistor-to-Transistor Logic.

twisted pair cable – Transmission medium consisting of two electrical conductors continuously twisted around one another.

twos complement – Notation system in which the complement of a bit string is formed by substituting 0 for all values of 1 and 1 for all values of 0.

Type I error – (1) Percentage of data items containing errors that will be incorrectly identified as error free. (2) Percentage of a sample or population that will be incorrectly identified as satisfying the null hypothesis.

Type II error – (1) Percentage of data items without errors that will be incorrectly identified as containing errors. (2) Percentage of a sample or population that will be incorrectly identified as not satisfying the null hypothesis.

unblocked – Storage of a single logical record within a physical record (block).

unconditional branch – Synonym of branch instruction.

undelete operation – Act of restoring a record or file by re-creating its index information (e.g., directory entry) and recovering its previously allocated storage locations.

underflow – (1) Condition that occurs when a value is too small to represent in floating point notation. (2) Overflow of a negative exponent in floating point notation.

Unicode – Standard 16-bit method of character coding.

uninterruptable power supply (UPS) – Device that provides electrical power when normal power inputs are interrupted.

unresolved reference – Reference within an object code module to a symbolic name not defined within that module.

unsigned integer – Integer lacking a sign bit (i.e., it can have only positive values).

update in place – Method of altering the contents of a record or field in which a new value overwrites the old value.

update operation – Process of modifying a set of data items to reflect the addition, change, and/or deletion of a data items.

UPS – Acronym of Uninterruptable Power Supply.

user – Person who directly or indirectly interacts with a program.

user authentication – Process of verifying user identity, usually by validating a user name and password.

utility program – Program that performs a commonly used function (e.g., file access, storage initialization, etc), typically provided by the operating system either as separate software tools or as part of the service layer.

validity checking – Any method of error detection in data transmission or storage.

variable length instruction – Instruction that can be in one of a set of instruction formats that vary in length.

VDT – Acronym of Video Display Terminal.

vector – (1) Line segment that has direction and length. (2) One-dimensional array.

vector list – List of line descriptions used for the input or output of graphic images.

vertical redundancy check (VRC) – Synonym of parity checking.

very large scale integration (VLSI) – Semiconductor devices (e.g., chips) containing tens of thousands of transistors or other primitive processing circuits.

victim – Memory page chosen to be swapped to secondary storage under virtual memory management (VMM).

video display terminal (VDT) – I/O device comprised of a keyboard and a video display surface.

virtual machine – Set of hardware resources visible to a user or application.

virtual memory management – Mode of operating system memory management in which secondary storage is used to extend the capacity of primary storage.

virtual resource – Resource visible to a user or program, but not necessarily available at all times or physically extant.

VLSI – Acronym of Very Large Scale Integration.

volatile storage – Storage devices that cannot retain their contents indefinitely (e.g., random access memory).

volatility – (1) Rate of change in a set of data items. (2) Lack of ability to permanently store data content.

von Neumann machine – Named after a principal member of its design team, John von Neumann, a computer system that contains a single processor, uses stored programs, and sequentially executes single instructions.

VRC – Acronym of Vertical Redundancy Check.

wait state – Idle processor cycle consumed while waiting for a response from another device.

WAN – Acronym of Wide Area Network.

Web site – Server that makes data content available to the Internet through various data exchange protocols, including hypertext transfer protocol and file transfer protocol.

wide area network (WAN) – Network that spans large physical distances (e.g., multiple buildings, cities, regions, or continents).

wideband – (1) Communication channel with high bandwidth. (2) Communication channel that implements frequency division multiplexing.

window manager – Operating system component that manages the placement of program windows and the routing of I/O to and from those windows.

window-based I/O – Form of user interface that uses windows and menus as a means of program interaction.

wiring closet – Room or closet where network hubs, switches, and/or wiring concentrators are located and to which transmission lines are routed from other rooms of a floor or building.

wiring concentrator – Synonym of network hub.

word – Unit of data processed by a single central processing unit (CPU) instruction.

working directory – Synonym of current directory.

working storage – (1) Synonym of registers or primary storage. (2) Set of data items manipulated by a program, process, or module.

workload model – Synonym of application demand model.

workstation – (1) Powerful microcomputer designed to support demanding numerical and/or graphical processing tasks. (2) A synonym of personal computer (PC) or microcomputer.

World Wide Web (WWW) – Set of all nodes (and their data content) accessible through the Internet.

WORM – Acronym of Write Once Read Many.

write once read many (WORM) – Type of optical storage media that is manufactured blank and can be written once.

write operation – Process of encoding and storing data within a storage device.

write protection – Protection of a set of data elements or a storage medium against accidental or malicious alteration.

XON/XOFF – Pair of control characters (ASCII 17 and 19) used to start and stop the flow of data in asynchronous data transmission.

XOR – Acronym of eXclusive OR.

zero crossing signal – Digital signal that has both positive and negative states.

INDEX

OR operations, 151
Pentium processors, 174
programming example, 156-158
sequence control operations, 153
store operations, 149-150
integer arithmetic, 110
integer data, 110-114
excess notation, 111-112
long, 114
negative and non-negative, 110
range and overflow, 113-114
signed and unsigned, 110
sign magnitude notation, 110
two's complement notation, 112-113
integrated circuits (ICs), 181-182
Intel processors, 83-86
4004, 183
memory address format, 130-131
Pentium. *See* Pentium processors
interactive processing, 46
interleaved execution, 505
International Alphabet Number 5 (IA5), 122
International Standards Organization (ISO), 122
Internet Control Message Protocol (ICMP), 369
Internet Packet Exchange (IPX), 623
Internet Protocol (IP), 368-370, 623
interpretation, 449
interpreters, 71, 449-451
resource utilization, 450-451
interprocess data communication, 502-504
interprocess synchronization, 499-502
interrupt(s), 259-264, 506-507
multiple, 260
processing, 507-508
stack processing, 261-262
timer, 512
interrupt code, 259
interrupt handlers, 260
interrupt register, 259
interrupt tables, 260
I/O. *See* input/output (I/O) entries
IP (Internet Protocol), 368-370, 623
IPX (Internet Packet Exchange), 623
ISO (International Standards Organization), 122

J

Java, 452-455
Java virtual machine (JVM), 453
Jaz drives, 228
JCL (job control language), 74, 611-613
job(s), 613
job control language (JCL), 74, 611-613
job steps, 613
job streams, 613
Joint Photographic Experts Group (JPEG), 392
journaling, 587
JPEG (Joint Photographic Experts Group), 392
jump operations, 153-154
JVM (Java virtual machine), 453

K

kernel, 489
operating systems, 74, 76-77
Kerr effect, 235
keyboard(s), 384-385
keyboard controllers, 384
kilobytes, 48

L

lands, CD-ROMs, 233
LANs. *See* local area networks (LANs)
laser(s), scanning, 409
laser printers, 405-406
late binding, 448-449
Latin-1, 126
L1 (level one) caches, 271
L2 (level two) caches, 271
LCDs (liquid crystal displays), 401
leakage, magnetic, 218
least significant byte, 212
least significant digit, 104
level one (L1) caches, 271
level two (L2) caches, 271
libraries, external subroutines. *See* support libraries
library calls, 445
linear address space, 250-252
linear velocity, constant, phase-change optical
technology, 234
line turnaround, 311
linked lists, 135-137
linkers, 446-447 (link editors)
linking (link editing), 446-449
dynamic and static, 448-449
link maps, 451
liquid crystal displays (LCDs), 401
lists, 135-137
linked, 135-137
little endian storage, 212
load balancing, 349
loaders, 213
load operations, 149-150
load sharing, 349
local area networks (LANs), 12, 356
bridges, 361
Ethernet, 373-374
location transparency, 624
locking
files, 579
records, 579-580
logical access, I/O ports. *See* input/output (I/O) ports
logical delete operations, 574
logical file structure, 546, 547
logical model, application development, 427-429
logical network interface driver, 622
logical records, 554-555
logical shift operations, 153
logic gates. *See* gates
logic instructions, 29
login, remote, 347-349

transmission media, 298-317
  bandwidth, 300-301
  channel sharing, 314-317
  electrical and optical cabling, 304-306
  full-duplex communication, 312
  half-duplex communication, 311-312
  parallel transmission, 312-313
  serial transmission, 313-314
  signal-to-noise ratio, 302-304
  simplex mode communication, 310
  speed and capacity, 299-300
  wireless data transmission, 306-309
transparent file access, 346-347
transparent file transfer, 346
transport layer, 365-366, 623
Transport Layer Interface (TLI), 623
tree structure directories, 560-563
truncation, real numbers, 118-119
TTL (transistor-to-transistor logic), 295-296
24-bit color, 390
twisted pair wire transmission media, 304
two's complement notation, 112-113
Type I errors, 322, 323
Type II errors, 322, 323

U

unconditional branches, 154
undelete operations, 575
underflow, real numbers, 118
Unicode, 127-128
uninterruptible power supplies (UPSs), 670-671
unresolved references, 447
unsigned integers, 110
update(s), lost, 578-580
update in place, 572
update operations, 572
upgrading network capacity, 327-328, 375-376
UPSs (uninterruptible power supplies), 670-671
user(s), 9-10
user authentication, 665
utility programs, 62-63
  software reuse, 63-64

V

value, absolute, 113
variable length instructions, 161
VDTs. See video display terminals (VDTs)
vectors, 392-394
vendors, Web sites, 16-17
vertical redundancy checking (VRC), 324
victims, 529
video display terminals (VDTs), 397-399
  full-screen input/output, 601-603
virtual memory management (VMM), 527-529
virtual resources, allocation, 494-495
VMM (virtual memory management), 527-529
volatile storage devices, 201
volatility, 36
von Neumann, John, 35
von Neumann machines, 35-36
VRC (vertical redundancy checking), 324

W

wait state, 166
WANs (wide area networks), bridges, 361
Web. See Web sites; World Wide Web (Web; WWW)
Web-based periodicals, as information source, 15-16
Web sites
  as information source, 16-17
  periodicals, 15-16
  vendors and manufacturers, 16-17
wide area networks (WANs), bridges, 361
window-based command interfaces, 619-621
window-based input/output, 605-607
window manager, 606
Windows, versions for business desktops, 518-519
Windows NT
  performance monitoring, 662-664
  scheduling, 515-517
Windows NTFS, 581-583
wireless data transmission, 306-309
wiring closets, 359
wiring concentrators, 359
word(s), definition, 170
word size, 170-172
  Pentium processors, 174
working directory, 561
workload models, 654-655
workload performance, 46-47
workstations, 38, 43. See also microcomputers
World Wide Web (Web; WWW). See also Web sites
  as computer technology information source, 15-19
  finding computer technology information, 17-19
  periodicals, 15-16
  vendor and manufacturer Web sites, 16-17
WORM (write once/read many) disks, 235
write caching, 268-269
write once/read many (WORM) disks, 235
write operations, 569-572
  direct access, 571
  indexed access, 571-572
  sequential access, 570-571
WWW. See Web sites; World Wide Web (Web; WWW)

X

XOR gates, 176
XOR (exclusive OR) operations, 151

Z

Zip drives, 228
zone networks, 356